Methods in Plant Biochemistry

Volume 9
Enzymes of Secondary Metabolism

METHODS IN PLANT BIOCHEMISTRY

Series Editors

P. M. DEY
Department of Biochemistry, Royal Holloway and Bedford New College, UK

J.B. HARBORNE
Plant Science Laboratories, University of Reading, UK

Methods in Plant Biochemistry

Series editors
P. M. DEY and J. B. HARBORNE

Volume 9
Enzymes of Secondary Metabolism

Edited by

P. J. LEA

Division of Biological Sciences, Lancaster University, UK

ACADEMIC PRESS
Harcourt Brace Jovanovich, Publishers
London   San Diego   New York
Boston   Sydney   Tokyo   Toronto

ACADEMIC PRESS LIMITED
24–28 Oval Road
London NW1 7DX

US edition published by
ACADEMIC PRESS INC
San Diego, CA 92101

Copyright © 1993, by
ACADEMIC PRESS LIMITED

ISBN 0–12–461019–6

Typeset by Colset Private Ltd, Singapore
Printed in Great Britain by The University Press, Cambridge

Contents

Contributors

W. R. Alonso, Institute of Biological Chemistry, Washington State University, Pullman, WA 99164–6340, USA.

H. Birecka, Department of Biological Sciences, Union College, Schenectady, NY 12308–2311, USA.

M. Birecki, Department of Biological Sciences, Union College, Schenectady, NY 12308–2311, USA.

P. M. Bramley, Department of Biochemistry, Royal Holloway and Bedford New College, Egham Hill, Surrey TW20 0EX, UK.

E. G. Brown, Biochemistry Research Group, School of Biological Sciences, University College of Swansea, Singleton Park, Swansea SA2 8PP, Wales, UK.

J. R. Coggins, Department of Biochemistry, University of Glasgow, Glasgow G12 8QQ, Scotland, UK.

R. Croteau, Institute of Biological Chemistry, Washington State University, Pullman, WA 99164–6340, USA.

V. De Luca, Plant Biology Research Institute, Department of Biological Sciences, University of Montréal, 4101 Sherbrooke Street East, Montréal, Québec, Canada H1X 2B2.

J. E. Graebe, Pflanzenphysiologisches Institut und Botanischer Garten der Universität Göttingen, Untere Karspüile 2, 2400 Göttingen, Germany.

W. T. Griffiths, Department of Biochemistry, University of Bristol, Bristol BS8 1TD, UK.

G. G. Gross, Universität Ulm, Abteilung Allgemeine Botanik (Biol II), Oberer Eselberg, D-7900 Ulm, Germany.

T. Hashimoto, Department of Agricultural Chemistry, Faculty of Agriculture, Kyoto University, Kyoto 606-01, Japan.

R. K. Ibrahim, Plant Biochemistry Laboratory, Concordia University, Sir George Williams Campus, 1455 De Maissoneuve Boulevard, West Montréal, Québec, Canada H3G 1M8.

F. Ikegami, Faculty of Pharmaceutical Sciences, Chiba University, Yayoi-cho 1–33, Inage-ku, Chiba 263, Japan.

F. Karp, Institute of Biological Chemistry, Washington State University, Pullman, WA 99164–6340, USA.

T. Lange, Pflanzenphysiologisches Institut und Botanischer Garten der Universität Göttingen, Untere Karspüile 2, 2400 Göttingen, Germany.

C. A. Mihaliak, Institute of Biological Chemistry, Washington State University, Pullman, WA 99164–6340, USA.

H.-P. Mock, Institut für Pharmazeutische Biologie der Technischen Universität Braunschweig, Mendelssohnstrasse 1, Braunschweig D-3300, Germany.

B. L. Møller, Plant Biochemistry Laboratory, Royal Veterinary and Agricultural University, Thorvaldsenvey 40, DK-1871, Frederiksberg C, Denmark.

D. M. Mousdale, Department of Biochemistry, University of Glasgow, Glasgow G12 8QQ, Scotland, UK.

I. Murakoshi, Faculty of Pharmaceutical Sciences, Chiba University, Yayoi-cho 1–33, Inage-ku, Chiba 263, Japan.

A. D. Parry, Department of Biological Sciences, University College of Wales, Aberystwyth, Dyfed SY23 3DA, Wales, UK.

J. E. Poulton, Department of Biological Sciences, The University of Iowa, Iowa City, IA 52242, USA.

A. G. Smith, Department of Plant Sciences, University of Cambridge, Downing Street, Cambridge CB2 3EA, UK.

D. Strack, Institut für Pharmazeutische Biologie der Technischen Universität Braunschweig, Mendelssohnstrasse 1, Braunschweig D-3300, Germany.

L. Varin, Plant Biochemistry Laboratory, Concordia University, Sir George Williams Campus, 1455 De Maissoneuve Boulevard, West Montréal, Québec, Canada H3G 1M8.

Y. Yamada, Department of Agricultural Chemistry, Faculty of Agriculture, Kyoto University, Kyoto 606-01, Japan.

S. F. Yang, Department of Vegetable Crops, University of California, Davis, CA 95616–8746, USA.

W.-K. Yip, Department of Vegetable Crops, University of California, Davis, CA 95616–8746, USA.

Preface to the Series

Scientific progress hinges on the continual discovery and extension of new laboratory methods and nowhere is this more evident than in the subject of biochemistry. The application in recent decades of novel techniques for fractionating cellular constituents, for isolating enzymes, for electrophoretically separating nucleic acids and proteins and for chromatographically identifying the intermediates and products of cellular metabolism has revolutionised our knowledge of the biochemical processes of life.

While there are many books and series of books on biochemical methods, volumes specifically catering for the plant biochemist have been few and far between. This is particularly unfortunate in that the isolation of DNA, enzymes or metabolites from plant tissues can often pose special problems not encountered by the animal biochemist. For a long time, the Springer series *Modern Methods in Plant Analysis*, which first appeared in the 1950s, provided the only comprehensive guide to experimental techniques for the investigation of plant metabolism and plant enzymology. This series, however, has never been completely updated; a second series has recently appeared but this is organised on a techniques basis and thus does not provide the comprehensive coverage of the first series. One of us (JBH) wrote a short guide to modern techniques of plant analysis *Phytochemical Methods* in 1976 (second edition, 1984) which showed the need for an expanded comprehensive treatment, but which by its very nature could only provide an outline of available methodology.

The time therefore seemed ripe to us to produce an entirely new multi-volume series on methods of plant biochemical analysis, which would be both thoroughly up-to-date and comprehensive. The success of *The Biochemistry of Plants*, edited by P. K. Stumpf and E. E. Conn and published by Academic Press, was an added stimulus to produce a complementary series on the methodology of the subject. With these thoughts in mind, we planned individual volumes covering: phenolics, carbohydrates, amino acids, proteins and nucleic acids, terpenoids, nitrogen and sulphur compounds, lipids, membranes and light receptors, enzymes of primary and secondary metabolism, plant molecular biology and biological techniques in plant biochemistry. Thus we have tried to cover all the major areas of current endeavour in phytochemistry and plant biochemistry.

The main aim of the series is to introduce to the scientist current knowledge of techniques in various fields of biochemically-related topics in plant research. It is also intended to present the historical background to each topic, to give experimental details of methods and analyses and appraisal of them, pointing out those methods that are

ix

most suitable for immediate application. Wherever possible illustrations and structures have been used and one or more case treatments presented. The compilation of known data and properties, where appropriate, is included in many chapters. In addition, the reader is directed to relevant references for further details. However, for the sake of clarity and completeness of individual reviews, some overlap between chapters of volumes has been allowed.

Finally, we extend our warmest thanks to our volume editors for undertaking the important task of organising each volume and cooperating in preparing the contents lists. Our special thanks go to the staff of Academic Press and to the many colleagues who have made this project a success.

P. M. DEY
J. B. HARBORNE

Preface to Volume 9

This book is a continuation of Volume 3, which was devoted to the enzymes of primary metabolism. It had been originally intended that the complete subject of primary and secondary metabolism would be contained in one volume. This has proved impossible and even now there are a number of subject areas that have not been covered. As with Volume 3, I have to apologise to those research workers whose subject areas have not been covered.

The book starts with the shikimate pathway which leads to the synthesis of the aromatic amino acids phenylalanine, tyrosine and tryptophan and to a range of secondary products. I have then attempted to cover as many secondary products as possible and finished with the key plant growth regulators. On two occasions it has been necessary to double the size of the chapters in order to include all the available data. Chlorophyll is included in this volume and the authors of the chapter have gone to great pains to tell me that this should be included under 'primary metabolism', a point that I have to agree with.

In my original letter to the authors I included the following statement. 'The chapter should include a brief historical background to the topic, full experimental details of the enzyme assays and method of purification. It should also include an appraisal of methods and modern techniques, pointing out those most suitable for specific purposes. Details of the properties of the enzymes should also be included, wherever possible illustrations and structures should be used'. Taking into account the confines of space, I feel these aims have been achieved. I have found it particularly interesting that in some instances it is only just possible to assay an enzyme in a crude extract, whilst in others, the enzyme has been purified to homogeneity and the gene isolated.

With a volume of this size it would be impossible to finish without expressing my thanks to a large number of people. Initially, I would like to thank Jeffrey Harborne, who originally suggested the idea of the volumes to me and was of considerable help in the early planning of the chapter titles and potential authors. I greatly appreciate the advice and encouragement received from Andrew Richford, James Gaussen and Carol Parr of Academic Press. I am of course totally indebted to all 29 authors, who produced their manuscripts within six months of the original submission date. Unfortunately, one author was unable to submit by the final closing date. During the last year I have received valiant service from my secretaries Alyson Allen and Annie Brotheridge. Finally, yet again, I am obliged to express my sincere and considerable thanks to Janice Turner for the vast amount of editorial work that she has carried out, on what is now her fifth book. She has worked tirelessly to check all the manuscripts for errors in spelling and in particular reference citation, as well as preparing the index. Without her help production of this book would not have been possible.

PETER J. LEA

1 The Shikimate Pathway

D. M. MOUSDALE and J. R. COGGINS

Department of Biochemistry, University of Glasgow, Glasgow G12 8QQ, Scotland, UK

I. INTRODUCTION

Seven enzymic steps convert a product of the Calvin cycle and/or pentose phosphate pathway (erythrose 4-phosphate, E4P) and a product of glycolysis (phospho*enol*

METHODS IN PLANT BIOCHEMISTRY Vol. 9
ISBN 0–12–461019–6

pyruvate, PEP) to chorismic acid, from which the aromatic amino acids and many other aromatic compounds are formed in plants and microorganisms. In sequence, the enzymes are: 3-deoxy-D-*arabino*-heptulosonic acid 7-phosphate (DAHP) synthase (systematic name 7-phospho-2-dehydro-3-deoxy-D-*arabino*-heptonate D-erythrose-4-phosphate lyase [pyruvate-phosphorylating], EC 4.1.2.15); 3-dehydroquinate (DHQ) synthase (7-phospho-2-dehydro-3-deoxy-D-*arabino*-heptulosonate phosphate lyase [cyclising], EC 4.6.1.3); 3-dehydroquinate dehydratase, DHQase (3-dehydroquinate hydro-lyase, EC 4.2.1.10); shikimate dehydrogenase, SDH (shikimate: NADP$^+$ oxidoreductase, EC 1.1.1.25); shikimate kinase (ATP:shikimate 3-phosphotransferase, EC 2.7.1.71); 5-*enol* pyruvylshikimate 3-phosphate (EPSP) synthase (phospho*enol* pyruvate 3-phosphoshikimate 1-carboxyvinyltransferase, EC 2.5.1.19); and chorismate synthase (O^5-[1-carboxyvinyl]-3-phosphoshikimate phosphate-lyase, EC 4.6.1.4).

The pathway is identical in plants and microorganisms, as evidenced by the demonstration of the individual reactions, the incorporation of intermediates and the purification of the relevant enzymes (Kishore and Shah, 1988; Coggins, 1989). The plant enzymes are monofunctional, with the exception of a bifunctional DHQase: SDH; apart from *Euglena*, no examples of *arom* pentafunctional polypetides incorporating the five enzymes from DHQ synthase to EPSP synthase are known in photosynthetic organisms (Patel and Giles, 1979). The shikimate pathway in plants is present in plastids and there is evidence for a second, independently regulated pathway in the cytosol (Jensen, 1986); these plastidic and soluble pathways in higher plants may function in aromatic amino acid biosynthesis *per se* and in secondary aromatic biosynthesis, respectively (Poulsen and Verpoorte, 1991). The quantitative importance of the shikimate pathway in plants is greatly magnified by the accumulation of many low molecular weight aromatics and of lignin; over a third of the total dry weight of a plant may be shikimate-derived (Boudet *et al.*, 1985). The enzymes of shikimate pathway in plants present challenges to the preparative biochemist because of their low tissue abundance and the necessity for separating them from the complex phytochemical mixtures often represented by plant extracts. Their presence as isozymes also hinders the application of high resolution techniques and reduces overall yields. The choice of plant source is important and in general cell cultures not requiring exogenous aromatic amino acid supplements would appear to offer the best starting material. Enzyme analyses in 'non-ideal' tissues and species may, however, be of more biological and agronomic interest as, for example, investigations of variation in herbicide-inhibitor sensitivity in natural weed populations. Further definition of chemical and physiological treatments to increase the expression of shikimate pathway enzymes in plant tissues generally is therefore of value both for the enzymologist and for the study of metabolic regulation.

II.   3-DEOXY-D-*arabino*-HEPTULOSONATE 7-PHOSPHATE SYNTHASE

A.   Assay

1.   *Colorimetric (periodate/thiobarbiturate) method*

DAHP synthase is most commonly assayed using the periodate/thiobarbiturate procedure for the determination of DAHP (Weissbach and Hurwitz, 1959; Srinivasan and Sprinson, 1959). This analytical method is replete with technical difficulties and the experimental aspects will therefore be considered in some detail. The test requires periodate oxidation of DAHP to yield formylpyruvate which then, on heating, produces a pink/red chromagen with thiobarbituric acid. 2-Keto-3-deoxy-sugars react to give derivatives with absorption maxima in the range 545–550 nm (Weissbach and Hurwitz, 1959). Deoxy-sugars can be oxidised by periodate to give malondialdehyde which produces a chromagen with an absorption maximum at 532 nm (Waravdekar and Saslaw, 1957). Both E4P and chorismate give positive reactions at > 1 mM (Jensen et al., 1973; Huisman and Kosuge, 1974). Shikimic and dehydroshikimic acids are detected by the assay (Millican, 1970) and it is likely that any related alicyclic intermediate will produce a similar coloration.

The following method (Schoner and Herrmann, 1976) reflects the changes in reagent composition and reaction times that have been adopted since the assay was first described:

> 400 μl test solution is added to 100 μl 0.2 M sodium periodate in 9 M ortho-phosphoric acid; after 20 min at 25°C, 400 μl 0.75 M sodium arsenite in 0.5 M sodium sulphate and 50 mM sulphuric acid is added and mixed thoroughly; 3 ml 40 mM thiobarbituric acid is added and the mixture incubated at 100°C for 15 min.

After cooling to room temperature, the absorbance can be measured at 549 nm. Opalescence is often a problem at this stage and can be avoided by maintaining the solution at 55–60°C and by reading the absorbance in a thermostatted cuvette (Jensen and Nester, 1966). Centrifugation can clarify cloudy reaction mixtures but if the protein or pigment content is high considerable chromagen can be co-precipitated (Mousdale and Coggins, 1985a). Alternatively the chromagen is extracted by shaking with an equal volume of cyclohexanone and the phases (cyclohexanone is the upper layer) separated by a brief centrifugation; this procedure intensifies the absorbance (Warren, 1959). The molar extinction coefficient, ε ($M^{-1} cm^{-1}$), has been reported to be 33 000–68 000. This range probably reflects the differing purities of DAHP used and uncorrected inhibition by dithiothreitol (DTT; Huisman and Kosuge, 1974) and β-mercaptoethanol (Catala et al., 1974). Reliable quantitative measurements require therefore the determination of ε under the actual reaction conditions employed. DAHP can be isolated (as the lithium salt) from selected Escherichia coli mutants requiring aromatic supplements and its purity assessed enzymatically (Lambert et al., 1985; Coggins et al., 1987); a chemical synthesis has also been described (Herrmann and Poling, 1975). The substrates E4P and PEP are commercially available; E4P may however be as little as 60% pure (DAHP synthase provides a convenient assay). Impurities in commercial batches of E4P inhibiting DAHP synthase have been suspected (Jensen et al., 1973). In general, the smaller the amount of DAHP to be measured, the less reliable is the straightforward measurement at 549 nm; cyclohexanone extraction gives an improved sensitivity together with greater accuracy. Appropriate controls, lacking either substrate or enzyme, must however be included and analysed in parallel with the complete incubation mix.

DAHP synthase has been measured with the colorimetric assay in extracts from many plant species. Table 1.1 summarises representative assay conditions together with enzyme activities (uniformly expressed as nkat per g fresh weight or nkat per mg protein). The final reaction volumes were 0.1–0.5 ml, the assays terminated by adding trichloroacetic acid (TCA) to a concentration of c. 5% and protein removed by

TABLE 1.1. Assay conditions for plant DAHP synthases.

Species	E4P (mM)	PEP (mM)	Metal ion	pH	T (°C)	DAHP synthase (nkat g⁻¹)	DAHP synthase (nkat mg⁻¹)	Reference
Sweet potato	1	5	Mg(II)	7.4	30	0.1		Minamikawa and Uritani (1967)
Mung bean shoot	2	4	Mg(II)	7.4	30	105	39.5	Minamikawa (1967)
Cauliflower	2	2	Mn(II)	6.6	25	2.5	0.08	Huisman and Kosuge (1974)
Pea cotyledon	2	1	Co(II)	7.6	37	8.3	2.0	Rothe et al. (1976)
Tea leaf	2	2	Mn(II)	6.6	30	0.04		Saijo and Takeo (1979)
Maize shoot	1	2	none	7.5	30		0.33	Graziana and Boudet (1980)
Pea leaf	2	1	Co(II)	7.6	37	1.7–3.3		Reinink and Borstlap (1982)
Carrot cells	3	7.5	Mn(II)	6.5	25		0.03–0.15	Suzich et al. (1984)
Mung bean shoot	0.5	2	Mn(II)	7.0	37		0.08	Rubin and Jensen (1985)
Mung bean shoot	4.0	2	Co(II)	8.8	37		0.06	Rubin and Jensen (1985)
C. succirubra cells	0.5	0.5	none	6.4	30		0.03–0.34	Schmauder et al. (1985)
Carrot root	3	7.5	Mn(II)	7.0	25	0.22	0.35	Suzich et al. (1985)
Tobacco cells	0.6	3	Mn(II)	8.0	37		0.19	Ganson et al. (1986)
Tobacco cells	6	2	Mg(II)/Co(II)	8.6	37		1.67	Ganson et al. (1986)
Cryptomeria cells	0.13	0.13	Co(II)/Mn(II)	7.5	30	1.2–4.8		Ishikura et al. (1986)
Perilla cells	0.13	0.13	Co(II)/Mn(II)	7.5	30	1.3–10		Ishikura et al. (1986)
Potato tuber	3	7.5	Mn(II)	7.0	25	0.80	0.33	Pinto et al. (1986)
Potato cells	3	7.5	Mn(II)	7.0	25		0.8–4.7	Pinto et al. (1988)
Parsley cells	0.6	3	Mn(II)	8.0	37		0.03–0.05	McCue and Conn (1990)
Parsley cells	6	2	Co(II)	8.6	37		0.01–0.02	McCue and Conn (1990)

DAHP synthase: nkat g⁻¹ = activity per g fresh weight; nkat mg⁻¹ = specific activity (per mg protein).

centrifugation. Incubation times of up to 30 min were used but it is seldom stated that the linearity with time of the DAHP formed was checked.

2. Direct spectrophotometric assay

A direct assay using the decline in absorbance at 232 nm (PEP $\varepsilon = 2800 \, M^{-1} cm^{-1}$; DAHP $\varepsilon = 415 \, M^{-1} cm^{-1}$) follows the utilisation of PEP (Schoner and Herrmann, 1976). This method suffers from two drawbacks when applied to crude and partly purified plant extracts: firstly, the absorbance in the absence of PEP is high and often unstable, and secondly, PEP breakdown by phosphatase action causes high 'background' assay rates. The assay has however been successfully used during the purification of cauliflower and potato DAHP synthases. The assay conditions were:

(1) in a final volume of 0.5 ml, 100 mM PIPES (pH 6.6), 5 mM DTT, 2 mM MnCl₂, 0.3 mg ml⁻¹ bovine serum albumin (BSA), 1 mM PEP and (to initiate assay) 1 mM E4P; the absorbance change at 240 nm (for PEP $\varepsilon = 1370 \, M^{-1} cm^{-1}$) was monitored over 3–5 min at 25°C; 0.3 mM fluoride was added to suppress phosphatase activity (Huisman and Kosuge, 1974);
(2) in a final volume of 1 ml, 50 mM bis-Tris propane (pH 7.0), 1 mM DTT, 150 μM PEP and (to initiate assay) 0.25 mM E4P; other additions were made to assess their effects on kinetic parameters (Suzich et al., 1985).

The problems with this assay can be avoided by the forward-coupling of the assay to DHQ synthase, DHQase and SDH so that the overall assay is followed by the utilisation of NADPH at 340 nm (Stuart, 1990).

B. Purification and Properties

Mn(II)-activated DAHP synthases have been purified to electrophoretic homogeneity from carrot root and potato tuber (Suzich et al., 1985; Pinto et al., 1986). The cauliflower enzyme was purified to > 95% purity (Huisman and Kosuge, 1974). Comparison of the purification profiles indicates that a purification of 1000- to 3000-fold over crude extracts is required (Table 1.2). The yields of enzyme protein were: cauliflower, 1500 μg from a starting 1400 g of tissue; carrot, 710 μg from 5000 g of tissue; and potato, 900 μg from 2000 g of tissue. The two unambiguously purified DAHP synthases have similar macromolecular properties. The carrot root enzyme has an apparent molecular weight of 103 000 and consists of two identical subunits of molecular weight 53 000 (Suzich et al., 1985); the values for the potato enzyme are 110 000 and 54 000 (Pinto et al., 1986).

No detailed study of the effects of extraction conditions or buffers has been published. β-Mercaptoethanol is essential for maximal stability (Ganson et al., 1986). This is not surprising, given the sensitivity of DAHP synthase to thiol-reactive compounds (Huisman and Kosuge, 1974). Since EDTA is also an inhibitor, this compound should be omitted from the buffers used in the purification schedule (Huisman and Kosuge, 1974). Poly-vinylpyrrolidone (PVP) has been included in extraction media to absorb reactive phenolic compounds (Suzich et al., 1985: Pinto et al., 1986).

TABLE 1.2. Purification of plant DAHP synthases.

Stage	Cauliflower				Carrot				Potato			
	A	SA	Y	P	A	SA	Y	P	A	SA	Y	P
Extract	3434	0.08	100	1	1090	0.35	100	1	1607	0.33	100	1
Protamine sulphate	3184	0.37	93	5								
0–60% (NH$_4$)$_2$SO$_4$	2450	0.45	71	6								
0–50% (NH$_4$)$_2$SO$_4$												
30–40% (NH$_4$)$_2$SO$_4$	1467	0.92	43	12								
24–36% (NH$_4$)$_2$SO$_4$									1249	1.33	78	4
DEAE cellulose	1050	9.7	31	121	275	8.8	25	25				
Phosphocellulose									808	32	50	97
Sulphopropyl Sephadex									432	253	27	767
Isoelectric focusing	343		10									
Matrix Gel Orange A (I)					238	45	22	129	247	273	15	827
Matrix Gel Orange A (II)					90	107	8	308				
Hydroxylapatite					83	118	8	337				
Ultracentrifugation	150	100	4	1250								

Data recalculated from: Huisman and Kosuge (1974), Suzich *et al.* (1985), Pinto *et al.* (1986).
A, activity (nkat); SA, specific activity (nkat mg^{-1} protein); Y, yield (%); P, purification (fold).

The range of substrate and cofactor concentrations in Table 1.1 partly reflect historical development, in particular the recognition of the isozymic status of higher plant DAHP synthases. The present knowledge can be summarised (Rubin et al., 1982; Rubin and Jensen, 1985; Ganson et al., 1986; Morris et al., 1989; McCue and Conn, 1990) as:

(1) there are at least two forms of DAHP synthase (resolvable by anion-exchange chromatography) which are widely distributed in higher plants;
(2) one isozyme (DS-Mn) requires the sulphydryl reductant DTT and Mn(II) for maximal activity;
(3) a second isozyme (DS-Co) is inhibited by DTT and requires a divalent metal ion cofactor, optimally Co(II);
(4) DS-Co has a higher pH optimum than DS-Mn;
(5) the plastidic isozyme is the DS-Mn form, while the DS-Co form is entirely cytosolic.

In the absence of an added metal ion, or with Co(II) as the exogenous cofactor, the 'DAHP synthase' measured is the sum of the DS-Co form and < 100% of the DS-Mn form (the precise amount being a function of the Mn(II)-activation of the latter isozyme). Three isozymes have been resolved in pea extracts by gradient ammonium sulphate fractionation (Rothe et al., 1976) and in carrot cell extracts by phospho-cellulose chromatography (Suzich et al., 1985). A complication is that in at least one species (potato) there is a deoxy-D-manno-octulosonate 8-phosphate synthase using PEP and arabinose 5-phosphate but which can also catalyse the DAHP synthase reaction (Morris et al., 1989).

The aspect of the kinetics of DAHP synthases which is of most biochemical interest is the feedback effect of phenylalanine, tyrosine or tryptophan. By analogy with microbial DAHP synthases, feedback inhibition (by the individual amino acids, or by their concerted/antagonistic action) might be expected (Herrmann, 1983). In only one instance (Cinchona succirubra cells) has a general inhibition by phenylalanine, tyrosine and tryptophan been reported (Schmauder et al., 1985). No feedback inhibition could be demonstrated with the DAHP synthases from cauliflower (Huisman and Kosuge, 1974), Camellia sinensis (Saijo and Takeo, 1979) or maize root (Graziana and Boudet, 1980). The DAHP synthase from maize shoot was however inhibited by tryptophan with an apparent K_1 of 65 μM (Graziana and Boudet, 1980). Prephenate, arogenate and (weakly) tryptophan inhibited the DS-Mn isozyme from mung bean while chorismate was an activator; the DS-Co form was inhibited by 3,4-dihydroxycinnamic acid (Rubin and Jensen, 1985). In contrast, the enzymes from carrot cells and potato tubers were activated by tryptophan and tyrosine (Suzich et al., 1985; Pinto et al., 1986). An important technical point is that tryptophan itself will inhibit the DAHP colorimetric reaction (Jensen et al., 1973) and appropriate controls must be included to assess the extent of any enzymic inhibition.

III. 3-DEHYDROQUINATE SYNTHASE

A. Assay

1. DAHP utilisation

DHQ synthase catalyses the cyclisation of DAHP to DHQ. The reaction can be followed by the loss of DAHP using the colorimetric methods described in Section II.A.1. The following protocol (used with pea shoot extracts) is typical (Pompliano et al., 1989):

in a final volume of 2 ml, 50 mM MOPS buffer (pH 7.5); 0.25 mM NAD; 0.66 mM DAHP; 0.25 mM $CoCl_2$; with enzyme (< 20 μl) or substrate to initiate the reaction at 15°C; 200 μl aliquots are withdrawn every 15–20 s up to 2 min and quenched immediately with 100μl 10% (v/v) TCA; after centrifugation the supernatant is assayed for DAHP (Gollub et al., 1970).

2. Inorganic phosphate liberation

Phosphate liberated during the cyclisation can be asayed by several sensitive techniques; that of Ames (1966) has been used for pea shoot extracts (Mehdi et al., 1987; Pompliano et al., 1989):

100 μl aliquots of the reaction mix are quenched immediately with 200 μl 1% (w/v) sodium dodecyl sulphate (SDS); 0.7 ml ascorbate-molybdate reagent is added and the samples incubated at 45°C for 20 min; after cooling to room temperature, the absorbance at 820 nm is read ($\varepsilon = 26\ 000\ \text{M}^{-1}\text{cm}^{-1}$).

3. Coupled enzyme assay

In the presence of excess DHQase, DHQ synthase can be continuously monitored by the increase in absorbance at 234 nm (Welch and Gaertner, 1976; Lambert et al., 1985). DHQase can be purified from wild-type or DHQase-overproducing E. coli (Lambert et al., 1985; Coggins et al., 1987). The following assay protocol was used with pea and spinach chloroplast preparations (Mousdale and Coggins, 1985a: Coggins, 1989):

in a final volume of 1 ml, 100 mM phosphate buffer (pH 7.0), 50 μM NAD, 0.1 mM $CoCl_2$, 0.4 mM DAHP, 0.5 nkat DHQase and < 50 μl enzyme; the reaction is initiated by the addition of substrate and the absorbance at 234 nm is monitored for 2–10 min at 25°C.

Both the DAHP and phosphate methods suffer from interference from phosphatases (and inorganic phosphate levels), especially in crude extracts. One possible correction is to determine the Co(II)- and NAD-dependent activity, although this may underestimate the total activity, depending on the quantitative effect of cofactor addition—see Section III.B. The coupled enzyme assay avoids these problems; the absorbance at 234 nm may, however, be very high in crude extracts, and the assay is best carried out by the addition of DAHP after a 2–5 min monitoring of the absorbance at 234 nm. Chloroplast lysates and derived fast protein liquid chromatography (FPLC) fractions offer excellent baseline stability.

B. Purification and Properties

DHQ synthase activity in a higher plant was first demonstrated in mung bean and pea seedlings (Ahmed and Swain, 1970). Subsequently the enzyme has been measured in extracts of: tobacco and the moss *Physcomitrella patens* (Berlyn *et al.*, 1970), mung bean (Yamamoto and Minamikawa, 1976; Yamamoto, 1977; Koshiba, 1979a; Yamamoto, 1980), sorghum, spinach, potato and sweet potato (Saijo and Kosuge, 1978), *Camellia sinensis* (Saijo and Takeo, 1979; Saijo, 1980), pea and spinach chloroplasts (Mousdale and Coggins, 1985a: Coggins, 1989) and potato cells cultures (Pinto *et al.*, 1988). The enzyme has been purified to electrophoretic homogeneity from mung bean and pea (Yamamoto, 1980; Pompliano *et al.*, 1989). The two procedures are compared in Table 1.3. The much higher initial activity in the mung bean extract is attributable to a higher tissue protein content and the higher assay temperature (30°C). The yields of enzyme protein were: 448 μg from a starting 10 000 g of mung bean seedlings and 300 μg from a starting 1300 g of pea shoots.

The purified DHQ synthase from mung bean has an apparent molecular weight of 67 000 and a subunit molecular weight of 43 000 (Yamamoto, 1980). The corresponding values for the pea enzyme are 66 000 and 33 500 (Pompliano *et al.*, 1989). The enzyme is probably therefore a homodimer. Plant DHQ synthases show a metal ion requirement, activity being stimulated by up to four-fold (Saijo and Kosuge, 1978; D. M. Mousdale and J. R. Coggins, unpubl. res.). The enzyme from sorghum was reported to be stimulated by Co(II); Mn(II), Cu(II) and Zn(II) had no effect; Co(II) also reversed the inhibition by EDTA (Saijo and Kosuge, 1978). The mung bean enzyme was activated by Co(II) and Cu(II); Cu(II) could be detected in the purified enzyme (Yamamoto, 1977, 1980). The DHQ synthase activity of the *arom* complex from *Neurospora crassa* is now thought to be Zn(II)-dependent (Lambert *et al.*, 1985). A practical consequence is that inclusion of Co(II) in buffers during enzyme extraction and purification is beneficial (Frost *et al.*, 1984; Pompliano *et al.*, 1989). In crude

TABLE 1.3. Purification of plant DHQ synthases.

Stage	Pea				Mung bean			
	A	SA	Y	P	A	SA	Y	P
Extract	33	0.002	100	1	3750	0.03	100	1
40–65% (NH$_4$)$_2$SO$_4$	42	0.005	127	2.5				
40–55% (NH$_4$)$_2$SO$_4$					5320	0.18	142	6
DEAE-cellulose (I)	55	0.04	167	20				
DEAE-cellulose					3310	1.13	88	38
DEAE-cellulose (II)	45	0.16	136	80				
Sephadex G-100					1960	2.54	52	85
Sephadex G-75	18	0.33	55	165				
Phosphocellulose					742	76.5	20	2550
Blue Sepharose	27	1.0	82	500				
Sephacryl S-200					676	188	18	6267
Dye Matrex Green	3	9.8	9	4900				
Blue Sepharose					364	813	10	27 100

Data recalculated from: Yamamoto (1980) and Pompiliano *et al.* (1989).
A, activity (nkat); SA, specific activity (nkat mg^{-1} protein); Y, yield (%); P, purification (fold).

extracts and chloroplast preparations exogenous NAD stimulates by up to 50% (Saijo and Kosuge, 1978; D. M. Mousdale and J. R. Coggins, unpubl. res.). There is recent evidence for a bound NAD coenzyme in *E. coli* DHQ synthase (also probably Zn(II)-containing) which may dissociate during the catalytic cycle (Bender *et al.*, 1989).

IV. DEHYDROQUINATE DEHYDRATASE:SHIKIMATE DEHYDROGENASE

A. Bifunctional DHQase:SDH Enzymes in Plants

In the blue-green alga *Anabaena variabilis* the DHQase and SDH activities are separable by sucrose density gradient centrifugation (Berlyn *et al.*, 1970); this has been confirmed with FPLC using the Mono-Q anion-exchanger (F. Macleod and J. R. Coggins, unpubl. res). In most higher photosynthetic organisms investigated, including a green alga, *Euglena*, a moss and dicotyledonous species, the two activities are not resolvable and co-purify (Berlyn *et al.*, 1970; Boudet, 1971; Boudet and Lecussan, 1974; Polley, 1978; Fiedler and Schultz, 1985; Mousdale *et al.*, 1987). In maize there is a second DHQase in a bifunctional enzyme with quinate dehydrogenase (Graziana *et al.*, 1980); this situation may occur in a range of monocotyledonous species (Boudet *et al.*, 1977).

B. Assay

1. Dehydroquinate dehydratase

DHQase is assayed by the increase in absorbance at 234 nm resulting from the dehydration of DHQ to form the unsaturated alicyclic acid dehydroshikimate, DHS; the extinction coefficient for this is $11\,900\,\mathrm{M^{-1}\,cm^{-1}}$ (Salamon and Davis, 1953; Mitsuhashi and Davis, 1954). The following protocol, used with extracts of pea, spinach, maize, carrot cell cultures, wheat, lettuce and *Arabidopsis thaliana* (Mousdale and Coggins, 1985a and unpubl. res.), is typical:

> in a final volume of 1 ml, 100 mM phosphate buffer (pH 7.0), enzyme and (to initiate assay) 0.8 mM DHQ; the increase in absorbance is monitored for 1–5 min at 25°C.

DHQ is prepared from the commercially available quinic acid and is isolated as the ammonium salt (Grewe and Haendler, 1966). The purity can be assessed enzymatically or by high performance liquid chromatography (HPLC) (Mousdale and Coggins, 1985b; Coggins *et al.*, 1987). High background absorbances at 234 nm are encountered in crude extracts and the stability of the baseline should be monitored before substrate addition. Centrifugation (at $80\,000 \times g$ or higher) greatly reduces the UV absorbance of tissue extracts (Mousdale *et al.*, 1987).

2. Shikimate dehydrogenase

SDH is the most commonly analysed enzyme of the shikimate pathway in plants because of the commercial availability of shikimic acid. A sensitive activity stain for SDH on starch or polyacrylamide gel has been widely used in studies of SDH isozymes and their use as markers for genetic variation, inheritance and species/cultivar differentiation (a method is given by Weeden and Gottlieb, 1980).

SDH is routinely assayed in the reverse (dehydrogenase) direction by the increase in absorbance at 340 nm accompanying NADP reduction. The following protocol (Mousdale and Coggins, 1985a), is typical:

> in a final volume of 1 ml, 100 mM sodium carbonate/bicarbonate buffer (pH 10.6), 2.0 mM NADP, enzyme and (to initiate assay) 4.0 mM shikimate; the increase in absorbance at 340 nm is followed for 1–5 min at 25°C.

The extinction coefficient for the reduction of NADP is often quoted as 6180–6220 $M^{-1}cm^{-1}$, but a more recent estimate is 6300 (used by Boehringer Mannheim UK, Lewes, East Sussex, UK). In crude tissue extracts, a positive assay rate is encountered before addition of the shikimate; this is often unstable and declines over 5 min. The background absorbance at 340 nm is greatly reduced by high speed centrifugation. The high pH in the assay is unsuitable with the enzyme from some species, for example spinach leaf, which requires a pH of 7.0–9.5 in phosphate or glycine-NaOH buffers (Fiedler and Schultz, 1985; D. M. Mousdale and J. R. Coggins, unpubl. res). The use of NADP in the assay allows a much higher sensitivity if fluorimetric detection is employed; typical conditions are λ_{ex} 340 nm/λ_{em} 455 nm (Nazfiger *et al.*, 1984). For the forward (DHS reductase) assay, the substrate must be prepared either chemically from quinic acid or shikimic acid (Haslam *et al.*, 1963) or enzymatically from DHQ (Coggins *et al.*, 1987). NADPH is unstable in aqueous solution and should be freshly prepared for kinetic studies.

C. Purification and Properties

Bifunctional DHQase:SDH enzymes have been purified to electrophoretic homogeneity from *Physcomitrella patens* (Polley, 1978), spinach chloroplasts (Fiedler and Schultz, 1985) and pea leaf and chloroplasts (Mousdale *et al.*, 1987). Considerable purifications may be required, several thousand-fold over whole-tissue extracts (Table 1.4). Ammonium sulphate has been reported to inhibit the spinach enzyme (Fiedler and Schultz, 1985). This inhibition can be encountered with SDH from other species but is entirely reversible by the repeated dialysis required to reduce the high conductivity of ammonium sulphate fractions before anion-exchange chromatography; this is a particular priority with pea DHQase:SDH which shows relatively poor binding to anion-exchange matrices (Mousdale and Coggins, 1985a; Mousdale *et al.*, 1987). The yields of purified DHQase:SDH obtained were: 200 μg from a starting 400 g fresh weight of *P. patens* protonema, 60 μg from 8000 g of spinach leaves and 14 μg from 250 g of pea shoots.

A rapid procedure purifies pea chloroplast DHQase:SDH by two tandem FPLC steps (Mousdale *et al.*, 1987). The chromatofocusing step on Mono-P is effective because of the relatively high isoelectric point (*c.* 5.9) and the early elution of the chloroplast isozymes; a much less dramatic purification is seen where the (apparent) isoelectric point is lower—for example, spinach DHQase:SDH with this procedure elutes at a pH below 5.0, even though the gel isoelectric focusing value (presumably for all four isozymes present) is reported to be 6.4 (Fiedler and Schultz, 1985).

TABLE 1.4. Purification of bifunctional DHQase:SDH from plants.

Stage	Physcomitrella				Spinach				Pea			
	A	SA	Y	P	A	SA	Y	P	A	SA	Y	P
Extract	7201	2.0	100	1	92	0.4	100	1	1471	1.0	100	1
Protamine sulphate	8135	2.3	113	1								
50–75% $(NH_4)_2SO_4$	7118	7.8	99	4								
35–55% $(NH_4)_2SO_4$									1331	2.6	91	3
DEAE-cellulose	5784	26.0	80	13								
DEAE-Sepharose					76	3.8	83	10				
DEAE Sephacel									757	11.5	52	12
Ultracentrifugation					66	3.8	72	10				
Sephacryl S-300					62	20.4	67	51				
Blue Sepharose	1250	567	17	284	53	63.5	58	159				
ADP-Sepharose									621	163	42	163
Hydroxylapatite	533	2667	7	1334	36	605	39	1513	389	229	26	229
Mono-Q									192	2400	13	2400
HPLC gel filtration									90	6443	6	6443

Data (SDH) recalculated from: Polley (1978), Fiedler and Schultz (1985) and Mousdale *et al.* (1987). A, activity (nkat); SA, specific activity (nkat mg^{-1} protein); Y, yield (100%); P, purification (fold).

Purified DHQase:SDH enzymes are monomeric with molecular weights (as determined by SDS-PAGE) of 48 000 (Polley, 1978), 59 000 (Fiedler and Schultz, 1985) and 60 000 (Mousdale *et al.*, 1987). The SDH:DHQase ratio is much lower for *P. patens* (3.5:1) than for spinach (9:1) and pea (10:1), although little variation is apparent at the various stages of the purifications (Polley, 1978; Fiedler and Schultz, 1985; Mousdale *et al.*, 1987). Two or three isozymes can be recognised in activity-stained gels of crude extracts (Weeden and Gottlieb, 1980; Linhart *et al.*, 1981; Gottlieb, 1981). The isozymes in spinach chloroplasts and pea shoot extracts have very similar SDH:DHQase ratios (Fielder and Schultz, 1985; Mousdale *et al.*, 1987 and unpubl. res.). The presence of isozymes does however limit the potential degree of purification obtainable with high resolution steps (for example HPLC/FPLC) as enzyme activity 'spreads' over more fractions than would be found with a single molecular species. DHQase and SDH both co-sediment with intact chloroplasts in density gradient centrifugation and exhibit latency (Mousdale and Coggins, 1985a; Mousdale *et al.*, 1987). There is, however, evidence for a dual localisation of SDH in the cytosol and plastids (Rothe *et al.*, 1983) and a minor isozyme observed in pea leaf appears to be absent from chloroplasts (Mousdale *et al.*, 1987).

V.   SHIKIMATE KINASE

A.   Assay

1. Fluorimetric coupled assay

Shikimate kinase activity was first demonstrated in plants using a coupled assay containing the enzymes EPSP synthase, chorismate synthase and anthranilate synthase provided by a crude extract of a *Neurospora crassa* mutant lacking shikimate kinase; the anthranilate formed from shikimate and ATP in the presence of $MgCl_2$, PEP, NADP and L-glutamine was measured fluorimetrically (Ahmed and Swain, 1970; Berlyn *et al.*, 1970). While this assay is primarily of historical interest, it does offer a very sensitive technique for use in both stopped-time and continuous assays if excess coupling enzymes can be provided.

2. Shikimate 3-phosphate determination

Using uniformly-labelled ^{14}C-shikimate as substrate, the product can be separated by precipitation as the barium salt. This method was developed for use with sorghum (Bowen and Kosuge. 1977) and has subsequently been applied to extracts of *Corydalis sempervirens* cell cultures (Smart *et al.*, 1985):

in a final volume of 200 μl, 50 mM PIPES buffer (pH 6.8) or 100 mM Tris (pH 9.0), 10 mM $MgCl_2$, 10 mM potassium phosphate, 10 mM sodium fluoride, 4 mM ATP, 1 mM [U-^{14}C] shikimate and (to initiate assay) 50 μl sample; after 30 min at 30°C the reaction is terminated by adding 50 μl 1.25 M TCA; 50 μl 1.25 M NaOH/50 mM ammonium sulphate, 50 μl 2 M barium acetate and 1.75 ml 95% ethanol are added and the mixture chilled for 10 min before vacuum-filtration through glass fibre filters (Whatman GF/C, 25 mm diameter); the filters are rinsed with 2 × 1 ml 50% ethanol and dried with acetone before liquid scintillation counting.

The substrate and product can be separated by thin-layer or ion-exchange column chromatography (Ahmed and Swain, 1970; Koshiba, 1978, 1979a, b; Bowen and Kosuge, 1979; Schmidt *et al.*, 1990). The following method is that most recently described (Schmidt *et al.*, 1990):

in a final volume of 60 μl, 100 mM glycine–NaOH buffer (pH 10.0), 10 mM MgCl$_2$, 5 mM DTT, 4 mM ATP, 1 mM [U-^{14}C] shikimate; after 10–60 min at 30°C the reaction is terminated by adding 10 μl of 10.5% (w/v) TCA; the precipitate is removed by centrifugation (10 min, 15 000 × g) and 50 μl of the supernatant fraction is transferred to a column containing 83 mg Dowex 1-X8 (Cl$^-$ form); shikimate is eluted with 0.3 mM acetic acid (7 × 200 μl) and shikimate 3-phosphate with 1 M NaCl (5 × 200 μl); the eluates are collected directly into scintillation vials and 5 ml scintillation fluid added before radioactivity counting.

Labelled shikimate can be obtained from New England Nuclear/Du Pont with a specific activity of up to 1.11 GBq mmol^{-1}. Potassium fluoride at 2.5 mM has been used to suppress phosphatase activity (Koshiba, 1978, 1979a, b; Bowen and Kosuge, 1979). Shikimate kinase has been measured using this approach in crude and purified extracts of mung bean, pea, sorghum and spinach chloroplasts.

3.   Coupled spectrophotometric assay

A convenient and continuous spectrophotometric assay follows the reaction by coupling ADP formation in the assay to NADH oxidation via pyruvate kinase (PK) and L-lactate dehydrogenase (LDH; Smith and Coggins, 1983; Coggins *et al.*, 1987). The following protocol has been used with chloroplast preparations from pea, spinach and lettuce (Mousdale and Coggins, 1985a; Coggins, 1989; D. M. Mousdale and J. R. Coggins, unpubl. res.):

in a final volume of 1 ml, 50 mM triethanolamine–HCl buffer (pH 7.0), 2.5 mM ATP, 1 mM PEP, 0.1 mM NADH, 2.5 mM MgCl$_2$, 50 mM KCl, 60 nkat PK, 50 nkat LDH and (to initiate assay) 1 mM shikimate; the absorbance at 340 nm is monitored for 1–5 min at 25°C.

Mixed coupling enzymes can be obtained from Boehringer Mannheim. The NADH solution is unstable and should be stored at −20°C.

The assay suffers from unworkably high background assay rates in crude and partially purified whole plant extracts in the absence of shikimate because of ATPase activity; this problem is much reduced in lysed and high-speed centrifuged chloroplast preparations and in shikimate kinase-containing fractions derived from these by HPLC or FPLC (Mousdale and Coggins, 1985a; Coggins, 1989).

4.   Shikimate utilisation

The disappearance of shikimate during the reaction can monitored by chemical assays for shikimate such as the alkaline periodate colorimetric assay (Gaitonde and Gordon, 1958). Unless the shikimate kinase activity is high, this approach requires care and precision for the measurement of the loss of shikimate from millimolar starting concentrations, but has been used with tobacco callus cultures (Beaudoin-Eagan and Thorpe, 1983).

B. Purification and Properties

Shikimate kinase has been purified to very close to electrophoretic homogeneity from spinach chloroplasts (Schmidt *et al.*, 1990) and partially purified from sorghum and mung bean (Bowen and Kosuge, 1979; Koshiba, 1979a). The procedures are outlined and compared in Table 1.5. The relative lack of success in purifying plant shikimate kinase is attributed to its instability and low tissue concentrations. Approximately 10 μg of the enzyme was obtained from a starting 1500 g of spinach leaves (Schmidt *et al.*, 1990).

The spinach chloroplast enzyme is a monomer of molecular weight 31 000 (Schmidt *et al.*, 1990). The purified enzyme is unstable and can be stabilised by added proteins, including oxidised or reduced thioredoxins (Schmidt and Schultz, 1987: Schmidt *et al.*, 1990). Magnesium is the optimum divalent metal cofactor with all sources of the enzyme. The optimum pH is alkaline, 8.6–9.6 (Bowen and Kosuge,1979; Koshiba, 1979a; Schmidt *et al.*, 1990). ADP is consistently found to be an inhibitor and this may have relevance to a regulation by the energy charge or the ADP:ATP ratio.

VI. EPSP SYNTHASE

A. Assay

1. Fluorimetric coupled assay

EPSP synthase activity was first demonstrated in plants using a coupled assay containing the enzymes chorismate synthase and anthranilate synthase in a procedure analogous to that described for shikimate kinase in Section V.A.1 (Ahmed and Swain, 1970; Berlyn *et al.*, 1970). This method was later adapted to use the purified coupling enzymes (Gaertner and DeMoss, 1970; Boocock and Coggins, 1983). The following protocol has been used with tomato cell extracts (Smith *et al.*, 1986):

> in a final volume of 1.5 ml, 50 mM MOPS buffer (pH 7.0), 500 μM PEP, 500 μM shikimate 3-phosphate (S-3-P), 100 μM NADPH, 10 μM flavin mononucleotide (FMN), 20 μM L-glutamine, 5 mM MgSO$_4$, 133 pkat chorismate synthase and 83 pkat anthranilate synthase; formation of anthranilate is monitored at 25°C using λ_{ex} 315 nm, λ_{em} 380 nm.

S-3-P can be isolated from an EPSP synthase-negative strain of *Klebsiella pneumoniae* and its purity assessed enzymatically and by HPLC (Coggins *et al.*, 1987).

2. Forward-coupled spectrophotometric assay

If, in the above procedure, anthranilate synthase is omitted, the formation of EPSP can be monitored by the formation of chorismic acid at 275 nm (Boocock and Coggins, 1983). The following method has been used with pea, spinach, maize and lettuce (Mousdale and Coggins, 1984, 1987):

> in a final volume of 1 ml, 50 mM triethanolamine buffer (pH 7.0), 50 mM KCl, 2.5 mM MgCl$_2$, 10 μM NADPH, 10 μM FMN, 500 μM S-3-P, 80 pkat chorismate synthase and (to initiate assay) 500 μM PEP; the reaction is monitored for 1–10 min at 275 nm.

TABLE 1.5. Purification of plant shikimate kinases.

Stage	Spinach				Sorghum				Mung bean			
	A	SA	Y	P	A	SA	Y	P	A	SA	Y	P
Extract	27	0.16	100	1	1.2	0.06	100	1	208 375	21.0	100	1
35–55% (NH$_4$)$_2$SO$_4$									143 862	46.5	69	2
DEAE-cellulose									95 686	146	46	7
Sephadex G-75	24	20.9	89	131								
Hydroxylapatite									20 671	265	10	13
Q-Sepharose	14	140	52	875								
Sephacryl S-200									2784	517	1.3	25
Blue Sepharose					1.2	0.95	100	16				
Dextran Blue Agarose	2.8	280	10	1750								

Data recalculated from: Bowen and Kosuge (1979), Koshiba (1979b) and Schmidt *et al.* (1990).
A, activity (nkat); SA, specific activity (nkat mg^{-1} protein); Y, yield (100%); P, purification (fold).

The assay is quantified using $\varepsilon_{275} = 2630\,\mathrm{M}^{-1}\,\mathrm{cm}^{-1}$ (Gibson, 1970). A very similar assay at pH 7.8 was used for *Corydalis sempervirens* and petunia cell cultures (Smart *et al.*, 1985; Steinrucken *et al.*, 1986). Because of the short wavelength used, this assay is often very difficult to apply to crude or partially purified plant extracts.

3. Reverse-coupled spectrophotometric assay

A convenient spectrophotometric assay follows the reverse reaction by coupling PEP formation in the assay to PK–LDH-coupled NADH oxidation (Lewendon and Coggins, 1983; Coggins *et al.*, 1987). The following protocol has been used with pea, spinach, maize and lettuce (Mousdale and Coggins, 1984, 1987):

> in a final volume of 1 ml, 100 mM potassium phosphate buffer (pH 7.0), 2.5 mM ADP. 0.1 mM NADH, 2.5 mM MgCl$_2$, 50 nkat PK, 42 nkat LDH and (to initiate assay) 50 µM EPSP; NADH oxidation is followed at 340 nm for 1–10 min at 25 °C.

EPSP can be prepared enzymatically using either purified *E. coli* EPSP synthase or *Neurospora crassa arom* (Coggins *et al.*, 1987) and isolated as its barium salt (Knowles *et al.*, 1970). The assay suffers from very high background assay rates in the absence of EPSP because of 'NADH oxidase' activity in crude whole plant extracts: this problem is much reduced after high-speed centrifugation, which removes considerable amounts of protein and pigment, and is not encountered after ammonium sulphate fractionation. Chloroplast preparations are readily assayable.

4. Phosphate release from PEP

S-3-P-dependent phosphate release from PEP has been used with the sensitive Malachite Green determination of inorganic phosphate (Lanzetta *et al.*, 1979) as a stopped-time assay method for EPSP synthase from tomato, *C. sempervirens*, carrot, tobacco and petunia (Nazfiger *et al.*, 1984; Rubin *et al.*, 1984; Schulz *et al.*, 1985; Smart *et al.*, 1985; Steinrucken *et al.*, 1986; Padgette *et al.*, 1987). Phosphatase interference is the major problem encountered with this procedure; ammonium heptomolybdate has been claimed to effectively inhibit phosphatase activity (Nazfiger *et al.*, 1984). The following procedure is typical (Steinrucken *et al.*, 1986):

> in a final volume of 100 µl, 50 mM HEPES buffer (pH 7.0), 1 mM S-3-P, 1 mM PEP, and 0.1 mM ammonium heptomolybdate; after a 5 min incubation at 30 °C, the reaction is initiated by the addition of enzyme; the reaction is terminated by adding 1.0 ml of the Malachite Green reagent, followed by 0.2 ml 34% citric acid after a further 60 s; control reactions are performed which either lack S-3-P or contain 10 mM glyphosate.

5. Other

S-3-P-dependent PEP utilisation has also been monitored in stopped-time assays using pyruvate kinase/lactate dehydrogenase-catalysed oxidation of NADH as a measure of residual PEP (Rubin *et al.*, 1984); phosphatase controls needed to be included. The preparation of [^{14}C]S-3-P and its use in a paper chromatography radiometric assay have been described (Sharps, 1984). Conversion of [^{14}C]PEP or [^{14}C]S-3-P to labelled

TABLE 1.6. Purification of plant EPSP synthases.

Stage	Pea				Sorghum				Petunia cells			
	A	SA	Y	P	A	SA	Y	P	A	SA	Y	P
Extract	37	0.03	100	1	1149	0.27	100	1	10 700	7.8	100	1
45–65% (NH$_4$)$_2$SO$_4$	34	0.10	92	3								
40–65% (NH$_4$)$_2$SO$_4$					1259	0.72	110	3	11 600	17	108	2
DEAE-Sephacel	29	0.72	78	24								
DEAE-Cellulose					908	2.8	79	10	8850	118	83	15
Hydroxylapatite					402	7.8	35	29	6140	165	57	21
Phenyl-Sepharose	24	6.5	65	217								
Phenyl-Agarose					366	90	32	333	5240	463	49	59
Cellulose phosphate	9	95	24	3167								
Sephacryl S-200					241	352	21	1304	3410	650	32	83

Data recalculated from: Mousdale and Coggins (1984), Steinrucken *et al.* (1986) and Ream *et al.* (1988).
A, activity (nkat); SA, specific activity (nkat mg^{-1} protein); Y, yield (100%); P, purification (fold).

EPSP, and their separation by HPLC, was a method devised for kinetic studies (Padgette *et al.*, 1987).

B. Purification and Properties

EPSP synthases have been purified to electrophoretic homogeneity from pea, *C. sempervirens*, petunia and sorghum (Mousdale and Coggins, 1984; Smart *et al.*, 1985; Steinrucken *et al.*, 1986; Ream *et al.*, 1988). Table 1.6 compares purifications from a dicotyledonous and a monocotyledonous species and from glyphosate-adapted, EPSP synthase-overexpressing cell cultures. The yields of EPSP synthase protein were: 6 μg from a starting 250 g fresh weight of pea shoots, 680 μg from 450 g of etiolated sorghum shoots and 5.2 mg from 360 g of petunia cells. The cDNA for the petunia enzyme has been cloned and expressed in *E. coli*, thus enabling the purification of relatively large amounts for mechanistic studies (Padgette *et al.*, 1987). Conversely, a rapid small-scale procedure has been described starting from chloroplast preparations which is applicable to a range of higher plant species (Mousdale and Coggins, 1986a, 1987).

The purified plant EPSP synthases are very similar structurally, being monomers of molecular weight 44 000–48 000. Their most significant kinetic parameter is a very potent inhibition by the herbicide glyphosate: this has been found with all higher plant EPSP synthases and is now recognised as the primary site of herbicide action (Kishore and Shah, 1988; Coggins, 1989). The much greater sensitivity of the plant as compared with microbial EPSP synthases became apparent as soon as comparative studies were undertaken (Duncan *et al.*, 1984; Schulz *et al.*, 1985). Tissue cultures adapted to glyphosate produce higher cellular levels of EPSP synthase, although the enzyme itself is still as susceptible to inhibition (Amrhein *et al.*, 1983; Nazfiger *et al.*, 1984; Smart *et al.*, 1985; Smith *et al.*, 1986; Steinrucken *et al.*, 1986). High resolution anion-exchange chromatography resolves a minor isozyme in extracts of pea shoot (Mousdale and Coggins, 1985a) and three isozymes in etiolated sorghum extracts (Ream *et al.*, 1988).

VII. CHORISMATE SYNTHASE

A. 'Aerobic' and 'Anaerobic' Chorismate Synthases

Chorismate synthase is the least studied enzyme of the shikimate pathway in plants. This is because of the technical difficulties which its assay presents. Two types of chorismate synthases can be defined: those which can only be assayed under anaerobic conditions—for example, *E. coli* and *Salmonella typhimurium* (Gollub *et al.*, 1967; Morell *et al.*, 1967)—and those for which no such precautions need to be taken, assay requiring only the addition of NADPH and FMN—for example *Neurospora crassa* and *Bacillus subtilis* (Welch *et al.*, 1974; Hasan and Nester, 1978). The essential difference between these two types of enzyme is that the latter contains a diaphorase, i.e. a flavoprotein catalysing the redox reaction between a reduced pyridine nucleotide and a suitable acceptor (usually a dye such as 2,6-dichlorophenol-indophenol). The diaphorase activity is associated with chorismate synthase either in a bifunctional polypeptide (White *et al.*, 1988) or as an enzyme complex (Hasan and Nester, 1978); monofunctional chorismate synthases lack this diaphorase and it is still unclear how their active site is maintained in a catalytically active reduced form. Both mono- and dicotyledonous species lack bifunctional chorismate synthases and enzyme activity has only been demonstrated in the presence of the powerful reducing agent dithionite (Mousdale and Coggins, 1986b and unpubl. res.; Schaller *et al.*, 1990).

B. Assay

In the presence of L-glutamine and anthranilate synthase, anthranilate formation from EPSP can be monitored by fluorimetric analysis. The two published assay protocols are very similar:

(1) in a final volume of 3 ml, 50 mM bis-Tris-HCl buffer (pH 7.0), 50 mM KCl, 5 mM $MgCl_2$, 10 mM glutamine, 10 μM FMN, and 180 pkat *Neurospora crassa* anthranilate synthase; sufficient solid sodium dithionite is added to decolorise the yellow solution and the assay is initiated by the addition of EPSP to a final concentration of 200 μM; after 30 min at 25 °C, the reaction is terminated with 50 μl 6 M HCl and the anthranilate extracted with 3 ml ethyl acetate; the layers are separated by a brief centrifugation and the fluorescence (λ_{ex} 315 nm, λ_{em} 390 nm) of the upper ethyl acetate layer measured (Mousdale and Coggins, 1986b).

(2) in a final volume of 250 μl, 50 mM triethanolamine–HCl (pH 8.0), 50 mM KCl, 2.5 mM $MgCl_2$, 10 mM glutamine, 10 μM FMN, 5 mM sodium dithionite, 50 pkat *Aerobacter aerogenes* anthranilate synthase and (to initiate assay) 80 μM EPSP; after 10–15 min at 30 °C, the assay is terminated with 20 μl 1 M HCl and the anthranilate measured fluorimetrically (λ_{ex} 340 nm, λ_{em} 400 nm) after extraction into 1 ml ethyl acetate (Schaller *et al.*, 1990).

The larger volume assay can also be used as a continuous assay with 3 ml fluorimetric cuvettes in a thermostatted fluorimeter. The anthranilate synthase coupling enzyme from *N. crassa* is partially purified and separated from chorismate synthase by chromatography on cellulose phosphate (Boocock and Coggins, 1983; Coggins *et al.*, 1987); *A. aerogenes* anthranilate synthase is prepared similarly (Schaller *et al.*, 1990). The assay can be quantified by calibration with anthranilate standards of known concentration. Aqueous solutions of dithionite are unstable but can be maintained for several hours on ice if made up in nitrogen-flushed water and sealed with a layer of petroleum.

Chorismate is easily separated from EPSP on the strong cation-exchange resins developed for organic acid analysis (Mousdale and Coggins, 1985a). A stopped-time assay for chorismate synthase uses this approach with UV (210 nm) detection of the chorismate (White *et al.*, 1987). While avoiding the need for coupling enzymes, this assay is relatively insensitive and plant extracts do not contain sufficient chorismate synthase to be accurately analysed. Purified and more concentrated fractions would, however, be suitable for this, as also for the Malachite Green procedure for the determination of the phosphate released from EPSP (Hawkes *et al.*, 1990).

C. Purification and Properties

Chorismate synthase has been purified to electrophoretic homogeneity from cultured cells of *Corydalis sempervirens* (Schaller *et al.*, 1990). An approximately 1000-fold purification was effected by ammonium sulphate fractionation and sequential chromatography on DEAE-cellulose, hydroxylapatite, cellulose phosphate and Blue Dextran-agarose. A yield of 400 μg purified enzyme was obtained from a starting 850 g fresh weight of cells. The purified enzyme is a dimer of subunit molecular weight 42 000; FMN is the preferred flavin cofactor over FAD (Schaller *et al.*, 1990). In pea seedlings chorismate synthase is found in intact chloroplasts separated by density gradient centrifugation (Mousdale and Coggins, 1986b).

REFERENCES

Ahmed, S. I. and Swain, T. (1970). *Phytochemistry* 9, 2287–2290.

Ames, B. N. (1966). *In* "Methods in Enzymology" (E. F.Neufeld and V.Ginsburg, eds), Vol. 8, pp. 115–118. Academic Press, New York and London.

Amrhein, N., Johanning, D., Schab, J. and Schulz, A. (1983). *FEBS Lett.* 157, 191–196.

Beaudoin-Eagan, L. D. and Thorpe, T. A. (1983). *Plant Physiol.* 73, 228–232.

Bender, S. L., Mehdi, S. and Knowles, J. R. (1989). *Biochemistry* 28, 7555–7560.

Berlyn, M. B., Ahmed, S. I. and Giles, N. H. (1970). *J. Bacteriol.* 104, 768–774.

Boocock, M. R. and Coggins, J. R. (1983). *FEBS Lett.* 154, 127–133.

Boudet, A. (1971). *FEBS Lett.* 14, 257–261.

Boudet, A. M. and Lecussan, R. (1974). *Planta* 119, 71–79.

Boudet, A. M., Boudet, A. and Bouyssou, H. (1977). *Phytochemistry* 16, 919–922.

Boudet, A. M., Graziana, A. and Ranjeva, R. (1985). In "Annual Proceedings of the Phytochemical Society of Europe" (C. F. Van Sumere and P. J. Lea, eds), Vol. 25, pp. 135–159. Clarendon Press, Oxford.

Bowen. J. R. and Kosuge, T. (1977). *Phytochemistry* 16, 881–884.

Bowen, J. R. and Kosuge, T. (1979). *Plant Physiol.* 64, 382–386.

Catala, F., Yapo, A., Patte, J. C. and Azerad, R. (1974). *Biochemie* 56, 1011–1023.

Coggins, J. R. (1989). *In* "Herbicides and Plant Metabolism" (A. D. Dodge, ed.), pp. 97–112. Cambridge University Press, Cambridge.

Coggins, J. R., Boocock, M. R., Chaudhuri, S., Lambert, J. M., Lumsden, J., Nimmo, G. A. and Smith, D. D. S. (1987). *In* "Methods in Enzymology" (S. Kaufman, ed.), Vol. 142, pp. 325–341. Academic Press, New York and London.

Duncan, K., Lewendon. A. and Coggins, J. R. (1984). *FEBS Lett.* 165, 121–127.

Fiedler, E. and Schultz, G. (1985). *Plant Physiol.* 79, 212–218.

Frost, J. W., Bender, J. I., Kadonaga, J. T. and Knowles, J. R. (1984). *Biochemistry* 23, 4470–4475.

Gaertner, F. H. and DeMoss, J. A. (1970). *In* "Methods in Enzymology" (H. Tabor and C. W. Tabor, eds), Vol. 17A, pp. 387–401. Academic Press, New York and London.

Gaitonde, M. K. and Gordon, M. W. (1958). *J. Biol. Chem.* 230, 1043–1050.

Ganson, R. J., D'Amato, T. A. and Jensen, R. A. (1986). *Plant Physiol.* 82, 203–210.

Gibson, F. (1970). *In* "Methods in Enzymology" (H. Tabor and C. W. Tabor, eds), Vol. 17A, pp. 362–364. Academic Press, New York and London.

Gottlieb, L. D. (1981). *Proc. Natl. Acad. Sci. USA* 78, 3726–3729.

Gollub, E., Zalkin, H. and Sprinson, D. B. (1967). *J. Biol. Chem.* 242, 5323–5328.

Gollub, E., Zalkin, H. and Sprinson, D. B. (1970). *In* "Methods in Enzymology" (H. Tabor and C. W. Tabor, eds), Vol. 17A, pp. 349–350. Academic Press, New York and London.

Graziana, A., Boudet, A. and Boudet, A. M. (1980). *Plant Cell Physiol.* 21, 1163–1174.

Graziana, A. and Boudet, A. M. (1980). *Plant Cell Physiol.* **21**, 793–802.
Grewe, R. and Haendler, H. (1966). *In* "Biochemical Preparations" (A. C. Maehly, ed.), Vol. 11, pp. 21–26. John Wiley and Sons, New York and London.
Hasan, N. and Nester, E. W. (1978). *J. Biol. Chem.* **253**, 4993–4998.
Haslam, E., Haworth, R. D. and Knowles, P. F. (1963). In "Methods in Enzymology" (S. P. Colowick and N. O. Kaplan, eds), Vol. 6, pp. 498–501. Academic Press, New York and London.
Hawkes, T. R., Lewis, T., Coggins, J. R., Mousdale, D. M., Lowe, D. J. and Thorneley, R. N. F. (1990). *Biochem. J.* **265**, 899–902.
Herrmann, K. M. (1983). In "Amino Acids: Biosynthesis and Genetic Regulation" (K. M. Herrmann and R. L. Somerville, eds), pp. 301–322. Addison-Wesley Publishing Co., Reading, MA.
Herrmann, K. M. and Poling, M. D. (1975). *J. Biol. Chem.* **250**, 6817–6821.
Huisman, O. C. and Kosuge, T. (1974). *J. Biol. Chem.* **249**, 6842–6848.
Ishikura, N., Teramoto, S., Takeshima, Y. and Mitsui, S. (1986). *Plant Cell Physiol.* **27**, 677–684.
Jensen, R. A. (1986). *Physiol. Plant.* **66**, 164–168.
Jensen, R. A. and Nester, E. W. (1966). *J. Biol. Chem.* **241**, 3365–3372.
Jensen, R. A., Calhoun, D. H. and Stenmark, S. L. (1973). *Biochim. Biophys, Acta* **293**, 256–268.
Kishore, G. M. and Shah, D. M. (1988). *Ann. Rev. Biochem.* **57**, 627–663.
Koshiba, T. (1978). *Z. Pflanzenphysiol.* **88**, 353–355.
Koshiba, T. (1979a). *Plant Cell Physiol.* **20**, 667–670.
Koshiba, T. (1979b). *Plant Cell Physiol.* **20**, 803–809.
Knowles, P. F., Levin, J. G. and Sprinson, D. B. (1970). *In* "Methods in Enzymology" (H. Tabor and C. W. Tabor, eds), Vol. 17A, pp. 360–362. Academic Press, New York and London.
Lambert, J. M., Boocock, M. R. and Coggins, J. R. (1985). *Biochem. J.* **226**, 817–829.
Lanzetta, P. A., Alvarez, L. J., Reinach, P. S. and Candia, O. A. (1979). *Anal. Biochem.* **100**, 95–97.
Lewendon, A. and Coggins, J. R. (1983). *Biochem. J.* **213**, 187–191.
Linhart, Y. B., Davis, M. L. and Mitton, J. B. (1981). *Biochem. Genet.* **19**, 641–646.
McCue, K. F. and Conn, E. E. (1990). *Plant Physiol.* **94**, 507–510.
Mehdi, S., Frost, J. W. and Knowles, J. R. (1987). *In* "Methods in Enzymology" (S. Kaufman, ed.), Vol. 142, pp. 306–314. Academic Press, New York and London.
Millican, R. C. (1970). *In* "Methods in Enzymology" (H. Tabor and C. W. Tabor, eds), Vol. 17A, pp. 352–354. Academic Press, New York and London.
Minamikawa, T. (1967). *Plant Cell Physiol.* **8**, 695–707.
Minamikawa, T. and Uritani, I. (1967). *J. Biochem.* **61**, 367–372.
Mitsuhashi, S. and Davis, B. D. (1954). *Biochem. Biophys. Acta* **15**, 54–61.
Morell, H., Clark, M. J., Knowles, P. F. and Sprinson, D. B. (1967). *J. Biol. Chem.* **242**, 82–90.
Morris, P. F., Doong, R.-L. and Jensen, R. A. (1989). *Plant Physiol.* **89**, 10–14.
Mousdale, D. M. and Coggins, J. R. (1984). *Planta* **160**, 78–83.
Mousdale, D. M. and Coggins, J. R. (1985a). *Planta* **163**, 241–249.
Mousdale, D. M. and Coggins, J. R. (1985b). *J. Chromatogr.* **329**, 268–272.
Mousdale, D. M. and Coggins, J. R. (1986a). *J. Chromatogr.* **367**, 217–222.
Mousdale, D. M. and Coggins, J. R. (1986b). *FEBS Lett.* **205**, 328–332.
Mousdale, D. M. and Coggins, J. R. (1987). *In* "Methods in Enzymology" (S. Kaufman, ed.), Vol. 142, pp. 348–354. Academic Press, New York and London.
Mousdale, D. M., Campbell, M. S. and Coggins, J. R. (1987). *Phytochemistry* **26**, 2665–2670.
Nazfiger, E. D., Widholm, J. M., Steinrucken, H. C. and Killmer, J. L. (1984). *Plant Physiol.* **76**, 571–574.
Padgette, S. R., Huynh, Q. K., Borgmeyer, J., Shah, D. M., Brand, L. A., Biest Re, D., Bishop, B. F., Rogers, S. G., Fraley, R. T. and Kishore, G. M. (1987). *Arch. Biochem. Biophys.* **258**, 564–573.
Patel, V. B. and Giles, N. H. (1979). *Biochim. Biophys. Acta* **567**, 24–34.

Pinto, J. E. B. P., Suzich, J. A. and Herrmann, K. M. (1986). *Plant Physiol.* **82**, 1040–1044.
Pinto, J. E. B. P., Dyer, W. E., Weller, S. C. and Herrmann, K. M. (1988). *Plant Physiol.* **87**, 891–893.
Polley. L. D. (1978). *Biochim. Biophys. Acta* **526**, 259–266.
Pompliano, D. L., Reimer, L. M., Myrvold, S. and Frost, J. W. (1989). *J. Am. Chem. Soc.* **111**, 1866–1871.
Poulsen, C. and Verpoorte, R. (1991). *Phytochemistry* **30**, 377–386.
Ream, J. E., Steinrucken, H. C., Porter, C. A. and Sikorksi, J. A. (1988). *Plant Physiol.* **87**, 232–238.
Reinink, M. and Borstlap, A. C. (1982). *Plant Sci. Lett.* **26**, 167–171.
Rothe, G. M., Maurer, W. and Mielke, C. (1976). *Ber. Deutsch. Bot. Ges.* **89**, 163–173.
Rothe, G. M., Hengst, G., Mildenberger, I., Scharer, H. and Utesch, D. (1983). *Planta* **157**, 358–366.
Rubin, J. L. and Jensen, R. A. (1985). *Plant Physiol.* **79**, 711–718.
Rubin, J. L., Gaines, C. G. and Jensen, R. A. (1982). *Plant Physiol.* **70**, 833–839.
Rubin, J. L., Gaines, C. G. and Jensen, R. A. (1984). *Plant Physiol.* **75**, 839–845.
Saijo, R. (1980). *Plant Cell. Physiol.* **21**, 989–998.
Saijo, R. and Kosuge, T. (1978). *Phytochemistry* **17**, 223–225.
Saijo, R. and Takeo, T. (1979). *Agric. Biol. Chem.* **43**, 1427–1432.
Salamon, I. I. and Davis, B. D. (1953). *J. Am. Chem. Soc.* **75**, 5567–5571.
Schaller, A., Windhofer, V. and Amrhein, N. (1990). *Arch. Biochem. Biophys.* **282**, 437–442.
Schmauder, H.-P., Groger, D., Koblitz, H. and Koblitz, D. (1985). *Plant Cell Rep.* **4**, 233–236.
Schmidt, C. L. and Schultz, G. (1987). *Physiol. Plant.* **70**, 65–67.
Schmidt, C. L., Danneel, H.-J., Schultz, G. and Buchanan, B. B. (1990). *Plant Physiol.* **93**, 758–766.
Schoner, R. and Herrmann, K. M. (1976). *J. Biol. Chem.* **251**, 5440–5447.
Schulz, A., Kruper, A. and Amrhein, N. (1985). *FEMS Microbiol. Lett.* **28**, 297–301.
Sharps, E. S. (1984). *Anal. Biochem.* **140**, 183–189.
Smart, C. C., Johanning, D., Muller, G. and Amrhein, N. (1985). *J. Biol. Chem.* **260**, 16338–16346.
Smith, C. M., Pratt, D. and Thompson, G. A. (1986). *Plant Cell Rep.* **5**, 298–301.
Smith. D. D. S. and Coggins, J. R. (1983). *Biochem. J.* **213**, 405–415.
Srinivasan, P. R. and Sprinson, D. B. (1959). *J. Biol. Chem.* **234**, 716–722.
Steinrucken, H. C., Schulz, A., Amrhein, N., Porter, C. A. and Fraley, R. T. (1986). *Arch. Biochem. Biophys.* **244**, 169–178.
Stuart. F. (1990). Ph.D. Thesis, University of Glasgow.
Suzich, J. A., Ranjeva, R., Hasegawa, P. M. and Herrmann, K. M. (1984). *Plant Physiol.* **75**, 369–371.
Suzich, J. A., Dean, J. F. D. and Herrmann, K. M. (1985). *Plant Physiol.* **79**, 765–770.
Waravdekar, V. S. and Saslaw, L. D. (1957). *Biochim. Biophys. Acta* **24**, 439.
Warren, L. (1959). *J. Biol. Chem.* **234**, 1971–1975.
Weeden, N. F. and Gottlieb, L. D. (1980). *Plant Physiol.* **66**, 400–403.
Weissbach, A. and Hurwitz, J. (1959). *J. Biol. Chem.* **234**, 705–709.
Welch, G. R. and Gaertner, F. H. (1976). *Arch. Biochem. Biophys.* **172**, 476–489.
Welch, G. R., Cole, K. W. and Gaertner, F. H. (1974). *Arch. Biochem. Biophys.* **165**, 505–518.
White, P. J., Mousdale, D. M. and Coggins, J. R. (1987). *Biochem. Soc. Trans.* **15**, 144–145.
White, P. J., Millar, G. and Coggins, J. R. (1988). *Biochem. J.* **251**, 313–322.
Yamamoto, E. (1977). *Plant Cell Physiol.* **18**, 995–1007.
Yamamoto, E. (1980). *Phytochemistry* **19**, 779–781.
Yamamoto, E. and Minamikawa, T. (1976). *J. Biochem.* **80**, 633–635.

2 Hydrolysable Tannins

GEORG G. GROSS

Universität Ulm, Abteilung Allgemeine Botanik, D-7900 Ulm, Germany

I. INTRODUCTION

Plant tannins have been used by mankind over several thousand years for the production of leather, inks, dyes and curatives (Larew, 1987), but it was only at the beginning of this century that chemical studies on the nature of these substances gave rise to an outpouring of important contributions. In particular the work of E. Fischer in Berlin was to set standards that are still valid, and to one of his disciples, K. Freudenberg (who received his eminent scientific reputation after changing his interest

METHODS IN PLANT BIOCHEMISTRY Vol. 9
ISBN 0–12–461019–6

to lignin, an even more puzzling and challenging class of phenolic natural products), we owe the classical subdivision into *condensed* and *hydrolysable* tannins (Freudenberg, 1920). However, with increasing recognition of the tremendous complexity of these plant constituents, interest in their structural elucidation waned considerably. A certain renaissance in this area took place in the 1950s, notably led by investigations of O. Th. Schmidt and W. Mayer in Heidelberg on ellagitannins. Highly active research on tannins was continued in England by E. C. Bate-Smith, T. Swain and E. Haslam (Sheffield), and in Japan by T. Okuda (Okayama) and I. Nishioka (Fukuoka) and their co-workers.

The results of these and many other efforts have provided detailed knowledge of the exact structures of innumerable hydrolysable tannins and related compounds, and also of their distribution in the Plant Kingdom (for recent relevant reviews see, e.g., Haslam, 1982, 1989a,b; Porter, 1989). Emerging from this solid basis, increasing attention has also been paid to related, far-reaching facets, e.g. potential medical applications (e.g. Haslam *et al.*, 1989; Okuda *et al.*, 1989a), chemico-biological interactions ranging from molecular aspects (cf. Haslam, 1988, 1989a) to ecological plant–herbivore relationships (e.g. Beart *et al.*, 1985; Scalbert and Haslam, 1987), and — last but not least — considerations on the biogenesis of these complex molecules (e.g. Haddock *et al.*, 1982a,b). Concerning this latter question, it has become common practice to tackle such problems using enzymatic studies. This chapter reports on the preparation, characterisation and assay of the enzymes involved in this biochemical pathway. Because tannins have a pronounced tendency to bind to proteins, thus causing their precipitation and/or denaturation including the effective inactivation of enzymes (cf. Haslam, 1988, 1989a), such efforts appeared initially very daring. Fortunately, these fears were unfounded — the enzymes isolated so far have been found to be remarkably resistant against their unfavorable substrates and products. Finally, it should be emphasised that the investigations reported below have already unravelled many principles of this pathway, but numerous details still await clarification before a final, conclusive picture can be drawn. It is thus one purpose of this chapter to stimulate readers to continue their own activities related to this area in the hope that they, too, will find these tannins not only frustrating but also fascinating (cf. Waterman, 1988).

II. BIOSYNTHESIS OF HYDROLYSABLE TANNINS: AN OVERVIEW

A. Classification and Structural Principles

As already mentioned, plant tannins are usually divided into *condensed* tannins, which are of flavonoid origin, and *hydrolysable* tannins, which are characterised by a central polyol moiety, usually β-D-glucopyranose. The hydroxyl groups of the latter are esterified with gallic acid (**1**); maximal substitution is reached with 1,2,3,4,6-penta-*O*-galloyl-β-D-glucose (**2**) which already exhibits pronounced tanning activity. This ester is regarded as the immediate precursor of both subclasses of hydrolysable tannins, i.e. *gallotannins* and *ellagitannins*. The former are characterised by the introduction of additional galloyl residues, linked to the pentagalloylglucose core via so-called *meta*-depside bonds (**3**) and reaching total substitution degrees of 10–12 galloyl units as has

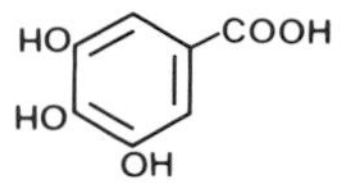

1

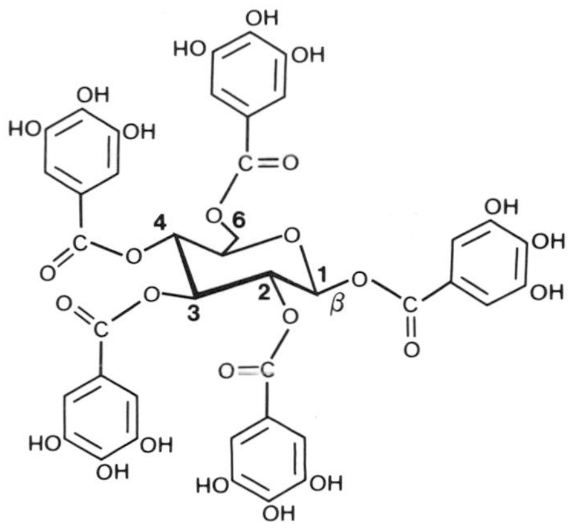

2

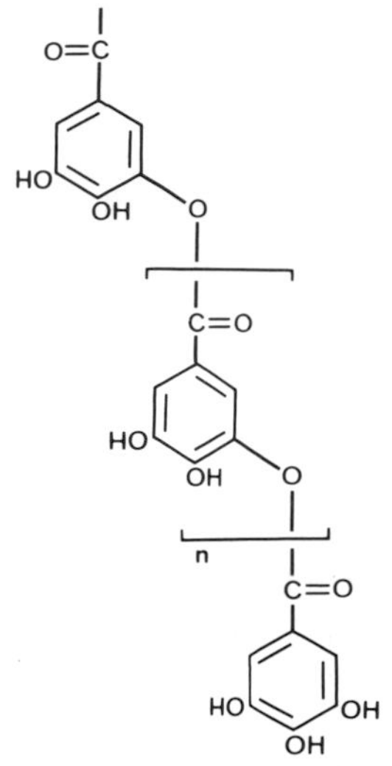

3

been shown for the tannins from *Rhus semialata* (Chinese gallotannin), *Quercus infectoria* (Turkish gallotannin), or *Paeonia albiflora* (*syn. lactiflora*) (Nishizawa *et al.*, 1980, 1982, 1983a, b). (It should be noted that in these investigations, evidence arose from NMR spectroscopy that gallotannins may be mixtures of *meta-* and *para*-depsides in Nature; however, this view still awaits confirmation from other laboratories). Ellagitannins, on the other hand, result from the introduction of secondary C–C linkages between spatially adjacent galloyl groups of pentagalloylglucose. In the case of the preferred $^4C_1$ configuration of D-glucose this event usually takes place between the residues at C-2/C-3 and C-4/C-6 (**4**)*; however, C-1/C-2 and C-1/C-6 coupling has also been observed (cf. Porter, 1989). Pentagalloylglucose may also adopt the

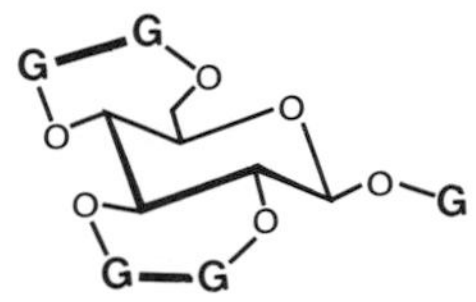

4 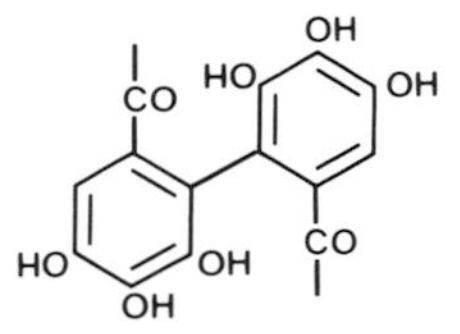   **5** 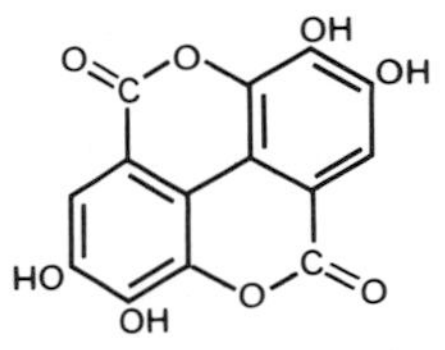   **6**

* The following abbreviations are used without further explanation: G, galloyl; G-G, 1,1′-dicarbonyl-3, 3′,4,4′,5,5′-hexahydroxydiphenoyl; Glc, glucose; HHDP, cf. G-G; HPLC, high performance liquid chromatography; MW, molecular weight; NMR, nuclear magnetic resonance; RP, reversed-phase; TLC, thin layer chromatography.

energetically less favourable $^1C_4$ conformation that remains fixed by galloyl coupling between C-2/C-4 and C-3/C-6. (R) or (S)-hexahydroxydiphenoyl (HHDP) residues (5) are formed in all these reactions; after hydrolysis, the resulting free acid rearranges spontaneously to the stable name-giving dilactone, ellagic acid (6). It should be mentioned finally that ellagitannins have a strong tendency to combine further to yield oligomeric derivatives coupled via aryl C—O—C linkages (cf. Haslam, 1982, 1989b; Porter, 1989).

B. Biosynthesis

On the basis of the structural relationships outlined above, together with an increasing knowledge of the structure and natural distribution of 'simple' esters (i.e. pentagalloylglucose and its partially substituted precursors), considerations on the biochemical pathways from gallic acid to complex hydrolysable tannins were developed (cf. Haddock et al., 1982a). However, as documented for instance by a review on that topic (Hillis, 1985), no experimental evidence was available until a few years ago. Fortunately this unsatisfactory situation has since changed considerably. Many earlier proposals were corroborated, but also modified or rejected as a consequence of enzyme studies, and in particular the nature and function of the 'energy-rich' intermediates required as acyl donors in these conversions was elucidated. To date, the entire biogenetic route may be conveniently divided into the following major sections: (1) the origin of the first specific intermediate, β-glucogallin (7), (2) the metabolic sequence from this monoester to pentagalloylglucose (2), and finally the secondary transformations of this pivotal intermediate to (3) gallotannins and (4) ellagitannins, respectively.

1. β-Glucogallin

There was, and is, no doubt as to the role of β-glucogallin (1-O-galloyl-β-D-glucose, 7) as the first specific intermediate in the pathway to hydrolysable tannins. The mechanism involved in the formation of this ester was unclear for a long time. By analogy to the ubiquitous plant phenolic chlorogenic acid (3-O-caffeoylquinic acid) and related depsides that were known to be formed via the coenzyme A (CoA) derivatives of their acyl moieties (for references see Gross, 1989), it was initially assumed that the biosynthesis of β-glucogallin should also occur according to this mechanism. Experiments with chemically synthesised galloyl-CoA (Gross, 1982a), however, demonstrated the incorrectness of this hypothesis (Gross, 1983a). Instead, a reaction of free gallic acid with activated glucose, in the form of UDP-glucose, was

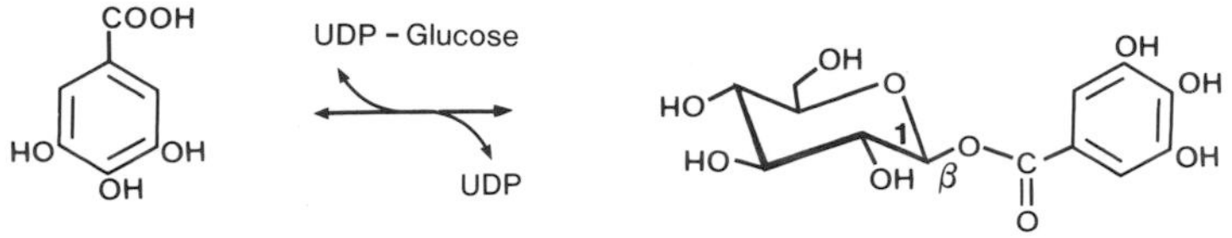

FIG. 2.1. Synthesis of β-glucogallin (7). UDP, uridine-5′-diphosphate.

found to be catalysed by an enzyme from oak leaves (Fig. 2.1; Gross, 1982b, 1983b) and this mechanism was consistent with other observations on the *in vitro* synthesis of 1-*O*-glucose esters of numerous phenolic acids (cf. Gross, 1989).

2. β-Glucogallin to pentagalloylglucose

In experiments on the biosynthesis of β-glucogallin it soon became apparent that this ester was not only the first intermediate in the pathway to pentagalloylglucose, but that the comparatively low group-transfer potential of its 1-*O*-acylester bond was still sufficiently high to permit subsequent transacylation reactions (Gross, 1983a). [Exact data for β-glucogallin do not exist; the comparable ester glucose-1-phosphate has a $\Delta G_0'$ of *c.* 5 kcal, whereas glucose-6-phosphate has a $\Delta G_0'$ of only about 2.5 kcal (Atkinson and Morton, 1960); the most common 'activated' acids, i.e. acyl-CoA thioesters, in contrast, have $\Delta G_0'$ values of 7–8 kcal.] In a first step (cf. Fig. 2.2), β-glucogallin was found to serve as both donor and acceptor yielding 1,6-di-*O*-galloylglucose (**8**) and free glucose (Gross, 1983a; Schmidt *et al.*, 1987); similar 'disproportionations' (cf. Strack *et al.*, 1984; Dahlbender and Strack, 1984) have meanwhile been found to be involved in the synthesis of 1,2-disinapoylglucose (Dahlbender and Strack, 1984, 1986) and of 3,5-dicaffeoylquinic acid (Kojima and Kondo, 1985; Villegas *et al.*, 1987).

Further studies with enzymes from oak or sumach leaves have also shown that the subsequent galloylation steps utilised β-glucogallin as acyl donor, thus specifically forming 1,2,6-tri-*O*-galloylglucose (**9**) (Gross and Denzel, 1991), 1,2,3,6-tetra-*O*-galloylglucose (**10**) (Hagenah and Gross, 1993), and finally 1,2,3,4,6-penta-*O*-galloylglucose (**2**) (Cammann *et al.*, 1989).

This clear and logical picture of an entirely β-glucogallin-dependent pathway was unfortunately blurred by the discovery of a parallel and apparently β-glucogallin-*independent* pathway. As depicted in Fig. 2.3, an enzyme was isolated from sumach

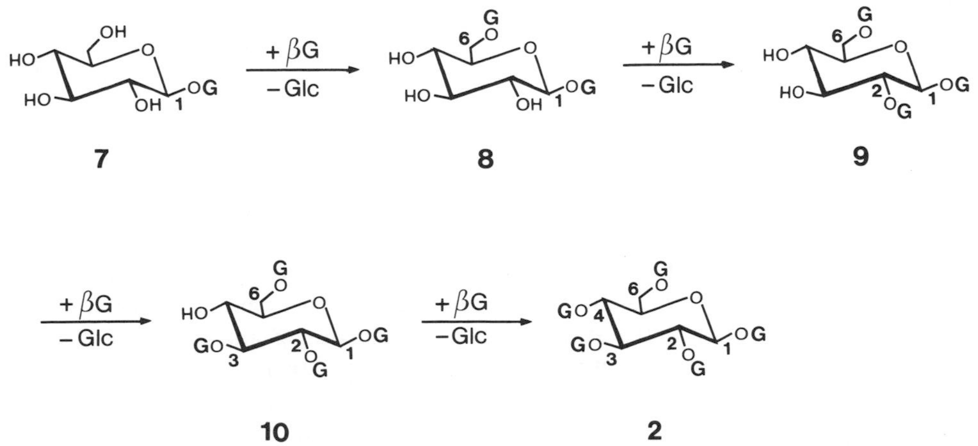

FIG. 2.2. Biosynthetic pathway from β-glucogallin (**7**) to 1,2,3,4,6-pentagalloylglucose (**2**). G, galloyl; βG, β-glucogallin; Glc, glucose.

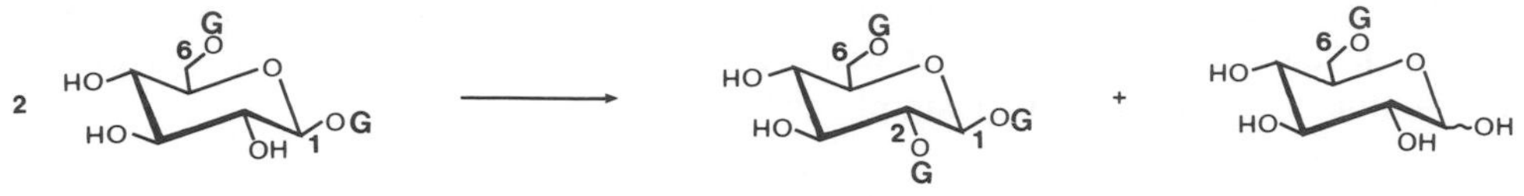

FIG. 2.3. Formation of 1,2,6-trigalloylglucose (9) by 'disproportionation' of two molecules 1,6-digalloyl-glucose (8), catalysed by 1,6-digalloylglucose: 1,6-digalloylglucose 2-O-galloyltransferase.

leaves that again catalysed a 'disproportionation' (see above), in this case, however, of two molecules of 1,6-digalloylglucose (8), yielding 1,2,6-trigalloylglucose (9) and anomeric 6-O-galloylglucose (11) (Denzel and Gross, 1991). Analogously, β-gluco-gallin could be replaced by 1,6-digalloylglucose as acyl donor in the conversion of 1,2,6-trigalloylglucose to 1,2,3,6-tetragalloylglucose (Hagenah and Gross, 1993). Moreover, tri- and tetragalloylglucoses bearing the essential 1-O-galloyl group also exhibited some activity as galloyl donors, a finding that was consistent with the recently discussed possibility (Haslam, 1989a, pp. 101–102) that the frequently occurring galloylglucoses with an unacylated anomeric position might originate from such reactions.

It should be mentioned finally that cell-free extracts from oak leaves were also found to contain a quite different type of acyltransferase that catalysed the following exchange reaction:

$$\beta\text{-Glucogallin} + {}^*\text{glucose} \rightleftharpoons \beta\text{-glucogallin} + ({}^*)\text{glucose}$$

where the asterisk symbolises an appropriate label (e.g. ^{14}C) to allow measurement of the reaction (Gross et al., 1986). The physiological significance of this enzyme is difficult to evaluate; however, it was successfully employed for the convenient preparation of labelled glucogallin and related esters (Denzel et al., 1988a; see Section III).

3. Gallotannins

Fortunately, the negative statement in a recent review (Gross, 1989) that 'virtually nothing is known about the formation of the characteristic meta-depside bonds' (3) of gallotannins was superseded by the observation that enzyme preparations from sumach (R. typhina) leaves catalysed the galloylation of 1,2,3,4,6-pentagalloylglucose, yielding a variety of hexa-, hepta- and nonagalloylglucoses (together with traces even of deca- and undecagalloylglucoses) as reaction products (Hofmann and Gross, 1990). By analogy to the major pathway to pentagalloylglucose, β-glucogallin was again found to serve as the specific acylating agent. Among the numerous reaction products, three hexagalloylglucoses have been purified to date to which structures 12–14 were assigned by means of ^{1}H and ^{13}C NMR spectroscopy (G. G. Gross, A. S. Hofmann and G. Schilling, unpubl. res.). These results are fully consistent with those obtained earlier for the hexagalloylglucoses produced in vivo in galls of the related species R. semialata (Chinese gallotannin; cf. Nishizawa et al., 1982).

4. Formation of ellagitannins

In spite of many attempts to elucidate the mechanism of the oxidative conversion of pentagalloylglucose to ellagitannins (reviewed by Gross, 1989) this question has remained highly enigmatic. The few preliminary indications of such a reaction mentioned by Gross (1989) have not led to any conclusive results. Sophisticated new strategies and techniques will be necessary to tackle this challenging problem.

III. PREPARATION OF SUBSTRATES

In most enzymatic investigations, problems are encountered regarding the required substrates, one major obstacle being commercial unavailability that enforces time-consuming and laborious chemical syntheses or isolation from natural sources, and this applies particularly to all tannins. The occurrence, isolation and characterisation of a vast number of such compounds is now well documented (for references see, e.g., the review articles of Haslam, 1982, 1989a, b; Haddock et al., 1982a, b, c; Gupta et al., 1982). Here in this section, only those procedures will be mentioned that have been recognised as most practicable for routine work on the basis of several years of experience in the author's laboratory.

A. β-Glucogallin

The particular importance of β-glucogallin (7) originates from its dual role as both primary intermediate and the predominant acyl donor in the entire biogenetic route to gallotannins (see above). For these reasons, this ester is required for enzyme studies not only in considerable amounts, but also in various labelled forms for special analytical applications. Given these requirements, both chemical and enzymatic syntheses have been developed, but isolation from natural sources can also be an alternative.

1. Chemical synthesis

The preparation of gramme quantitites of β-glucogallin is best accomplished by chemical means; the method summarised below is intended to be practicable for chemically less experienced biochemists or biologists. As a prerequisite, the free OH-groups of the reactants must be properly protected; we decided to use acetyl residues for this purpose because of their facile elimination. Thus, 2,3,4,6-tetra-O-acetyl-β-D-glucose was prepared by debromination of commercially available α-acetobromo-glucose with silver carbonate (Fischer and Hess, 1912). [High purity of the precursor was found essential for good recoveries; if necessary, recrystallisation, e.g. from diethyl ether (Koenigs and Knorr, 1901), is recommended]. Gallic acid was converted to the triacetyl derivative (Lesser and Gad, 1926) from which the acylchloride was prepared with PCl_5 (Fischer et al., 1918). Esterification of these two components (Fischer and Bergmann, 1918) gave hepta-acetylgalloylglucose whose protecting groups were finally removed with sodium methylate (Schmidt and Reuss, 1961), yielding analytically pure β-glucogallin. Upon prolonged storage, the reaction product was found to undergo slight degradation. If necessary, purification was achieved by boiling with diethyl ether and crystallisation of the solid residue in a minimum of hot water.

2. Enzymatic syntheses

Enzymatic syntheses are normally applicable only for small-scale preparations which, nevertheless, are sufficient for many biochemical investigations. A major advantage is the avoidance of all problems related to the introduction and elimination of protecting groups. For both reasons, such procedures are particularly attractive for the synthesis of radioactively labelled compounds.

The reported enzymatic synthesis of unlabelled β-glucogallin and related esters (Weisemann et al., 1988) is thus of only limited interest in this chapter. Of greater importance is the possibility of synthesising [14C]glucosyl-labelled β-glucogallin from UDP-D-[14C]glucose (Gross, 1983a). A serious disadvantage of this procedure as initially reported, however, arose from the enormous costs of the labelled substrate. The problem was overcome by adapting the above-mentioned β-glucogallin-glucose exchange reaction (see Section II.B.2) for preparative purposes. In this case free [14C]D-glucose, 50 times cheaper to purchase, could be employed. This justified the comparatively low yields achieved by this latter method (Gross et al., 1986). [14C-glucosyl]β-Glucogallin was economically synthesised on this basis and conveniently purified by chromatography on octadecyl-substituted (RP-18) silica gel, followed, if desired, by semipreparative HPLC to remove traces of residual gallic acid (Denzel et al., 1988a).]

For other applications, the use of β-glucogallin labelled in the galloyl moiety was required. Again, this ester was accessible by an enzymatic method utilising UDP-glucose and free gallic acid which, however, is difficult to obtain in radioactively labelled form. Sophisticated chemical and enzymatic procedures have been elaborated for this purpose (e.g. Kozák et al., 1978; Ishikura et al., 1984; Schildknecht and Milde, 1987); alternatively, a convenient and inexpensive method based on the photo-assimilation of $^{14}CO_2$ by sumach leaves in the presence of the herbicide glyphosate

[*N*-(phosphonomethyl)glycine] was developed recently (N. Amrhein, pers. comm.; see also Amrhein *et al.*, 1984; Krüper *et al.*, 1990).

3. Isolation from natural tannin

β-Glucogallin normally occurs only as a minor constituent of tannins, e.g. in roots of rhubarb (*Rheum*) or in leaves of sumach (*Rhus*). Its isolation from such sources is thus not very advisable except for those cases where it has been enriched as by-product from the purification of more abundantly occurring galloylglucoses (see below).

B. Multiple-galloylated Glucose Esters

Due to difficulties in obtaining suitable partially protected glucoses, chemical syntheses play no role; only 1,2,3,4,6-pentagalloylglucose is easily obtained from free glucose. Instead, isolation from natural tannin sources is recommended, but efficient enzymatic syntheses have also been developed.

1. 1,6-Di-O-galloyl-β-D-glucose (8)

This rare ester was extracted from rhubarb roots (*Rhizoma rhei*). After prepurification by repeated chromatography on Sephadex LH-20 (Pharmacia) as described by Nonaka and Nishioka (1983), it was found advantageous to proceed with chromatography on RP-18 silica gel (40–60 μm particle size), followed by semipreparative RP-HPLC and a final crystallisation step from water. Yields up to 180 mg of pure diester were thus obtained from 3 kg plant material (Denzel *et al.*, 1988b). It must be emphasised, however, that the composition and contents of the components of rhubarbs of different origin were found to vary so significantly (Nonaka *et al.*, 1985) that the suitability of unknown plant drugs has to be carefully estimated before working them up.

The above laborious and time-consuming isolation procedure was replaced by an efficient enzymatic method using β-glucogallin: β-glucogallin 2-O-galloyltransferase (see Section II.B.2) that had been immobilised on a Phenyl-Sepharose column. By simply cycling a buffered solution of β-glucogallin through this 'enzyme reactor', followed by adsorption of the reaction product on a subsequent column with RP-18 silica gel, 60 mg of pure 1,6-digalloylglucose could be conveniently synthesised within 5 days (including the time required for the preparation of enzyme) in 60–90% yield (Gross *et al.*, 1990).

2. 1,2,6-Trigalloyl-β-D-glucose (9)

The purification of this ester from rhubarb has been described by Nonaka *et al.* (1981). The plant material available to us, however, was found to contain only traces of this compound. Better results were achieved with sumach leaves from which it was obtained as a by-product from the purification of pentagalloylglucose (see below). Later, when sufficient amounts of enzymatically synthesised 1,6-digalloylglucose were available (cf. Section III.B.1), this strategy was also applied for the preparation of 1,2,6-trigalloylglucose (Hagenah and Gross, 1993). Briefly, β-glucogallin (250 mg), 1,6-digalloyl-glucose (15 mg) and enzyme [prepared as described by Denzel *et al.* (1988b), the acid-precipitated pellet being suspended in sodium citrate buffer, pH 5.0] were incubated under stirring in a total volume of 45 ml at 40°C. After 5 h, the suspension was centrifuged and the sedimented protein resuspended in fresh substrate solution. The combined supernatants from five incubations were adsorbed on RP-18 silica gel (column 6 × 2 cm i.d.) and purified by sequential elution with water (affording gallic acid and β-glucogallin), 10% aqueous methanol (6-galloylglucose), 20% methanol (1,6-digalloylglucose), and finally 25% methanol, yielding *c.* 15 mg of 1,2,6-trigalloyl-glucose of 90–95% purity.

3.   *1,2,3,6-Tetra-O-galloyl-β-D-glucose* (10)

A good source for this ester was commercially available tannin. Starting with only 30 g of this crude product, 240 mg of pure ester were obtained after chromatography on Sephadex LH-20 in ethanol (cf. Nishizawa *et al.*, 1980) and isocratic semipreparative RP-HPLC (for details, see Cammann *et al.*, 1989).

4.   *1,2,3,4,6-Penta-O-galloyl-β-D-glucose* (2)

This ester is known as an abundant constituent of many plant tannins. Acetone extracts of such sources (sumach leaves, for instance, are easily available) are conveniently worked up by column chromatography on Sephadex LH-20 with the ethanol–water–acetone solvent system developed by Nishizawa *et al.* (1980). However, chemical methods provide an interesting alternative because unprotected glucose can be used. By this strategy, gramme quantities of pentagalloylglucose were prepared by reacting triacetylgalloylchloride (cf. Section III.A.1) with β-D-glucose (Fischer and Bergmann, 1918) and subsequent hydrolysis of the protecting acetyl groups from penta(triacetyl-galloyl)-β-D-glucose (Fischer and Bergmann, 1919). The crude product was purified by chromatography on Sephadex LH-20 as above. This chemical method was also well suited for the small-scale preparation of [^{14}C]glucosyl-labelled pentagalloylglucose (G. G. Gross, unpubl. res.).

C.   Gallotannins

Detailed procedures have been published for the purification of gallotannins from *R. semialata* (Nishizawa *et al.*, 1982), *Q. infectoria* (Nishizawa *et al.*, 1983b) and *P. lactiflora* (Nishizawa *et al.*, 1983a). The method comprises acetone extraction of dried plant material, partitioning against ethyl acetate, prepurification on Sephadex LH-20 (Nishizawa *et al.*, 1980) and final purification by preparative RP-HPLC with acetonitrile–water mixtures (18–24% acetonitrile, depending on the compounds to be purified; cf. Hofmann and Gross, 1990) that are supplemented with oxalic acid (2–3 g l^{-1}) to achieve sharp peaks; the acid is then easily removed from the eluates by passage through a column with anion-exchange resin (OH$^-$ form) before lyophilisation.

IV. ENZYME ASSAY TECHNIQUES

The unequivocal identification and quantification of hydrolysable tannins has always been a major problem because of many closely related structures and isomeric forms. The analysis and characterisation of these compounds has been the subject of recent comprehensive treatises, e.g. by Porter (1989; see Vol. 1, Ch. 11 of this series), Haslam (1989a) and Okuda *et al.* (1989b). Among the wide repertoire of available methods only a few are suitable for the specific requirements of enzyme assays, i.e. rapid and facile operation, specificity, sensitivity, and reproducibility. Colour tests, for instance, usually lack sensitivity, in both qualitative and quantitative terms, while sophisticated techniques with a high potential of evidence (e.g. NMR spectroscopy) are not practicable for an enzymologist's daily routine. There remain chromatographic procedures, among which HPLC has attained outstanding importance. In particular, RP-HPLC on octadecyl-substituted matrices has become the method of choice as it perfectly fulfills the above prerequisites, but other techniques that require no expensive equipment can also be useful tools for certain applications.

A. Thin Layer Chromatography

This simple and inexpensive method was successfully applied for the separation of UDP-glucose, free glucose, gallic acid, β-glucogallin and digalloylglucose, i.e. the substrates and products involved in the first steps of the pathway to gallotannins (Gross, 1982b, 1983a, b; Gross *et al.*, 1986; Schmidt *et al.*, 1987). Initially, unexpected problems due to insufficient separation and pronounced tailing were encountered. Satisfactory results were achieved with silica gel-coated plastic sheets (Sil N-HR, Macherey-Nagel, Düren, Germany) and the solvent system ethyl acetate–ethyl methyl ketone–formic acid–water (5:3:1:1; v/v). Aromatic compounds were easily located on the chromatograms by the UV absorbance of previously added carriers. Reliable quantification was best achieved by using ^{14}C-labelled substrates; their radioactivity was determined by a TLC scanner or — much more quickly and more precisely — by simply cutting out the respective spots and liquid scintillation counting in cocktails based on toluene as solvent (constant counting efficiencies are assured by the insolubility of the above compounds in toluene).

Disadvantages of this technique were the unusually long developing times for the chromatograms of 3.5–4 h (however, this was partially compensated by the possibility of chromatographing dozens of samples simultaneously) and the comparatively low resolution potential that prevented the identification of isomers and the separation of higher galloylated glucoses.

B. Low Pressure Column Chromatography

The critical time factor of the above TLC technique was overcome with the introduction of C_{18}-substituted silica gels on the market that allowed facile and fast separations by low-pressure RP-chromatography on small disposable columns. Thus, the determination of [^{14}C]β-glucogallin formed by β-glucogallin–glucose exchange (Gross *et al.*, 1986) was completed within 10 min by simply adsorbing the deproteinised enzyme assay on RP-18 silica gel, rinsing with water and eluting the reaction product with methanol, followed by liquid scintillation counting of its radioactivity (Denzel *et al.*, 1988b). Analogous separations were developed for 1,6-digalloylglucose (Schmidt *et al.*, 1987) and 1,2,6-trigalloylglucose (cf. Section III.B.2) by adequately modifying the elution conditions.

C. High Performance Liquid Chromatography

Regardless of the merits of the above techniques, HPLC has now developed as the predominating analytical method (in spite of the considerable cost of the necessary equipment) and particularly the enormous versatility of RP-HPLC has significantly contributed to this situation. With respect to gallotannins, both normal-phase HPLC (on unmodified silica gel) and RP-HPLC (in most cases with C_{18}-substituted supports) have been used.

Normal-phase HPLC had the disadvantage that application of aqueous samples had to be avoided, thus preventing the direct analysis of enzyme assay mixtures, and also its resolution capacity was rather limited. On the other hand, this method allowed the efficient separation of gallotannins in the order of their substitution degrees (see Fig. 2.4). Modifications of the complex solvent system developed by Nishizawa *et al.* (1980) for this purpose (consisting of *n*-hexane, methanol, tetrahydrofuran, formic acid and oxalic acid) were applied to analyse 'simple' galloylglucoses (Krajci and Gross,

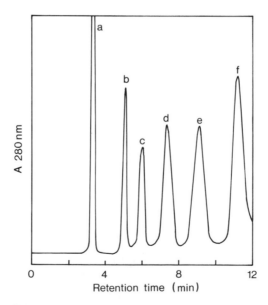

FIG. 2.4.   Normal-phase HPLC of galloylglucoses. (a) Gallic acid (**1**); (b) β-glucogallin (**7**); (c)–(f) di-, tri-, tetra- and pentagalloylglucose. Chromatography conditions: μPorasil (Waters), particle size 10 μm, column 300 × 4 mm i.d.; solvent: *n*-hexane–methanol–tetrahydrofurane–formic acid (63:27:9:1; v/v) plus 320 mg oxalic acid per 1000 ml (Gross, 1983a; modified from Nishizawa *et al.*, 1980); flow rate 2 ml min^{-1}.

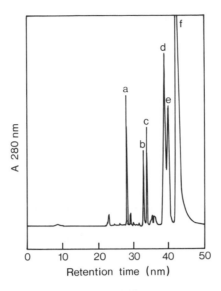

FIG. 2.5. RP-HPLC of galloylglucoses. (a) 1,6-Digalloylglucose (**8**); (b) 1,2,6-trigalloylglucose (**9**); (c) 1,3,6-trigalloylglucose; (d) 1,2,3,6-tetragalloylglucose (**10**); (e) 1,2,4,6-tetragalloylglucose; (f) 1,2,3,4,6-pentagalloylglucose (**2**). Chromatography conditions: LiChrosorb RP-18 (Merck), particle size 5 μm, column 180 × 3 mm i.d.; solvent: acetonitrile–aqueous H_3PO_4 gradient as given by Haddock *et al.* (1982a, b); flow rate 1 ml min^{-1}.

1987) and oligomeric ellagitannins (Okuda *et al.*, 1989b). Of particular importance was the observation that plots of galloylation degrees against the logarithm of the corresponding retention times resulted in straight lines; thus limited numbers of available references also allowed predictions on the nature of unknown tannins (Krajci and Gross, 1987).

As in most other fields, RP-HPLC is now the most common chromatography technique for the analysis of hydrolysable tannins. This must be attributed not only to the excellent results achieved by this method, but certainly also to the usually simple solvent systems, their facile adaptation to many different separation problems, and the convenient application of gradient systems to further improve resolution capacities. In most cases, aqueous solutions of methanol or acetonitrile are used to which small amounts (0.05%) of acetic or phosphoric acid are added to suppress ionisation. For the chromatography of gallotannins, the presence of higher concentrations of oxalic acid (0.2–0.3%) in the solvents is favourable (cf. Nishizawa *et al.*, 1982, 1983a, b). Based on these few components, a host of different isocratic systems and gradients of various shapes has been developed for the analysis of enzyme reaction products or the preparation of substrates (for relevant details the reader is referred to the experimental data of the publications cited in Sections III and V). The enormous potential of RP-HPLC is illustrated by the example depicted in Fig. 2.5, demonstrating that even closely related isomers are effectively separated by this method.

V. ENZYMES OF THE PATHWAY TO HYDROLYSABLE TANNINS

A. General Preparative Methods

1. Plant materials

In initial experiments, young leaves from oak trees growing in the open air were collected during spring and early summer. Although in theory this enzyme source was suitable, serious problems were encountered with leaves that were older than about three months as these contained abundant amounts of highly viscous mucilage that made it impossible to employ the usual enzyme techniques. Identical problems were observed with leaves from greenhouse grown plantlets of *Q. robur* or *Q. rubra*, and it was found that this effect bore no relation to the actual season but depended only on leaf age. Elimination of the disturbing contaminants by adsorption/desorption of enzymes on $CaPO_4$-gel (cf. Gross, 1983b) was unsatisfactory due to excessive losses of enzyme activity. The problem was finally solved by simply cutting the stems of older plantlets to enforce the subsequent development of new shoots and fresh leaves. By this means, an almost continuous supply of suitable plant material was ensured; unavoidable gaps could be bridged by storage of leaves at $-20°C$ after shock-freezing in liquid nitrogen and sealing in evacuated plastic bags.

Leaves of sumach (*R. typhina*), which were used in many experiments, were also collected from trees grown in the open air or from plants raised in the greenhouse. In the latter case, it was found indispensable to break the dormancy of seeds prior to sowing, e.g. by soaking in conc. H_2SO_4 for 3 h (Rolston, 1978). In contrast to oak, autumnal leaf abscission happened with sumach even in the greenhouse; however, leaf material of this species could be stored frozen without problems.

After several years of practical experience it appears that these two enzyme sources should be sufficient for most investigations as these plants are not only easily accessible but also produce a wide array of gallotannins and ellagitannins. However, this does not preclude the use of other plants or of plant organ and cell cultures. Although increasing information is available on the production of tannins by the latter (e.g. De Langhe *et al.*, 1987; Krajci and Gross, 1987; Yazaki and Okuda, 1989, 1990; Ishimaru *et al.*, 1990), cost-benefit estimations make it questionable whether these will become an alternative to native plants as enzyme sources.

2. Preparation of cell-free extracts

One of the most crucial steps in enzymology is the preparation of cell-free extracts. Particular problems are encountered with plant materials as these not only require strong mechanical forces to disrupt their rigid cell walls, but also often contain high concentrations of organic acids and phenolics that effectively denature the released enzymes. It is obvious that these negative factors apply particularly to plants that actively produce tannin. However, with the precautions of the standard procedure described below, these problems have effectively been eliminated.

An appropriate quantity of frozen leaves (typically 80 g) is suspended in liquid nitrogen and ground to a fine powder (0.5 mm particle size) in a precooled ultra-centrifugal mill. The still frozen powder is mixed with an equal weight of prewashed insoluble polyvinylpyrrolidone (polyclar AT) to bind phenols (Loomis and Bataille, 1966) and, while thawing in an ice bath, extracted under mechanical stirring for 30 min with 100 ml borate buffer (0.1 M, pH 7.5) and 150 ml Tris-HCl buffer (1 M, pH 8), both buffers being supplemented with 5–10 mM 2-mercaptoethanol as antioxidant. In spite of the high buffer concentration it is important to readjust the pH of the brei occasionally with Tris buffer to compensate for the release of acids. (Smaller amounts of leaf material may be homogenised analogously in a chilled mortar with quartz sand.) The mixture is then squeezed through four layers of muslin, followed by centrifugation of the filtrate (30 000 × g, 20 min). The supernatant crude extract is depleted of residual acids and phenolics by stirring for 15 min with 8 g anion-exchange resin (Lam and Shaw, 1970), e.g. Dowex 1X4, 50–100 mesh (BO_3^{3-}-form). After filtering through glass wool, a clear cell-free extract is obtained from which the individual enzymes described below can be purified.

B. Properties of Individual Enzymes

1. UDP-Glucose: vanillate 1-O-glucosyltransferase (EC 2.4.1.-)

UDPG + phenylcarboxylic acid → 1-O-phenylcarboxyl-β-D-glucose + UDP

The enzyme was detected in leaves of *Q. robur* (Gross, 1982b) and purified 45-fold (10% yield) from cell-free extracts of *Q. rubra* (Gross, 1983b). The glucosyltransferase had a MW of 68 kDa, a pH optimum at 6.5–7.0, and a temperature optimum of 40°C. UDP-Glucose (K_m = 2.3 mM) was the exclusive donor, while a variety of benzoic and, at significantly lower rates, cinnamic acids was utilised as acceptor molecules. The above systematic name was proposed according to the best substrate, vanillate; however, it was concluded that the physiological role of the enzyme is the formation of β-glucogallin. The preparation and characterisation of several related esters by this enzyme has been reported by Weisemann *et al.* (1988).

2. 1-O-Benzoylglucose: D-glucose 1-O-benzoyltransferase (EC 2.3.1.-)

1-O-Phenylcarboxyl-β-D-glucose + D-glucose → D-glucose

+ 1-O-phenylcarboxyl-β-D-glucose

Evidence for an enzyme catalysing this unusual exchange reaction was obtained with extracts from *Q. robur* (Gross, 1983a) from which it was purified 53-fold with 20% yield; it had a MW of 380 kDa, a pH optimum of 6.6, and a temperature optimum of 30°C (Gross *et al.*, 1986). The transferase was active with a variety of 1-O-acylglucoses, the best being 1-O-benzoylglucose, while D-glucose was the exclusive acceptor (Denzel *et al.*, 1988a). The exchange reaction, whose physiological role is still unclear, was used for the economic preparation of [^{14}C-glucosyl]β-glucogallin (see Section III.A.1).

3. *β-Glucogallin: β-glucogallin 6-O-galloyltransferase* (EC 2.3.1.90)

$$\beta\text{-Glucogallin} + \beta\text{-glucogallin} \rightarrow 1,6\text{-di-}O\text{-galloylglucose} + \text{D-glucose}$$

This early postulated enzyme (Gross, 1983a) was purified 31-fold from leaves of *Q. robur* (recovery 14%). It had a MW of *c.* 400 kDa, a temperature optimum at 30°C, and a pH optimum of 6.5; it was stable between pH 4.5 and 6.0 (Schmidt *et al.*, 1987). β-Glucogallin ($K_m = 67$ mM) was the best substrate among several related 1-*O*-acylglucoses. Immobilised on Phenyl-Sepharose, the enzyme was employed for the preparation of 1,6-digalloylglucose (Gross *et al.*, 1990; see Section III.B.1).

4. *β-Glucogallin: 1,6-di-O-galloyl-β-D-glucose 2-O-galloyl-transferase* (EC 2.3.1.-)

$$\beta\text{-Glucogallin} + 1,6\text{-di-}O\text{-galloylglucose} \rightarrow 1,2,6\text{-tri-}O\text{-galloylglucose} + \text{D-glucose}$$

This enzyme was detected in cell-free extracts from leaves of *R. typhina*, from which it was partially purified by precipitation with acetic acid (Denzel *et al.*, 1988b), but was found to occur also in oak leaves. Mainly because of solubility problems with enzyme prepared this way, a different purification procedure was developed (Gross and Denzel, 1991). The enzyme had an unusually high MW of *c.* 700 kDa and a temperature optimum of 50°C; the optimal pH was at 5.0–5.5, and maximal stability was between pH 3.4 and 5.8. Surprisingly, addition of penta- or hexagalloylglucose caused a pronounced stimulation of enzyme activity. Besides several unphysiological substrates, β-glucogallin and its dihydroxy analogue, 1-*O*-protocatechuoylglucose, acted as potent acyl donors. For the preparation of 1,2,6-trigalloylglucose with this enzyme, see Section III.B.2.

5. *1,6-Di-O-galloyl-β-D-glucose: 1,6-di-O-galloyl-β-D-glucose 2-O-galloyltransferase* (EC 2.3.1.-)

$$1,6\text{-Di-}O\text{-galloylglucose} + 1,6\text{-di-}O\text{-galloylglucose} \rightarrow 1,2,6\text{-tri-}O\text{-galloylglucose}$$

$$+ 6\text{-}O\text{-galloylglucose}$$

In studies on the above enzyme (cf. Section V.B.4), a related galloyltransferase was discovered in sumach that again catalysed the formation of 1,2,6-trigalloylglucose but showed no requirement for the acyl donor β-glucogallin (Denzel and Gross, 1991). This new enzyme was purified almost 1700-fold (yield 85%). It had a MW of 56 kDa, a K_m value of 12 mM, was stable between pH 4.5 and 6.5, and most active at pH 5.9 and 40°C. Unlike its β-glucogallin-dependent analogue (see above), it was drastically inhibited by added pentagalloylglucose. Also 1,6-diprotocatechuoylglucose was accepted as substrate; moreover, tri- and tetragalloylglucoses could act as both donors and acceptors.

6. *β-Glucogallin: 1,2,6-trigalloyl-β-D-glucose 3-O-galloyltransferase* (EC 2.3.1.-)

$$\beta\text{-Glucogallin} + 1,2,6\text{-tri-}O\text{-galloylglucose} \rightarrow 1,2,3,6\text{-tetra-}O\text{-galloylglucose}$$

$$+ \text{D-glucose}$$

This enzyme from sumach leaves is currently being investigated (Hagenah and Gross, 1993). It had a MW of 380 kDa, a temperature optimum of 55°C, and a pH optimum at 5.5–6.0. The standard acyl donor, β-glucogallin ($K_m = 0.5$ mM), could be replaced by 1,6-digalloylglucose with about 50% relative activity. The acceptor, 1,2,6-trigalloyl-glucose, and its unphysiological 1,3,6-analogue had nearly identical K_m values (5.0 and 5.4 mM, respectively); however, an about 10-fold higher reaction rate was determined for the latter.

7. *β-Glucogallin: 1,2,3,6-tetragalloyl-β-D-glucose 4-O-galloyltransferase*
   (EC 2.3.1.-)

β-Glucogallin + 1,2,3,6-tetra-O-galloylglucose → 1,2,3,4,6-penta-O-galloylglucose

+ D-glucose

The enzyme was purified 22-fold from oak leaves in 37% yield. It dependend strictly on β-glucogallin as acyl donor ($K_m = 2.3$ mM) and the acceptor 1,2,3,6-tetragalloyl-glucose ($K_m = 1$ mM), whereas the 1,2,4,6-isomer was inactive. The transferase (MW 260 kDa) was stable between pH 5.0 and 6.5; highest reactivities were observed at pH 6.3 and 40°C (Cammann *et al.*, 1989).

8. *Synthesis of gallotannins*

The characterisation of the enzymes involved in the conversion of pentagalloylglucose to gallotannins is currently being investigated. It has been shown that β-glucogallin again served as the principal acyl donor for the *meta*-depsidic galloyl groups of *in vitro* synthesised hexa-, hepta-, octa-, nona-, and eventually decagalloylglucoses (Hofmann and Gross, 1990), but no information is to date available on details of this reaction sequence.

VI. CONCLUSIONS AND PERSPECTIVES

This chapter describes, with special emphasis on methodological aspects, the current status in the investigation of the pathways to gallotannins and ellagitannins. Consider-ing the fatal effects exerted by these natural products on proteins, it is surprising that so many properties of the enzymes from these reaction sequences have been clarified in such a comparatively short time. Nevertheless, these efforts represent only a beginning. There still remain many unanswered questions, for example on the mechanism of the formation of ellagitannins. Moreover, every success is known to create immediate new challenges. The observation, for example, that the pentagalloyl-glucose-producing and -metabolising enzymes are so remarkably resistant against these agressive tannins raises the question as to whether these proteins might possess unique structural features. This new problem demands, after purifying such enzymes to homogeneity, the elucidation of their molecular structures. As one prerequisite for such studies, it is important to show that now tannins can be investigated successfully by enzymatic means, and it is hoped that this chapter provides a stimulus for further and intensive research on the basis of these techniques.

ACKNOWLEDGEMENTS

I am indebted to the co-workers who contributed to the research reported from my laboratory, and to the Deutsche Forschungsgemeinschaft and the Fonds der Chemischen Industrie for financial support.

REFERENCES

Amrhein, N., Topp, H. and Joop, O. (1984). *Plant Physiol.* **75** (Suppl.), 18.
Atkinson, M. R. and Morton, R. K. (1960). *In* "Comparative Biochemistry, Vol. II: Free Energy and Biological Function" (M. Florkin and H. S. Mason, eds), pp. 1–95. Academic Press, New York–London.
Beart, J. E., Lilley, T. H. and Haslam, E. (1985). *Phytochemistry* **24**, 33–38.
Cammann, J., Denzel, K., Schilling, G. and Gross, G. G. (1989). *Arch. Biochem. Biophys.* **273**, 58–63.
Dahlbender, B. and Strack, D. (1984). *J. Plant Physiol.* **116**, 375–379.
Dahlbender, B. and Strack, D. (1986). *Phytochemistry* **25**, 1043–1046.
De Langhe, J., Van Maele, A., Nollet, L. and Poma, K. (1987). *Med. Fac. Landbouww. Rijksuniv. Gent* **52**, 1479–1488.
Denzel, K. and Gross, G. G. (1991) *Planta* **184**, 285–289.
Denzel, K., Weisemann, S. and Gross, G. G. (1988a). *J. Plant Physiol.* **133**, 113–115.
Denzel, K., Schilling, G. and Gross, G. G. (1988b). *Planta* **176**, 135–137.
Fischer, E. and Bergmann, M. (1918). *Ber. Deutsch. Chem. Ges.* **51**, 1760–1804.
Fischer, E. and Bergmann, M. (1919). *Ber. Deutsch. Chem. Ges.* **52**, 829–854.
Fischer, E. and Hess, K. (1912). *Ber. Deutsch. Chem. Ges.* **45**, 912–915.
Fischer, E., Bergmann, M. and Lipschitz, W. (1918). *Ber. Deutsch. Chem. Ges.* **51**, 45–79.
Freudenberg, K. (1920). "Die Chemie der natürlichen Gerbstoffe", p. 4. Springer-Verlag, Berlin.
Gross, G. G. (1982a). *Z. Naturforsch.* **37c**, 778–783.
Gross, G. G. (1982b). *FEBS Lett.* **148**, 67–70.
Gross, G. G. (1983a). *Z. Naturforsch.* **38c**, 519–523.
Gross, G. G. (1983b). *Phytochemistry* **22**, 2179–2182.
Gross, G. G. (1989). *In* "Plant Cell Wall Polymers: Biogenesis and Biodegradation" (N. G. Lewis and M. G. Paice, eds), pp. 108–121. ACS Symp. Ser. Vol. 399, Washington, DC.
Gross, G. G. and Denzel, K. (1991). *Z. Naturforsch.* **46C**, 389–394.
Gross, G. G., Schmidt, S. W. and Denzel, K. (1986). *J. Plant Physiol.* **126**, 173–179.
Gross, G. G., Denzel, K. and Schilling, G. (1990). *Z. Naturforsch.* **45c**, 37–41.
Gupta, R. K., Al-Shafi, S. M. K., Layden, K. and Haslam, E. (1982). *J. Chem. Soc., Perkin Trans. I*, 2525–2534.
Haddock, E. A., Gupta, R. K., Al-Shafi, S. M. K., Layden, K., Haslam, E. and Magnolato, D. (1982a). *Phytochemistry* **21**, 1049–1062.
Haddock, E. A., Gupta, R. K., Al-Shafi, S. M. K. and Haslam, E. (1982b). *J. Chem. Soc., Perkin Trans. I*, 2515–2524.
Haddock, E. A., Gupta, R. K. and Haslam, E. (1982c). *J. Chem. Soc., Perkin Trans. I*, 2535–2545.
Hagenah, S. and Gross, G. G. (1993) *Phytochemistry*, in press.
Haslam, E. (1982). *Fortschr. Chem. Org. Naturst.* **41**, 1–46.
Haslam, E. (1988). *J. Soc. Leather Technologists & Chemists* **72**, 45–64.
Haslam, E. (1989a). "Plant Polyphenols. Vegetable Tannins Revisited". Cambridge University Press, Cambridge.
Haslam, E. (1989b). *In* "Natural Products of Woody Plants I. Chemicals Extraneous to the Lignocellulosic Cell Wall" (J. W. Rowe, ed.), pp. 399–438. Springer-Verlag, Berlin.
Haslam, E., Lilley, T. H., Cai, Y., Martin, R. and Magnolato, D. (1989). *Planta Med.* **55**, 1–8.

Hillis, W. E. (1985). *In* "Biosynthesis and Biodegradation of Wood Components" (T. Higuchi, ed.), pp. 325–347. Academic Press, Orlando, FL.

Hofmann, A. S. and Gross, G. G. (1990). *Arch. Biochem. Biophys.* **283**, 530–532.

Ishikura, N., Hayashida, S. and Tazaki, K. (1984). *Bot. Mag. Tokyo* **97**, 355–367.

Ishimaru, K., Hirose, M., Takahashi, K., Koyama, K. and Shimomura, K. (1990). *Phytochemistry* **29**, 3827–3830.

Koenigs, W. and Knorr, E. (1901). *Ber. Deutsch. Chem. Ges.* **34**, 957–981.

Kojima, M. and Kondo, T. (1985). *Agric. Biol. Chem.* **49**, 2467–2469.

Kozák, I., Kronrád, L. and Procházka, M. (1978). *J. Labelled Comp. Radiopharm.* **15**, 401–405.

Krajci, I. and Gross, G. G. (1987). *Phytochemistry* **26**, 141–143.

Krüper, A., Gehrke, P. and Amrhein, N. (1990). *J. Labelled Comp. Radiopharm.* **28**, 713–718.

Lam, T. H. and Shaw, M. (1970). *Biochem. Biophys. Res. Commun.* **39**, 965–968.

Larew, H. G. (1987). *Economic Botany* **41**, 33–40.

Lesser, R. and Gad, G. (1926). *Ber. Deutsch. Chem. Ges.* **59**, 233–236.

Loomis, W. D. and Battaile, J. (1966). *Phytochemistry* **5**, 423–438.

Nishizawa, M., Yamagishi, T., Nonaka, G. and Nishioka, I. (1980). *Chem. Pharm. Bull.* **28**, 2850–2852.

Nishizawa, M., Yamagishi, T., Nonaka, G. and Nishioka, I. (1982). *J. Chem. Soc., Perkin Trans. I,* 2963–2968.

Nishizawa, M., Yamagishi, T., Nonaka, G., Nishioka, I., Nagasawa, T. and Oura, H. (1983a). *Chem. Pharm. Bull.* **31**, 2593–2600.

Nishizawa, M., Yamagishi, T., Nonaka, G. and Nishioka, I. (1983b). *J. Chem. Soc., Perkin Trans. I,* 961–965.

Nonaka, G. and Nishioka, I. (1983). *Chem. Pharm. Bull.* **31**, 1652–1658.

Nonaka, G., Nishioka, I., Nagasawa, T. and Oura, H. (1981). *Chem. Pharm. Bull.* **29**, 2862–2870.

Nonaka, G., Nishioka, I., Nishizawa, M. and Yamagishi, T. (1985). *J. Med. Pharm. Soc. for Wakan-Yaku* **2**, 37–47.

Okuda, T., Yoshida, T. and Hatano, T. (1989a). *Planta Med.* **55**, 117–122.

Okuda, T., Yoshida, T. and Hatano, T. (1989b). *J. Nat. Prod.* **52**, 1–31.

Porter, L. J. (1989). *In* "Methods in Plant Biochemistry, Vol. 1: Plant Phenolics" (J. B. Harborne, ed.), pp. 389–419. Academic Press, London.

Rolston, M. P. (1978). *Bot. Rev.* **44**, 365–396.

Scalbert, A. and Haslam, E. (1987). *Phytochemistry* **26**, 3191–3195.

Schildknecht, H. and Milde, R. (1987). *Carbohydrate Res.* **164**, 23–31.

Schmidt, O. Th. and Reuss, H. (1961). *Liebigs Ann. Chem.* **649**, 137–148.

Schmidt, S. W., Denzel, K., Schilling, G. and Gross, G. G. (1987). *Z. Naturforsch.* **42c**, 87–92.

Strack, D., Dahlbender, B., Grotjahn, L. and Wray, V. (1984). *Phytochemistry* **23**, 657–659.

Villegas, R. J. A., Shimokawa, T., Okuyama, H. and Kojima, M. (1987). *Phytochemistry* **26**, 1577–1581.

Waterman, P. G. (1988). *Phytochemistry* **27**(12), xii–xiii.

Weisemann, S., Denzel, K., Schilling, G. and Gross, G. G. (1988). *Bioorg. Chem.* **16**, 29–37.

Yazaki, K. and Okuda, T. (1989). *Plant Cell Rep.* **8**, 346–349.

Yazaki, K. and Okuda, T. (1990). *Phytochemistry* **29**, 1127–1130.

3 Hydroxycinnamic Acids and Lignins

DIETER STRACK and HANS-PETER MOCK

*Institut für Pharmazeutische Biologie der Technischen Universität,
D-3300 Braunschweig, Germany*

I. INTRODUCTION

The term 'hydroxycinnamic acids' (HCAs) covers a group of chemical compounds built up as open-chain phenylpropanoids (3-phenyl-2-propenoic acids) with different

METHODS IN PLANT BIOCHEMISTRY Vol. 9
ISBN 0–12–461019–6

patterns of ring hydroxylation and methoxylation. They belong to the class of 'phenolic compounds' characterised by at least one aromatic ring (C_6) bearing one or more hydroxyl constituents, including their derivatives. The HCAs (C_6–C_3), along with hydroxybenzoic acids (C_6–C_1), represent the two common families of 'phenolic acids' found in plants (Gross, 1981).

Unsubstituted cinnamic acid was identified in cinnamon oil by Dumas and Peligot in 1834. Later the substituted cinnamic acids were found: sinapic acid (4-hydroxy-3,5-dimethoxycinnamic acid) in mustard by von Babo and Hirschbrunn in 1852; 4-coumaric acid (4-hydroxycinnamic acid) and ferulic acid (4-hydroxy-3-methoxy-cinnamic acid) in aloe by Hlasiwetz in 1865 and Hlasiwetz and Barth in 1866, respectively; and caffeic acid (3,4-dihydroxycinnamic acid) by Hlasiwetz in 1867.

	R^1	R^2	R^3
Cinnamic acid	H	H	H
4-Coumaric acid	H	OH	H
Caffeic acid	OH	OH	H
Ferulic acid	OCH_3	OH	H
Sinapic acid	OCH_3	OH	OCH_3

Hydroxycinnamic acids are now well known to occur ubiquitously in higher plants, caffeic acid being the commonest one (Bate-Smith, 1962; Harborne, 1966). The great diversity of several thousand known phenylpropanoid structures in plants is principally based on alterations of the phenylpropenoic acid side-chain leading to the various classes of phenylpropanoids (Barz et al., 1985), e.g. the most widespread and important flavonoids (Hahlbrock, 1981) and lignins (Grisebach, 1981).

On the other hand, HCAs accumulate as intact moieties in combined forms as conjugates, of which several hundred individual compounds are known. They are formed by liberation of H_2O mostly as esters and amides, rarely as glycosides, preserving the C_6–C_3 structure, with a vast number of different compounds, e.g. carbohydrates, proteins, lipids, amino acids, amines, carboxylic acids, terpenoids, alkaloids or flavonoids (Herrmann, 1978; Harborne, 1980). In addition, HCAs occur insolubly bound to polymers such as cutins and suberins (Riley and Kolattukudy, 1975), lignins (Higuchi et al., 1967; Scalbert et al., 1985; Smith, 1955) or polysaccharides in cell wall fractions (Fry, 1986).

Both these conjugates and the diverse HCA derivatives are of considerable biological significance, see for instance their importance as support materials (e.g. suberins, lignins) or as ecological agents (e.g. phytoalexins, allelopathics, repellents, attractants, UV light protectants). The various areas of phenylpropanoid metabolism are dependent on the physiology of the plant's primary metabolism (e.g. dependence on physiological factors such as light, temperature, water and mineral nutrition, or phytohormones). Phenylpropanoids probably also perform active roles in plant growth and general metabolism (Harborne, 1980).

The ultimate goal in research on secondary plant products, such as the phenylpropanoids, is the understanding of the regulation of the temporal metabolic activities during plant development and the spatial distribution in certain tissues or cell types, leading finally to the fascinating field of molecular genetics (Hahlbrock and Scheel, 1989; Dixon and Lamb, 1990). One of the starting points for such research activities is a thorough knowledge of the enzymology of phenylpropanoid derivatives. Within the last decade both new sophisticated techniques in structural elucidation of natural compounds and isolation of enzymes involved in their metabolism (biosynthesis, interconversion, degradation) have been developed and have led to a vast number of well-defined pathways.

In this chapter, we will summarise briefly the biochemistry of HCAs and their derivatives, with special emphasis on the general phenylpropanoid metabolism (Hahlbrock and Grisebach, 1975) and the biosynthesis of HCA conjugates and lignins. Degradative reactions were not considered. The main sections contain methods of enzymology of general interest including some selected protocols for extraction, assay and purification. In some cases the amount (weight or volume of solution) of protein is not given. This is either because these amounts are variable, or because no amount is given by the author. Common practices, e.g. incubation at 30°C, and some basic tools are not reported; it is not intended to be exhaustive. We will rather point to examples of well-known and widely used methods in enzymic studies of HCAs and lignins either from earlier references or from more recent publications. For further general discussions and reports of properties about the enzymes described, the reader should consult the complementary series *The Biochemistry of Plants*, Vol. 7 (ed. E. E. Conn, 1981) and the references cited in the present contribution.

II. PATHWAYS

A. General Phenylpropanoid Pathway

The carbon skeleton of HCAs is supplied by L-phenylalanine by way of the shikimate/arogenate pathway (Jensen, 1986). The key reaction leading to the structures of the secondary phenylpropanoids is the non-oxidative deamination of phenylalanine in the formation of *E*-cinnamic acid catalysed by phenylalanine ammonia-lyase (PAL; EC 4.3.1.5) (Koukol and Conn, 1961). In some cases, e.g. grasses and certain fungi, the enzyme may also act on L-tyrosine (Hanson and Havir, 1981). The PAL activity seems to determine the extent to which phenylalanine is withdrawn from primary metabolism entering the general phenylpropanoid metabolism. Induction of PAL and regulation of its activity play an important role in plant development (Camm and Towers, 1977; Jones, 1984). The supply of phenylalanine, however, may also be rate-limiting for phenylpropanoid biosynthesis (Jones, 1984). The enzymes involved in the general phenylpropanoid pathway belong to the possibly coordinatedly induced group I (Hahlbrock *et al.*, 1976).

With respect to the coordination of phenylalanine fluxes into protein and phenylpropanoid biosyntheses, a recently suggested hypothesis of Hrazdina and Jensen (1990) is noteworthy. These authors discussed multiple parallel pathways in the phenylpropanoid metabolism with two locations of the shikimate/arogenate pathway in the plant cell including several isoenzymes. These locations are the chloroplast, in which the aromatic amino acids are produced mainly for protein biosynthesis, and the cytoplasm, in which, most probably membrane associated, phenylalanine is produced for the formation of the secondary phenylpropanoids. The latter includes pathway channelling (Hrazdina and Wagner, 1985a). This idea is supported by data indicating coordinate enzyme induction (Hahlbrock et al., 1976; Hrazdina and Wagner, 1985b).

The product of PAL activity, E-cinnamic acid, is subject to a series of hydroxylation and methylation reactions, resulting in the sequential formation of the common 4-coumaric, caffeic, ferulic and sinapic acids. Hydroxyferulic acid, which has recently been found esterified with cell wall components in Zea mays and Hordeum vulgare (Ohashi et al., 1987), is apparently the substrate for sinapic acid formation.

The HCAs enter various pathways (Barz et al., 1985) of four predominant types of side-chain reactions. Important reactions are: (1) condensation (side-chain elongation) with three malonyl-CoAs accompanied by liberation of three molecules CO_2 leading to the flavonoids; (2) degradation (side-chain shortening) by removal of an acetic acid unit leading to the hydroxybenzoic acids; (3) reduction (NADPH-dependent) leading to the lignin precursors hydroxycinnamyl alcohols; and (4) conjugation through attachment to an hydroxyl- or amino-group-bearing molecule with liberation of H_2O leading to esters, amides or glycosides.

B. Hydroxycinnamic Acid Conjugation

It is a characteristic feature of secondary plant products, especially the soluble phenolics, that they occur in the form of conjugates. Conjugation reactions are important because they alter the chemical/physical properties of compounds which may result in a site of accumulation different from that occupied by the aglycone. This may greatly influence whether a compound is converted into a metabolically inactive final endproduct (permanent storage) or into a transiently accumulating product which may be subject to further temporal metabolic pathways (e.g. further substitution or conjugation, interconversion, or degradation).

The distribution of HCA conjugates shows in several cases close correlation with the systematic arrangement of plant families, for example caffeic acid esters (Molgaard and Ravn, 1988) such as caffeoyldisaccharides are present mainly in Scrophulariaceae and Oleaceae, 5-O-caffeoylquinic acid (chlorogenic acid) in Asteraceae, Solanaceae and Rubiaceae, and caffeoyldihydroxyphenyllactic acid (rosmarinic acid) in Lamiaceae and Boraginaceae. Other examples are the characteristic occurrence of sinapoylcholine (sinapine) in Brassicaceae (Schultz and Gmelin, 1953), feruloylsucrose in Liliaceae (Meurer et al., 1984), di-HCA-spermidine amide in families of the Fagales, Juglandales and Myricales (Meurer et al., 1988) or tricoumaroylspermidine amide in Rosaceae (Strack et al., 1990a).

Besides UDP-glucose-dependent transferases involved in the formation of HCA acylglucose esters and glucosides, there are three types of transferases which have been described to date catalysing the formation of HCA conjugates: (1) HCA-CoA thioester-, (2) 1-O-HCA-acylglucose- and (3) HCA O-ester-dependent transferases. Further conjugation reactions in which HCAs act as acceptor molecules are possible, see for example the recently described glutathione S-transferase which catalyses the conjugation of cinnamic acid with glutathione (Edwards and Dixon, 1991a).

The alternative pathways shown so far for the HCA-ester formations are obviously not dependent on the nature of the conjugating moiety, but rather on the source of the enzyme used, i.e. the plant investigated. The biosynthesis of HCA-glucaric acid ester is a good example of converging lines for the formation of HCA esters. In *Secale cereale* leaves it proceeds via the HCA-CoA thioester (Strack *et al.*, 1987b), in *Cestrum elegans* leaves via the 1-*O*-HCA-acylglucoside (Strack *et al.*, 1987c), and in *Lycopersicon esculentum* leaves via the *O*-HCA-quinic acid ester (chlorogenic acid) (Strack and Gross, 1990). The biosynthesis of chlorogenic acid is another example of such alternative pathways. It can be formed via caffeoyl-CoA, e.g. in *Nicotiana alata* cell cultures (Stöckigt and Zenk, 1974) and various other systems (Ulbrich and Zenk, 1979), or via the 1-*O*-caffeoylglucose in *Ipomoea batatas* (Kojima and Villegas, 1984). It will be of great interest to search for further alternative acyltransferase reactions in plants and to evaluate the role of such biochemical convergence in the metabolism of HCA conjugates.

C. Lignin Formation

Lignins constitute a family of cell wall-located heterogeneous phenylpropane polymers arising from peroxidase-initiated and chemically driven dehydrogenative polymerisation of hydroxycinnamyl alcohols (Sarkanen and Ludwig, 1971). Gymnosperm lignin is composed mainly of coniferyl alcohol, whereas angiosperm lignin is built up by both guaiacyl and syringyl units. Lignins show a wide structural variety in different plant tissues and are always associated with polysaccharides. Monties (1989) gives a thorough account of the historical background and general structure and properties of lignins as well as methods of analysis.

In the inital reactions leading to lignins, the precursors, HCAs, are reduced to the hydroxycinnamyl alcohols via the activities of the HCA-CoA reductase and hydroxy-cinnamaldehyde dehydrogenase. The alcohols are glucosylated (4-*O*-β-glucosides) into an intercellular transport form and secreted into the lignifying cell wall. Here the glucose is removed by a β-glucosidase and the liberated alcohol enters the lignin pathway. Lignin biosynthesis starts with the formation of free radicals by peroxidase activity and H_2O_2. Interconversion of the mesomeric free radicals leads to random coupling, followed by intramolecular reaction of a quinone methide-type dimer. Further intermolecular reactions with other phenolics and polysaccharides in the wall form stable cross-links. For further reading and references on lignin structure and biosynthesis, see Freudenberg and Neish (1968), Grisebach (1981), Gross (1985), Fry (1988) and Lewis and Yamamoto (1990).

III. ENZYMES OF THE GENERAL PHENYLPROPANOID PATHWAY

A. Phenylalanine Ammonia-lyase (PAL; EC 4.3.1.5)

(a) *Reaction*

$$\text{L-Phenylalanine} \longrightarrow E\text{-Cinnamic acid} + NH_3$$

The elimination of ammonia from amino group-bearing molecules is catalysed by ammonia-lyases (carbon–nitrogen lyases). The respective enzyme in the general phenylpropanoid pathway is the L-phenylalanine ammonia-lyase (PAL) which has become the most intensively studied enzyme in phenylpropanoid metabolism. Phenylalanine ammonia-lyase catalyses the non-oxidative elimination of NH_3 from L-phenylalanine including the *pro*-3*S* hydrogen (Hanson and Rose, 1975) with the formation of a double bond to give *E*-cinnamic acid. This enzyme is ubiquitously distributed in higher plants and occurs also in certain bacteria, algae and fungi. It has been the subject of numerous biochemical and physiological investigations (Camm and Towers, 1977; Jones, 1984).

(b) Extraction for assay. Extraction of PAL is usually accomplished either by homogenisation of plant material in sodium borate buffer (100 mM, pH 8.5–8.8) or by way of the 'acetone powder' (Koukol and Conn, 1961; Neish, 1961) extracted with borate buffer. The acetone powder is prepared by homogenisation of freshly chopped material (e.g. 1–2 min in a Waring blender) in cold acetone ($-10°C$ to $-20°C$). The insoluble material containing the precipitated protein is washed several times with cold acetone. Additional washings with diethyl ether remove any remaining lipophilic components. The powder is dried at room temperature and stored in the cold. The PAL activity is quite stable in this powder (Havir and Hanson, 1968; Zucker, 1965), however, 20–40% loss of activity has been reported when stored for a week at $-15°C$ (Marsh *et al.*, 1968).

Protein solutions may lose 50% of the inital PAL activity at 4°C within 10 days (Zimmermann and Hahlbrock, 1975). This can be partially avoided by the addition of a thiol reagent, such as 1 mM dithiothreitol (DTT) or 1–5 mM 2-mercaptoethanol with only 25% loss of activity.

(c) Assay

PAL activity from Petroselinum crispum (P. hortense) (*Zimmermann and Hahlbrock, 1975*):

- 10 mM L-phenylalanine
- 100 mM sodium borate buffer (pH 8.8)
- protein in a final volume of 1 ml

The PAL-catalysed reaction does not require cofactors. Assays found in the literature are essentially according to the original description given by Koukol and Conn in 1961. The reaction can easily be measured spectrophotometrically by following the initial rate of *E*-cinnamic acid formation at 290 nm (Zucker, 1965). The amount of product formed can be calculated from the increase in absorbance using an extinction coefficient for *E*-cinnamic acid at pH 8.8 of $1 \times 10^7 \, cm^2 \, mol^{-1}$. This method may cause problems in the case of absorbance in assays with high protein concentrations. Problems due to possible side reaction by phenylalanine transaminase (Erez, 1973), which might be active in the presence of contaminating 2-oxocarboxylic acids (strong absorbance of phenylpyruvate–*enol* tautomer–borate complex at 290 nm; Knox and Pitt, 1957), are eliminated by acetone powder preparations or filtration of crude extracts through Sephadex G-25 or ion-exchangers.

Low PAL activities may be detected by radiotracer methods with ^{14}C-labelled phenylalanine. The amount of ^{14}C-labelled cinnamic acid formed can be determined by liquid scintillation counting after separation by thin layer chromatography (TLC) of a diethyl ether extract of the assay including cinnamic acid as a marker (Brödenfeldt and Mohr, 1986; see also Havir *et al.*, 1971) for TLC separation. As an alternative the acidified mixture (e.g. 0.2 ml 3 N H_2SO_4 to a 3 ml assay; Amrhein and Zenk, 1971) may be extracted with a diethyl ether–hexane mixture (1:1; v/v) or toluene (Koukol and Conn, 1961) and the organic phase subjected to ^{14}C counting without chromatographic separation of the product.

Other techniques for PAL activity determination, combined with product identification, are gas chromatography (GC) (Czichi and Kindl, 1975) and high performance liquid chromatography (HPLC) (Murphy and Stutte, 1978). Amrhein *et al.* (1976) described a procedure which permits activity measurement in intact plant cells based on ^3HOH release (recovered by sublimation) via N^3HH_2 from L-[3-^3H]phenylalanine. The specificity of the ^3H release, however, has to be investigated in each case (Holländer and Amrhein, 1981). Difficulties in adapting PAL assay to determination of TAL activity (4-coumaric acid production from L-tyrosine) are discussed by Hanson and Havir (1981).

(d) Purification and properties. Phenylalanine ammonia-lyase has been purified from a large number of plant tissues. Its properties have been thoroughly described by Hanson and Havir (1981). Purification of PAL is not always easy. During the purification procedures the enzyme undergoes various changes (Hanson and Havir, 1981). Purification procedures may drastically change kinetic characteristics, such as K_m values or cooperativity. Zimmermann and Hahlbrock (1975) described a column chromatographic purification procedure which includes the commonly used DEAE-cellulose, Sephadex G-200 and hydroxyapatite which proved to be superior to other described methods (Table 3.1).

The enzyme was finally purified by preparative polyacrylamide gel electrophoresis (PAGE) and exhibited typical properties, such as M_r 300 000 ($\pm 10\%$) with four probably identical subunits of M_r 83 000. This PAL showed characteristic negative cooperativity with respect to phenylalanine, i.e. two K_m values, one at low and the other at high phenylalanine concentrations (32 and 240 μM) and a pH optimum in the range 8–9. For discussions on the use of PAL inhibitors, such as the highly effective D- and L-2-aminooxy-3-phenylpropionic acid (D- and L-AOPP; Amrhein and Gödeke, 1977), see Hanson and Havir (1981). Recently, Zon and Amrhein (1992) introduced a novel excellent inhibitor of PAL, 2-aminoindan-2-phosphonic acid (AIP), which is a chemically stable compound whose synthesis and handling (e.g. autoclaving) is much simpler than those of the other PAL inhibitors.

TABLE 3.1. Purification of PAL from *Petroselinum crispum* cell cultures.

Purification step	Protein (mg)	Specific activity (units^a $\times 10^{-3}$ mg^{-1})	Purification (-fold)	Yield (%)
Initial crude extract	13 200	2.9	1	100
Dowex 1X2 supernatant	12 200	3.0	1.1	99
$(NH_4)_2SO_4$ (40–55%)	1 183	22.6	8	71
DEAE-cellulose	92	116	41	28
Sephadex G-200	13	515	180	17
Hydroxyapatite	4.7	1250	440	16

^{a} Amount of enzyme that catalyses the formation of 1 μmol cinnamic acid in 1 min.
From Zimmermann and Hahlbrock (1975).

B. Hydroxylases

The sequence of hydroxylation reactions from cinnamic acid to 4-coumaric, caffeic and 5-hydroxyferulic acid via ferulic acid, is catalysed by monooxygenases, referred to as mixed-function oxidases. There are two types of these enzymes, the membrane-bound (microsomal) cytochrome P-450-dependent oxygenases, involved in the formation of 4-coumaric and 5-hydroxyferulic acids, and the soluble phenolase which catalyses the introduction of a second hydroxyl group into a monophenol, as in caffeic acid formation. The preferred electron donors are NADPH and ascorbic acid. The cytochrome P-450-linked oxygenases are characterised by the requirement of O_2 and light-reversible CO inhibition. Inhibition studies with cytochrome c and various cytochrome P-450-specific inhibitors, such as prochloraz, ketoconazole and metapyrone, have also been used to characterise hydroxylases involved in other metabolic areas, e.g. a microsomal protopine 6-hydroxylase (Tanahashi and Zenk, 1990).

1.   Cinnamic acid 4-hydroxylase (CA4H; EC 1.14.13.11)

(a)   Reaction

$$\text{Cinnamic acid} + O_2 + AH + H^+ \longrightarrow \text{4-Coumaric acid} + H_2O + A$$

Cinnamic acid 4-hydroxylase (CA4H) specifically introduces a hydroxyl group in the 4-position of E-cinnamic acid with a concomitant NIH shift (Zenk, 1967; Reed et al., 1973), i.e. an intramolecular migration of the proton which is being displaced to an adjacent ortho position on the aromatic ring (Haslam, 1974, and literature cited therein). The CA4H is inactive with the Z-isomer (Pfändler et al., 1977). The enzyme was first detected in microsomal preparations from Pisum sativum seedlings (Russell and Conn, 1967; Russell, 1971) and has subsequently been described in various other plants. Hydroxylation at the 2-position (Gesteiner and Conn, 1974) involved in coumarin biosynthesis has also been reported from several plants (Gross, 1985) and again was found to be membrane bound, exhibiting similar properties to those of CA4H.

(b)   Extraction for assay.

Extraction of CA4H from Zea mays seedlings (Oba and Conn, 1988) is accomplished by homogenisation of plant material in Tris-HCl buffer (50 mM, pH 8.0) containing 330 mM sorbitol, 4 mM $MgCl_2$, 4 mM EDTA and 2 mM 2-mercaptoethanol (essential in crude extracts). After filtration and centrifugation (10 min at $10\,000 \times g$), the supernatant was passed through a Sephadex G-25 column. The protein eluate was further centrifuged at $100\,000 \times g$ for 1.5 h (microsomal fraction). The resulting pellet was resuspended with a Teflon-glass homogeniser in Tris-HCl buffer (50 mM, pH 8.0) containing 2 mM 2-mercaptoethanol.

Hahlbrock et al. (1971) isolated CA4H from Petroselinum crispum cell cultures by homogenisation (1 min) of a mixture of 1 g of cells with 2 ml potassium phosphate buffer (200 mM, pH 7.5) containing 2 mM 2-mercaptoethanol. After centrifugation at $10\,000 \times g$ for 15 min, the supernatant was passed (centrifuged) through c. 4 g of Sephadex G-25 (equilibrated with the extraction buffer).

(c) Assay

CA4H activity from Petroselinum crispum (*Hahlbrock* et al., *1971*):

- 60 μM [1-^{14}C]cinnamic acid (850 Bq assay^{-1})
- 500 μM NADP$^+$
- 1 mM glucose 6-phosphate
- 0.1 enzyme unit glucose 6-phosphate dehydrogenase
- 2 mM 2-mercaptoethanol
- 200 mM potassium phosphate buffer (pH 7.5)
- 100–200 mg protein in a total volume of 500 μl

A common assay, similar to that used originally by Russell (1971), contains a NADPH-regenerating system and ^{14}C-labelled cinnamic acid. The reaction was terminated after 60 min with 50 μl acetic acid containing 500 μg each of cinnamic and 4-coumaric acids as markers. After TLC on silica gel in benzene–acetic acid (9:1; v/v). 4-coumaric acid is detected under UV light, scraped off and subjected to liquid scintillation counting (LSC).

Lamb and Rubery (1975) described an alternative method using a rapid spectrophotometric assay for CA4H, based on the characteristic spectra of cinnamic acid and 4-coumaric acid at 340 nm after acidification (destruction of NADPH) of the reaction mixture and subsequent readjustment to pH 11 (bathochromic shift of UV absorption of 4-coumaric acid). This method is not applicable to low enzyme activities (Billett and Smith, 1978).

(d) Purification and properties.   The CA4H has not been purified to homogeneity because of rapid loss of activity due to the complexity of the protein system (multienzyme system). Total loss of activity from *Triticum aestivum* has been reported to occur within 12 h at 4°C (Maule and Ride, 1983), similar to the original enzyme preparation of Russell (1971). The latter reported that storage of the enzyme for one week at −20°C in the presence of 10% glycerol (v/v) resulted in no loss of activity. The enzyme is usually enriched by ultracentrifugation, i.e. preparation of microsomal fractions. Further subfractionation may be achieved by centrifugation on sucrose gradients (e.g. Benveniste *et al.*, 1977, 1978; Billett and Smith, 1978; Postius and Kindl, 1978; Fujita and Asahi, 1985). The enzyme system is a membrane-bound haemoprotein linked with a cytochrome P-450 system. The preferred cosubstrate is NADPH + H$^+$. It has been found that NADH or a NADH-generating system cannot replace NADPH. However, it may have a synergistic effect on hydroxylation (Benveniste *et al.*, 1977; Billett and Smith, 1978).

The reaction optimum is usually found at pH 7.5, which is used in the common assay buffers. Higher pH optima, however, may be found, e.g. 8.0 from *Ipomoea batatas* (Tanaka *et al.*, 1974). 2-Mercaptoethanol, in millimolar concentrations, is required for optimal activity (Russel, 1971). It could not be replaced by other thiol reagents. However this cannot be generalised, as shown with an enzyme preparation from *Cucumis sativus* (Billett and Smith, 1978) that required glutathione for highest activity, whereas 2-mercaptoethanol was inhibitory at concentrations higher than 0.4 mM. Powerful tools for specificity and elucidation of the physiological role of different plant monooxygenases are inhibitors with differential inactivation of cytochrome P-450 enzymes, such as 1-aminobenzotriazole (suicide substrate) for selective CA4H inhibition (Reichhart *et al.*, 1982).

2. *4-Coumaric acid hydroxylase* (Phenolase; EC 1.14.18.1)

(a) Reaction

$$\text{4-Coumaric acid} + O_2 + AH + H^+ \longrightarrow \text{Caffeic acid} + H_2O + A$$

This phenolase reaction catalyses the introduction of a hydroxyl group in the 3-position of 4-coumaric acid in the formation of caffeic acid. Because of various reports about low substrate specificity and the observation of a second reaction leading to quinones, it is a matter of dispute to what extent phenolases are involved in the biosynthetic route of HCAs (Gross, 1985). The more recent findings of specific hydroxylases, acting at the level of HCA-conjugates (see below), support this notion.

Phenolase activities have to be distinguished from those of polyphenol oxidase (PPO), which show a broad substrate specificity towards monophenols in the presence of electron donors such as ascorbic acid. They catalyse a second reaction, the oxidase reaction, leading to production of quinones. Various examples of such unspecific reactions cast serious doubts on the role of phenolases in the HCA pathway (Gross, 1985). This agrees with the recent finding of Kojima and Takeuchi (1989), who found a new membrane-bound specific phenolase in *Phaseolus mungo* (*Phaseolus radiatus*) after complete elimination of PPO activity in tentoxin-treated plants (Kojima and Takeuchi, 1989). Tentoxin, a chlorosis-inducing fungal toxin, has been found to inhibit development of PPO (Duke and Vaughn, 1982). The new phenolase was inhibited by FAD and FMN, in contrast to some other analogous enzymes which require FAD in addition to NAD(P)H (Kamsteeg *et al.*, 1981; Boniwell and Butt, 1986). The enzyme activity described by Boniwell and Butt (1986) requires caffeic acid in trace quantities as an initiator. Caffeic acid initiates the hydroxylation of 4-coumaric acid, which is similar to the hydroxylation of other monophenols (Vaughan and Butt, 1970). These authors suggest that *o*-dihydroxy phenols interact with the enzymic copper to produce a reactive species that affects the hydroxylation of 4-coumaric acid.

(b) Extraction for assay.

The phenolase from *Phaseolus radiatus* (Kojima and Takeuchi, 1989) was extracted essentially by the method of Duke and Vaughn (1982). Samples were homogenised in a chilled mortar in 10 mM citrate–phosphate buffer (pH 5.3), containing 700 mM D-sorbitol and 1 mM EDTA. The homogenate was centrifuged (15 min at 13 000 × g) and the precipitate resuspended in the extraction buffer, containing 700 mM D-sorbitol. The specific enzyme activity was recovered in the precipitate, whereas PPO activity remained primarily in the supernatant. Location of PPO including phenolase in higher plants is still controversial. It is either easily solubilised, bound to various membranes or associated to high molecular weight complexes (Stafford and Dresler, 1972; Stafford and Bliss, 1973).

(c) Assay

Assay for phenolase activity from Beta vulgaris (*Vaughan and Butt, 1969*):

- 3.3 mM 4-coumaric acid
- 3.3 mM ascorbic acid

- 500 mM $(NH_4)_2SO_4$
- citrate (13.3 mM)–phosphate (33.3 mM) buffer (pH 5.3)
- protein in a total volume of 3 ml

The assay used by Vaughan and Butt (1969) was run in a shaking incubator. Of various reductants tested, only NADPH was less active. However, Kojima and Takeuchi (1989) used the more effective NADPH instead of ascorbic acid as electron donor with a phenolase from *Phaseolus radiatus*. The ammonium sulphate activates enzyme activity. Catalase may be added to the assay mixture (Duke and Vaughn, 1982; Kojima and Takeuschi, 1989) to prevent peroxidase-related hydroxylation. In addition the PPO inhibitor diethyldithiocarbamate may be included (Duke and Vaughn, 1982).

The reaction can be stopped by the addition of 10% (w/v) trichloroacetic acid (Vaughan and Butt, 1969) or a mixture of 12 N HCl and 4.1 M $(NH_4)_2SO_4$ (Kojima and Takeuchi, 1989). Enzyme activity is measured by caffeic acid production (Vaughan and Butt, 1969). The caffeic acid may be measured spectrophotometrically (320 nm) after chromatographic isolation from an ethyl acetate fraction (Kojima and Takeuchi, 1989) or after chemical modification in the visible absorption range (Vaughan and Butt, 1969; Duke and Vaughn, 1982).

(d) Purification and properties. The phenolase was originally purified from *Beta vulgaris* leaves (Table 3.2) by Vaughan and Butt (1969). The specific phenolase from *Phaseolus radiatus* (Kojima and Takeuchi, 1989) has not been purified. The differences between its activity and other hydroxylase activities, however, clearly indicate the presence of a new enzyme. Table 3.3 summarises some properties compared to other hydroxylases.

TABLE 3.2. Purification of 4-coumaric acid hydroxylase from *Beta vulgaris* leaves.

Purification step	Protein (mg ml^{-1})	Activity (units[a] $\times 10^{-3}$ mg^{-1})	Purification (-fold)	Yield (%)
Initial crude extract	49 200	0.68	1	100
$(NH_4)_2SO_4$ (35–70%)	2 320	6.4	9.4	44
Heat treatment (10 min at 60°C), dialysis	488	25	37	36
DEAE-Cellulose	168	73	107	36
$(NH_4)_2SO_4$, dialysis	79	111	162	26
CM-Cellulose	3	727	1074	7

[a] Amount of enzyme that catalyses the uptake of 1 μmol O_2 in 1 min.
From Vaughan and Butt (1969).

3. Ferulic acid hydroxylase

(a) Reaction

Ferulic acid + O_2 + NADPH + H$^+$ $\longrightarrow$ 5-Hydroxyferulic acid + H_2O + NADP$^+$

Ferulic acid hydroxylase (FAH) catalyses the introduction of a hydroxyl group in the 5-position of ferulic acid. It has been found in *Populus* x *euramericana* stems (Grand,

TABLE 3.3. Selected hydroxylases from various sources.

Source	Substrate ⟶ product	pH optimum	Cofactor requirement	Remarks	Reference
Beta vulgaris	4-Coumaric acid ⟶ caffeic acid	$(5.3)^a$	Ascorbate, less active with NADPH; most active with a pteridine compound	Soluble; low specificity high catechol oxidase activity	Vaughan and Butt (1969)
Phaseolus radiatus	4-Coumaric acid ⟶ caffeic acid	5.0	NADPH, less active with ascorbate	Membrane bound; inhibition by 2-mercaptoethanol and diethyldithiocarbamate as well as FAD and FMN	Kojima and Takeuschi (1989)
Petroselinum crispum	4-Coumaroyl-CoA ⟶ caffeoyl-CoA	6.5	Zn^{2+}; ascorbate	Soluble	Kneusel *et al.* (1989)
Silene dioica	4-Coumaroyl-CoA ⟶ caffeoyl-CoA	8.1^b	NADPH; FAD	Soluble; strong inhibition by chloromercuribenzoate and Zn^{2+}	Kamsteeg *et al.* (1981)
Petroselinum crispum	5-O-(4-Coumaroyl)-shikimate ⟶ 5-O-caffeoylshikimate	7.4	NADPH	Microsomal cytochrome P-450-dependent; elicitor inducible	Heller and Kühnl (1985)

[a] Assay pH (optimum not determined).
[b] In phosphate buffer (pH 8.6 in glycylglycine buffer).

1984) and is involved in the biosynthesis of sinapic acid. The enzyme is a membrane-bound cytochrome P-450-dependent mixed-function monooxygenase, distinct from the cinnamic acid 4-hydroxylase. Activity requires NADPH. NADH exhibits synergistic effects.

(b) Extraction for assay. Extraction of FAH from *Populus* stems (Grand, 1984) was achieved by homogenisation of the plant material in 20 mM Tris-HCl buffer (pH 7.5) containing 10 mM 2-mercaptoethanol, 500 mM mannitol, 0.1% bovine serum albumin (BSA), 0.5% polyethylene glycol and 2 g polyvinylpolypyrrolidone (PVP). The enzyme was active in a subcellular fraction which precipitated between 10 000 and 100 000 × g (microsomal fraction). The activity was unstable (90% loss of activity after 24 h at 4°C) but could be stabilised by the addition of 15% polyethylene glycol (45% of the original activity after 1 week at −20°C).

(c) Assay

Assay for FAH activity from Populus *x* euramericana *stems (Grand, 1984):*

- 1 mM [OCH$_3$-^{14}C]ferulic acid (18.5 kBq assay^{-1})
- 1.5 mM NADP$^+$
- 2 mM glucose 6-phosphate
- 0.1 units glucose 6-phosphate dehydrogenase
- 1 mM 2-mercaptoethanol
- 500 μl microsomal fraction containing BSA (1 mg ml^{-1})
- 25 mM Tris-HCl buffer (pH 7.5) containing 15% polyethylene glycol
- protein in a total volume of 1 ml

The FAH activity was determined by LSC as follows: (1) reaction stopped after 20 min by adding 3 ml of 3 M HCl; (2) precipitated proteins removed by centrifugation; (3) supernatant extracted (×3) with 10 ml ethyl acetate; (4) extract evaporated to dryness and residue dissolved in 0.2 ml ethanol; (5) paper chromatography with chloroform–acetic acid–water (2:1:1; v/v); (6) 5-hydroxyferulic acid localised with a radioactivity scanner; (7) 5-hydroxyferulic acid zone cut out and dipped in liquid scintillation cocktail; (8) measurement of radioactivity (LSC).

(d) Properties. Ferulic acid hydroxylase has not been purified. Attempts to solubilise the enzyme resulted in total loss of activity. The pH optimum was found to be 7.5. Glutathione has the most stimulating effect of the thiol reagents tested. Ferulic and cinnamic acids were accepted. However, the enzyme was inactive with 4-coumaric acid.

4. 4-Coumaroyl-CoA hydroxylase

(a) Reaction

Silene dioica:

4-Coumaroyl-CoA + O$_2$ + NADPH + H$^+$ ⟶ Caffeoyl-CoA + H$_2$O + NADP$^+$

Petroselinum crispum:

$$4\text{-Coumaroyl-CoA} + O_2 + \text{ascorbic acid} \longrightarrow \text{Caffeoyl-CoA} + H_2O$$
$$+ \text{ dehydroascorbic acid}$$

Hydroxylase activities have been detected in *Silene dioica* (Kamsteeg *et al.*, 1981) and in *Petroselinum crispum* (Kneusel *et al.*, 1989) that specifically catalyse the hydroxylation in the 3-position of 4-coumaroyl-CoA in the formation of caffeoyl-CoA. This points again to the discussion about the role of the phenolase-catalysed hydroxylations at the level of the free acids (see above).

(b) Extraction for assay.   The enzyme from *Silene dioica* (*Silene* enzyme; Kamsteeg *et al.*, 1981) was extracted by homogenisation (glass Potter Elvehjem) of petals in 50 mM potassium–sodium phosphate buffer (pH 7.8) containing 20 mM 2-mercapto-ethanol, 5% soluble PVP and 0.1% Triton X-100. After centrifugation at $38\,000 \times g$ for 30 min the clear supernatant can be used as the source of enzyme activity. Enzyme purification from this extract is shown in Table 3.4.

TABLE 3.4.   Purification of 4-coumaroyl-CoA hydroxylase from *Silene dioica* petals.

Purification step	Protein (mg ml^{-1})	Activity (units$^a \times 10^{-3}$ mg^{-1})	Purification (-fold)	Yield (%)
Initial crude extract	6.3	10.8	1	100
Supernatant crude extract	4.6	12.2	1.1	73
Supernatant $38\,000 \times g$	4.4	12.1	1.1	64
Precipitate $38\,000 \times g$	2.5	2.3	0.21	0.6
Dowex 1X2 eluate	2.15	70	6.5	262
Sephadex G-150	0.36	417	39	259

^{a} Amount of enzyme which consumes 1 nmol oxygen in 1 min at 20°C.
From Kamsteeg *et al.* (1981).

Kneusel *et al.* (1989) extracted the hydroxylase from *Petroselinum crispum* (*Petroselinum* enzyme) cell cultures by homogenisation of frozen cells in a mortar with Dowex 1X2 and 50 mM 2-(*N*-morpholino)ethanesulphonic acid (pH 6.5). The homogenate was filtered through a nylon screen and the filtrate centrifuged at $20\,000 \times g$ for 15 min. The supernatant was passed through Sephadex G-25, equilibrated with the extraction buffer.

(c) Assay

Assay for the Silene *enzyme (Kamsteeg et al., 1981):*

- 83 μM 4-coumaroyl-CoA
- 83 μM NADPH
- 8.3 μM FAD
- 6.7 mM potassium–sodium phosphate buffer (pH 8.0)
- 2.7 mM 2-mercaptoethanol
- 0.8 ml protein solution in total volume of 1.2 ml

Assay for the Petroselinum *enzyme (Kneusel et al., 1989):*

- 150 mM [2-^{14}C]4-coumaroyl-CoA (1.4 MBq assay^{-1})

- 50 mM ascorbic acid
- 500 μM ZnSO$_4$
- 20 μg protein in 50 mM 2-(N-morpholino)ethanesulfonate buffer (pH 6.5) in a total volume of 50 μl

Kamsteeg *et al.* (1981) described three different methods for determination of hydroxylase activity: (1) measurement of absorbance of NADPH (oxidation) at 340 nm (extinction coefficient of 6×10^6 cm^2 mol^{-1}); (2) measurement of absorbance of FAD (reduction; extinction coefficient 11.3×10^6 cm^2 mol^{-1}); (3) measurement of oxygen consumption. The *Petroselinum* enzyme was measured by a radiotracer method using [2-^{14}C]4-coumaroyl-CoA (Kneusel *et al.*, 1989). The latter was analysed by TLC (location and ^{14}C-measurement by TLC analyser) or reversed-phase HPLC (C$_{18}$) with 50 mM sodium citrate (pH 4.6)–methanol (65:35; v/v).

(d) Purification and properties. The *Silene* enzyme has been purified (39-fold; see Table 3.4). Both the *Silene* and *Petroselinum* enzymes are soluble and specific for the hydroxylation of 4-coumaroyl-CoA, although they differ remarkably in some of their properties (see Table 3.3).

5. *Hydroxycinnamic acid O-ester hydroxylase*

(a) Reaction

4-Coumaric acid *O*-ester + NADPH + H$^+$ $\longrightarrow$ 4-Caffeic acid *O*-ester
$$+ \, H_2O + NADP^+$$

The *O*-ester-dependent hydroxylase introduces a hydroxyl group in the 3-position of a 4-coumaric acid *O*-ester in the formation of the respective caffeic acid *O*-ester. This type of enzyme was first detected in *Petroselinum crispum* cell cultures challenged with an elicitor from *Phytophthora megasperma* in the specific formation of 5-*O*-caffeoylshikimic acid (Heller and Kühnl, 1985).

Hydroxylation which leads to the caffeoyl moiety in HCA-conjugates may also occur in other plants. For example, based on kinetic trapping experiments with radioactive labelled cinnamic acid as well as enzymatic studies on substrate specificities (CoA ligase), a possible hydroxylase-catalysed formation of chlorogenic acid via 5-*O*-(4-coumaroyl)quinic acid has been discussed (e.g. Nagels and Parmentier, 1976). This is in agreement with the recent discovery of a light-induced enzyme acting on 5-*O*-(4-coumaroyl)quinic acid/shikimic acid which is thought to be involved in chlorogenic acid biosynthesis in *Daucus carota* suspension cultures (Kühnl *et al.*, 1987).

This type of hydroxylase might also be involved in other analogous pathways. There is some evidence that the formation of caffeoyltartronic acid in *Phaseolus radiatus* also proceeds via 4-coumaroyltartronate (Strack *et al.*, 1986), which could possibly be proved with the enzyme preparation described by Kojima and Takeuchi (1989). It is very likely, however, that there are different biosynthetic routes, depending on the actual metabolic pool sizes and the plant investigated (Zenk, 1979).

(b) Extraction for assay. The *O*-ester specific hydroxylase challenged in *Petroselinum crispum* cell cultures (Heller and Kühnl, 1985) was extracted by homogenisation in a mortar with Dowex 1X2 in 100 mM potassium phosphate buffer (pH 7.0).

After centrifugation at $40\,000 \times g$ for 10 min the supernatant was passed through Sephadex G-25. Microsomal fractions were prepared after extraction in 100 mM potassium phosphate buffer (pH 7.5) containing 14 mM 2-mercaptoethanol and 10% (w/v) sucrose either by ultracentrifugation ($100\,000 \times g$) or Mg^{2+} precipitation ($40\,000 \times g$) (Britsch *et al.*, 1981; Hagmann *et al.*, 1983). Specific activity in the microsomal fractions prepared by ultracentrifugation was lower ($20\,\mu$kat kg^{-1}) than in microsomal fractions prepared by Mg^{2+} precipitation ($74\,\mu$kat kg^{-1}).

(c)   Assay

Assay for 5-O-(4-coumaroyl)-shikimic acid hydroxylase from Petroselinum crispum *(Heller and Kühnl, 1985):*

- 50 μM 5-*O*-(4-coumaroyl)-[G-^{14}C]shikimic acid (300 Bq assay^{-1})
- 1 mM NADPH
- 10 mM potassium phosphate buffer (pH 7.4)
- 1.4 mM 2-mercaptoethanol
- 1% (w/v) sucrose
- 8–24 μg protein (microsomal fraction) in a total volume of 50 μl

The reaction (at 20°C in the dark) was terminated after 20 min by adding 5 μl acetic acid to the assay. Phenolic compounds were extracted with ethyl acetate and separated on cellulose plates (3% aq. formic acid). Compounds were localised by radioscanning and the enzyme activity determined by peak integration.

(d)   Properties.   The enzyme was stable over a period of 2 months when stored at −70°C. At 4°C it had a half-life of 25 min. Some properties are summarised in Table 3.3. The enzyme accepted neither the *Z*-isomer nor free 4-coumaric, 4-coumaroyl-CoA or 5-*O*-(4-coumaroyl)quinic acid. There is a strict NADPH dependence, although NADH enhances enzyme activity. Inhibition studies with cytochrome and carbon monoxide suggested a cytochrome P-450-dependent mixed-function monooxygenase.

C.   *O*-Methyltransferases

O-Methylation of caffeic and hydroxyferulic acids to yield ferulic and sinapic acids, respectively, are important reactions not only in lignin biosynthesis, but also in the formation of various HCA conjugates. However, as discussed above for hydroxylation reactions, *O*-methylations may also occur at the level of HCA-CoAs, depending on the plant investigated. The methyl donor *S*-adenosyl-L-methionine (SAM) is converted during the methyltransfer to *S*-adenosyl-L-homocysteine (SAH), which may be the substrate for a specific hydrolase in favour of the overall formation of the methoxylated products (Poulton and Butt, 1976).

1.   Caffeic and 5-hydroxyferulic acid O-methyltransferases

(a)   Reaction

Caffeic acid + SAM ⟶ Ferulic acid + SAH

5-Hydroxyferulic acid + SAM ⟶ Sinapic acid + SAH

The O-methyltransferases (OMTs) considered here mediate the O-methylation of caffeic acid (EC 2.1.1.68) and 5-hydroxyferulic acid, specifically at the 3-position. It is not certain whether one or two specific enzymes are involved in these reactions. The activities have been classified in relation to lignin biosynthesis on the basis of substrate specificities into three groups: (1) gymnosperm-type OMT catalysing predominantly the formation of guaiacyl units; (2) angiosperm-type OMT catalysing predominantly the formation of syringyl units and with lower activity (c. 30%) guaiacyl units; (3) the bamboo-type OMT catalysing both units with similar activities. The functions of OMTs are discussed not only for lignin formation but also for the biosynthesis of all groups of phenylpropanoids (literature in Kuroda *et al.*, 1981; Poulton, 1981).

(b) Extraction for assay. Since the discovery of OMT in cambial tissue of the apple tree (Finkle and Nelson, 1963), a large number of publications on this enzyme have appeared. One of these (Shimada *et al.*, 1970), dealing with *Populus nigra* shoots, will be cited as a representative method. Young shoots (100 g) were homogenised in a Waring blender in an equal weight of 100 mM Tris-HCl buffer (pH 7.3) containing 10 mg DTT and 40 mg $NaBH_4$. The filtrate was centrifuged (20 min at $5500 \times g$) and the supernatant brought to 80% $(NH_4)_2SO_4$ saturation. The precipitated protein was redissolved in a small volume of 10 mM Tris-HCl buffer (pH 7.5) and the solution passed through Sephadex G-25 to remove excess $(NH_4)_2SO_4$. The eluate was used as the source of enzyme activity.

(c) Assay

OMT activity from Brassica oleracea *leaves (De Carolis and Ibrahim, 1989):*

- 100 μM HCA (caffeic or 5-hydroxyferulic acid)
- 100 μM SAM (1.85 GBq assay^{-1})
- 50 mM Tris-HCl buffer (pH 7.6)
- up to 50 μg protein in a final volume of 100 μl

For a convenient radiochemical method of activity determination the reaction mixture contains [methyl-^{14}C]SAM. The reaction was terminated after 30 min by the addition of 10 μl 6 N HCl to 100 μl assay volume. The methylated products formed were extracted in 250 μl ethyl acetate. Aliquots of the organic phase were counted for radioactivity by LSC. Identification of the reaction products may be achieved by co-chromatography with reference compounds on TLC (cellulose-silica, 1:1; w/w) using benzene–acetic acid–water (2:3:1; v/v) as solvent system (De Carolis and Ibrahim, 1989) or alternatively on silica plates in two dimensions (Edwards and Dixon, 1991b) using initially petroleum ether–ethyl acetate–methanol (10:10:1; v/v) and then ethyl acetate–methyl ethyl ketone–formic acid–water (5:3:1:1; v/v). Compounds can be visualised by UV light and detected by autoradiography. Product identification can be achieved by HPLC (Elkind *et al.*, 1990).

(d) Purification and properties. OMTs have been enriched from various plants (Poulton, 1981). Only recently have they been purified to homogeneity (see Tables 3.5 and 3.6) from *Brassica oleracea* leaves (De Carolis and Ibrahim, 1989) and from *Medicago sativa* suspension cultures (Edwards and Dixon, 1991b). The most effective

TABLE 3.5. Purification of HCA O-methyltransferase (formation of ferulic acid, FA) and 5-hydroxyferulic acid O-methyltransferase (formation of sinapic acid, SA) from *Brassica oleracea* leaves.

Purification step	Protein (mg)	Activity (pkat mg^{-1})		Purification (-fold)	Yield
		SA	FA		
Initial crude extract[a]	600	1.2	0.4	1	100
Sephadex G-100	140	4.4	1.5	3.7	86
DEAE-Sephacel	9.4	12.5	4.2	10	17
Mono Q					
Peak I	2.5	19.0	10.8	16	9
Peak II	2.1	16.0	6.1	13	5
Adenosine-agarose					
Peak I	<0.025	>6.8 × 10^3	>3.9 × 10^3	>5660	<24
Peak II	<0.025	No activity			

[a] Stirred with Dowex 1X2.
From De Carolis and Ibrahim (1989).

TABLE 3.6. Purification of caffeic acid O-methyltransferase from *Medicago sativa* suspension cultures.

Purification step	Protein (mg)	Activity (pkat mg^{-1})	Purification (-fold)	Yield
Initial crude extract	1470	0.7	1	100
$(NH_4)_2SO_4$ (45–85%)	436	1.3	1.9	59
DEAE Sepharose CL-6B	128	4.8	6.9	63
SAH-agarose affinity				
Peak I	0.24	139	198	3
Peak II	0.22	145	207	3
Hydropore HIC				
Peak I	0.11	187	268	2
Peak II	0.11	255	322	3

From Edwards and Dixon (1991b).

purification step applies affinity column chromatography on adenosine-agarose or SAH-agarose. Although an attempt was made to answer the question of the specificity of O-methylation in the formation of ferulic and sinapic acids (De Carolis and Ibrahim, 1989), and indeed two isoforms were isolated, no definite conclusion could be drawn (see Table 3.5 for sinapic acid and ferulic acid activities). Edwards and Dixon (1991b) were also able to isolate two OMT isoforms, although both forms had equal affinities towards caffeic acid with K_ms of 53 and 59 μM. A detailed investigation on the kinetics of the *B. oleracea* enzyme indicated an ordered bi bi mechanism which is consistent with a caffeoyl-CoA O-methyltransferase (Pakusch and Matern, 1991), although the order of binding and release is different (see below). Thus the first enzyme binds SAM and then releases the SAH product, the second enzyme first binds caffeoyl-CoA and then releases feruloyl-CoA.

The OMTs have usually been reported to be soluble enzymes. However, microsomal forms (Stafford, 1974) and cell wall-associated, membrane-bound forms have been described (Monroe and Johnson, 1984).

For a summary of the general features of *meta*-directed OMTs see the thorough review of Poulton (1981). The purified OMTs are similar with respect to pH optimum (7–8), enzyme stabilisation by sulphydryl reagents and some kinetic characteristics such as competitive inhibitions by SAH. Their M_rs were 41 000 (*M. sativa*) and 42 000 (*B. oleracea*). This is in agreement with other M_rs reported for OMTs; however, some other appreciably higher values have been reported and may arise from heteromeric interactions with other proteins (Edwards and Dixon, 1991b).

2. *Caffeoyl-CoA O-methyltransferases*

(a) Reaction

$$\text{Caffeoyl-CoA} + \text{SAM} \longrightarrow \text{Feruloyl-CoA} + \text{SAH}$$

As discussed in the previous section on hydroxylases, OMTs may also catalyse *O*-methylation with HCA-CoAs. The general question as to whether the free HCAs are the actual *in vivo* substrates, or rather the HCA thioesters or even their *O*-esters, must await further studies.

(b) Extraction for assay. To date two sources of caffeoyl-CoA OMTs are known. These are suspension cultures of *Daucus carota* (Kühnl *et al.*, 1989) and *Petroselinum crispum* (Pakusch *et al.*, 1989, 1991). Enzymes from *Daucus carota* have been prepared as described for 5-*O*-(4-coumaroyl)shikimic acid hydroxylase from *Petroselinum crispum* culture (Heller and Kühnl, 1985).

(c) Assay

Caffeoyl-CoA OMT from Daucus carota *(Kühnl et al., 1989):*

- 0.2 mM caffeoyl-CoA
- 50 μM [^{14}C]SAM (380 Bq assay^{-1})
- 1 mM MgCl$_2$
- 100 mM potassium phosphate buffer (pH 7.5)
- *c.* 15 μg protein in a total volume of 50 μl

The presence of Mg^{2+}, as well as a sulphhydryl reagent, markedly stimulates enzyme activity. This was also observed with the enzyme from *Petroselinum crispum*. Pakusch *et al.* (1989) included DTT and glycerol in their enzyme assays as activating and stabilising factors. In both studies activities were detected by radiotracer methods. The amount of ^{14}C-labelled product formed can be determined by LSC after alkaline hydrolysis of the CoA esters, acidification (HCl) and extraction of the HCAs in ethyl acetate. Product identification can be carried out by cellulose TLC in various solvent systems as described by Kühnl *et al.* (1989) and Pakusch *et al.* (1989).

(d) Purification and properties. A procedure for the purification of homogeneous caffeoyl-CoA OMT is described by Pakusch *et al.* (1991). These authors were unable to purify the enzyme by affinity chromatography, commonly used for OMT

TABLE 3.7. Purification of caffeoyl-CoA O-methyltransferase from *Petroselinum crispum* suspension cultures.

Purification step	Protein (mg)	Activity (pkat mg^{-1})	Purification (-fold)	Yield (%)
Initial crude extract	15 940	19 ·	1	100
(NH$_4$)$_2$SO$_4$ (35–60%)	5 250	50	2.6	88
Q-Sepharose	1 200	108	5.9	44
Blue Sepharose CL-6B	94	428	23	14
Hydroxyapatite	30	676	36	7
Isoelectric focusing	1.7	522	28	0.3

From Pakusch *et al.* (1991).

purifications. They achieved complete purification but had to cope with substantial loss of the enzyme (Table 3.7). Investigation of the properties of this enzyme has been done in a previous study with an 82-fold enriched enzyme activity and a yield of 1.7% (Pakusch *et al.*, 1989). The enzyme was highly specific for caffeoyl-CoA, but also accepted various other caffeic acid esters with low affinities.

D. Coenzyme A Ligases

(a) Reaction

$$\text{HCA} + \text{CoA} + \text{ATP} \longrightarrow \text{HCA-CoA} + \text{AMP} + \text{diphosphate}$$

The enzymes involved in the transformation of HCAs into their CoA-thioesters are CoA ligases; e.g. 4-coumaroyl-CoA:ligase (EC 6.2.1.12). The reaction proceeds via HCA-AMP as intermediate. Since there are different V_{max}/K_m ratios of ligase isoforms for the different HCAs (Grisebach, 1981), it is very likely that there are specific ligases for specific pathways, e.g. in flavonoid, lignin, and ester formation. Thus these enzymes are classified as HCA:CoA ligases (Gross and Zenk, 1974). These HCA-CoAs are key precursors of several classes of phenylpropanoids (Zenk, 1979) and enter various reactions involved in side-chain modification. However, it is still unclear whether there is a relationship between the various CoA ligase isoforms and the specific pathways of phenylpropanoid metabolism (Heller and Forkmann, 1988).

There is an often mentioned discrepancy between some ligase substrate specificities and phenolic substitution patterns of the products. However, possible secondary phenolic substitution reactions which occur with HCA-CoAs as substrates, such as hydroxylations or O-methylations (see above), may explain these differences.

(b) Extraction for assay.
Problems with ligase preparations arising from enzyme inhibition by interfering phenolic compounds were effectively circumvented in a study on ligase from *Taxus bacata* (Gross *et al.*, 1975) by using Dowex anion-exchange resin together with PVP, which gave ligase activities approximately twice as high as the control without PVP and Dowex. Most of the published enzyme extractions include at least PVP together with a sulphhydryl reagent.

Walton and Butt (1970) extracted CoA ligase activity from frozen (liquid nitrogen) powdered *Beta vulgaris* leaves by stirring the powder in the presence of an equal weight of PVP in 60 ml 100 mM potassium phosphate buffer (pH 8.0) containing 0.5 mM EDTA and freshly added 20 mM ascorbic acid. The mixture was stirred for 10–15 min. The filtrate was adjusted to pH 8.0 with KOH and then centrifuged (28 000 × g for 15 min). The supernatant was used for assaying enzyme activity.

(c) Assay

CoA ligases from anthers (Sütfeld and Wiermann, 1974):

- 1.8 mM HCA (dissolved in 200 mM Tris-HCl buffer (pH 8.5), 1% KHCO$_3$)
- 2.8 mM ATP
- 7.4 mM ascorbic acid
- 0.5 mM MgCl$_2$
- 1.8 mM 2-mercaptoethanol
- 163 mM Tris-HCl buffer (pH 8.5)
- 1 ml protein solution in a total volume of 5.4 ml

Enzyme activity can easily be determined spectrophotometrically at the absorption maxima of the respective HCA-CoAs (Gross and Zenk, 1966; Stöckigt and Zenk, 1975; Zenk, 1979). Product identification can be achieved after separation of the assay mixture on Sephadex G-15 columns followed by alkaline hydrolysis of the HCA-CoAs and chromatography of the liberated HCAs with reference material. HPLC can also be used to identify the HCA-CoAs together with the produced AMP (Knogge *et al.*, 1981).

(d) Purification and properties. A recent procedure for the isolation of two homogeneous ligase isoforms from *Petroselinum crispum* to determine their primary structures by sequencing cDNAs, corresponding to the ligase genes, has been published by Lozoya *et al.* (1988) and is shown in Table 3.8. The application of Mono Q-FPLC is the most effective purification step for the two *Petroselinum* ligases.

Another efficient method for isolation of ligase isoforms was published by Grand *et al.* (1983). A crude extract from *Populus* x *euramericana* stems was chromatofocused on a polybuffer exchanger column PBE 94 (equilibrated with 25 mM histidine buffer, pH 6.5) which resulted in the separation of three isoforms which exhibited different

TABLE 3.8. Purification of two isoenzymes of HCA:CoA ligase from *Petroselinum crispum* suspension cultures.

Purification step	Protein (mg)	Activity (μkat kg^{-1})	Purification (-fold)	Yield (%)
Initial crude extract	516	112	1	100
(NH$_4$)$_2$SO$_4$ (40–60%)	138	90	0.8	21
Blue Sepharose	1.82	5 075	45	15
Red Agarose	0.319	16 487	147	9
Mono Q				
Peak I	0.002	531 200	4743	2
Peak II	0.003	320 900	2865	2

From Lozoya *et al.* (1988).

substrate specificities. This indicated that ligase isoforms could play a role in the control of the monomeric composition of lignin.

With a few exceptions, the various ligases so far described showed similar properties (Gross, 1985). The pH optima were found to be near 7.5 and the M_r values reported most often were near 55 000. For a summary of HCA-CoA ligase properties from various sources, see Zenk (1979) and Heller and Forkmann (1988).

V. ENZYMES OF HYDROXYCINNAMIC ACID CONJUGATION

A. Glucosyltransferases

(a) Reaction

$$\text{UDP-glucose} + \text{HCA} \longrightarrow \text{HCA-glucose} + \text{UDP}$$

The involvement of UDP-glucose as glucosyl donor in the formation of various phenylpropanoid glucosides is well established (Hösel, 1981). The same mechanism has been found in the formation of phenolic glucose esters, catalysed e.g. by a UDP-glucose:sinapic acid glucosyltransferase (EC 2.4.1.120; Strack, 1980, and references cited therein; Nurmann and Strack, 1981; Halaweish and Dougall, 1990).

Not much is known about the formation of HCA glycosides. It was earlier assumed that the same enzyme which catalyses the UDP-glucose-dependent 1-*O*-acylglucoside formation was also responsible for the formation of the HCA glucoside (EC 2.4.1.126; Fleuriet *et al.*, 1980). It was recently shown with *Tulipa* anthers (Bäumker *et al.*, 1987), however, that there are two different glucosyltransferases catalysing the formation of either the HCA-acylglucoside (glucose attached to the HCA carboxylic group at C-9) or the HCA glucoside (glucose attached to a phenolic hydroxyl group). Glucosylation of 2-hydroxycinnamic acid from UDP-glucose and the free acid, a reaction possibly involved in coumarin biosynthesis, has earlier been shown by protein extracts from *Melilotus alba* (Kleinhofs *et al.*, 1967). The well-known UDP-glucose:hydroxycinnamyl alcohol glucosyltransferases (Grisebach, 1981) will be covered below.

(b) Extraction for assay. The following extraction procedure is described for the sinapic acid glucosyltransferase from *Raphanus sativus* seedlings (Strack, 1980; Nurmann and Strack, 1981). Young seedlings (2 days old) were treated with an Ultra Turrax homogeniser at 0°C in the presence of insoluble PVP in 100 mM potassium phosphate buffer (pH 7.0), containing 10 mM DTT. The suspension was centrifuged at 3000 × *g* for 15 min and the filtered supernatant fractionated by $(NH_4)_2SO_4$ (30–60% saturation). The redissolved protein was passed through Sephadex G-25.

(c) Assay

Glucosyltransferase from Raphanus sativus *(Strack, 1980):*

- 1 mM HCA
- 1 mM UDP-glucose
- 9.3 mM DTT

- 93 mM potassium phosphate buffer (pH 7.0)
- 100 μl protein solution in a total volume of 2.91 ml

Glucosyltransferase from Chenopodium rubrum *cultures (Bokern et al., 1991a):*

- 400 μM HCA
- 1 mM UDP-glucose
- 20 mM citrate–phosphate buffer (pH 6.0)
- 10 μl protein solution in a total volume of 50 μl

Enzyme activities can be measured by HPLC determination of the produced glucose esters, enabling product identification by UV detection (absorption of the HCA moieties) and by means of radio detection, e.g. transfer of radioactive labelled glucose from UDP-glucose to the phenolic acids (Strack, 1980).

High performance liquid chromatography analysis of glucosyltransferase assays is a recommendable technique to distinguish between formation of glucose esters and glucosides (Bäumker *et al.*, 1987). See also this work for further UV-spectroscopic, chemical and PC/TLC chromatographic characterisation of both products.

(d) Purification and properties. An enzyme which catalyses the formation of cinnamoylglucose has been purified from *Ipomoea batatas* (Shimizu and Kojima, 1984). The purification protocol is shown in Table 3.9. It was necessary to include phenylmethylsulphonylfluoride (PMSF) and 2-mercaptoethanol during the purification procedures to avoid loss of enzyme activity. Storage of the partially purified enzyme at −20°C resulted in total loss of activity in one day. The enzyme could be stabilised in 10% sorbitol. The M_r of the enzyme was near 45 000 and the pH optimum at 5.8. Besides cinnamic acid as the best acceptor, the enzyme showed activities with various other phenolic acids.

TABLE 3.9. Purification of UDP-glucose:cinnamic acid glucosyltransferase from *Impomoea batatas* roots.

Purification step	Protein (mg)	Activity (units[a] mg^{-1})	Purification (-fold)	Yield (%)
Initial crude extract	19 272	0.13	1	100
(NH$_4$)$_2$SO$_4$ (40–70%)	3 045	0.37	2.8	44
DEAE-Sephacel	903	0.90	6.9	32
Sephadex G-100	304	3.32	26	40
DEAE-Toyopearl	107	4.15	32	17
Ultrogel AcA 34	14.1	18.8	144	10
Hydroxyapatite	0.97	70.1	539	3

[a] Amount of enzyme that catalyses the formation of 1 μmol cinnamoylglucose in 1 min.
From Shimizu and Kojima (1984).

Bäumker *et al.* (1987) published a purification procedure, including chromatofocusing and anion-exchange FPLC, which allowed the separation of two different glucosyltransferases, one involved in the formation of the glucose ester and the other of the glucoside, with M_rs of 45 000 (ester formation) and 25 000 (glucoside formation).

There are indications that some of the glucosyltransferase reactions are freely reversible. UDP-glucose is formed from HCA-glucose and UDP (Strack, 1980; Nurmann and Strack, 1981; Bokern *et al.*, 1991a; Mock and Strack, 1992). The free reversibility, which might be of physiological significance in possible catabolic pathways by preserving the hydrolytic energy in the form of UDP-glucose, has also been shown for coniferin (Schmid and Grisebach, 1982) and for flavonol glucosides (Sutter and Grisebach, 1975; Kleinehollenhorst *et al.*, 1982; Heilemann and Strack, 1991).

B. Acyltransferases

1. Coenzyme A thioester acyltransferases

(a) Reaction

$$\text{HCA-CoA} + \text{HO-acceptor} \longrightarrow \text{HCA } O\text{-conjugate} + \text{CoA}$$

$$\text{HCA-CoA} + \text{NH}_2\text{-acceptor} \longrightarrow \text{HCA } N\text{-conjugate} + \text{CoA}$$

These enzymes catalyse the transfer of HCA from HCA-CoA to an HO- or NH_2-bearing acceptor molecule in the syntheses of *O*-esters and amides, respectively. This type of HCA transferase was initially shown with *Nicotiana alata* cell cultures in the biosynthesis of *O*-esters, e.g. chlorogenic acid (Stöckigt and Zenk, 1974). The reaction involved in HCA amide formation has been found in *Hordeum vulgare* in the biosynthesis of 4-coumaroylagmatine (Bird and Smith, 1981, 1983). Thereafter, various other transferase reactions proceeding via the HCA-CoAs have been described (Table 3.10).

(b) Extraction for assay

(i) *HCA O-conjugates.* The enzyme involved in the biosynthesis of chlorogenic acid, HCA-CoA:quinic acid HCA-transferase (CQT), in fruits of *Lycopersicon esculentum* (Rhodes and Wooltorton, 1977), was extracted in 100 mM Tris-HCl buffer (pH 8.0) containing 500 mM sucrose, 1 mM EDTA, 2 mM DTE and PVP as 3.2% of the weight of plant material. The extract was filtered through Miracloth and the filtrate centrifuged at $10\,000 \times g$ for 10 min. The supernatant was treated with 85% $(\text{NH}_4)_2\text{SO}_4$ saturation and the precipitate taken up in 100 mM Tris-HCl buffer (pH 7.45) containing 0.1 mM DTE. This solution was desalted on Sephadex G-25 and used for the assay of HCA-transferase as well as some other phenylpropanoid-related enzymes, i.e. PAL, alcohol dehydrogenase and ligase.

(ii) *HCA N-conjugates.* The extraction of an enzyme involved in amide formation from *Hordeum vulgare* is described as an example, i.e. 4-coumaroyl-CoA:agmatine 4-coumaroyltransferase (Bird and Smith, 1981). Young *H. vulgare* seedlings were homogenised in 4 volumes of extraction medium containing 100 mM Tris-HCl buffer (pH 8.5), 2 mM EDTA, 10 mM 2-mercaptoethanol, 1% bovine serum albumin (BSA) and 1% PVP. After centrifugation at $10\,000 \times g$ for 20 min the supernatant was dialysed against 100 mM Tris-HCl buffer (pH 8.5) and 10 mM 2-mercaptoethanol.

TABLE 3.10. Hydroxycinnamic acid-ester and -amide formations via hydroxycinnamoyl-CoA thioesters.

HCA-conjugate	Plant	Reference
HCA-ester		
(1) 5-*O*-Caffeoylquinic acid (chlorogenic acid)	*Nicotiana alata*	Stöckigt and Zenk (1974)
	Fagopyrum esculentum	Ulbrich *et al.* (1976)
	Solanum tuberosum	Lamb (1977), Rhodes *et al.* (1979)
	Lycopersicon esculentum	Rhodes and Wooltorton (1976)
(2) 5-*O*-(4-Coumaroyl)- and	Various plants and cell cultures	Ulbrich and Zenk (1979)
5-*O*-caffeoylquinic acids		Murakoshi *et al.* (1977)
(3) 4-Coumaroyl- and feruloyllupinine	*Lupinus luteus*	Strack *et al.* (1991)
(4) 4-Coumaroyl-3-*O*-	*Pisum sativum*	Saylor and Mansell (1977)
triglucosylkaempferol		
(5) 5-*O*-(4-Coumaroyl)-shikimic acid	Various plants and cell cultures	Ulbrich and Zenk (1980)
(6) 4-Coumaroyl- or caffeoyl-3-*O*-rutinosyl-	*Silene dioica*	Kamsteeg *et al.* (1980)
or 3-*O*-rutinosyl-5-*O*-glucosylanthocyanidin		
(7) 4-Coumaroyl- and caffeoyltartronic acids	*Phaseolus radiatus*	Strack *et al.* (1986)
(8) Caffeoylisocitric acid	*Amaranthus cruentus*	Strack *et al.* (1987a)
(9) HCA-glucuronate, -glucarate and -galactarate	*Secale cereale*	Strack *et al.* (1987b)
(10) 4-Coumaroyl- or caffeoyl-3-*O*-glucosyl- or -3-*O*-	*Matthiola incana*	Teusch *et al.* (1987)
xylosylglucosylanthocyanidin		
(11) Caffeoyldihydroxyphenyllactic acid (rosmarinic	*Coleus blumei*	Petersen and Alfermann (1988)
acid)		
(12) Cinnamoyl-(13-hydroxylupanine)	*Lupinus angustifolius*	Strack *et al.* (1991)
(13) 4-Coumaroyl-*myo*-inositol	*Taxus baccata*	Heilemann *et al.* (1990)
HCA-amide		
(14) 4-Coumaroylagmatine	*Hordeum vulgare*	Bird and Smith (1983)
(15) Feruloyltyramine	*Nicotiana tabacum*	Negrel and Martin (1984),
		Fleurence and Negrel (1989)
	Nicotiana glutinosa	Villegas and Brodelius (1990)
	Eschscholtzia californica	Villegas and Brodelius (1990)
(16) Caffeoylputrescine	*Nicotiana tabacum*	Meurer-Grimes *et al.* (1989),
		Negrel (1989)
(17) 4-Coumaroylspermidine	*Nicotiana tabacum*	Negrel *et al.* (1991)

(c) Assay

(i) HCA O-conjugates

CQT from various plants (Ulbrich and Zenk, 1979):
Forward reaction:

- 20 μM 4-coumaroyl-CoA (chosen because of its greater stability)
- 4 mM quinic acid
- 100 mM potassium phosphate buffer (pH 6.5)
- 0.75–2.4 μg protein in a total volume of 0.5 ml

Reverse reaction:

- 0.2 mM chlorogenic acid
- 0.4 mM CoA
- 100 mM potassium phosphate buffer (pH 7.0)
- *c.* 20 μg protein in a total volume of 0.5 ml

It has been repeatedly demonstrated that the HCA-CoA-dependent transferase reactions are freely reversible (Rhodes and Wooltorton, 1976; Ulbrich and Zenk, 1979, 1980; Strack *et al.*, 1986). This facilitates the assay and activity determination and it is therefore not always a prerequisite to synthesise the HCA-CoAs for measurement of the respective enzymes.

Both forward and reverse reactions can be photometrically measured by the method developed by Zenk and co-workers (Gross and Zenk, 1966; Stöckigt and Zenk, 1975), or by means of HPLC (Strack *et al.*, 1986, 1987a, b). The latter includes a convenient method of product identification, the HCA *O*-ester in the forward and the HCA-CoA in the reverse reaction.

Enzyme activities were measured with the optical assays (Ulbrich and Zenk, 1979) at an incubation temperature of 35°C, the forward reaction by a decrease in absorbance at 342 nm (extinction of 0.01 corresponds to the formation of 0.735 nmol 4-coumaroylquinic acid) and the reverse reaction by an increase in absorbance at 360 nm (extinction of 0.01 corresponds to the formation of 0.769 nmol caffeoyl-CoA). Both reactions were linear over a period of 5 min.

(ii) HCA N-conjugates

4-Coumaroyl-CoA:agmatine 4-coumaroyltransferase from Hordeum vulgare *(Bird and Smith, 1983)*:

- 10 μM 4-coumaroyl-CoA
- 200 μM agmatine sulphate
- 1 mM EDTA
- <3 mM 2-mercaptoethanol (decrease due to reaction with 4-coumaroyl-CoA)
- 100 mM Tris-HCl buffer (pH 7.5)
- protein in a final volume of 1 ml

The activity of this enzyme was measured in an optical assay as described above. The reaction was started by the addition of agmatine and the decrease in absorbance at 333 nm was recorded for 4–5 min. The inital rate was calculated by substracting the rate of loss of absorption due to 2-mercaptoethanol. Product identification was achieved by using ^{14}C-labelled agmatine and TLC of the assay products (Bird and Smith, 1981). Activity measurement including product identification can also be performed by HPLC, as shown for the formation of caffeoylputrescine (Meurer-Grimes *et al.*, 1989).

(d) Purification and properties

(i) HCA O-conjugates. Rhodes *et al.* (1979) have published a rapid method for the CQT purification from *Solanum tuberosum* tubers which includes affinity chromatography on Blue Sepharose, the enzyme being eluted with a CoA gradient. The protocol of the purification is shown in Table 3.11. The pH optimum for the three enzyme peaks was at 6.25 and the activity was stimulated by the inclusion of BSA. The M_r was near 41 500.

The purified CQT activity has an absolute specificity for quinate. In crude extracts, however, there is an accompanying activity with shikimic acid (Zenk, 1979) which has also been found with *Cestrum elegans* leaves (Strack *et al.*, 1988). Ulbrich and Zenk (1979) were able to separate clearly a specific enzyme involved in shikimic acid O-ester formation and furthermore showed the wide distribution of this activity in plants which do not accumulate HCA-shikimic acid (Ulbrich and Zenk, 1980). Table 3.12

TABLE 3.11. Purification of CQT from *Solanum tuberosum* tubers, stored at 0°C.

Purification step	Protein (mg)	Activity (nkat mg^{-1})	Purification (-fold)	Yield (%)
Initial crude extract	519	0.20	1	100
$(NH_4)_2SO_4$ (30–85%)	413	0.22	1.1	87
Sephadex G-25	329	0.33	1.7	113
DEAE-Cellulose	28.5	2.50	13	69
Dialysis	28.1	1.94	9.7	52
Blue Sepharose[a]				
Peak I	0.155	279	1440	42
Peak II[b]	0.228	49.6	248	11
Peak III	1.77	5.5	28	9

[a] Elution of peaks I, II and III at 220, 290 and 390 μM CoA, respectively.
[b] This peak shows also a small activity towards shikimic acid which is thought to be due to an underlying peak of a shikimic acid-specific enzyme.
From Rhodes *et al.* (1979).

TABLE 3.12. Purification of 4-hydroxycinnamoyl-CoA:shikimic acid 4-coumaroyltransferase from *Cichorium endivia*.

Purification step	Protein (mg)	Activity (nkat mg^{-1})	Purification (-fold)	Yield (%)
Initial crude extract	511	0.15	1	100
$(NH_4)_2SO_4$ (35–60%)	199	0.34	2.2	87
Ultrogel AcA 44	38.5	1.68	11	85
DE-52-Cellulose	12.5	3.77	25	62
Hydroxyapatite	4.2	9.2	61	51
PAGE	0.4	18.9	126	10

From Ulbrich and Zenk (1980).

shows the purification protocol of this enzyme. In contrast to the CQT isolated by Ulbrich and Zenk (1979) from *Nicotiana alata*, this enzyme was not inibited by Tris-HCl buffer. The pH optimum was at 6.5. The M_r of 58 000 was of the same order of magnitude as those values for CQT from *Solanum* (see above), *Nicotiana* (75 000) and *Stevia rebaudiana* (45 000).

The obscure role of the shikimic acid HCA-transferase needs to be investigated. It is possible that HCA-shikimic acid serves as acyl donor in as yet unknown transacylation reactions (Ulbrich and Zenk, 1980), a reaction which has recently been shown with chlorogenic acid in the formation caffeoylglucaric acid in *Lycopersicon esculentum* leaves (Strack *et al.*, 1987d; Strack and Gross, 1990).

For a summary of early work on *O*-acylconjugate formation see Zenk (1979). Additional and more recent publications are listed in Table 3.10.

(ii) HCA N-conjugates. The first *N*-acylating enzyme, the 4-coumaroyl-CoA: agmatine *N*-4-coumaroyltransferase (ACT), has been purified by affinity chromatography on AC-Ultrogel AcA 34-agmatine (Table 3.13), the enzyme eluted with a agmatine gradient. There was a strict specificity towards agmatine, although cinnamoyl- and other HCA-CoAs were accepted. ACT activity was maximal at pH 7.5 and was about twice as active in Tris-HCl buffer as in Bicine buffer, in contrast to the CQT inhibition observed by Ulbrich and Zenk (1979). The M_r of the enzyme was 40 000. The inclusion of a thiol reagent is essential for maximal enzyme activity, although one should be aware of reactions between 4-coumaroyl-CoA and thiols which result in a decrease of the absorbance in the optical assay (see above). This possible risk in optical assays can be circumvented by using HPLC (Meurer-Grimes *et al.*, 1989).

One should be aware of the general problem of assaying CoA-thioester-dependent transferases, e.g. contaminating thiol esterase activities which can be eliminated during purification procedures. Other possible problems may be encountered with phosphohydrolysis activities (Negrel and Smith, 1984) which can be inhibited by inorganic diphosphate, sodium fluoride and purine nucleotides.

Table 3.10 lists some other *N*-acylating enzymes that have been described as also being involved in the formation of tyramine, putrescine and spermidine amides. These enzymes have rather alkaline pH optima (8–10); e.g. a HCA-CoA:putrescine HCA-transferase from *Nicotiana tabacum* suspension cultures has a pH optimum at pH 8.8

TABLE 3.13. Purification of 4-coumaroyl-CoA:agmatine N-4-coumaroyltransferase from *Hordeum vulgare* shoots.

Purification step	Protein (mg)	Activity (nkat mg^{-1})	Purification (-fold)	Yield (%)
Initial crude extract	81.9	0.20	—	100
Protamine sulphate	44.9	0.38	1.9	104
(NH$_4$)$_2$SO$_4$ (0–60%)	37.4	0.42	2.1	95
Bio-Gel A	12.0	1.55	7.6	112
AC Ultrogel AcA 34-agmatine[a]	0.53	20.4	100	66

[a] Affinity chromatography with an agmatine gradient (enzyme elution between 1 and 2 mM). From Bird and Smith (1983).

with 4-coumaroyl-CoA, 10.0 with caffeoyl- and sinapoyl-CoA and 10.3 with feruloyl-CoA (Meurer-Grimes *et al.*, 1989).

Negrel *et al.* (1991) were able to separate putrescine- and spermidine HCA-transferases (PHT and SHT) extracted from *Nicotiana tabacum* callus cultures. This was achieved by ion-exchange chromatography on DEAE Trisacyl. Whereas the SHT acted only on spermidine, the PHT was not specific for putrescine and could use other diamines (mainly cadaverine and diaminopropane).

2. 1-O-Acylglucose acyltransferase

(a) Reaction

$$\text{1-}O\text{-HCA-acylglucose} + \text{acceptor} \longrightarrow \text{HCA-conjugate} + \text{glucose}$$

The acylglucose-dependent *O*-transferase catalyses the formation of an *O*-ester by way of transfer of the HCA moiety of 1-*O*-HCA-acylglucose to an HO-bearing acceptor molecule. Harborne and Corner proposed as early as 1961 that the widely occurring phenolic glucose esters (1-*O*-acyl-β-glucosides) may be essential intermediates in phenylpropanoid metabolism. This has often been discussed by other authors (Stafford, 1974; Schlepphorst and Barz, 1979; Hanson and Havir, 1979) and it was found that glucose esters can be metabolically active (Harborne and Corner, 1961; Schlepphorst and Barz, 1979; Kojima and Uritani, 1972, 1973; Strack, 1977; Molderez *et al.*, 1978). Indeed, a glucose ester-dependent transferase reaction was first discovered in plant hormone metabolism. Michalczuk and Bandurski found in 1980 that the biosynthesis of the indolylacetic acid(IAA)-*myo*-inositol ester in kernels of sweet corn (*Zea mays*) proceeds via the 1-*O*-IAA-acylglucoside. In the same year, Tkotz and Strack (1980) described the same type of reaction for the first time in phenylpropanoid metabolism, i.e. the biosynthesis of sinapoylmalate in radish (*Raphanus sativus*). It has recently been shown (Mock and Strack, 1992) that the 1-*O*-acyl-β-glucosides have a high group-transfer potential.

Meanwhile there is increasing evidence that HCA-ester formation via the 1-*O*-HCA-acylglucosides plays an important role as an alternative to the CoA-dependent transferase reactions. The same mechanism has been described for leaves of *Quercus robur*, in which 1-*O*-galloylglucose is involved in gallotannin biosynthesis (see Gross, Chapter 2, this volume). Recently, the first example of an acylglucose-dependent acylation of anthocyanins was described (Glässgen and Seitz, 1991) as an alternative to the known CoA thioester-dependent one (Kamsteeg *et al.*, 1980; Teusch *et al.*, 1987). Table 3.14 lists the transferase reactions that proceed via 1-*O*-HCA-acylglucose as acyl donor.

(b) Extraction for assay.

Extraction of an acylglucose-dependent enzyme involved in chlorogenic acid (5-caffeoylquinic acid) biosynthesis in *Ipomoea batatas* roots (Villegas and Kojima, 1985) was achieved using acetone powder (see above). This was extracted with 100 mM phosphate buffer (pH 7.0). The extract was fractionated with $(NH_4)_2SO_4$ (40–70% saturation) and the precipitate dissolved in 25 mM phosphate buffer (pH 6.3) containing 13 mM 2-mercaptoethanol. The solution was passed through Sephadex G-25 to remove excess $(NH_4)_2SO_4$ and the eluate was used as source of enzyme activity.

A routine extraction of an enzyme catalysing the formation of sinapine (*O*-sinapoylcholine) via 1-*O*-sinapoylglucose was used in a study on the distribution

TABLE 3.14. Hydroxycinnamic acid-ester formation via 1-O-hydroxycinnamic acid-acylglucosides.

HCA-ester	Plant	Reference
(1) Sinapoylmalate	*Raphanus sativus*	Tkotz and Strack (1980)
	Brassica napus	Strack *et al.* (1990b)
	Brassica rapa (=B. campestris)	Mock *et al.* (1992)
	Arabidopsis thaliana	
(2) Sinapoylcholine (sinapine)	Various Brassicaceae	Strack *et al.* (1983), Regenbrecht and Strack (1985), Gräwe and Strack (1986)
(3) 1,2-Di-O-sinapoylglucose	*Raphanus sativus*	Dahlbender and Strack (1984, 1986)
(4) 5-O-(4-Coumaroyl)- and -caffeoylquinic acids	*Ipomoea batatas*	Kojima and Villegas (1984), Villegas and Kojima (1985, 1986)
(5) 4-Coumaroyltartaric acid	*Spinacia oleracea*	Strack *et al.* (1987c)
(6) Caffeoylglucaric acid	*Cestrum elegans*	Strack *et al.* (1988)
(7) 4-Coumaroyl- and feruloylbetanin (lampranthin I, II)	*Lampranthus sociorum*	Bokern and Strack (1988)
(8) 4-Coumaroyl- and feruloylamaranthin (celosianin I, II)	*Chenopodium rubrum*	Bokern and Strack (1988)
(9) Feruloylgomphrenin I (gomphrenin III), lampranthin II, celosianin II	*Gomphrena globosa*	Bokern *et al.* (1992)
(10) Lampranthin II	*Beta vulgaris*, *Mirabilis jalapa*	Bokern *et al.* (1992)
(11) Celosianin II	*Celosia argentea*	Bokern *et al.* (1992)
(12) Lampranthin II, celosianin II	*Iresine lindenii*	Bokern *et al.* (1992)
(13) HCA 3-O-glucosyllathyroside-cyanidin	*Daucus carota*	Glässgen and Seitz (1991)

of this enzyme in members of the Brassicaceae (Regenbrecht and Strack, 1985). Brassicaceae seeds were frozen with liquid nitrogen and ground in a mortar together with quartz sand, PVP and 100 mM phosphate buffer (pH 7.0). The homogenate was allowed to stand with continuous stirring for 1 h at 4°C and subsequently filtered through Miracloth. After centrifugation at 48 000 × g the enzyme activity was prepared from the supernatant by $(NH_4)_2SO_4$ precipitation (30–70% saturation) followed by chromatography on Sephadex G-25.

(c)   Assay

1-O-(4-coumaroyl)-glucose:quinic   acid   4-coumaroyltransferase   from Ipomoea batatas *roots (Villegas and Kojima, 1986):*

- 1.7 mM 1-*O*-(4-coumaroyl)-glucose
- 23 mM quinic acid
- 11 mM $MgCl_2$
- 17 mM phosphate buffer (pH 6.0)
- protein in a final volume of 760 µl

1-O-sinapoylglucose:choline   sinapoyltransferase   from   Brassicaceae   seeds (Regenbrecht and Strack, 1985):

- 2 mM 1-*O*-sinapoylglucose
- 250 mM choline chloride
- 200 mM potassium phosphate buffer (pH 7.0)
- protein in a final volume of 100 µl

The reaction with protein from *Ipomoea batatas* roots (Villegas and Kojima, 1986) was terminated after 30 min by adding 3 ml ethanol and 2 ml acetone. The mixture was left at room temperature for 45 min and then centrifuged. The supernatant was concentrated and purified by paper chromatography (Whatman No. 1; development with *n*-butanol–pyridine–water, 14:3:3; v/v). The UV-quenching product (5-*O*-[4-coumaroyl]-quinic acid, R_f 0.3) was cut out and eluted with ethanol–water (1:1; v/v). The amount of product formed was calculated from absorbance at 315 nm using an extinction coefficient of $19 \times 10^6 \, cm^2 \, mol^{-1}$. The addition of $MgCl_2$ was essential for efficient precipitation of protein and inorganic salts after the termination of the reaction with ethanol and acetone. The reaction product, chlorogenic acid, has also been measured by HPLC (Villegas and Kojima, 1985). The paper eluate was applied to a C_{18} column and the product eluted with 35% aq. methanol containing 10 mM acetic acid.

The reactions with protein extracted from Brassicaceae seeds, which catalyses the formation of sinapine via 1-*O*-sinapoylglucose, were stopped after 2 h by transfer to −20°C. The enzymic activities were analysed by reversed-phase HPLC on C_8 developed isocratically using 30% acetonitrile, 15% acetic acid, 1% phosphoric acid and 0.05% SDS in water. However, C_{18} column material without including SDS in the solvent system is to be preferred (D. Strack, unpubl. res.). For further references on this type of transferase, see Table 3.14.

(d)   Purification and properties.   The transferase from the root of *Ipomoea batatas* has been purified 160-fold to apparent homogeneity (Table 3.15; Villegas and Kojima,

TABLE 3.15. Purification of 1-O-HCA-glucose:quinic acid HCA-transferase from *Ipomoea batatas* roots.

Purification step	Protein (mg)	Activity (units^a mg^{-1})	Purification (-fold)	Yield (%)
Initial crude extract	27 900	0.023	1	100
(NH$_4$)$_2$SO$_4$ (40–70%)	20 300	0.034	1.5	107
Sephadex G-25	19 800	0.055	2.4	169
DEAE-Toyopearl	2 100	0.12	5.1	38
CM-Sephadex	130	0.66	29	13
Mono Q	33	1.7	74	9
Mono S	2.0	3.5	152	1
Superose 12	0.56	3.6	157	0.3

^{a} Amount of enzyme that catalyses the formation of 1 nmol 4-coumaroylquinic acid in 1 min.
From Villegas and Kojima (1986).

TABLE 3.16. Purification of 1-O-sinapoylglucose:choline sinapoyltransferase from *Raphanus sativus* seeds.

Purification step	Protein (mg)	Activity (nkat mg^{-1})	Purification (-fold)	Yield (%)
Initial crude extract	1091	0.019	1	100
(NH$_4$)$_2$SO$_4$ (35–65%)	416	0.038	2.0	75
Heat (10 min at 65°C)	224	0.068	3.5	73
CM-Sepharose	13.7	0.5	26	32
Ultrogel AcA 44	0.47	8.1	420	18

From Gräwe and Strack (1986).

1986). The enzyme was stable in a precipitated form. The (NH$_4$)$_2$SO$_4$ precipitate could be stored in ice for 1 month without loss of enzyme activity. There was no cofactor requirement and the M_r was near 25 000. The isoelectric point was at 8.5 and the optimal pH at 6.0.

The 1-O-sinapoylglucose:choline sinapoyltransferase has been partially purified from *Raphanus sativus* and *Sinapis alba* seedlings (Table 3.16 for the *Raphanus* enzyme; Gräwe and Strack, 1986). Highest activities were found at pH 7.2. As with the *Ipomoea* enzyme, there was no cofactor requirement. However the activity was inhibited by Mg^{2+}, EDTA as well as by sulphhydryl reagents such as DTT. The purified enzyme was quite stable. When stored at −20°C for 14 days, repeatedly thawed and refrozen, no loss of activity was observed. However, the activity in the crude extract lost *c.* 20% when stored at −20°C.

3. O-*Ester acyltransferases*

(a) *Reaction*

$$\text{HCA-}O\text{-ester}_1 + \text{acceptor}_1 \rightarrow \text{ester moiety}_1 + \text{HCA-}O\text{-ester}_2$$

This enzyme catalyses the HCA-transfer from an HCA O-ester to an HO-bearing acceptor molecule (transesterification). For example, chlorogenic acid may act as acyl donor for caffeoyltransferases. This has been demonstrated for the formation of 3,5-dicaffeoylquinic acid ('disproportionation') in *Ipomoea batatas* roots (Villegas *et al.*, 1987) and caffeoylglucaric acid in *Lycopersicon esculentum* leaves (Strack and Gross, 1990). The disproportionation reaction has also been shown with 1-*O*-HCA-acylglucose in the formation of 1,2-di-*O*-sinapoylglucose (Dahlbender and Strack, 1984, 1986) and with 1-*O*-galloylglucose in the formation of 1,6-di-*O*-galloylglucose (Gross *et al.*, 1990; Gross, chapter 2, this volume).

(b) Extraction for assay. The enzyme from *Ipomoea batatas* (Kojima and Kondo, 1985) was extracted from acetone powder (see above) with 100 mM phosphate buffer (pH 7.0) containing 0.1% 2-mercaptoethanol. The protein precipitated at 70% $(NH_4)_2SO_4$ saturation was collected and redissolved in 25 mM phosphate buffer (pH 6.4). Excess $(NH_4)_2SO_4$ was removed by filtration through Sephadex G-25.

The enzyme from *Lycopersicon esculentum* (Strack and Gross, 1990) was prepared by homogenisation of frozen leaves in 100 mM Tris-HCl buffer (pH 8.0) containing 1 mM EDTA, 10 mM DTT and PVP. The homogenate was passed through Miracloth and the filtrate centrifuged. The activity was precipitated by $(NH_4)_2SO_4$ (30–80% saturation) and redissolved in 20 mM potassium phosphate buffer (pH 5.7). Excess $(NH_4)_2SO_4$ was removed by filtration through Sephadex G-25.

(c) Assay

Chlorogenic acid:chlorogenic acid caffeoyltransferase from Ipomoea batatas *roots (Villegas* et al., *1987):*

- 5.4 mM chlorogenic acid
- 182 μM MgCl$_2$
- 36 mM acetate buffer (pH 5.0)
- protein in a total volume of 550 μl

Chlorogenic acid:glucaric acid caffeoyltransferase from Lycopersicon esculentum *leaves (Strack and Gross, 1990):*

- 20 mM chlorogenic acid
- 4 mM glucaric acid
- 20 mM potassium phosphate buffer (pH 5.7)
- protein in a total volume of 50 μl

Activity of the *Ipomoea* caffeoyltransferase was determined by a similar method to that described above for the acylglucose-dependent quinic acid *O*-ester formation (Villegas and Kojima, 1986). The caffeoyltransferase reaction from *Lycopersicon* (Strack and Gross, 1990) was stopped after 30 min by transferring the mixture into liquid nitrogen. The activity was determined by reversed-phase HPLC as follows: gradient elution with 1.5% phosphoric acid in water as solvent A and 1.5% phosphoric acid, 20% acetic acid and 25% acetonitrile in water as solvent B.

(d) Purification and properties. The purification protocols of the two enzymes described are listed in Tables 3.17 and 3.18. The *Ipomoea* enzyme (Villegas *et al.*, 1987) showed an M_r of 25 000. The isoelectric point was at pH 4.6 and the optimum pH of reaction was 5.0. There was no cofactor requirement.

TABLE 3.17. Purification of chlorogenic acid caffeoyltransferase from *Ipomoea batatas* roots.

Purification step	Protein (mg)	Activity (units^a mg^{-1})	Purification (-fold)	Yield (%)
Initial crude extract	2380	0.26	1	100
$(NH_4)_2SO_4$ (40–70%)	725	0.37	1.4	40
Sephadex G-25	470	0.36	1.4	28
DEAE-Toyopearl	200	1.2	4.6	29
DEAE-Sephacel	12	9.58	37	19
Hydroxyapatite	2.9	17	66	8
Superose 12	0.38	76	290	5
PAGE	0.015	930	3600	2

^{a} Amount of enzyme that catalyses the formation of 1 nmol 3,5-*O*-caffeoylquinic acid in 1 min.
From Villegas *et al.* (1987).

TABLE 3.18. Purification of chlorogenic acid:glucaric acid caffeoyltransferase from *Lycopersicon esculentum*.

Purification step	Protein (mg)	Activity (nkat mg^{-1})	Purification (-fold)	Yield (%)
Initial crude extract	1245	0.098	1	100
$(NH_4)_2SO_4$ (30–80%)	658	0.114	1.2	62
DEAE-Sephacel	35.7	1.32	14	39
CM-Sepharose	1.3	17.5	179	19
Ultrogel AcA 44	0.21	51.4	524	9
Ultropac TSK 3000 CW (×2)	0.022	236	2412	4

From Strack and Gross (1990).

The *Lycopersicon* enzyme (Strack and Gross, 1990) gave an M_r of 40 000. Highest activity was found at pH 5.7. The isoelectric point of the enzyme was at pH 5.75. The addition of Mg^{2+} and Ca^{2+} gave 125 and 150% activity of the control.

C. Preparation of Acyl Donors

1. Preparation of hydroxycinnamoyl-CoA thioesters

Preparation of HCA-CoA thioesters is essential for studies of the various CoA-dependent biosyntheses of phenylpropanoids. Thus, much effort has been put into the elaboration of protocols for thioester preparations. Initially the problem of side reactions in the chemical synthesis of HCA-CoA thioesters was circumvented by enzymatic preparations with beef liver mitochondria (Gross and Zenk, 1966) or from a strain of *Pseudomonas putida* (Zenk *et al.*, 1980). However, excellent chemical methods were developed and it is now easy to synthesise the CoA esters chemically by the ester-exchange reaction via the acyl *N*-hydroxysuccinimide esters (Stöckigt and Zenk, 1975). A simplified method for the synthesis of the hydroxysuccinimic esters of HCAs is described by Negrel and Smith (1984). Following the work of Zenk and co-workers, various other CoA esters have been synthesised, e.g. with tropic acid (Gross and Koelen, 1980), gallic acid (Gross, 1982), retinoic acid (Kutner *et al.*, 1986) or piperic acid (Semler *et al.*, 1987). Purifications of the CoA esters can be achieved by open column chromatography, e.g. on DEAE-cellulose (Stöckigt and Zenk, 1975; Gross and Koelen, 1980), Sephadex types (Mieyal *et al.*, 1974; Gross, 1982) and polyamide (Zenk *et al.*, 1980; Strack *et al.*, 1987b; Strack *et al.*, 1991), or by (semi)preparative HPLC on reversed-phase material such as the most popular silica-C_{18} as has been shown with piperoyl-CoA thioester (Semler *et al.*, 1987). The latter may easily be applied to purification of the HCA-CoAs. Whereas the intermediary succinimide ester has usually to be isolated, it is possible to isolate the end products, with 20% recoveries, in scaled-down reactions from the crude reaction mixture from which only the precipitated N,N'-dicyclohexyl urea has been removed by filtration (Semler *et al.*, 1987).

Open column chromatography on polyamide of the complete reaction mixture is a convenient method for isolation and purification of the HCA-CoAs on a preparative scale (Strack *et al.*, 1987b, 1991). The following protocol gave recoveries of 30–40%, as estimated by UV spectroscopy (Stöckigt and Zenk, 1975) and by HPLC (Strack *et al.*, 1991). The reaction mixture (see above) was concentrated under reduced pressure to remove most of the organic solvent. The remaining aqueous phase was filtered through glass wool directly onto a H_2O-equilibrated polyamide column (24 cm × 2 cm i.d.). Then the following stepwise gradient was applied: 300–350 ml each of H_2O, MeOH, and 0.15, 0.44, 0.73, 1.5, 4.4 and 7.3% aq. NH_3 in MeOH. Whereas the 0.15–0.44% NH_3 and 4.4–7.3% NH_3 contained the free HCAs and free CoA, respectively, the HCA-CoAs were eluted in the range 0.73–1.5% NH_3, depending on the substitution pattern of the HCA moiety. The polyamide used for these purifications has to be carefully freed from fine particles and thoroughly washed with 1 M aqueous NH_3 in MeOH, followed by H_2O, 1 M aqueous HCl and finally H_2O.

2. Preparation of 1-O-acylglucosides

The 1-*O*-HCA-acyl-β-glucosides, which may be used for acylglucose-dependent HCA-transferases, can be obtained in quite different ways: (1) as natural products from intact plants (Birkofer *et al.*, 1961; Harborne and Corner, 1961) (e.g. 4-coumaroyl- and feruloylglucoses from *Antirrhinum majus* petals or sinapoylglucose from *Raphanus sativus* seedlings) (Linscheid *et al.*, 1980) or tissue cultures (e.g. feruloylglucose shown with *Chenopodium rubrum* or *Beta vulgaris* cell cultures) (Bokern *et al.*, 1991a, b); (2) as detoxification products from cell cultures (biotransformation; glucose ester production from free HCAs; see e.g. Bokern *et al.*, 1991a); and (3) as chemically synthesised products.

The chemical synthesis of 1-*O*-(4-coumaroyl)-glucose has been detailed by Shimizu and Kojima (1984) following protocols described by Birkofer *et al.* (1966) (see also Birkofer *et al.*, 1969, for NMR spectroscopy of acylglucosides). The following will be treated as an example of this kind of chemical synthesis. Synthesis proceeded as follows: 4-coumaric acid (5.2 g) was dissolved in 25 ml pyridine and 4.8 g acetic anhydride added. After incubation for 22 h at room temperature the mixture was poured into 500 ml water and kept in ice overnight to allow crystallisation of 4-acetylcoumaric acid (m.p. 200–205°C). The dried 4-acetylcoumaric acid (4.2 g) was mixed with 25 g thionylchloride and refluxed at 112°C for 3 h. After removal of the remaining thionyl chloride *in vacuo* the product (4-acetylcoumaroylchloride) was purified by recrystallisation from benzene (m.p. 115–116°C). 2,3,4,6-Tetraacetyl-α-glucopyranosylbromide was prepared by a method described by Lemifux (1963) and converted to β-glucose-2,3,4,6-tetraacetate by the method of McCliskey and Coleman (1955).

The conjugation product 1-*O*-acetyl-(4-coumaroyl)-2,3,4,6-tetraacetyl-β-glucose can be obtained by the method of Birkofer *et al.* (1961). A mixture (*c.* 1:1 of 6–11 g each) of the coumaroyl and the glucose derivatives was dissolved in 25 ml chloroform and 3.7 g pyridine. After a short initial warming and 2 days incubation at room temperature, another 25 ml of chloroform was added and the mixture was washed with 2 M aqueous sulphuric acid, sodium bicarbonate solution and water. The product was obtained from the remaining chloroform phase by crystallisation (addition of ethyl ether in the cold). The final product was obtained by deacetylation according to Birkofer *et al.* (1966) to give 1-*O*-(4-coumaroyl)-β-glucose.

V. ENZYMES OF THE LIGNIFICATION PATHWAY

A. Reductase (EC 1.2.1.44)

(a) Reaction

$$\text{HCA-CoA} + \text{NADPH} + \text{H}^+ \longrightarrow \text{Hydroxycinnamaldehyde} + \text{CoA} + \text{NADP}^+$$

The enzyme, HCA-CoA:NADPH oxidoreductase, catalyses the reduction of HCA-CoAs to the corresponding aldehydes. The overall reaction from ferulic acid to coniferyl alcohol, proceeding via CoA-ligase, HCA-CoA reductase and hydroxycinnamyl alcohol dehydrogenase, was first demonstrated with a protein preparation from *Salix alba* (Mansell *et al.*, 1972). The proof of the enzymatic reduction of HCA-CoAs to hydroxycinnamaldehydes came from studies with *Forsythia suspensa* (Gross and Kreiten, 1975) and *Glycine max* (Wengenmayer *et al.*, 1976). The enzyme is highly specific for HCA-CoAs and NADPH. It belongs to the class B type (H_s) of dehydrogenases (Gross and Kreiten, 1975). Substrate specificities were compared between purified enzymes form *Glycine max* and *Picea abies* (Lüderitz and Grisebach, 1981) and showed that sinapoyl-CoA was only a good substrate for the angiosperm enzyme.

(b) Extraction for assay. Frozen cambial sap from *Picea abies* (Lüderitz and Grisebach, 1981) was thawed in Tris-HCl buffer (pH 7.5) and filtered through a nylon net. The residue was extracted by grinding in a mortar with buffer and the slurry was pressed though the nylon net. The filtrates were combined and centrifuged (20 min, $14\,000 \times g$). The resulting extract was treated with Dowex and a precipitation with polyethyleneimine was carried out before $(NH_4)_2SO_4$ precipitation, which may be necessary to detect enzyme activity. However, with the *Glycine max* enzyme (from cell suspension cultures) activity could be detected after dialysis of the crude extract (Wengenmayer *et al.*, 1976). The buffers used should contain stabilising components (2-mercaptoethanol, ethylene glycol).

(c)   Assay

Reduction activity with HCA-CoAs (Lüderitz and Grisebach, 1981):

- 41 μM feruloyl-CoA
- 300 μM NADPH
- 100 mM phosphate buffer (pH 6.25)
- protein in a total volume of 1 ml

Oxidation activity with hydroxycinnamaldehydes (Lüderitz and Grisebach, 1981):

- 100 μM coniferaldehyde dissolved in 20 μl methoxyethanol
- 250 μM NADP$^+$
- 335 μM CoA
- 200 mM Tris-HCl buffer (pH 7.8)
- protein in a total volume of 1 ml

For activity determinations changes in absorbance at 366 nm were monitored (decrease in reduction activity and increase in oxidation activity). With the extinction coefficients at 366 nm of feruloyl-CoA (14.9×10^6 cm^2 mol^{-1}), NADPH (3.36×10^6 cm^2 mol^{-1}) and coniferaldehyde (10.1×10^6 cm^2 mol^{-1}) the reduction activity may be calculated using a resulting coefficient of 8.16×10^6 cm^2 mol^{-1}. With 4-coumaroyl-CoA and sinapoyl-CoA, the coefficients are 8.16×10^6 and 7.06×10^6, respectively. The oxidation activity with coniferaldehyde may be calculated using a value of 5.86×10^6 cm^2 mol^{-1}.

(d)   Purification and properties.   HCA-CoA reductase was purified to apparent homogeneity from *Glycine max*, *Picea abies* (Lüderitz and Grisebach, 1981) and *Populus* x *euramericana* (Sarni *et al.*, 1984). Table 3.19 shows the purification protocol for the reductase from *Populus*. The enzyme from *Glycine max* (Wengenmayer *et al.*, 1976) was purified by a procedure including affinity chromatography using CoA-hexane-agarose or 5′-AMP bound to Sepharose, the enzyme eluted specifically with NADP$^+$.

The pH optima of reductase activities vary between 6.0 and 6.5 (Wengenmayer *et al.*, 1976; Lüderitz and Grisebach, 1981; Sarni *et al.*, 1984), although values between 7.4 and 7.8 were also reported (Gross and Kreiten, 1975). The enzyme is a monomeric protein with M_r values between 33 000 (Lüderitz and Grisebach, 1981) and 40 000 (Gross and Kreiten, 1975).

TABLE 3.19.   Purification of HCA-CoA reductase from *Populus* x *euramericana* stems.

Purification step	Protein (mg)	Activity (nkat mg^{-1})	Purification (-fold)	Yield (%)
(NH$_4$)$_2$SO$_4$ (40–70%)[a]	51.6	0.05	1	100
DEAE-Sephacel	2.2	0.76	15	50
Sephacryl S-200	0.052	11.5	220	23
Blue Sepharose	0.006	29.7	595	7

[a] No activity detected in crude extracts.
From Sarni *et al.* (1984).

B. Dehydrogenases (EC 1.1.1.194/195)

(a)   Reaction

Hydroxycinnamaldehyde + NADPH + H$^+$ ⟶ Hydroxycinnamyl alcohol + NADP$^+$

This enzyme catalyses the reduction of the hydroxycinnamaldehyde to the respective alcohol. It has an absolute requirement for NADPH belonging to the class A (H$_R$) of dehydrogenases (Mansell *et al.*, 1974). Determination of the equilibrium constant and K_m values for aldehydes and alcohols showed that the formation of alcohols is favoured under physiological conditions (Wyrambik and Grisebach, 1975; Sarni *et al.*, 1984). Enzyme preparations from gymnosperms exhibited poor activity with sinapaldehyde compared to coniferaldehyde and 4-coumaraldehyde, whereas in angiosperms all three aldehydes are accepted equally.

Screening of a vast number of species from different taxonomic groups revealed highest activities in the cambial regions of woody dicots and gymnosperms, whereas in mosses only low activities were detected (Mansell *et al.*, 1974), thus confirming its role in lignin biosynthesis. In cell suspension cultures of *Phaseolus vulgaris* a fungal elicitor induces the *de novo* synthesis of the enzyme (Grand *et al.*, 1987) in agreement with the process of lignification as a defence mechanism.

(b)   Extraction for assay.

In the extraction procedure used for *Forsythia suspensa* and many other species (Mansell *et al.*, 1974, 1976) the tissue is frozen in liquid nitrogen and powdered. After addition of pre-wet PVP and 100 mM borate buffer (pH 7.8) containing 20 mM 2-mercaptoethanol the extract is stirred for 1 h at 4°C, squeezed through cheesecloth and centrifuged (25 min at 48 000 × *g*). The enzyme from *Forsythia* was very unstable below pH 6.8. Stability may be enhanced by the addition of ethylene glycol (Wyrambik and Grisebach, 1975) or glycerol (Kutsuki *et al.*, 1982).

(c)   Assay

Reduction activity with aldehydes (Wyrambik and Grisebach, 1975):

- 34 μM coniferaldehyde dissolved in 20 μl methoxyethanol
- 200 μM NADPH
- 200 mM potassium phosphate buffer (pH 6.5)
- protein up to 0.8 μg in a total volume of 1 ml

Oxidation activity with alcohols (Mansell et al., 1974):

- 200 μM coniferyl alcohol
- 200 μM NADP$^+$
- 100 mM Tris-HCl buffer (pH 8.8)
- protein in a total volume of 500 μl

The enzyme activities can be determined in both the forward and reverse reaction. For measurements of the forward reaction (aldehyde reduction) the change in absorbance at 340 nm is monitored. The rate of alcohol formation is calculated according to the formula (ΔA_{340})/(sum of extinction coefficients of coniferaldehyde and NADPH) with the coefficient of coniferaldehyde of $15.8 \times 10^6 \, \text{cm}^2 \, \text{mol}^{-1}$. The coefficients for the other aldehydes are 23.5×10^6 for 4-coumaraldehyde and 20.8×10^6 for sinapaldehyde.

The formation of coniferaldehyde (reverse reaction) is monitored photometrically at 400 nm (extinction coefficient of $18.5 \times 10^6 \, \text{cm}^2 \, \text{mol}^{-1}$). As an alternative the reduction of $NADP^+$ may be followed at 340 nm (Lewis *et al.*, 1987).

Preparation of the substrates may be carried out through enzymatic hydrolysis (Mansell *et al.*, 1974) of isolated syringin (Freudenberg *et al.*, 1951), the corresponding glucoside. Other alcohols may be synthesised from the HCAs (Mansell *et al.*, 1974). The aldehydes may be synthesised according to Pearl and Darling (1957).

(d) Purification and properties. The enzyme was first purified and characterised from *Forsythia suspensa* (Mansell *et al.*, 1974) and *Glycine max* (Wyrambik and Grisebach, 1975, 1979). From experiments with thiol reagents it was concluded that a sulphhydryl group is necessary for enzyme activity (Mansell *et al.*, 1974; Wyrambik and Grisebach, 1979). The enzyme is a zinc-containing dimer with identical subunits of M_r 40 000 (Wyrambik and Grisebach, 1979; Lüderitz and Grisebach, 1981; Sarni *et al.*, 1984). Multiple forms were found in a few species, e.g. *Glycine max* (Wyrambik and Grisebach, 1975) or *Salix* (Mansell *et al.*, 1976). Purification to apparent homogeneity has been reported for *Glycine max* (Wyrambik and Grisebach, 1979), *Picea abies* (Lüderitz and Grisebach, 1981), *Pinus thunbergii* (Kutsuki *et al.*, 1982) and *Populus* x *euramericana* (Sarni *et al.*, 1984).

The purification protocol given for *Picea abies* (Lüderitz and Grisebach, 1981) in Table 3.20 may be used also for HCA:CoA-ligase, HCA-CoA reductase and HCA alcohol glucosyltransferase. Wyrambik and Grisebach (1979) purified an isoform from *Glycine max* about 3760-fold to apparent homogeneity including affinity chromatography on a $NADP^+$-agarose column with elution of the enzyme with a $NADP^+$ gradient.

The pH optima of the aldehyde reduction are in the range 6–7.6, those of the alcohol oxidation are markedly higher, e.g. 8.8 (Mansell *et al.*, 1974) or 9.2 (Lüderitz and Grisebach, 1981). The isoelectric point of the enzyme from *Populus* x *euramericana* was found to be at pH 5.6 (Sarni *et al.*, 1984). Possible isoenzymes (Wyrambik and Grisebach, 1975, 1979) exhibited strict substrate specificities, thus they can be discriminated in crude extracts by the use of different substrates.

TABLE 3.20. Purification of hydroxycinnamyl alcohol dehydrogenase from cambial sap (5 kg fresh weight) of *Picea abies*.

Purification step	Protein (mg)	Activity (nkat mg^{-1})	Purification (-fold)	Yield (%)
Initial crude extract	26 150	0.70	1	100
Polyethyleneimine, $(NH_4)_2SO_4$ (35–75%)	9 900	1.4	2.0	75
DEAE-Cellulose	2 150	5.3	7.6	62
Hydroxyapatite	327	33.4	48	60
Blue Sepharose	23	461	659	58
Red Agarose	8.8	1076	1537	52

From Lüderitz and Grisebach (1981).

C. Glucosyltransferases (EC 2.4.1.111)

(a) Reaction

UDP-Glucose + hydroxycinnamyl alcohol ⟶ Hydroxycinnamyl 4-O-β-glucoside
                                                                                     + UDP

The enzyme catalyses the transfer of glucose from UDP-glucose to the 4-OH of hydroxycinnamyl alcohols with the formation of hydroxycinnamyl 4-O-β-glucoside. This enzyme was first detected in cell suspension cultures of Paul's Scarlet Rose (Ibrahim and Grisebach, 1976) in the glucosylation of coniferyl and sinapoyl alcohol. Investigation of a wide range of species from different taxa showed that in ferns and mosses low activities were also present (Ibrahim, 1977).

The purified enzyme from *Picea abies* exhibited a pronounced specificity for UDP-glucose as donor and coniferyl alcohol as acceptor. The reaction was shown to be freely reversible and from the kinetic data a mono-iso ordered bi bi reaction mechanism was postulated (Schmid and Grisebach, 1982).

In the hypocotyl of 10-day-old *Picea* seedlings, the enzyme is located predominantly in the epidermal and subepidermal layer and in the vascular bundles (Schmid *et al.*, 1982). This correlates well with the findings that during seedling development lignification starts in these cells and the coniferin-specific β-glucosidase is present (Marcinowski *et al.*, 1979).

(b) Extraction for assay.

Extraction from cambial sap is performed as described for the HCA-CoA reductase (Lüderitz and Grisebach, 1981). In crude extracts, enzyme stability is increased by the addition of 2-mercaptoethanol (42 mM) and a high buffer concentration (200 mM Tris-HCl, pH 7.5). For dilute solutions, the addition of glycerol or ethylene glycol is recommended (Schmid and Grisebach, 1982). Dialysis of the crude extract may be necessary to detect enzyme activity.

(c) Assay

UDP-glucose:coniferyl alcohol glucosyltransferase from Picea abies *(Schmid and Grisebach, 1982)*:

- 2.5 mM coniferyl alcohol dissolved in 5 μl 2-methoxyethanol
- 2.5 mM UDP-[^{14}C]glucose (4.6 kBq assay^{-1})
- 25 mM Tris-HCl buffer (pH 7.5)
- protein in a total volume of 80 μl

After 15 min the reaction was stopped by the addition of 5 μl acetic acid and 10 μl coniferin (2 mg ml^{-1} 2-methoxyethanol). The mixture was centrifuged and the supernatant applied to Whatmann No. 1 paper. Descending chromatography was carried out with the upper phase of *n*-butanol–acetic acid–water (4:1:5; v/v; the lower phase was used to saturate the chromatographic tank). The coniferin spot ($R_f = 0.62$) was detected under UV light (254 nm), cut out, and radioactivity determined by LSC.

Concerning the E/Z isomerisation, only the E-hydroxycinnamyl alcohols were considered to be involved in lignification (Yamamoto *et al.*, 1989). However, this must not be generalised, since in recent studies of Lewis and co-workers an acceptor specificity for the Z-forms in the glucosyltransferases has been found, e.g. formation of Z-coniferin in protein preparations from *Fagus grandifolia* (Yamamoto *et al.*, 1990). The standard assay for such a specific enzyme reaction is as follows:

UDP-Glucose:coniferyl alcohol glucosyltransferase from Fagus grandifolia *(Yamamoto* et al.*, 1990):*

- 1.02 mM *E*- or *Z*-coniferyl alcohol
- 26 µM UDP-glucose (9.25 KBq UDP-[U-^{14}C]glucose assay^{-1})
- 7.5 mM DTT
- 75 mM Tris-HCl buffer (pH 7.2)
- protein solution (40 µl) in a total volume of 80 µl

After 20 min the reaction was terminated by the addition of 5 µl 50% aq. acetic acid. Then unlabelled *E*- (77.5 nmol in 10 µl) and *Z*-coniferin were added as carrier (141.6 nmol in 20 µl). The resulting suspension was centrifuged and the supernatant passed through a filter. The resulting filtrate was applied onto reversed-phase HPLC (C$_{18}$). Details of HPLC analyses, as well as substrate preparations, are described in Lewis *et al.* (1989) and Yamamoto *et al.* (1990). The most interesting result of these studies by Lewis and co-workers was the unusual substrate specificity for *Z*- and not for *E*-hydroxycinnamyl alcohols. The *Z*-form was readily converted into its glucoside, e.g. *Z*-coniferyl alcohol into *Z*-coniferin, and no *E*-coniferin was produced. In an assay containing *E*-coniferyl alcohol there was only a poor conversion to *E*-coniferin, but interestingly a significant conversion into *Z*-coniferin occurred.

In the *Fagus* bark the *Z*-hydroxycinnamyl alcohols and their glucosides are accumulated exclusively (Lewis *et al.*, 1988; Yamamoto *et al.*, 1990). The authors obtained preliminary evidence for the existence of a novel *E/Z*-hydroxycinnamyl alcohol isomerase, possibly involved in the *Z*-formation of the substrates for gluco-syltransferase activities. Experiments with radiolabelled precursors showed that this isomerisation takes place at the monolignol (hydroxycinnamyl alcohol) level (Lewis *et al.*, 1987). Involvement of these monolignols in lignification has not yet been shown (Lewis and Yamamoto, 1990), but *in vitro* a synthetic dehydrogenatively polymerised lignin is formed by the action of horseradish peroxidase (Morelli *et al.*, 1986).

TABLE 3.21. Purification of UDP-glucose:coniferyl alcohol glucosyltransferase from cambial sap of *Picea abies*.

Purification step	Protein (mg)	Activity (nkat mg^{-1})	Purification (-fold)	Yield (%)
Initial crude extract	9462	0.024	1	100
Polyethyleneimine, (NH$_4$)$_2$SO$_4$ (35–80%)	3658	0.095	4.0	100
DEAE-Cellulose	605	0.299	13	79
Hydroxyapatite	216	0.667	28	63
Phenyl Sepharose	94	1.18	49	49
Red Agarose	11	6.37	265	30
Orange Agarose	0.75	40.7	1694	13

From Schmid and Grisebach (1982).

(d) Purification and properties. The protocol developed for the enzyme extracted from cambial sap of *Picea abies* (Table 3.21; Schmid and Grisebach, 1982) may be used in combination with the procedures for HCA-CoA reductase and hydroxycinnamyl alcohol dehydrogenase from the same source (Lüderitz and Grisebach, 1981).

The enzyme from *Picea* is a monomeric protein with an M_r of 50 000. Divalent cations had no influence on its activity, but a sulphhydryl reagent is essential (Schmid and Grisebach, 1982). Similar characteristics were given for the enzyme from *Forsythia orata* (Ibrahim, 1977) and Paul's Scarlet rose (Ibrahim and Grisebach, 1976).

D.   Glucosidases

(a) Reaction

Hydroxycinnamyl 4-*O*-β-glucoside $\longrightarrow$ Hydroxycinnamyl alcohol + glucose

The β-glucosidase activity is thought to be responsible for the hydrolysis of hydroxycinnamyl 4-*O*-β-glucosides which are secreted from the cytoplasm into the cell wall. Cell wall-bound β-glucosidases using hydroxycinnamyl alcohol glucosides as substrates were first characterised from *Picea abies* (Marcinowski and Grisebach, 1978), *Cicer arietinum* (Hösel *et al.*, 1978) and *Glycine max* (Hösel and Todenhagen, 1980). Purification of the enzymes enabled the preparation of antibodies which were used for localisation in immunofluorescence studies. In cross-sections from the hypocotyl of developing *Picea* seedlings the β-glucosidase activity was localised at the inner layer of the secondary cell wall; it was present in all cells, but the fluorescence seemed to be stronger in the epidermal layer and in the vascular bundles (Marcinowski *et al.*, 1979). Similar investigations with *Cicer* showed that the β-glucosidase is located in the cell walls of tracheary elements and of other tissues known to contain either lignin or other polymers derived from phenylpropanoids (Burmeister and Hösel, 1981).

(b) Extraction for assay. The extraction procedure for *Cicer arietinum* cell suspension cultures (Hösel *et al.*, 1978) starts with the preparation of a cell wall fraction by homogenisation in 200 mM citrate–sodium phosphate buffer (pH 5.0). The homogenate is centrifuged and the resulting pellet washed several times to obtain a crude cell wall preparation. The β-glucosidase is solubilised from the pellet by treatment with 500 mM NaCl (3 ×). A similar procedure has been given for *Picea abies* (Marcinowski and Grisebach, 1978).

(c) Assay

    β-Glucosidase from Cicer arientinum *cell suspension cultures (Hösel et al., 1978):*

- 2 mM coniferin
- 50 mM NaCl
- 50 mM citrate–sodium phosphate buffer (pH 5.0)
- protein in a total volume of 500 μl

After 30 min the reaction is stopped by the addition of 500 μl NaCO₃ (1 M) and the liberated coniferyl alcohol is photometrically determined at 325 nm (extinction coefficient of $7 \times 10^6 \, cm^2 \, mol^{-1}$). Unspecific β-glucosidase activities may be measured using 4-nitrophenyl glucoside as substrate.

(d) Purification and properties. Coniferin-specific β-glucosidase may be purified from NaCl extracts of cell wall. Table 3.22 lists the purification procedure from *Cicer arietinum* cell-suspension cultures (Hösel *et al.*, 1978). Separation from other cell wall-located β-glucosidases can be monitored by measuring activities towards 4-nitrophenyl glucoside.

TABLE 3.22. Purification of coniferin β-glucosidase from *Cicer arietinum* cell suspension cultures.

Purification step	Protein (mg)	Activity[a] (units[b] mg^{-1})	Purification (-fold)	Yield (%)
NaCl solubilized fraction from crude particulate pellets[c] $(NH_4)_2SO_4$ (0–65%)	3300	0.08	1	100
Sephadex G-25	420	0.43	5.4	68
Sephadex G-200	79	1.3	16	38
Dialysis	78	1.05	13	31
CM Sephadex	18	2.7	34	18

[a] Activity towards 4-nitrophenyl β-glucoside was separated from coniferin specific activity.
[b] Amount of enzyme that catalyses the formation of 1 μmol in 1 min.
[c] Crude particulate fraction before and after NaCl treatment showed activities towards coniferin of 130 and 50% compared to the NaCl solubilised fraction.
From Hösel *et al.* (1978).

In the presence of 500 mM NaCl the enzyme was stable for several months at 4°C. The CM-Sephadex fraction (Table 3.22) was further separated into different activities by isoelectric focusing (Hösel *et al.*, 1978). For *Picea abies* two activities were separated during the purification procedures, which differed in their isoelectric points (Marcinowski and Grisebach, 1978). The function of these multiple forms remains unclear. The pH optima are in the range 4.5–6.0. In *Picea* the enzyme was a monomer with an M_r of 58 500, whereas in *Cicer* a dimeric form was found with an M_r of 110 000 with subunits of 63 000 and 43 000. The β-glucosidase from both sources turned out to be glycoproteins. The β-glucosidases purified from *Glycine max* (Hösel and Todenhagen, 1980) had an M_r of 45 000. It showed high activities towards coniferin and syringin, which may be correlated with the higher content of sinapyl alcohol in angiosperm lignin.

E. Peroxidases (EC 1.11.1.7)

(a) Reaction

$$AH + AH + H_2O_2 \longrightarrow A\text{-}A + 2\,H_2O$$

Coniferyl and related alcohols are polymerised to form lignin. These reactions involve oxidations catalysed by peroxidase and H_2O_2. The free radicals produced couple to dilignols and, after reoxidation, finally form lignin macromolecules with hetero-geneous mixtures of C–C and C–O bonds (Gross, 1978; Grisebach, 1981).

The generation of H_2O_2 is also a peroxidase-catalysed process involving the action of a cell wall-bound malate dehydrogenase (Gross, 1978). In addition, H_2O_2 needed for cross-linking events may be provided by the action of a cell wall-bound polyamine oxidase (Angelini and Federico, 1989).

Peroxidases are present in a large number of isoforms in different compartments and with different biochemical roles (Gaspar *et al.*, 1982; van Huystee, 1987), e.g. lignifica-tion (Gross, 1978; Grisebach, 1981) and lignin degradation (Sarkanen *et al.*, 1991), suber-isation (Espelie *et al.*, 1986), cross-linking of extensin precursors (Everdeen *et al.*, 1988), cross-linking of pectic polysaccharides with phenolic acids (Fry, 1983), chlorophyll degradation (Matile, 1980) and indolylacetic acid (IAA) oxidation (Ray, 1960). Correla-tion of a certain peroxidase isoform with a specific biochemical role is often uncertain (Xu *et al.*, 1990). Attribution may be facilitated with the use of specific antibodies (Kim *et al.*, 1988; Lang *et al.*, 1990; Xu *et al.*, 1990) and molecular probes (Lagrimini *et al.*, 1987).

In *Nicotiana tabacum* acidic isoperoxidases are mainly located in the cell wall and the basic isoforms group C1 is found in the vacuole (Schloss *et al.*, 1987). Lignification was therefore attributed to the acidic peroxidases. In *Zinnia elegans* mesophyll cells differentiating into tracheary elements, an induction-specific cationic isoperoxidase was found and designated as lignoperoxidase (Church and Galston, 1988). Elicitation of lignin biosynthesis in suspension cultures of *Ricinus communis* led to a change in the activity of one anionic and one cationic extracellular peroxidases, as well as to the appearance of three new cationic extracellular peroxidases (Bruce and West, 1989).

(b) Extraction for assay

(i) Extraction from Nicotiana tabacum *leaves (Mäder, 1980).*   Leaves of *Nicotiana tabacum* were homogenised in a Waring blender in 10 mM sodium–potassium phosphate buffer (pH 7.5) and the homogenate pressed through four layers of muslin. This filtrate is also used for the purification of the isoforms group A1 (see below).

(ii) Extraction of isoperoxidases from cultured Zinnia elegans *mesophyll cells induced for tracheary element differentiation (Masuda* et al., *1983).*   Formation of tracheary elements induced in isolated mesophyll cells is a convenient model system to investigate cytodifferentiation (Fukuda and Komamine, 1985). Formation of lignin, changes in enzyme activities correlated to lignin biosynthesis, and changes in peroxidase patterns have been investigated (Fukuda and Komamine, 1982; Masuda *et al.*, 1983; Church and Galston, 1988).

Zinnia elegans cells and medium (containing 'extracellular peroxidases') are separated by vacuum filtration. The cells are washed and then homogenised in 5 mM Tris-HCl buffer (pH 7.0). After centrifugation the supernatant contains the 'soluble peroxidases'. The pellet is washed in buffer to obtain a cell wall fraction. 'Ionically bound peroxidases' are extracted from the cell walls with 200 mM $CaCl_2$. The pellet is then digested enzymatically with 1% Cellulase Onozuka R-10 and 0.5% Macero-enzyme R-10 to obtain finally 'enzyme-released' and 'tightly bound' peroxidases.

(iii) Electrophoretic separation of peroxidase isoforms.   After extraction, frac-tionation or during purification, peroxidase activity may be determined in total (see

the following section), but usually the isoform pattern is of interest. Separation may be performed with starch gel electrophoresis (e.g. van den Berg and van Huystee, 1984), which allows the simultaneous separation of cationic and anionic enzymes. By disc electrophoresis in polyacrylamide gels, cationic and anionic isoforms are separated in different buffer systems (e.g. Mäder and Bopp, 1976). As an alternative, isoelectric focusing may be used (e.g. Ros Barcelo *et al.*, 1987); differences in the number of isoforms have been monitored by comparing separation with isoelectric focusing and disc electrophoresis (Mäder and Bopp, 1976).

(c) Assay

(i) Spectrophotometrical activity determination

Peroxidase activity from Pisum sativum *(Siegel and Galston, 1967) and* Nicotiana tabacum *(Mäder et al., 1975) with guaiacol as substrate:*

- 5 mM guaiacol
- 5 mM H_2O_2
- 200 mM sodium potassium phosphate buffer (pH 5.8)
- protein

Peroxidase activity from Populus x euramericana *with syringaldazine (syringaldazine oxidase; Imberty et al., 1985):*

- 41.6 μM syringaldazine (stock solution: 10 mM syringaldazine dissolved in methanol and then mixing with 2 volumes of dioxane)
- 1.1 mM H_2O_2
- 100 mM sodium potassium phosphate buffer (pH 6.0)
- 50 or 100 μl protein solution in a final volume of 4 ml

Enzyme activities are determined photometrically, the guaiacol activity at 470 nm (extinction coefficient for tetraguaiacol of 26.6×10^6 cm^2 mol^{-1}; George, 1953) and the syringaldazine activity at 530 nm. The syringaldazine oxidase activity has been correlated with lignification (Imberty *et al.*, 1985), but differences in substrate specificity have been found between histochemical (*in situ*) and biochemical (*in vitro*) investigations (Pang *et al.*, 1989). Syringaldazine oxidase was also detected *in vitro* in extracts from non-lignifying tissues.

Other frequently used hydrogen donors for assaying peroxidase activity are eugenol (Liu and Gibson, 1977), tetramethylbenzidine, *p*-phenylenediamine-pyrocatechol (Imberty *et al.*, 1984), 4-aminoantipyrine (van den Berg *et al.*, 1983) and *o*-dianisidine (Church and Galston, 1988). Assays with coniferyl alcohol (Ferrer *et al.*, 1990) and fluorinated analogues of ferulic acid and coniferyl alcohol (Goldberg *et al.*, 1988) have been described.

(ii) Histochemical localization of peroxidase with syringaldazine (Harkin and Obst, 1973; Imberty et al., 1985).   Tissue sections are prepared in 100 mM Tris-HCl buffer (pH 7.6) and incubated for 5 min in 0.1% ethanolic syringaldazine solution containing 1.25 mM H_2O_2, then rinsed with water and observed by light microscopy. Controls are run without H_2O_2. Syringaldazine oxidase has also been histochemically localised

in non-lignifying tissues (Grison and Pilet, 1985). As in *in vitro* assays, a wide range of other hydrogen donors were also used.

(iii) Zymogram technique. After separation of peroxidase isoforms with native gel electrophoresis or isoelectric focusing, visualisation of enzyme activities is accomplished by immersing the gels in a staining solution containing buffer, H_2O_2 and a hydrogen donor. In principle the staining solution may be identical with the mixture used for the *in vitro* assays, but the buffer capacity must be high enough to counteract the pH conditions of the gel.

Staining is performed for 5–20 min according to the enzyme activity (Mäder *et al.*, 1975). The gel is then washed with water and the banding pattern is recorded. The use of different hydrogen donors often leads to different isoform patterns. The sequential application of hydrogen donor and then H_2O_2 allows detection of more bands than with simultaneous application (Kay and Basile, 1987).

(d) Purification and properties. Peroxidase isoenzymes have been purified from total cell homogenates but differential extraction of aploplastic peroxidases can be obtained from intercellular washing fluids (Rathmell and Sequeira, 1974; van den Berg and van Huystee, 1984; Li and McClure, 1990) and from cell wall preparations (Masuda *et al.*, 1983; Ros Barcelo *et al.*, 1987; Church and Galston, 1988; Li and McClure, 1990). Table 3.23 summarises a purification protocol of the isoform group A1 (formerly G_1) from filtered homogenate (see above extraction for assay) of *Nicotiana tabacum* (Mäder, 1980; Lang *et al.*, 1990). Purification steps are checked by native PAGE and SDS-PAGE. Chromatography was performed at room temperature and after each step 5 mM $CaCl_2$ was added to the fractions containing peroxidase activities in order to stabilise the enzyme.

TABLE 3.23. Purification of isoenzyme group A1 from *Nicotiana tabacum* leaves.

Purification step	Protein (mg)	Activity (units[a] mg^{-1})	Purification (-fold)	Yield (%)
Initial crude extract	590	25	1	100
Acetone precipitation	110	99	4.0	74
DE-52-Cellulose (stepwise)	24.6	365	15	61
DE-52-Cellulose (gradient)	6.2	1402	56	59
Sephadex G-75	3.0	2310	92	47
TSK Phenyl-5-PW	0.14	7786	310	7

[a] Photometrical determination (ΔA at 470 nm in 1 min; see peroxidase assay above).
From Lang *et al.* (1990).

The same purification scheme was used for the purification of the isoform groups A2 and C1 from *Nicotiana* cell suspension cultures (Lang *et al.*, 1990). The group A1 exhibited a high affinity towards 4-coumaryl- and coniferyl alcohol (Mäder *et al.*, 1977) and is therefore possibly involved in lignification (Grisebach, 1981). The isoelectric points of these isoforms are in the range 4.0–4.5 (Mäder and Bopp, 1976) and the M_r was determined to be near 38 000 (Lang *et al.*, 1990).

ACKNOWLEDGEMENTS

Work of the senior author has been supported by the Deutsche Forschungsgemeinschaft und the Fonds der Chemischen Industrie.

REFERENCES

Amrhein, N. and Gödeke, K.-H. (1977). *Plant Sci. Lett.* **8**, 313–317.
Amrhein, N. and Zenk, M. H. (1971). *Z. Pflanzenphysiol.* **64**, 145–168.
Amrhein, N., Gödeke, K.-H. and Gerhardt, J. (1976). *Planta* **131**, 33–40.
Angelini, R. and Federico, R. (1989). *J. Plant Physiol.* **135**, 212–217.
Bäumker, P. A., Jütte, M. and Wiermann, R. (1987). *Z. Naturforsch.* **42c**, 1223–1230.
Barz, W., Köster, J., Weltring, K.-M. and Strack, D. (1985). *In* "Annual Proceedings of the Phytochemical Society of Europe", Vol. 25 (C. F. van Sumere and P. J. Lea, eds), pp. 307–347. Clarendon Press, Oxford.
Bate-Smith, E. C. (1962). *J. Linn. Soc.* (Bot.) **58**, 95–173.
Benveniste, I., Salaun, J.-P. and Durst, F. (1977). *Phytochemistry* **16**, 69–73.
Benveniste, I., Salaun, J.-P. and Durst, F. (1978). *Phytochemistry* **17**, 359–363.
Billett, E. E. and Smith, H. (1978). *Phytochemistry* **17**, 1511–1516.
Bird, C. R. and Smith, T. A. (1981). *Phytochemistry* **20**, 2345–2346.
Bird, C. R. and Smith, T. A. (1983). *Phytochemistry* **22**, 2401–2403.
Birkofer, L., Kaiser, C., Nouvertne, W. and Thomas, U. (1961). *Z. Naturforsch.* **16b**, 249–251.
Birkofer, L., Kaiser, C., Kosmol, H., Donike, M. and Michaelis, G. (1966). *Liebigs Ann. Chem.* **699**, 223–231.
Birkofer, L., Kaiser, C., Hillges, B. and Becker, F. (1969). *Liebigs Ann. Chem.* **725**, 196–202.
Bokern, M. and Strack, D. (1988). *Planta* **174**, 101–105.
Bokern, M., Wray, V. and Strack, D. (1991a). *Planta* **184**, 261–270.
Bokern, M., Heuer, S., Wray, V., Witte, L., Macek, T., Vanek, T. and Strack, D. (1991b). *Phytochemistry* **30**, 2361–3265.
Bokern, M., Heuer, S. and Strack, D. (1992). *Botanica Acta*, **105**, 146–151.
Boniwell, J. M. and Butt, V. S. (1986) *Z. Naturforsch.* **41c**, 56–60.
Britsch, L., Heller, W. and Grisebach, H. (1981). *Z. Naturforsch.* **36c**, 742–750.
Brödenfeldt, R. and Mohr, H. (1986). *Z. Naturforsch.* **41c**, 61–68.
Bruce, R. J. and West, C. A. (1989). *Plant Physiol.* **91**, 889–897.
Burmeister, G. and Hösel, W. (1981). *Planta* **152**, 578–586.
Camm, E. L. and Towers, G. H. N. (1977). *Progr. Chem.* **4**, 169–188.
Church, D. L. and Galston, A. W. (1988). *Plant Physiol.* **88**, 679–684.
Conn, E. E. (ed.) (1981). "The Biochemistry of Plants", Vol. 7: Secondary Plant Products. Academic Press, New York.
Czichi, U. and Kindl, H. (1975). *Hoppe-Seyler's Z. Physiol. Chem.* **356**, 457–485.
Dahlbender, B. and Strack, D. (1984). *J. Plank Physiol.* **116**, 375–379.
Dahlbender, B. and Strack, D. (1986). *Phytochemistry* **25**, 1043–1046.
De Carolis, E. and Ibrahim, R. K. (1989). *Biochem. Cell Biol.* **67**, 763–769.
Dixon, R. A. and Lamb, C. J. (1990). *In* "Annual Proceedings of the Phytochemical Society of Europe", Vol. 30 (B. V. Charlwood and M. J. C. Rhodes, eds), pp. 101–116. Clarendon Press, Oxford.
Duke, S. O. and Vaughn, K. C. (1982). *Physiol. Plant.* **54**, 381–385.
Dumas, J. and Peligot, E. (1834). *J. Liebig's Ann. Chem. Pharm.* **12**, 24–25.
Edwards, R. and Dixon, R. A. (1991a). *Phytochemistry* **30**, 79–84.
Edwards, R. and Dixon, R. A. (1991b). *Arch. Biochem. Biophys.* **287**, 372–379.
Elkind, Y., Edwards, R., Mavandad., M., Hedrick, S. A., Ribak, O., Dixon, R. A. and Lamb, C. J. (1990). *Proc. Natl. Acad. Sci. USA*, **87**, 9057–9061.
Erez, A. (1973). *Plant Physiol.* **51**, 409–411.
Espelie, K. E., Franceschi, V. R. and Kolattukudy, P. E. (1986). *Plant Physiol.* **81**, 487–492.

Everdeen, D. S., Kiefer, S., Willard, J. J., Muldoon, E. P., Dey, P. M., Li, X.-B. and Lamport, D. T. A. (1988). *Plant Physiol.* **87**, 616–621.

Ferrer, M. A., Pedreno, M. A., Calderon, A. A., Munoz, R. and Ros Barcelo, A. (1990). *Physiol. Plant.* **79**, 610–616.

Finkle, B. J. and Nelson, R. F. (1963). *Biochem. Biophys. Acta* **78**, 747–749.

Fleurence, J. and Negrel, J. (1989). *Phytochemistry* **28**, 733–736.

Fleuriet, A., Macheix, J. J., Suen, R. and Ibrahim, R. K. (1980). *Z. Naturforsch.* **35c**, 967–972.

Freudenberg, K. and Neish, A. C. (1968). "Constitution and Biosynthesis of Lignin". Springer-Verlag, Berlin.

Freudenberg, K., Kraft, R. and Heimberger, W. (1951). *Chem. Ber.* **84**, 472–476.

Fry, S. C. (1983). *Planta* **157**, 111–123.

Fry, S. C. (1986). *Ann. Rev. Plant Physiol.* **37**, 165–186.

Fry, S. C. (1988). "The Growing Plant Cell Wall: Chemical and Metabolic Analysis". Longman Scientific & Technical, John Wiley & Sons, New York.

Fujita, M. and Asahi, T. (1985). *Plant Cell Physiol.* **26**, 389–395.

Fukuda, H. and Komamine, A. (1982). *Planta* **155**, 423–430.

Fukuda, H. and Komamine, A. (1985). *In* "Cell Culture and Somatic Cell Genetics of Plants", Vol. 2 (I. K. Vasil, ed), pp. 149–212. Academic Press, London.

Gaspar, T., Penel, C., Thorpe, T. and Greppin, H. (1982). "Peroxidases 1970–1980. A Survey of their Biochemical and Physiological Roles in Higher Plants". University of Geneva.

George, P. (1953). *J. Biol. Chem.* **201**, 413–434.

Gesteiner, B. and Conn, E. E. (1974). *Arch. Biochem. Biophys.* **163**, 617–624.

Glässgen, W. E. and Seitz, H. U. (1991). *Planta* **186**, 582–585.

Goldberg, R., Pang, A., Pierron, M., Catesson, A.-M., Czaninski, Y., Francesch, C. and Roland., C. (1988). *Phytochemistry* **27**, 1647–1651.

Gräwe, W. and Strack, D. (1986). *Z. Naturforsch.* **41c**, 28–33.

Grand, C. (1984). *FEBS Lett.* **169**, 7–11.

Grand, C., Boudet, A. and Boudet, A. M. (1983). *Planta* **158**, 225–229.

Grand, C., Sarni, F. and Lamb, C. J. (1987). *Eur. J. Biochem.* **169**, 73–77.

Grison, R. and Pilet, P.-E. (1985). *J. Plant Physiol.* **118**, 201–208.

Grisebach, H. (1981). *In* "The Biochemistry of Plants", Vol. 7 (E. E. Conn, ed.), pp. 457–478. Academic Press, London.

Gross, G. G. (1978). *Recent Adv. Phytochem.* **12**, 177–220.

Gross, G. G. (1981). *In* "The Biochemistry of Plants", Vol. 7 (E. E. Conn, ed.), pp. 301–316. Academic Press, London.

Gross, G. G. (1982). *Z. Naturforsch.* **37c**, 778–783.

Gross, G. G. (1985). *In* "Biosynthesis and Degradation of Wood Components" (T. Higuchi, ed.), pp. 229–271. Academic Press, New York.

Gross, G. G. and Koelen, K. J. (1980). *Z. Naturforsch.* **35c**, 363–366.

Gross, G. G. and Kreiten, W. (1975). *FEBS Lett.* **54**, 259–262.

Gross, G. G. and Zenk, M. H. (1966). *Z. Naturforsch.* **21b**, 683–690.

Gross, G. G. and Zenk, M. H. (1974). *Eur. J. Biochem.* **42**, 453–459.

Gross, G. G., Mansell, R. L. and Zenk, M. H. (1975). *Biochem. Physiol. Pflanzen* **168**, 41–51.

Gross, G. G., Denzel, K. and Schilling, G. (1990). *Z. Naturforsch.* **45c**, 37–41.

Hagmann, M., Heller, W. and Grisebach, H. (1983). *Eur. J. Biochem.* **134**, 547–554.

Hahlbrock, K. (1981). *In* "The Biochemistry of Plants", Vol. 7 (E. E. Conn, ed.), pp. 425–456. Academic Press, London.

Hahlbrock, K. and Grisebach, H. (1975). *In* "The Flavonoids" (J. B. Harborne, T. J. Mabry, and H. Mabry, eds), pp. 866–915. Chapman and Hall, London.

Hahlbrock, K. and Scheel, D. (1989). *Ann. Rev. Plant Physiol. Plant Mol. Biol.* **40**, 347–369.

Hahlbrock, K., Ebel, J., Ortmann, R., Sutter, A., Wellmann, E. and Grisebach, H. (1971). *Biochim. Biophys. Acta* **244**, 7–15.

Hahlbrock, K., Knobloch, K. H., Kreuzaler, F., Potts, J. R. M. and Wellmann, E. (1976). *Eur. J. Biochem.* **61**, 199–206.

Halaweish, F. and Dougall, D. K. (1990). *Plant Sci.* **71**, 179–184.

Hanson, K. R. and Havir, E. A. (1979). *In* "Recent Advances in Phytochemistry", Vol. 12:

Biochemistry of Plant Phenolics (T. Swain, J. B. Harborne and C. F. van Sumere, eds), pp. 91-137. Plenum Press, New York.

Hanson, K. R. and Havir, E. A. (1981). *In* "The Biochemistry of Plants", Vol. 7 (E. E. Conn, ed), pp. 577-625. Academic Press, London.

Hanson, K. R. and Rose, I. A. (1975). *Acc. Chem. Res.* **8**, 1-10.

Harborne, J. B. (1966). *Z. Naturforsch.* **21b**, 604-605.

Harborne, J. B. (1980). *In* "Encyclopedia of Plant Physiology", New Series, Vol. 8: Secondary Plant Products (E. A. Bell and B. V. Charlwood, eds), pp. 329-402. Springer-Verlag, Berlin, Heidelberg, New York.

Harborne, J. B. and Corner, J. J. (1961). *Biochem. J.* **81**, 242-250.

Harkin, J. M. and Obst, J. R. (1973). *Science* **180**, 296-298.

Haslam, E. (1974). "The Shikimate Pathway". Butterworth & Co., London.

Havir, E. A. and Hanson, K. R. (1968). *Biochemistry* **7**, 1896-1903.

Havir, E. A., Reid, P. D. and Marsh Jr., H. V. (1971). *Plant Physiol.* **14**, 130-163.

Heilemann, J., Wray, V. and Strack, D. (1990). *Phytochemistry* **29**, 3487-3489.

Heilemann, J. and Strack, D. (1991). *Phytochemistry* **30**, 1773-1776.

Heller, W. and Forkmann, G. (1988). *In* "The Flavonoids. Advances in Research Since 1980" (J. B. Harborne, ed.), pp. 399-425. Chapman and Hall, New York.

Heller, W. and Kühnl, T. (1985). *Arch. Biochem. Biophys.* **253**, 453-460.

Herrmann, K. (1978). *Fortschr. Chem. Organ. Naturstoffe* **35**, 73-132.

Higuchi, T., Ito, Y., Shimada, M. and Kawamura, I. (1967). *Phytochemistry* **6**, 1551-1556.

Hlasiwetz, H. (1865). *J. Liebig's Ann. Chem. Pharm.* **136**, 31-36.

Hlasiwetz, H. (1867). *J. Liebig's Ann. Chem. Pharm.* **142**, 219-245.

Hlasiwetz, H. and Barth, L. (1866). *J. Liebig's Ann. Chem. Pharm.* **138**, 61-67.

Hösel, W. (1981). *In* "The Biochemistry of Plants", Vol. 7 (E. E. Conn, ed.), pp. 725-753. Academic Press, London.

Hösel, W. and Todenhagen, R. (1980). *Phytochemistry* **19**, 1349-1353.

Hösel, W., Surholt, E. and Borgmann, E. (1978). *Eur. J. Biochem.* **84**, 487-492.

Holländer, H. and Amrhein, N. (1981). *Planta* **152**, 374-378.

Hrazdina, G. and Jensen, R. A. (1990). *In* "Structural and Organizational Aspects of Metabolic Regulation" (P. A. Srere, M. E. Jones and C. K. Mathews, eds), pp. 27-42. Wiley-Liss, New York.

Hrazdina, G. and Wagner, G. J. (1985a). *In* "Annual Proceedings of The Phytochemical Society of Europe", Vol. 25 (C. F. van Sumere and P. J. Lea, eds), pp. 119-133. Clarendon Press, Oxford.

Hrazdina, G. and Wagner, G. J. (1985b). *Arch. Biochem. Biophys.* **237**, 88-100.

Ibrahim, R. K. (1977). *Z. Pflanzenphysiol.* **85**, 253-262.

Ibrahim, R. K. and Grisebach, H. (1976). *Arch. Biochem. Biophys.* **176**, 700-708.

Imberty, A., Goldberg, R. and Catesson, A.-M. (1984). *Plant Sci. Lett.* **35**, 103-108.

Imberty, A., Goldberg, R. and Catesson, A.-M. (1985). *Planta* **164**, 221-226.

Jensen, R. A. (1986). *Physiol. Plant.* **66**, 164-168.

Jones, D. H. (1984). *Phytochemistry* **23**, 1349-1359.

Kamsteeg, J., Van Brederode, J., Hommels, C. H. and Van Nigtevecht, G. (1980). *Biochem. Physiol. Pflanzen* **175**, 403-411.

Kamsteeg, J., Van Brederode, J., Verschuren, P. M. and Van Nigtevecht, G. (1981). *Z. Pflanzenphysiol.* **102**, 435-442.

Kay, L. E. and Basile, D. V. (1987). *Plant Physiol.* **84**, 99-105.

Kim, S.-H., Terry, M. E., Hoops, P., Dauwalder, M. and Roux, S. J. (1988). *Plant Physiol.* **88**, 1446-1453.

Kleinehollenhorst, G., Behrens, H., Pegels, G., Srunk, N. and Wiermann, R. (1982). *Z. Naturforsch.* **37c**, 587-599.

Kleinhofs, A., Haskins, F. A. and Gorz, H. J. (1967). *Phytochemistry* **6**, 1313-1318.

Kneusel, R. E., Matern, U. and Nicolay, K. (1989). *Arch. Biochem. Biophys.* **269**, 455-462.

Knogge, W., Weissenböck, G. and Strack, D. (1981). *Z. Naturforsch.* **36c**, 197-199.

Knox, W. E. and Pitt, B. M. (1957). *J. Biol. Chem.* **225**, 675-688.

Kojima, M. and Kondo, T. (1985). *Agric. Biol. Chem.* **49**, 2467-2469.

Kojima, M. and Takeuchi, W. (1989). *J. Biochem.* **105**, 265–270.
Kojima, M. and Villegas, R. J. A. (1984). *Agric. Biol. Chem.* **48**, 2397–2399.
Kojima, M. and Uritani, I. (1972). *Plant Cell Physiol.* **13**, 311–319.
Kojima, M. and Uritani, I. (1973). *Plant Physiol.* **51**, 768–771.
Koukol, J. and Conn, E. E. (1961). *J. Biol. Chem.* **236**, 2692–2698.
Kühnl, T., Koch, U., Heller, W. and Wellmann, E. (1987). *Arch. Biochem. Biophys.* **258**, 226–232.
Kühnl, T., Koch, U., Heller, W. and Wellmann, E. (1989). *Plant Sci.* **60**, 21–25.
Kuroda, H., Shimada, M. and Higuchi, T. (1981). *Phytochemistry* **20**, 2635–2639.
Kutner, A., Renstrom, B., Schnoes, H. K. and DeLuca, H. F. (1986). *Proc. Natl. Acad. Sci. USA* **83**, 6781–6784.
Kutsuki, H., Shimada, M. and Higuchi, T. (1982). *Phytochemistry* **21**, 19–23.
Lagrimini, L. M., Burkhart, W., Moyer, M. and Rothstein, S. (1987). *Proc. Natl. Acad. Sci.* **84**, 7542–7546.
Lamb, C. J. (1977). *FEBS Lett.* **75**, 37–40.
Lamb, C. J. and Rubery, P. H. (1975). *Anal. Biochem.* **68**, 554–561.
Lang, S., Hilgenfeldt, U. and Mäder, M. (1990). *J. Plant Physiol.* **136**, 494–498.
Lemifux, R. U. (1963). *In* "Methods in Carbohydrate Chemistry" (L. Whistler and M. L. Wolfrom, eds), Vol. 2, pp. 221–222. Academic Press, New York.
Lewis, N. G. and Yamamoto, E. (1990). *Ann. Rev. Plant Physiol. Plant Mol. Biol.* **41**, 455–496.
Lewis, N. G., Dubelstein, P., Eberhardt, T. L., Yamamoto, E. and Towers, G. H. N. (1987). *Phytochemistry* **26**, 2729–2734.
Lewis, N. G., Inciong, M. E. J., Ohashi, H., Towers, G. H. N. and Yamamoto, E. (1988). *Phytochemistry* **27**, 2119–2121.
Lewis, N. G., Inciong, M. E. J., Dhara, K. P. and Yamamoto, E. (1989). *J. Chromatogr.* **479**, 345–403.
Li, Z.-C. and McClure, J. W. (1990). *J. Plant Physiol.* **136**, 398–403.
Linscheid, M., Wendisch, D. and Strack, D. (1980). *Z. Naturforsch.* **35c**, 907–914.
Liu, E. H. and Gibson, D. M. (1977). *Anal. Biochem.* **79**, 597–601.
Lozoya, E., Hoffmann, H., Douglas, C., Schultz, W., Scheel, D. and Hahlbrock, K. (1988). *Eur. J. Biochem.* **176**, 661–667.
Lüderitz, T. and Grisebach, H. (1981). *Eur. J. Biochem.* **119**, 115–124.
Mäder, M. (1980). *Z. Pflanzenphysiol.* **96**, 283–296.
Mäder, M. and Bopp, M. (1976). *Planta* **128**, 247–253.
Mäder, M., Meyer, Y. and Bopp, M. (1975). *Planta* **122**, 259–268.
Mäder, M., Nessel, A. and Bopp, M. (1977). *Z. Pflanzenphysiol.* **82**, 247–260.
Mansell, R. L., Stöckigt, J. and Zenk, M. H. (1972). *Z. Pflanzenphysiol.* **68**, 286–288.
Mansell, R. L., Gross, G. G., Stöckigt, J., Franke, H. and Zenk, M. H. (1974). *Phytochemistry* **13**, 2427–2435.
Mansell, R. L., Babbel, G. R. and Zenk, M. H. (1976). *Phytochemistry* **15**, 1849–1853.
Marcinowski, S. and Grisebach, H. (1978). *Eur. J. Biochem.* **87**, 37–44.
Marcinowski, S., Falk, H., Hammer, D. K., Hoyer, B. and Grisebach, H. (1979). *Planta* **144**, 161–165.
Marsh Jr., H. V., Havir, E. A. and Hanson, K. R. (1968). *Biochemistry* **7**, 1915–1918.
Masuda, H., Fukuda, H. and Komamine, A. (1983). *Z. Pflanzenphysiol.* **112**, 417–426.
Matile, P. (1980). *Z. Pflanzenphysiol.* **99**, 475–478.
Maule, A. J. and Ride, J. P. (1983). *Phytochemistry* **22**, 1113–1116.
McCliskey, C. M. and Coleman, G. H. (1955). *In* "Organic Synthesis", Vol. 3 (E. C. Horning, ed.), pp. 434–436. John Wiley & Sons, New York.
Meurer, B., Strack, D. and Wiermann, R. (1984). *Planta Med.* **50**, 376–380.
Meurer, B., Wiermann, R. and Strack, D. (1988). *Phytochemistry* **27**, 823–828.
Meurer-Grimes, B., Berlin, J. and Strack, D. (1989). *Plant Physiol.* **89**, 488–492.
Michalczuk, L. and Bandurski, R. S. (1980). *Biochem. Biophys. Res. Commun.* **93**, 588–592.
Mieyal, J. J., Webster, T. and Siddiqui, U. S. (1974). *J. Biol. Chem.* **249**, 2633–2640.
Mock, H.-P. and Strack, D. (1992). *Phytochemistry*, in press.
Mock, H.-P., Vogt, T. and Strack, D. (1992). *Z. Naturforsch*, in press.
Molderez, M., Nagels, L. and Parmentier, F. (1978). *Phytochemistry* **17**, 1747–1750.

Molgaard, P. and Ravn, H. (1988). *Phytochemistry* **27**, 2411–2421.

Monroe, S. H. and Johnson, M. A. (1984). *Phytochemistry* **23**, 1541–1543.

Monties, B. (1989). *In* "Methods in Plant Biochemistry", Vol. 1: Plant Phenolics (J. B. Harborne, ed.), pp. 113–157. Academic Press, London.

Morelli, E., Rej, R. N., Lewis, N. G., Just, G. and Towers, G. H. N. (1986). *Phytochemistry* **25**, 1701–1705.

Murakoshi, I., Ogawa, M., Toriizuka, K., Haginiwa, J., Ohmiya, S. and Otomasu, H. (1977). *Chem. Pharm. Bull.* **25**, 527–529.

Murphy, B. J. and Stutte, C. A. (1978). *Anal. Biochem.* **86**, 220–228.

Nagels, L. and Parmentier, F. (1976). *Phytochemistry* **15**, 703–706.

Negrel, J. (1989). *Phytochemistry* **28**, 477–481.

Negrel, J. and Martin, J. (1984). *Phytochemistry* **23**, 2797–2801.

Negrel, J. and Smith, T. A. (1984). *Phytochemistry* **23**, 31–34.

Negrel, J., Javelle, F. and Paynot, M. (1991). *Phytochemistry* **30**, 1089–1092.

Neish, A. C. (1961). *Phytochemistry* **1**, 1–24.

Nurmann, G. and Strack, D. (1981). *Z. Pflanzenphysiol.* **102**, 11–17.

Oba, K. and Conn, E. E. (1988). *Phytochemistry* **27**, 2447–2450.

Ohashi, H., Yamamoto, E., Lewis, N. G. and Towers, G. H. N. (1987). *Phytochemistry* **26**, 1915–1916.

Pakusch, A.-E. and Matern, U. (1991). *Plant Physiol.* **96**, 327–330.

Pakusch, A.-E., Kneusel, R. E. and Matern, U. (1989). *Arch. Biochem. Biophys.* **271**, 488–494.

Pakusch, A.-E., Matern, U. and Schiltz, E. (1991). *Plant Physiol.* **95**, 137–143.

Pang, A., Catesson, A.-M., Francesch, C., Rolando, C., and Goldberg, R. (1989). *J. Plant Physiol.* **135**, 325–329.

Pearl, I. A. and Darling, S. F. (1957). *J. Org. Chem.* **22**, 1266–1267.

Petersen, M. and Alfermann, A. W. (1988). *Z. Naturforsch.* **43c**, 501–504.

Pfändler, R., Scheel, D., Sandermann, H. and Grisebach, H. (1977). *Arch. Biochem. Biophys.* **178**, 315–316.

Postius, C. and Kindl, H. (1978). *Z. Naturforsch.* **33c**, 65–69.

Poulton, J. (1981). *In:* "The Biochemistry of Plants", Vol. 7 (E. E. Conn, ed.), pp. 667–723. Academic Press, New York.

Poulton, J. E. and Butt, V. S. (1976). *Arch. Biochem. Biophys.* **172**, 135–142.

Rathmell, W. G. and Sequeira, L. (1974). *Plant Physiol.* **53**, 317–318.

Ray, P. M. (1960). *Arch. Biochem. Biophys.* **87**, 19–30.

Reed, D. J., Vimmerstedt, J., Jerina, D. M. and Daly, J. W. (1973). *Arch. Biochem. Biophys.* **154**, 642–647.

Regenbrecht, J. and Strack, D. (1985). *Phytochemistry* **24**, 407–410.

Reichhart, D., Simon, A., Durst, F., Mathews, J. M. and de Montellano, P. R. O. (1982). *Arch. Biochem. Biophys.* **216**, 522–529.

Rhodes, M. J. C. and Wooltorton, L. S. C. (1976). *Phytochemistry* **15**, 947–951.

Rhodes, M. J. C. and Wooltorton, L. S. C. (1977). *Phytochemistry* **16**, 655–659.

Rhodes, M. J. C., Wooltorton, L. S. C. and Lourenco, E. J. (1979). *Phytochemistry* **18**, 1125–1129.

Riley, R. G. and Kolattukudy, P. E. (1975). *Plant Physiol.* **56**, 650–654.

Ros Barcelo, A., Munoz, R. and Sabater, F. (1987). *Physiol. Plant.* **71**, 448–454.

Russell, D. W. (1971). *J. Biol. Chem.* **246**, 2870–3878.

Russell, D. W. and Conn, E. E. (1967). *Arch. Biochem. Biophys.* **122**, 256–258.

Sarkanen, K. V. and Ludwig, C. H. (1971). *In* :"Lignins: Occurrence, Formation, Structure, and Reactions" (K. V. Sarkanen and C. H. Ludwig, eds), pp. 1–18. Wiley Interscience, New York.

Sarkanen, S., Razal, R. A., Piccariello, T., Yamamoto, E. and Lewis, N. G. (1991). *J. Biol. Chem.* **266**, 3636–3642.

Sarni, F., Grand, C. and Boudet, A. M. (1984). *Eur. J. Biochem.* **139**, 259–265.

Saylor, M. H. and Mansell, R. L. (1977). *Z. Naturforsch.* **32c**, 765–768.

Scalbert, A., Monties, B., Lallemand, J.-Y., Guittet, E. and Rolando, C. (1985). *Phytochemistry* **24**, 1359–1362.

Schlepphorst, R. and Barz, W. (1979). *Planta Med.* **36**, 333-342.
Schloss, P., Walter, C. and Mäder, M. (1987). *Planta* **170**, 225-229.
Schmid, G. and Grisebach, H. (1982). *Eur. J. Biochem.* **123**, 363-370.
Schmid, G., Hammer, D. K., Ritterbusch, A. and Grisebach, H. (1982). *Planta* **156**, 207-212.
Schultz, O. E. and Gmelin, R. (1953). *Z. Naturforsch.* **8b**, 151-156.
Semler, U., Schmidtberg, G. and Gross, G. G. (1987). *Z. Naturforsch.* **42c**, 1070-1074.
Shimada, M., Ohashi, H. and Higuchi, T. (1970). *Phytochemistry* **9**, 2463-2470.
Shimizu, T. and Kojima, M. (1984). *J. Biochem.* **95**, 205-212.
Siegel, B. Z. and Galston, A. W. (1967). *Plant Physiol.* **42**, 221-226.
Smith, D. C. C. (1955). *Nature* **176**, 267-268.
Stafford, H. A. (1974). *Rec. Adv. Phytochem.* **8**, 53-79.
Stafford, H. A. and Bliss, M. (1973). *Plant Physiol.* **52**, 453-458.
Stafford, H. A. and Dresler, S. (1972). *Plant Physiol.* **49**, 590-595.
Stöckigt, J. and Zenk, M. H. (1974). *FEBS Lett.* **42**, 131-134.
Stöckigt, J. and Zenk, M. H. (1975). *Z. Naturforsch.* **30c**, 352-358.
Strack, D. (1977). *Z. Pflanzenphysiol.* **84**, 139-145.
Strack, D. (1980). *Z. Naturforsch.* **35c**, 204-208.
Strack, D. and Gross, W. (1990). *Plant Physiol.* **92**, 41-47.
Strack, D., Knogge, W. and Dahlbender, B. (1983). *Z. Naturforsch.* **38c**, 21-27.
Strack, D., Ruhoff, R., Gräwe, W. (1986). *Phytochemistry* **25**, 833-837.
Strack, D., Leicht, P., Bokern, M., Wray, V. and Grotjahn, L. (1987a). *Phytochemistry* **26**, 2919-2922.
Strack, D., Keller, H. and Weissenböck, G. (1987b). *J. Plant Physiol.* **131**, 61-73.
Strack, D., Heilemann, J., Boehnert, B., Grotjahn, L. and Wray, V. (1987c). *Phytochemistry* **26**, 107-111.
Strack, D., Gross, W., Wray, V. and Grotjahn, L. (1987d). *Plant Physiol.* **83**, 475-478.
Strack, D., Gross, W., Heilemann, J., Keller, H. and Ohm, S. (1988). *Z. Naturforsch.* **43c**, 32-36.
Strack, D., Eilert, U., Wray, V, Wolff, J. and Jaggy, H. (1990a). *Phytochemistry* **29**, 2893-2896.
Strack, D., Ellis, B.E., Gräwe, W. and Heilemann, J. (1990b). *Planta* **180**, 217-219.
Strack, D., Becher, A., Brall, S. and Witte, L. (1991). *Phytochemistry* **30**, 1493-1498.
Sütfeld, R. and Wiermann, R. (1974). *Z. Pflanzenphysiol.* **72**, 163-171.
Sutter, A. and Grisebach, H. (1975). *Arch. Biochem. Biophys.* **167**, 444-447.
Tanahashi, T. and Zenk, M. H. (1990). *Phytochemistry* **29**, 1113-1122.
Tanaka, Y., Kojima, M. and Uritani, I. (1974). *Plant & Cell Physiol.* **15**, 843-854.
Teusch, M., Forkmann, G. and Seyffert, W. (1987). *Phytochemistry* **26**, 991-994.
Tkotz, N. and Strack, D. (1980). *Z. Naturforsch.* **35c**, 835-837.
Ulbrich, B. and Zenk, M. H. (1979). *Phytochemistry* **18**, 929-933.
Ulbrich, B. and Zenk, M. H. (1980). *Phytochemistry* **19**, 1625-1629.
Ulbrich, B., Stöckigt, J. and Zenk, M. H. (1976). *Naturwissenschaften* **63**, 484.
van den Berg, B. M., Chibbar, R. N. and van Huystee, R. B. (1983). *Plant Cell Rep.* **2**, 304-307.
van den Berg, B. M. and van Huystee, R. B. (1984). *Physiol. Plant.* **60**, 299-304.
van Huystee, R. B. (1987). *Ann. Rev. Plant Physiol.* **38**, 205-219.
Vaughan, P. F. T. and Butt, V. S. (1969). *Biochem. J.* **113**, 109-115.
Vaughan, P. F. T. and Butt, V. S. (1970). *Biochem. J.* **119**, 89-94.
Villegas, M. and Brodelius, E. (1990). *Physiol. Plant.* **78**, 414-420.
Villegas, R. J. A. and Kojima, M. (1985). *Agric. Biol. Chem.* **49**, 263-265.
Villegas, R. J. A. and Kojima, M. (1986). *J. Biol. Chem.* **261**, 8729-8733.
Villegas, R. J. A., Shimokawa, T., Okuyama, H. and Kojima, M. (1987). *Phytochemistry* **26**, 1577-1581.
von Babo, L. and Hirschbrunn, M. (1852). *J. Liebig's Ann. Chem. Pharm.* **84**, 10-32.
Walton, E. and Butt, V. S. (1970). *J. Exp. Bot.* **21**, 887-891.
Wengenmayer, H., Ebel, J. and Grisebach, H. (1976). *Eur. J. Biochem.* **65**, 529-536.
Wyrambik, D. and Grisebach, H. (1975). *Eur. J. Biochem.* **59**, 9-15.
Wyrambik, D. and Grisebach, H. (1979). *Eur. J. Biochem.* **97**, 503-509.

Xu, Y., Hu, C. and van Huystee, R. B. (1990). *J. Exp. Bot.* **41**, 1479–1488.

Yamamoto, E., Bokelman, G. H. and Lewis, N. G. (1989). *In*: "Plant Cell Wall Polymers. Biogenesis and Biodegradation", Am. Chem. Soc. Symposium Series 399 (N. G. Lewis and M. G. Paice, eds), pp. 68–88. American Chemical Society, Washington, DC.

Yamamoto, E., Inciong, M. E. J., Davin, L. B. and Lewis, N. G. (1990). *Plant Physiol.* **94**, 209–213.

Zenk, M. H. (1967). *Z. Pflanzenphysiol.* **57**, 477–478.

Zenk, M. H. (1979). *In*: "Recent Advances in Phytochemistry", Vol. 12 (T. Swain, J. B. Harborne and C. F. van Sumere, eds), pp. 139–176. Plenum Press, New York.

Zenk, M. H., Ulbrich, B., Busse, J. and Stöckigt, J. (1980). *Anal. Biochem.* **101**, 182–187.

Zimmermann, A. and Hahlbrock, K. (1975). *Arch. Biochem. Biophys.* **166**, 54–62.

Zon, J. and Amrhein, N. (1992). *Liebigs Ann. Chem.* 625–628.

Zucker, M. (1965). *Plant Physiol.* **40**, 779–784.

4 Flavonoid Enzymology

RAGAI K. IBRAHIM and LUC VARIN

*Plant Biochemistry Laboratory, Department of Biology,
Concordia University, Montreal, Canada H3G 1M8*

This chapter is dedicated to the late Professor Hans Grisebach for his valuable contribution to flavonoid enzymology. Professor Grisebach had originally agreed to write this chapter.

I. INTRODUCTION

Flavonoid compounds are characterised by their ubiquitous occurrence and diversity of structural patterns. They are classified according to the oxidation level of ring C which results in a number of major classes including chalcones, flavanones, flavones, isoflavones, flavonols and anthocyanidins (Scheme 4.1). These aglycones may undergo further hydroxylation, methylation, glycosylation, acylation, prenylation or sulpha-

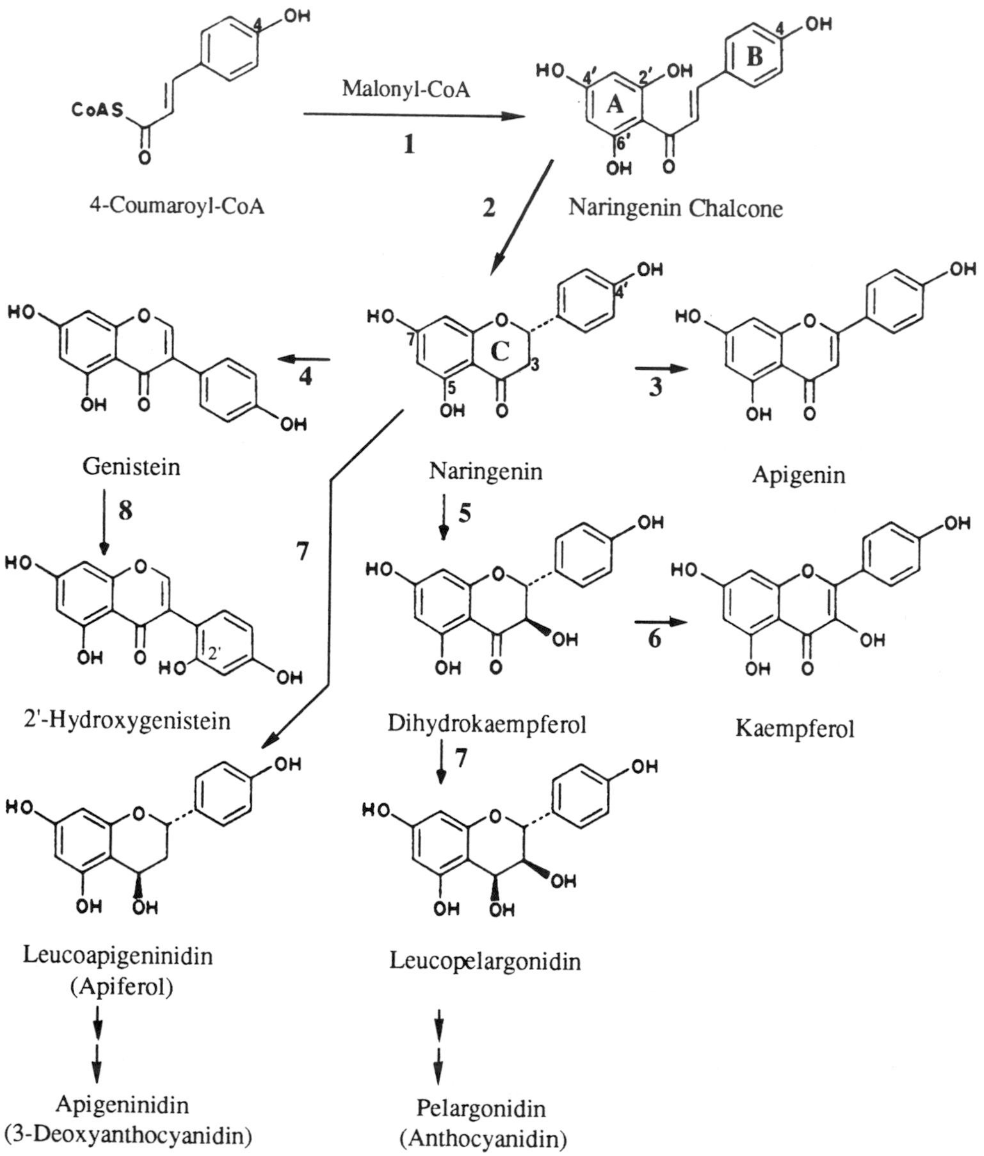

SCHEME 4.1. Enzymatic synthesis of flavonoids with 5,7-hydroxy A-rings. Key: 1, chalcone synthase; 2, chalcone isomerase; 3, flavone synthase; 4, isoflavone synthase; 5, flavanone 3-hydroxylase; 6, flavonol synthase; 7, dihydroflavone/dihydroflavonol 4-reductase; 8, isoflavone 2′-hydroxylase.

tion. Previous studies of the phytochemistry, biosynthesis and enzymology of flavonoids have been summarised in recent reviews (Harborne and Mabry, 1982; Harborne, 1988; Stafford, 1990).

This chapter deals mainly with the methods involved in protein purification and enzyme assays, as well as the properties of the major flavonoid enzymes, especially those enzymes which have been purified and characterised. Where appropriate, methods for the preparation of commercially unavailable enzyme substrates will be described. General reviews of protein separation techniques are to be found in several books (Freifelder, 1982; Scopes, 1987; Janson and Rydén, 1989) and reviews (Jakoby, 1984; Deutscher, 1990).

Since flavonoid enzymes occur in low abundance, in the presence of relatively high amounts of phenolic compounds and other secondary metabolites, special care must be taken during protein extraction in order to minimise enzyme inactivation. A commonly used procedure for the extraction of cytosolic enzymes consists of reducing the liquid N_2-frozen tissue to a powder, to which is added Polyclar AT and the appropriate extraction buffer containing 5 mM each of diethylammonium diethyldithiocarbamate and dithiothreitol. Usually all extraction and purification steps are carried out at temperatures between 2 to 4°C. After tissue homogenisation, the mixture is filtered through nylon cloth and the filtrate centrifuged at 15 000 × g for 15 min. The supernatant is then stirred for 10 min with Dowex 1×2 and filtered. This crude protein extract may be used for further enzyme purification, or passed through a Sephadex G-25 column and used directly for enzyme assays. Where necessary, modifications to the above extraction procedure will be indicated in the following sections.

II. ENZYMES INVOLVED IN THE FORMATION OF FLAVONOID SKELETONS

A. Chalcone Synthase

Chalcone synthase (CHS; EC 2.3.1.74), which was earlier referred to as the flavanone synthase (Kreuzaler and Hahlbrock, 1975), is the first and key enzyme of flavonoid biosynthesis. It catalyses the stepwise condensation of three molecules of malonyl-CoA with one molecule of hydroxycinnamoyl-CoA to form naringenin chalcone (2',4,4', 6'-tetrahydroxychalcone), the central intermediate common to all flavonoids with a 5,7-dihydroxy A-ring (Scheme 4.1). The chalcone formed may, however, undergo spontaneous isomerisation *in vitro* to yield the corresponding flavanone, naringenin. Very recently, an NADPH-dependent reductase activity that coacts with CHS has been shown to account for the production of 6'-deoxynaringenin chalcone, isoliquiritigenin (Scheme 4.2) (Ayabe *et al.*, 1988; Welle and Grisebach, 1988). Whereas 4-coumaroyl-CoA is the physiological substrate in most plants (Hrazdina *et al.*, 1976; Spribille and Forkmann, 1982a), CHS from petals of *Dianthus caryophyllus* (Spribille and Forkmann, 1982b) and *Verbena hybrida* (Stotz *et al.*, 1984), as well as from flowers and cell cultures of *Daucus carota* (Hinderer *et al.*, 1983), soybean cell culture (Welle and Grisebach, 1988) and rye leaves (Peters *et al.*, 1988), have been reported to utilise caffeoyl-CoA. On the other hand, petals of *Cosmos sulphureus* (Sütfeld and Wiermann, 1981), tulip anthers (Sütfeld *et al.*, 1978; Sütfeld and Wiermann, 1981), spinach leaves (Beerhues and Wiermann, 1985), buckwheat hypocotyls (Hrazdina *et al.*, 1986) and rye leaves (Peters *et al.*, 1988) have been reported to utilise feruloyl-CoA as well as caffeoyl-CoA, but to a lesser degree than the monohydroxy analogue.

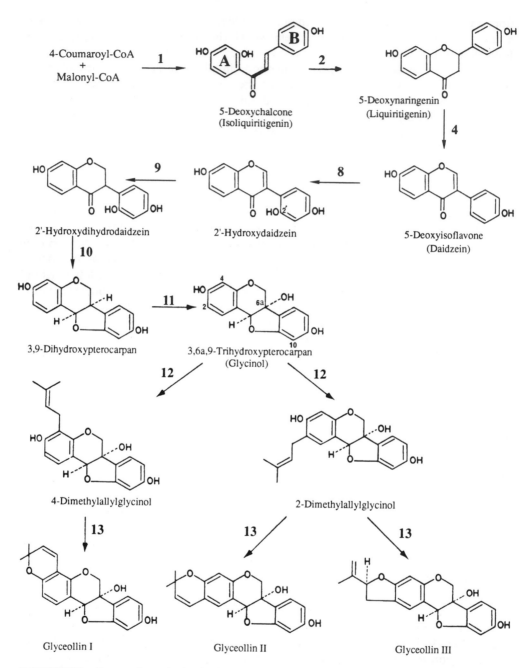

SCHEME 4.2. Enzymatic synthesis of 5-deoxyisoflavonoids and pterocarpans. Key: 1, chalcone synthase; 2, chalcone isomerase; 4, isoflavone synthase; 8, isoflavone 2'-hydroxylase; 9, isoflavone oxidoreductase; 10, pterocarpan synthase; 11, pterocarpan 6a-hydroxylase; 12, prenyltransferase; 13, prenylated pterocarpan cyclase.

1. Synthesis of substrates

Hydroxycinnamoyl-CoA esters are synthesised by the method of Stöckigt and Zenk (1975), in which the cinnamic acid is reacted with *N*-hydroxysuccinimide in the presence of dicyclohexylcarbodiimide. The succinimide ester formed is purified by preparative silica gel thin layer chromatography (TLC) in chloroform–ethanol (4:1, v/v) before conjugation with CoA. The CoA ester formed is further purified, to remove unreacted CoA, lyophilised and resuspended in water acidified to pH 3.5 with dilute HCl, then stored at $-70°C$.

2. Assay

Chalcone synthase is assayed by determining the incorporation into naringenin chalcone of radioactivity from $[2\text{-}^{14}C]$malonyl-CoA $(10 \,\mu M)$ in the presence of 4-coumaroyl-CoA $(10 \,\mu M)$ and the enzyme protein (in 0.1 M phosphate buffer, pH 8.0). However, the presence of chalcone isomerase in enzyme preparations of varying purity, and the non-enzymic conversion of chalcone to flavanone, reduce the yield of the chalcone product. Furthermore, the addition of 2-mercaptoethanol to the assay mixture results in the formation of short-chain release products, originating from the condensation of 4-coumaroyl-CoA with one or two molecules of malonyl-CoA (Hrazdina *et al.*, 1976). The enzyme reaction, which is linear for 20 min, is terminated by the addition of $20 \,\mu l$ of methanol containing $10 \,\mu g$ naringenin and its chalcone. Adjusting the pH of the reaction mixture to pH 4.0 by the addition of 0.1 M citric acid prevents the non-enzymic cyclisation of chalcone to flavanone since the former is stable under these conditions (Sütfeld and Wiermann, 1980). Recent modification of the enzyme assay involves the use of N_2-saturated phosphate buffer (pH 6.8), and the addition of sodium ascorbate $(20 \,mM)$ and cysteine $(10 \,mM)$ instead of 2-mercapto-ethanol. The modified assay conditions resulted in >95% of $[^{14}C]$naringenin as the reaction product (Britsch and Grisebach, 1985).

3. Purification

Chalcone synthase has been partially purified from some 20 species (see Heller and Forkmann, 1988; Stafford, 1990), including one gymnosperm (Rolf and Kindl, 1984) and, to apparent homogeneity, from carrot cell culture (Ozeki *et al.*, 1985), buckwheat hypocotyls (Hrazdina *et al.*, 1986), spinach leaves and cell culture (Beerhues and Wiermann, 1988), and rye leaves (Peters *et al.*, 1988). Since the enzyme activity is inhibited by Tris-HCl buffer, the protein is extracted in 0.1 M phosphate buffer, pH 8.0 and precipitated with solid ammonium sulphate (50–80% saturation) or polyethylene glycol (7–30% in 30 mM $MgCl_2$). Addition of 20 mM sodium ascorbate, instead of 2-mercaptoethanol, to the N_2-saturated extraction buffer appears to reduce enzyme loss during purification, and improves product yield while avoiding the formation of short-chain release products (Britsch and Grisebach, 1985). Further purification steps include gel filtration and successive chromatography on ion-exchange and hydroxy-apatite columns. The enzyme usually binds to anion exchangers (DEAE-cellulose, DEAE-Sephacel or DEAE-Trisacryl), except for the carrot enzyme (Ozeki *et al.*, 1985) which bound only to CM-Sepharose CL-6B. The final steps of purification, to apparent homogeneity, may involve chromatofocusing (Beerhues and Wiermann, 1988) or high performance liquid chromatography (HPLC) on hydroxyapatite (Hrazdina *et al.*, 1986), but are not without extensive loss in recovery. Nevertheless, 70- to 3000-fold purification may be achieved with specific activities of 16–300 pkat/mg^{-1} protein and recoveries of 0.3–20%.

4. Properties

Chalcone synthase is considered a cytosolic enzyme (Beerhues *et al.*, 1988), although small amounts of the enzyme may be associated with membranes (Hrazdina and Wagner, 1985), particularly the cytoplasmic side of the endoplasmic reticulum (ER) (Hrazdina *et al.*, 1987). The enzyme is a dimer with an apparent mol. wt. of 78–88 kDa and 40–44 kDa for the subunits as determined by SDS-PAGE. Two forms of CHS have been reported in spinach with different native molecular weights although with similar subunit size and extensive homology between the two proteins (Beerhues and Wiermann, 1988). The pH optimum of the enzyme reaction, in phosphate buffer, is 7.5–8.5 and 6.5–6.8 for 4-coumaroyl-CoA and caffeoyl-CoA, respectively, and the pI of the enzyme is 5.2. Chalcone synthase from most sources is highly specific for 4-coumaroyl-CoA, although the enzymes from *Cosmos* petals and *Tulipa* anthers utilise 4-coumaroyl-CoA, caffeoyl-CoA and feruloyl-CoA with similar efficiency (Sütfeld and Wiermann, 1981). Buckwheat CHS accepts caffeoyl-CoA and feruloyl-CoA at 20% and 80% of that of 4-coumaroyl-CoA (Hrazdina *et al.*, 1986). The K_m values for the enzyme range from 0.6 to 1.6 μM for 4-coumaroyl-CoA, 1.5 to 7.7 μM for caffeoyl-CoA and 1.0 to 35 μM for malonyl-CoA. Chalcone synthase is inhibited by higher concentration of both the substrate and cosubstrate (>10 and 50 μM, respectively), CoASH, acetyl CoA, and the reaction products as well as their structural analogues, flavanone and flavone. While the enzyme activity is also inhibited by >50% in the presence of 10 μM of either NEM, PCMB or iodoacetamide, it is not affected by EDTA since it has no requirement for divalent cations. Chalcone synthase is relatively unstable with a half-life of 1–4 days at 0–4°C, but can be stabilised in the presence of 20% glycerol at −20°C for up to 4 weeks.

B. Chalcone Isomerase

Chalcone isomerase (CHI; EC 5.5.1.6), the second enzyme in flavonoid biosynthesis, catalyses the stereospecific isomerisation of chalcones to their corresponding (−)flavanones, thus establishing their (−)2S configuration (Scheme 4.1). The latter are the only substrates utilised by subsequent enzymes in the flavonoid pathway. Two types of isomerases have been recognised. One type cyclises chalcones with phloroglucinol-type A-ring only, as in parsley (Kreuzaler and Hahlbrock, 1975) and tulip anthers (Chmiel *et al.*, 1983), and the other cyclises both types of A-ring substitution (6′-hydroxy- and 6′-deoxychalcones) as that of bean cell culture (Dixon, *et al.*, 1982). The substrate specificity of the two isomerase types seems to correlate with the naturally occurring flavonoids in these tissues.

1. Assay

Different chalcones are prepared by the Claisen–Schmidt condensation of their corresponding acetophenones and 4-hydroxybenzaldehyde in alcoholic KOH (Geissman and Clinton, 1946). The resulting chalcones are crystallised from aqueous EtOH and purified on polyamide column using 95% EtOH as eluant. Chalcone isomerase activity is determined spectrophotometrically by following the decrease in absorbance at λ_{max} (385–405 nm) of the respective chalcones (20–40 μM) in the presence of the enzyme in 50 mM Tris-HCl buffer, pH 7.6. The addition of 40 mM KCN to the reaction mixture inhibits any contaminating chalcone peroxidase activity (Wilson and Wong, 1976). Correction is made for any decrease in absorbance which occurs by spontaneous cyclisation in the absence of the enzyme. Enzyme activity is calculated at 390 nm using the molar extinction coefficient of the respective chalcone (e.g. 29 400 M^{-1} cm^{-1} for 2′,4′, 4-trihydroxychalcone) (Bednar and Hadcock, 1988). Non-enzymic cyclisation and peroxidase activity are considered major problems for the accurate measurements of CHI activity. Measurements of enzyme activity at acidic pH (5.5–6.5) and the addition of 2 mg ml^{-1} bovine serum albumin (BSA) have been reported to reduce nonenzymic cyclisation (Mol et al., 1985).

2. Purification

Chalcone isomerase has been purified from several sources (Heller and Forkmann, 1988; Stafford, 1990 and references therein). Tissues are usually extracted in presence of PVPP and Dowex 1x2, with 0.05–0.1 M phosphate or Tris-HCl buffers, pH 7.5–8.5, containing 1–4 mM 2-mercaptoethanol, 1 mM EDTA and 1 mM phenylmethylsulfonyl-fluoride, and centrifuged. In order to eliminate endogenous peroxidase, the supernatant is brought to pH 4–5 for 10–20 min, readjusted to pH 8–8.5, and the precipitates removed by centrifugation. The protein is fractionated with solid ammonium sulphate (50–75% saturation) and the resulting pellet desalted by gel filtration. Enzyme purification is carried out by chromatography on DEAE-cellulose and eluted with a gradient of 10 to 60 mM phosphate buffer, pH 7.5 (Dixon et al., 1982; Robbins and Dixon 1984), or 0 to 150 mM KCl in 15 mM bis-Tris-HCl, pH 6.8 (Bednar and Hadcock 1988). The latter step separates CHI activity, which elutes at 50–60 mM KCl, from chalcone peroxidase activity, which elutes with the breakthrough protein and also later in the gradient (c. 130 mM). Further purification of CHI can be achieved by affinity chromatography on one of the following ligands: (a) p-aminobenzamidohexyl-Sepharose 4B linked to 2′,4′-dimethoxy-4-hydroxychalcone and elution with 30 mM phosphate buffer, pH 8.0, containing 0.5 mg ml^{-1} of the chalcone substrate (Dixon et al., 1982); (b) Blue Dextran–Sepharose eluted with 0.1 M phosphate buffer, pH 7.4, containing 1 mg ml^{-1} naringenin chalcone (Van Tunen and Mol, 1987); or (c) Amicon Matrex Orange A eluted with 15 mM bis-Tris-HCl, pH 6.5, containing 50 μM naringenin (Bednar and Hadcock, 1988). These purification steps resulted in an increase of specific activity of 400-fold for the Phaseolus cell culture (Dixon et al., 1982), 200-fold for the Petunia petal (Van Tunen and Mol, 1987) and 11 000-fold for the soybean seed (Bednar and Hadcock, 1988) enzymes. Whereas the french bean

(Dixon *et al.*, 1982), the soybean (Bednar and Hadcock, 1988) and parsley (Kreuzaler and Hahlbrock, 1975) enzymes appear as single bands on SDS-PAGE, that of *Petunia* petals gives a number of components on isoelectric focusing (IEF) (Van Tunen and Mol, 1987) which were correlated to the expression of different CHI genes (Dixon *et al.*, 1988).

3. Properties

Chalcone isomerase is a monomer with a mol. wt. of 24–30 kDa, although large variations have been reported for enzymes from different sources (Dixon *et al.*, 1988). The pH optimum is around 7.5 and the pI varies between 5 and 5.7. The K_m values for tetra- and pentahydroxychalcones are in the micromolar range (8–50 μM). Except for soybean seed CHI which exhibits no substrate inhibition up to 10 times the K_m (Bednar and Hadcock, 1988), other CHI enzymes are inhibited by both the substrates and products (15–30 μM), and competitively inhibited by a number of putative isoflavonoid phytoalexins (Dixon *et al.*, 1982) as well as by various flavones and flavonols (Boland and Wong, 1975). The enzyme activity is severely inhibited by SH group reagents, especially PCMB (50% inhibition at 1 μM) , but is neither affected by divalent cations nor EDTA. An activation energy of 17.56 kJ mol^{-1} (Dixon *et al.*, 1982) and equilibrium constants of 15 and 7.6 for spontaneous and enzymatic reactions, respectively (Bednar and Hadcock, 1988), have been determined. Chalcone isomerase is stable for one year when stored at a concentration of 1 mg ml^{-1} at −80°C, but loses 30% of its activity when stored at 4°C for 3 months. Dilute enzyme preparations are unstable, especially on vortexing in enzyme assays.

C. Chalcone Cyclase

Another enzyme, which has been purified from immature grapefruit, acts on the chalcone–flavanone equilibrium reaction (Raymond and Maier, 1977). The enzyme requires, for activity, neohesperidose at C-4′ of the chalcone-A-ring and a free hydroxyl group at C-4 of ring B. Chalcone cyclase (CHC) does not accept either homoeriodictyol neohesperidose or prunin chalcone, suggesting that 7-*O*-glycosylation of flavanones occurs prior to heterocyclic ring formation, whereas ring B *O*-methylation occurs after cyclisation. The enzyme exhibits a pH optimum of 7.0, and does not revert naringin (naringenin 7-neohesperidoside) to the corresponding chalcone between pH 4 and 10. CHC has a mol. wt. of 53 kDa, is not affected by EDTA, PCMB or NaN$_3$, but is totally inhibited by KCN (Raymond and Maier, 1977).

III. ENZYMES INVOLVED IN THE MODIFICATION OF RING C

A. Flavanone 3-Hydroxylase

Flavanone 3β-hydroxylase (F3H) [(2*S*)-flavanone, 2-oxoglutarate:oxygen oxidoreductase, (3*R*)-hydroxylating; EC 1.14.11.9] catalyses the stereospecific hydroxylation of (2*S*)-flavanones to (2*R*,3*R*)-dihydroflavonols (Scheme 4.1). The latter compounds are intermediates in the biosynthesis of flavonols, anthocyanidins and catechins (Heller and Forkmann, 1988). This cytosolic, 2-oxoglutarate-dependent dioxygenase was first characterised from crude extracts of *Matthiola incana* (Forkmann *et al.*, 1980) and from illuminated parsley cell cultures (Britsch *et al.*, 1981). F3H was highly purified from *Petunia hybrida* (Britsch and Grisebach, 1986) and, more recently, purified to apparent homogeneity from the same source (Britsch, 1990a).

1. Substrate preparation

A radiolabelled enzyme assay utilises (2*S*)-[4a,6,8-^{14}C]naringenin as the substrate (Britsch *et al.*, 1981). The latter is prepared enzymatically using a partially purified preparation containing both CHS and CHI (Sections II.A,B) in the presence of 4-coumaroyl-CoA and [2-^{14}C]malonyl-CoA as substrates. (2*S*)-[^{14}C]Eriodictyol and 5,7,3′,4′,5′-pentahydroxyflavanone may be prepared from [^{14}C]naringenin with a microsomal preparation containing flavanone 3′,5′-hydroxylase (Section IV.A) from the blue-flowering mutant of *Petunia hybrida* (Stotz and Forkmann, 1982). The labelled enzyme products are purified by TLC using 15% HOAc as solvent.

2. Assay

The standard assay system (Britsch and Grisebach, 1986) contains 100 pmol labelled flavanone, 8 μmol sodium glycinate (pH 8.0), 0.2 mg catalase, 1 μmol sodium ascorbate, 10 nmol ferrous ion, 25 nmol 2-oxoglutarate and the enzyme protein, all of which are added in this sequence in a final volume of 100 μl. The mixture is incubated for 10 min at 37°C in an open vial. The reaction is terminated by the addition of 10 μl of saturated aqueous EDTA and the flavonoid products extracted twice with 40 μl ethyl acetate. The organic layer is co-chromatographed with reference compounds on cellulose TLC using CHCl$_3$–HOAc–H$_2$O (10:9:1; v/v/v) or 15% aqueous HOAc. The flavonoids are detected by their fluorescence in UV light (254 nm), and by spraying with 0.1% aqueous fast Blue B salt, followed by exposure to ammonia vapour. Dihydroflavonols may also be detected by treatment of the air-dried TLC plates with zinc dust followed by spraying with 6 M HCl (Barton, 1968). This colour reaction may be used for rapid detection of enzyme activity during purification. For a semi-quantitative but rapid enzyme assay, unlabelled naringenin (50 nmol) may be used as substrate and the resulting product, dihydrokaempferol, is then scraped off the TLC plate, eluted with a small aliquot of methanol, and its absorbance determined at 292 nm. This assay is one-tenth as sensitive as the radiolabelled assay (Britsch and Grisebach, 1986).

3. Purification

Flavonone 3-hydroxylase has been purified to apparent homogeneity from a red-flowering mutant of *Petunia hybrida* (Britsch, 1990a). Due to rapid loss of activity during purification, the enzyme is stabilised by the addition of 20 mM sodium ascorbate, 0.2 mM 2-oxoglutarate and 0.1 mM ferrous sulphate to the extraction buffers, which are degassed and kept under a stream of N$_2$ at 4°C. Frozen petals (750 g) are ground in a mortar with 10 g sodium ascorbate, 300 g Dowex 1x2 and 1.5 l Tris-HCl buffer, pH 7.3 and 10% glycerol. After centrifugation, the supernatant is fractionated with solid ammonium sulphate (40–80% saturation) and the protein desalted on Sephadex G-25 using 50 mM imidazole-HCl buffer, pH 6.8, containing 10% glycerol (buffer A). The desalted protein is chromatographed on DEAE-Sepharose CL-6B and the enzyme eluted with a linear gradient of 0 to 0.5 M NaCl in buffer A. The active fractions are passed through Sephacryl S-200 using 25 mM of buffer A, before being chromatographed on hydroxyapatite and eluted with a linear gradient of 5 to 200 mM phosphate buffer, pH 6.8, containing 10 mM sodium ascorbate and 10% glycerol. The partially purified enzyme is then subjected to chromatofocusing on Mono P column, followed by hydrophobic interaction chromatography on phenyl-Sepharose column. In the latter step, the active Mono P fractions are adjusted to 1.3 M ammonium sulphate, and passed through the phenyl-Sepharose column. The enzyme is eluted by a descending gradient of 1.3 to 0.0 M ammonium sulphate. This elaborate procedure resulted in an apparently homogenous preparation, although the specific activity and the recovery were low (Britsch, 1990a), in comparison with a previous protocol which yielded a partially purified enzyme with a higher specific activity (Britsch and Grisebach, 1986).

4. Properties

The native, catalytically active F3H has a mol. wt. of 70–75 kDa which, during extensive purification or SDS-PAGE, gives rise to species with lower molecular weights (34–38 kDa) (Britsch and Grisebach, 1986; Britsch, 1990a). The enzyme exhibits a pH optimum of 8.5 and pI values of 4.8 and 5.0 on chromatofocusing and isoelectric focusing, respectively. Flavanone 3-hydroxylase has an absolute requirement for molecular oxygen, 2-oxoglutarate, Fe^{2+} and ascorbate as cofactors. Catalase has a prominent stimulating effect on F3H activity, as is the case with other 2-oxoglutarate-dependant dioxygenases (Kondo et al., 1981). The latter treatment results in a 10-fold increase in enzyme activity with crude extracts and 3- to 5-fold increase with the purified enzyme. Flavanone 3-hydroxylase is stereospecific for (2S), but not (2R), enantiomers. It accepts naringenin (K_m 5.6 μM) and eriodictyol (K_m 12 μM), but not 5,7,3′,4′,5′-pentahydroxyflavone, and its K_m for 2-oxoglutarate is 20 μM. The enzyme is competitively inhibited by pyridine 2,4-dicarboxylate (K_i 1.2 μM) and 2,5-dicarboxylate (K_i 40 μM) with respect to 2-oxoglutarate. Some product inhibition (15–40%) is also observed at concentrations of 5–100 μM of (+)-dihydrokaempferol. Despite the various additives used for enzyme stabilisation and the use of FPLC to reduce the time of purification, F3H is very unstable as it suffers from partial proteolytic degradation. However, the enzyme from the ion-exchange step may be stored, at −70°C in presence of 20 mM ascorbate, for 6 months without loss of activity (Britsch and Grisebach, 1986).

B. Flavone Synthase

Flavone synthase (FS) catalyses the conversion of (2S)-flavanones to the corresponding flavones (Scheme 4.1). The enzyme was first observed in cell-free extracts of primary leaves and irradiated cell cultures of parsley (Sutter et al., 1975). Whereas further studies in parsley cell cultures (Britsch et al., 1981) indicated that FS is a cytosolic 2-oxoglutarate-dependent dioxygenase (FS I), in Antirrhinum majus flowers (Stotz and Forkmann, 1981) and osmotically stressed soybean cell cultures (Kochs and Grisebach,

1987) it occurs as an NADPH-dependent microsomal enzyme (FS II). This is a case in which the same metabolic product is formed by two distinctly different enzyme reactions.

1. Flavone synthase I (EC 1.14.11)

(a) Extraction and assay. For the preparation of a crude enzyme extract, 10–20 g tissue is homogenised in a chilled mortar with 10 g quartz sand, 2 g Dowex 1x2 and 10–20 ml of 0.1 M Tris-HCl buffer, pH 7.5, containing 14 mM 2-mercaptoethanol (Britsch *et al.*, 1981). After centrifugation of the homogenate (10 000 × g, 10 min), the supernatant is desalted on a Sephadex G-25 column and may be assayed directly for FS I activity. The same fraction may also serve as a source of the membrane-associated 3′-hydroxylase activity (Section IV.A) for the preparation of eriodictyol. The membrane pellet is prepared by the addition of 30 mM $MgCl_2$ and centrifugation at 40 000 × g for 20 min. After washing the pellet with extraction buffer containing 2.8 mM 2-mercaptoethanol, it is suspended in 1–2 ml of the buffer and homogenised in a glass homogeniser.

The synthesis of labelled naringenin, and the incubation mixture for the FS I assay, are the same as previously described for F3H (Section III.A). The reaction products are extracted with ethyl acetate and co-chromatographed with apigenin and/or luteolin, as reference compounds, on cellulose TLC plates in 30% aqueous HOAc or $CHCl_3$-HOAc-H_2O (10:9:1; v/v/v), or on silica gel plates in $CHCl_3$-MeOH-H_2O (93:6:1; v/v/v). Flavonoids are detected in UV light, scanned for radioactivity (or autoradiographed) and the radioactive zones are scraped off the plate for liquid scintillation counting.

(b) Purification. More recently, FS I has been purified to apparent homogeneity from illuminated parsley cell cultures by a six-step procedure (Britsch, 1990b) . Due to rapid loss of enzyme activity, the extraction procedure has been modified as described for F3H (Section III.A). The enzyme was purified by precipitation with solid ammonium sulphate (40–80% saturation), desalting on Sephadex G-25 and successive chromatography on Q-Sepharose, Sephacryl S-200, Mono-P and finally on phenyl-Superose. The chromatofocusing step was essential for the elimination of other 2-oxoglutarate-dependant dioxygenases, mainly flavanone 3-β-hydroxylase and flavonol synthase. This step, which may be replaced by chromatography on hydroxy-apatite, resulted in a 24-fold purification with a specific activity of 210 pkat mg^{-1} and 2% recovery, which declined to 12-fold, 95 pkat mg^{-1} and 0.05% after chromatography on phenyl-Superose.

(c) Properties. As for F3H, FS I has an absolute requirement for molecular oxygen, 2-oxoglutarate, Fe^{2+} ions and ascorbate. Fe^{2+} could not be replaced by other divalent cations, although 20 mM Cu^{2+} or Zn^{2+} completely inhibited enzyme activity (Britsch, 1990b). Ascorbate could not be replaced by reductants such as 2-mercaptoethanol or dithiothreitol. In fact, the addition of 1.4 or 14 μM of 2-mercaptoethanol to the buffers, instead of ascorbate, resulted in irreversible inactivation of enzyme activity. On the other hand, the enzyme activity was stimulated, up to 10-fold, by addition to the assay of enzymatically active catalase (1–2 mg ml^{-1}, 65 000 U mg^{-1}) which is a

general characteristic of 2-oxoglutarate-dependant dioxygenases (Kondo *et al.*, 1981). The enzyme has a pI of 4.8 and exhibits activity over a broad range from pH 5 to a maximum at 8.5 followed by a steep decline of activity at pH 9.0. The enzyme utilises (2S)-naringenin and (2S)-eriodictyol but neither of the (2R)-enantiomers, 2-hydroxy-naringenin, 5,7,3′,4′,5′-pentahydroxyflavanone nor dihydroflavonols. The K_m and V_{max} values for the two former substrates were 5 and 8 μM and 0.23 and 0.065 nmol l^{-1} s^{-1}, respectively. The apparent K_m values for 2-oxoglutarate in the presence of 20 μM (2S)-naringenin was 16 μM. The enzyme activity was competitively inhibited by prunin (K_i 57 μM), as well as by analogues of 2-oxoglutarate (K_i 1.8 μM). The mol. wt. of the native FS I is 48 kDa, as determined by gel filtration, with a subunit mass of 24–25 kDa by SDS-PAGE.

It was postulated earlier that the conversion of flavanones to flavones by FS I (Britsch *et al.*, 1981) and FS II (Section III.B.2) proceeds by hydroxylation of the 2-position of flavanones and subsequent elimination of water, thus resulting in the C_2–C_3 double bond of flavones. However, it has recently been suggested that double bond formation may be catalysed by direct abstraction of vicinal hydrogen atoms (Britsch, 1990b). The fact that FS I does not utilise the hypothetical intermediate (2-hydroxyflavanone) as substrate, coupled with the lack of a dehydratase activity, support this mechanism. Consequently, FS I belongs to a novel subgroup referred to as 2-oxoglutarate-dependent desaturases, in which the 2-hydroxyflavanone inter-mediates are apparently not freely dissociable, and are not capable of entering the active site of the enzyme to compete with the flavanone (Britsch, 1990b).

2. Flavone synthase II

(a) Extraction. The tissue homogenate is prepared as described for FSI (Section III.B.1) and centrifuged at 8000 × g for 10 min. A membrane pellet is prepared by MgCl$_2$ (30 mM) precipitation and centrifugation at 90 000 × g for 30 min (Stotz and Forkmann, 1981) or by centrifugation of the supernatant at 160 000 × g for 80 min (Kochs and Grisebach, 1987). The pellet is suspended, at a concentration of 5–10 mg protein ml^{-1}, in 80 mM phosphate buffer, pH 7.5, containing 10% sucrose and 14 mM 2-mercaptoethanol.

(b) Assay. The assay mixture consists of 10 μmol phosphate buffer, 80–100 pmol (2S)-[^{14}C]naringenin (or eriodictyol), 100 nmol NADPH and 30–50 μg microsomal protein in a total volume of 100 μl. The radiolabelled substrates are prepared enzyma-tically as previously described (Section III.A). The reaction mixture without NADPH is equilibrated for 5 min at 10°C, and the reaction is started by addition of the cofactor and incubation for 15 min at 10°C with shaking in open vials. The low incubation temperature is preferred because of enzyme instability at 30°C and to prevent degradation of NADPH (Kochs and Grisebach, 1987). The enzyme reaction is terminated by addition of 50 μl ethyl acetate containing 10 μg each of naringenin and apigenin (or eriodictyol and luteolin). The flavonoid products are extracted twice with 100 μl each of ethyl acetate and chromatographed on cellulose plates in 30% HOAc or *t*-BuOH–HOAc–H$_2$O (3:1:1; v/v/v), visualised in UV light, and scanned for radioactivity as described in Section III.B.1.

(c) Properties. Flavone synthase II is a microsomal enzyme which has an absolute requirement for NADPH and O_2 and its activity is reduced by more than 80% if NADPH is replaced by NADH. The enzyme has the typical properties of a cytochrome P-450-dependant monooxygenase, as indicated by CO inhibition and its reversal by light, strong inhibition by typical cytochrome P-450 inhibitors, inhibition by KSCN, and correlation with the inhibition of NADPH-cytochrome reductase (Kochs and Grisebach, 1987). The optimum pH of the reaction is 7.5 with half-maximum activity at pH 7.0 and 8.4. It is partially inhibited by 0.5 mM diethylpyrocarbonate (50%), 5 mM KCN (30%), and 0.1 mM PCMB (25%). Flavone synthase II is inhibited in presence of 25 and 50 μM cytochrome c by 47 and 72%, respectively, and is competitively inhibited by $NADP^+$. It exhibits strict stereospecificity for (2R)-naringenin and (2S)-eriodictyol, but is not inhibited by an equimolar mixture of the (2S)-enantiomer. The apparent K_m values for naringenin and NADPH are 10 and 40 μM, respectively (Kocks and Grisebach, 1987).

C. Flavonol Synthase

This enzyme catalyses the conversion of dihydroflavonols to flavonols (Scheme 4.1). Its activity was first detected in cell-free extracts of parsley cell cultures (Britsch *et al.*, 1981) and later in young flower buds of *Matthiola incana* (Spribille and Forkmann, 1984) and *Petunia hybrida* (Forkmann *et al.*, 1986).

1. Extraction and assay

The preparation of protein extracts and enzyme assays are similar to those described for FS I (Section III.B.1). The radiolabelled substrates, dihydrokaempferol and dihydroquercetin, are prepared enzymatically from (2S)-[^{14}C]naringenin using a flavanone 3-hydroxylase (Section III.A) and from (2R,3R)-[^{14}C]dihydrokaempferol using a flavonoid 3'-hydroxylase (Section IV.A), respectively.

2. Properties

Flavonol synthase is a cytosolic, 2-oxoglutarate-dependent dioxygenase. The enzyme exhibits the same properties as those of FS I including inhibition by EDTA, KCN, diethyldithiocarbamate and PCMB (Spribille and Forkmann, 1984). The mechanism of 2,3-double bond formation of flavonols is postulated in a manner similar to that of flavones (Section III.B.1).

D. Isoflavone Synthase

This enzyme catalyses the conversion of (2S)-naringenin to the isoflavone genistein (Scheme 4.1). Isoflavone synthase (IFS) has been characterised from microsomal preparations of elicitor-treated soybean cell cultures (Hagmann and Grisebach, 1984; Kochs and Grisebach, 1986) and the roots of soybean seedlings infected with *Phytophtora megasperma* (Bonhoff *et al.*, 1986).

1. Extraction and assay

The tissue homogenate is prepared as described for FS II (Section III.B.2) but using 0.1 M Tris-HCl buffer, pH 7.5, containing 14 mM 2-mercaptoethanol and 20% sucrose. After centrifugation for 10 min at $8000 \times g$, the supernatant is further centrifuged for 80 min at $160\,000 \times g$, and the pellet washed twice with 80 mM phosphate buffer, pH 8.5, containing 14 mM 2-mercaptoethanol and 20% sucrose, then homogenised. The enzyme assay is the same as described for the FS II (Section III.B.2), except that 0.5 mM glutathione is used instead of 2-mercaptoethanol. Because of enzyme instability, incubation is carried out at 10°C for 15 min. Genistein (10 μg in MeOH) is added as carrier during the extraction of reaction products. The latter are chromatographed on silica gel (F-254) plates in $CHCl_3$–acetone–NH_4OH (70:60:1; v/v/v), and genistein is detected in UV light or by spraying with a 0.2% solution of fast Blue B salt and exposure to ammonia vapours. Labelled products are scanned or autoradiographed, and counted for radioactivity.

2. Properties

Isoflavone synthase is an NADPH-dependent monooxygenase that requires the participation of cytochrome P-450. It is relatively unstable even in the presence of 14 mM 2-mercaptoethanol and 20% sucrose. The pH optimum for isoflavone formation is 8–8.6 with half-maximal activity at pH 7.0 and 9.0. The enzyme utilises (2S)-naringenin (K_m 8.7 μM) and (2S)-5,4'-dihydroxyflavanone as substrates but not their (2R)-enantiomers. At a saturating concentration of (2S)-naringenin, enzyme activity is inhibited 30% by the addition of an equimolar amount of the (2R)-enantiomer. Except for strong inhibition by 2-mercaptoethanol and dithiothreitol, other properties of IFS are similar to those of FS II (Section III.B.2), including those characteristic of cytochrome P-450-dependent monooxygenases (Kochs and Grisebach, 1986).

E. Dihydroflavone/dihydroflavonol 4-Reductase

Dihydroflavone/dihydroflavonol 4-reductase (DHFR) catalyses the stereospecific reduction of (+)-dihydroflavonols and (2S)-dihydroflavones (flavanones) to *cis*-flavan 3,4-diols and flavan 4-ols, respectively (Scheme 4.1). The latter compounds are intermediates in the biosynthesis of anthocyanidins, proanthocyanidins and (+)catechins (Stafford, 1990). Dihydroflavone/dihydroflavonol 4-reductase activity has been demonstrated in cell cultures of Douglas fir (Stafford and Lester, 1982), tissue cultures of *Ginkgo biloba* (Stafford and Lester, 1985), *Matthiola incana* flowers (Heller *et al.*, 1985), barley grains (Kristiansen, 1986) and *Sinningia cardinalis* flowers (Stich and Forkmann, 1988). The enzyme has been partially purified from *Cryptomeria* (a gymnosperm) cell cultures (Ishikura *et al.*, 1988) and, to apparent homogeneity, from *Dahlia variabilis* flowers (Fischer *et al.*, 1988).

1. Assay

The labelled substrates dihydrokaempferol, dihydroquercetin and dihydromyricetin are synthesised sequentially from (2S)-[4a,6,8-^{14}C]naringenin using a partially purified flavanone 3-hydroxylase (Section III.A.1), flavonoid 3'-, and 3',5'-hydroxylase preparations (Section IV.A), respectively. The enzyme assay (50 μl) consists of 50 nmol of the labelled flavonoid substrate, 150 nmol NADPH (in 10 μl water) and 15–50 μg protein in 0.1 M imidazole buffer, pH 6.0, containing 2.8 mM 2-mercaptoethanol and 10% glycerol (Heller *et al.*, 1985). After incubation for 20 min at 25°C, the mixture is extracted with ethyl acetate and the leucoanthocyanidins chromatographed on cellulose TLC in $CHCl_3$–MeOH–H_2O (10:9:1; v/v/v) for determination of their radioactivity. Leucopelargonidin, leucocyanidin and leucodelphinidin migrate at R_f values of 0.3, 0.14 and 0.04, as compared with their respective dihydroflavonols which migrate at R_f values of 0.71, 0.44 and 0.19. Other solvent systems for the separation of different substrates and products have also been reported (Heller *et al.*, 1985).

2. Purification

Due to instability of the enzyme, all buffer solutions are degassed, equilibrated with N_2 gas and further degassed. All buffers contained 10% glycerol as well as 20–40 mM sodium ascorbate, and 14 mM 2-mercaptoethanol or 1 mM dithiothreitol (Fischer *et al.*, 1988). The following protocol was used for purification of the enzyme from Scarlet star *Dahlia* flowers. Frozen petals are homogenised with 0.2 M Tris-HCl buffer, pH 7.5, containing 10% Dowex (w/v) and then centrifuged. The protein is precipitated with solid ammonium sulphate (40–80% saturation) and the pellet dissolved under N_2 in a minimal amount of 50 mM imidazole-HCl buffer, pH 6.8, and desalted on Sephadex G-25 using the same buffer. The enzyme is purified by chromatography on a Q-Sepharose CL-6B column and elution with a linear gradient of 0 to 0.5 M NaCl in 0.1 M imidazole-HCl buffer. Further purification of the enzyme protein is achieved by affinity chromatography on Blue-Sepharose CL-6B, in 50 mM phosphate buffer, pH 6.6, and elution with 5 mM $NADP^+$ in the same buffer. Despite the low binding capacity of the latter column (*c.* 3 mg protein ml^{-1} of Blue-Sepharose), this protocol resulted in a 150-fold purification with a 5% recovery. Final purification is achieved on a Mono Q column and elution with a linear gradient (0 to 0.4 M) of NaCl in phosphate buffer. Although the latter step gave an apparently homogenous protein, it reduced the specific activity and recovery of the enzyme to 65-fold and 1.2%, respectively (Fischer *et al.*, 1988).

3. Properties

Dihydroflavone/dihydroflavonol 4-reductase consists of one polypeptide with a mol. wt. of about 44 kDa as estimated by its elution volume from a Superose 12 column and SDS-PAGE. The enzyme reaction has a pH optimum of 6.8 for dihydroflavonols and 6.0 for dihydroflavones. DHFR converts (+)dihydrokaempferol and (+)dihydroquercetin equally well to their corresponding leucoanthocyanidins with K_m values of 10 and 15 μM, respectively and, at a lower rate, dihydromyricetin to leucodelphinidin. It also catalyses the reduction of (2S)-naringenin (K_m 2.3 μM) and (2S)-eriodictyol (K_m 2.0 μM) to the 3-deoxyproanthocyanidins (flavan 4-ols), apiferol and luteoferol, respectively, although with lower V/K_m values than those for the dihydroflavonols. Compared with dihydroquercetin and naringenin, the Lineweaver–Burk plots for dihydrokaempferol and eriodictyol are curvilinear at concentrations above 60 and 10 μM, respectively, suggesting inhibition by these substrates. Dihydroflavone/dihydroflavonol 4-reductase requires NADPH for activity with a K_m of 42 μM, whereas the rate of reaction with NADH is about 20%. In contrast, the *Sinningia* enzyme utilises both NADPH and NADH equally well (Stich and Forkmann, 1988). Dihydroflavone/dihydroflavonol 4-reductase is completely inhibited by 0.1 μM PCMB, whereas EDTA (5 mM) or 1,10-phenanthrolin (0.1 mM) were not inhibitory. Hydride ion transfer has been shown to involve the pro-*S*(B-site) of NADPH to the carbonyl group which binds to the enzyme with its *syn*-conformation. Therefore, the enzyme belongs to the pro-*S*(B-specific) dehydrogenases (Fischer *et al.*, 1988).

The substrate specificity of DHFR varies in different sources and, therefore, determines the composition of anthocyanidins in the respective plant species. The enzyme from *Petunia hybrida* reduces both dihydroquercetin and dihydromyricetin, but not dihydrokaempferol, whereas that from *Zea mays* utilises predominantly dihydrokaempferol and dihydroquercetin. Transfer of the DHFR gene from maize to *Petunia* resulted in the formation of leucopelargonidin and the ultimate production of a new brick-red flower pigment, pelargonidin, in this species (Meyer *et al.*, 1987).

F. Isoflavone Oxidoreductase

Isoflavone oxidoreductase (IFOR) catalyses the stereospecific reduction of 5-deoxy-2′-hydroxyisoflavones to their (3*R*)-isoflavanone derivatives (Scheme 4.2). The latter are the substrates for the pterocarpan synthase (Section III.G). Isoflavone oxidoreductase was first detected in chickpea cell cultures (Tiemann *et al.*, 1987), CuCl$_2$-treated pea seedlings (Preisig *et al.*, 1990) and, more recently, purified to apparent homogeneity from elicitor-challenged soybean cell cultures (Fischer *et al.*, 1990a).

1. Synthesis of substrates

The enzyme substrate 7,2′,4′-trihydroxyisoflavone (2′-hydroxydaidzein) is prepared according to Dewick (1977). Briefly, 2,4-dibenzoyloxybenzaldehyde is heated with 2-hydroxy-4-benzoyloxyacetophenone in presence of NaOH. The 2,4,4′-tribenzoyloxy-2′-hydroxychalcone formed is converted to the tribenzoyloxyisoflavone which, in turn, is debenzoylated to 7,2′,4′-trihydroxyisoflavone. The latter is extracted with ethyl acetate and the organic layer is concentrated *in vacuo*. The precipitated 2′-hydroxydaidzein is crystallised from aqueous MeOH. The acetylated product can be hydrogenated catalytically with Pd-charcoal (Woodward, 1980) and deacetylated to 2′-hydroxydihydrodaidzein, which is used as a reference compound to the enzyme reaction product, as well as substrate for the pterocarpan synthase (Section III.G).

2. Assay

The oxidoreductase assay mixture consists of 50 μM 2′-hydroxydaidzein, 1 mM NADPH and the enzyme protein (1–80 μg) in a total volume of 100 μl. The mixture is incubated for 10 min at 30°C, extracted with 200 μl of ethyl acetate and analysed by TLC or HPLC. Thin layer chromatography is performed on Nano silica HPTLC plates using toluene–CHCl$_3$–acetone (1:1:2; v/v/v), petroleum ether–ethyl acetate–

MeOH (10:10:1; v/v/v), toluene–acetone (1:1; v/v) or chloroform–MeOH (20:1; v/v), and the isoflavonoids are detected by spraying with a 0.2% solution of Fast Blue B salt and exposure to ammonia vapours. High performance liquid chromatography is carried out on Lichrosorb RP$_{18}$, 5 μm, 60 × 4 mm using MeOH–H$_2$O (40:60 and 52:48, v/v). The R_t values, in mins, are 3.2, 2.7, 11.1 and 8.5 for 2′-hydroxy-daidzein, 2′-hydroxydihydrodaidzein, 2′-hydroxyformononetin, and 2′-hydroxydihydroformononetin, respectively (Fischer *et al.*, 1990a).

3. Purification

Isoflavone oxidoreductase, which is present in very small amounts in soybean cell cultures, can be induced by treatment for 24 h with a glucan elicitor from *Phytophtora megasperma* f. sp. *glycinea* (Fischer *et al.*, 1990a). The filtered cells (2.5 kg) are homogenized with 250 g Dowex 1x2 and 1.7 l of 0.2 M Tris-HCl buffer, pH 7.5. All buffers contain 10% glycerol and 14 mM 2-mercaptoethanol. The slurry is filtered through nylon net and the filtrate centrifuged for 20 min at 10 800 × g. A 10% solution of polyethyleneimine, adjusted to pH 7.5, is added to a final concentration of 0.02% and the precipitate removed by centrifugation. The protein is then precipitated with 60% ammonium sulphate and the pellet dissolved in 150 ml of the same buffer (but at 0.1 M). After desalting on Bio-gel P column, the protein is transferred to a Q-Sepharose CL-6B column, washed with the above buffer and eluted with a gradient (0 to 0.6 M) of NaCl in the same buffer. The active fractions are then chromatographed on a Blue Sepharose CL-6B affinity column, washed with 0.1 M NaCl in 0.1 M Tris-HCl buffer, and the protein eluted with 40 ml of Tris buffer containing 3.5 mM NADP$^+$. The enzyme fractions are finally chromatographed on a Sephacryl S-200 HR column using 0.1 M Tris-HCl buffer, then concentrated by ultrafiltration. This protocol results in 28-fold purification with a 1.5% recovery and a specific activity of 1110 pkat mg^{-1} for the apparently homogenous protein (Fischer *et al.*, 1990a).

4. Properties

Isoflavone oxidoreductase has a pH optimum of 7.0 in phosphate buffer and a pI of 5.0. Its mol. wt. is 34 kDa as determined by elution from Sephacryl S-200 and SDS-PAGE, and consists of one polypeptide. The oxidoreductase reaction, which requires NADPH, is stereospecific and results in a 3R configuration at C-3 of the dihydroiso-flavone. It converts 2′-hydroxydaidzein and 2′-hydroxyformononetin to their dehydro derivatives; however, it does not accept biochanin A, daidzein or genistein, as sub-strates. The apparent K_m values for the 2′-hydroxyisoflavones and NADPH are about 50–60 μM (Fischer *et al.*, 1990a).

G. Pterocarpan Synthase

Pterocarpan synthase (PTS) catalyses the NADPH-dependent conversion of 2′-hydroxyisoflavanones to their corresponding pterocarpans (Scheme 4.2). The enzyme was first detected in chickpea cell cultures which, when treated with yeast extract, accumulated the pterocarpans, medicarpin and maackiain (Bless and Barz, 1988). It has since been partially purified from elicitor-treated soybean cell cultures which accumulate glyceollins as pterocarpan phytoalexins (Fischer *et al.*, 1990b).

1. Assay

The isoflavonoid substrate, 2'-hydroxydihydrodaidzein is synthesised as described previously (Section III.F). It can also be synthesised enzymatically as the (3R)-enantiomer (Fischer *et al.*, 1990a, Section III.F). The assay mixture consists of 75 μM (RS)2'-hydroxyisoflavanone, 1 mM NADPH and 20–40 μg protein in 0.2 M phosphate buffer, pH 6.0, and is incubated for 10 min at 30°C. The reaction product is extracted in 200 μl ethyl acetate, evaporated to dryness, and dissolved in 30 μl MeOH. Product identification is carried out by TLC on Nano silica HPTLC plates in toluene–CHCl$_3$–acetone (1:1:2; v/v/v), CHCl$_3$–MeOH (20:1; v/v) or CHCl$_3$–acetone–NH$_4$OH (70:60:1; v/v/v), and detection with 0.2% Fast Blue B salt solution and ammonia vapour. The pterocarpan product can also be identified by chromatography on HPLC (Lichrosorp RP$_{18}$, 5 μM, 60 × 4 mm) with MeOH–H$_2$O (2:3 and 3:2; v/v) with R_t of 2.7 and 4.6 min for 2'-hydroxydihydrodaidzein and 3,9-dihydroxypterocarpan, respectively (Fischer *et al.*, 1990b).

2. Purification

Pterocarpan synthase from soybean cultures is inducible after challenge with glucan elicitor from *Phytophtora megasperma* f. sp. *glycinea* or yeast extract (Ebel, 1986). Preparation of crude extracts, ammonium sulphate precipitation, and anion-exchange chromatography on Q-Sepharose are similar to those described for the 2'-hydroxyiso-flavone oxidoreductase (Section III.F). Pterocarpan synthase is separated from the oxidoreductase by chromatography on Blue Sepharose, which retains the latter enzyme, while PTS is recovered in the wash buffer. Pterocarpan synthase activity is further purified by chromatography on Sephacryl S-200 HR. The enzyme preparation does not bind to other dye ligands and its activity is lost on chromatofocusing. This protocol results in a 7-fold purification with a specific activity of 1000 pkat mg^{-1} protein and a 5.7% recovery (Fischer *et al.*, 1990b). The chickpea enzyme has been enriched 3.5-fold, with a 10.5% recovery, and a specific activity of 166 pkat mg^{-1} protein (Bless and Barz, 1988).

3. Properties

Pterocarpan synthase has a mol. wt. of 29 000 as determined by elution from Superose 12, and a pH optimum of 6.0 with half-maximum activity at pH 4.8 and 7.25. The enzyme which is stereospecific for the (3R)-enantiomer of isoflavanones, has only 50% yield with (RS)-2'-hydroxydihydrodaidzein (prepared by catalytic reduction of the corresponding isoflavone). In contrast with the chickpea PTS, which converts the 4'-methoxy compound (vestitone), the soybean enzyme accepts isoflavanones with 4'-hydroxy- and 4'-methoxy substitutions for conversion to their corresponding pterocarpans. The K_m values for NADPH and the 2'-hydroxydihydrodaidzein are 45 and 75 μM, respectively. The enzyme reaction is competitively inhibited by NADP$^+$

but not affected by divalent cations (Mg^{2+}, Ca^{2+}, or Zn^{2+}), EDTA (up to 40 mM) or o-phenanthrolin (up to 100 μM) (Fischer et al., 1990b).

IV. ENZYMES INVOLVED IN FLAVONOID SUBSTITUTION

The diversity of naturally occurring flavonoid compounds is the result of various substitutions within each flavonoid class. These substitutions which include hydroxylation, glycosylation, methylation, acylation, prenylation and sulphation, are catalysed by substrate-specific, position-oriented enzymes. Partially or highly purified enzymes, with specificity towards one or another flavonoid class, are described in the following sections.

A. Hydroxylases

Substitution of flavonoids may occur at the C_{15} stage by specific hydroxylases. A number of flavonoid-specific hydroxylases have been shown to hydroxylate positions 3' of flavanones, flavones, dihydroflavonols and flavonols (Fritsch and Grisebach, 1975; Forkmann et al., 1980; Stotz and Forkmann, 1981; Spribille and Forkmann, 1982b; Stotz et al., 1984, 1985; Larson and Bussard, 1986), 3' and 5' of flavanones and dihydroflavonols (Stotz and Forkmann, 1982; Stotz et al., 1984), 2' and 3' of isoflavones (Hinderer et al., 1987), and 6a of pterocarpans (Scheme 4.2) (Hagmann et al., 1984).

1. Extraction

Several methods of protein extraction have been described for the preparation of flavonoid hydroxylases. For the isolation of flavonoid 3',5'-hydroxylase, the tissue (5 g) is homogenised in presence of 2.5 g Dowex 1x2, 1.5 g quartz sand, with 15 ml of 0.1 M phosphate, pH 7.5, containing 28 mM 2-mercaptoethanol (Stotz and Forkmann, 1982). For the extraction of 3'-hydroxylase, 7.5 g of maize seedlings are homogenised with 40 ml of 0.05 M Bicine buffer, pH 8.5, containing 0.8 M sucrose and 5.0 mM dithiothreitol in the presence of 1.8 g of Polyclar AT (Larson and Bussard, 1986). For the isoflavone 2'- and 3'-hydroxylases, chickpea cell suspension cultures were challenged with yeast extract for 16 h prior to preparation of the microsomal preparation. The crude extract is then mixed with 1 M $MgCl_2$ (to a final concentration of 30 mM), and kept on ice for 10 min before centrifugation at 17 000 $\times$ g for 20 min to recover the microsomal pellet. Alternatively, the microsomal pellet can be prepared by ultracentrifugation of the crude extract at 90 000 $\times$ g for 75 min.

2. Assay

Labelled naringenin and eriodictyol can be prepared as described in Section III.A, and labeled dihydrokaempferol, as described in Section III.E.

For the 3',5'-hydroxylase, the assay mixture (100 μl) consists of 20 μmol phosphate buffer, pH 7.5, containing 0.4–2.5 μmol 2-mercaptoethanol, 0.1–0.3 nmol of radioactive substrate, 0.1 μmol NADPH, and 20–60 μg protein (Stotz and Forkmann, 1982). The mixture is incubated at 30°C for 10–30 min and incubation terminated by the addition of 20 μl methanol. The reaction products are extracted twice with ethyl acetate and chromatographed on cellulose TLC plates with $CHCl_3$–HOAc–H_2O (10:9:1; v/v/v). Radioactive spots are localised by scanning the plates and counted for radioactivity.

For the 3′-hydroxylase, the assay mixture (200 μl) consists of 50 mM Bicine buffer, pH 8.5, containing 2.0 mM dithiothreitol, 0.35 mM kaempferol, 0.37 mM NADPH, and about 200 μg of protein (Larson and Bussard, 1986). The reaction mixture is incubated at 30°C for 30 min and terminated by the addition of 800 μl of a mixture (2:1; v/v) of $CHCl_3$–MeOH (containing 1% HCl). After centrifugation, the reaction products are recovered in the methanolic upper phase. Separation and quantitative determination of the reaction products is carried out by HPLC as described by Larson and Bussard (1986).

For the isoflavone 2′-/3′-hydroxylase, the reaction mixture consists of 100 mM phosphate buffer, pH 7.5, containing 400 mM sucrose, 1 mM NADPH, 50 μM of the isoflavone substrate, and 0.4–0.6 mg of enzyme protein in a final volume of 2 ml (Hinderer et al., 1987). The reaction is incubated at 25°C for 30 min and stopped by extraction with ethyl acetate. The organic layer is then evaporated to dryness and redissolved in 100 μl of methanol. Separation of the substrate and products is performed by HPLC as described by Köster et al. (1983).

For the pterocarpan 6a-hydroxylase, the assay mixture consists of 4.8 μmol phosphate buffer, pH 7.5, containing 14 mM 2-mercapto-ethanol and 20% sucrose, 400 pmol (±)3,9-dihydroxy[6,11a-^{3}H]pterocarpan (prepared by reduction of 7,2′,4′-trihydroxyisoflavone with sodium boro[^{3}H]hydride according to Dewick and Steele, 1982), 5–10 μg microsomal protein and 100 nmol NADPH in a total of 100 μl. After shaking in open vials for 10 min at 30°C, the reaction is terminated by addition of 5 mM EDTA, and the products extracted with ethyl acetate (Hagmann et al., 1984).

3. Properties

Hydroxylases are microsomal enzymes with an absolute requirement for NADPH and molecular oxygen. Their microsomal localisation, cyanide resistance, inhibition by cytochrome c, and by CO which is reversible in presence of light, suggest that they are cytochrome P-450-dependent monooxygenases. Different flavonoid hydroxylases exhibit pH optima in the range of 7.0 to 8.5. In parsley and maize, the flavonoid 3′-hydroxylase catalyses the hydroxylation of naringenin to eriodictyol, dihydrokaempferol to dihydroquercetin, kaempferol to quercetin and apigenin to luteolin (Hagmann et al., 1983; Larson and Bussard, 1986). K_m values of the maize enzyme for kaempferol and NADPH are 7.15 and 5.8 μM, respectively (Larson and Bussard, 1986). The pterocarpan 6a-hydroxylase activity is increased by 32% when both FMN and FAD (5 μM) are added to NADPH (Hagmann et al., 1984). In Verbena hybrida, the flavonoid ring B hydroxylase catalyses the hydroxylation of naringenin and dihydrokaempferol to 5,7,3′,4′,5′-pentahydroxyflavanone and dihydromyricetin, respectively (Stotz and Forkmann, 1982). The enzyme also accepts the 3,4′-hydroxylated flavonoids, eriodictyol and dihydroquercetin to give rise to their 5′-hydroxy derivatives. The chickpea hydroxylases catalyse the hydroxylation of 4′-

methoxyisoflavones (biochanin A and formononetin) to their 2'- and 3'-hydroxy derivatives, whereas 4'-hydroxyisoflavones such as genistein and daidzein were not accepted as substrates.

B. Glycosyltransferase

A multitude of O-glycosyltransferases (GT; EC 2.4.1.-) with different specificities toward the sugar donor and flavonoid acceptor have been characterised over the past 20 years (Hösel, 1981; Ibrahim *et al.*, 1986; Heller and Forkmann, 1988; Stafford, 1990). Most of these enzymes catalyse single glycosylation steps of various flavonoid classes, including the less common 2'- and 5'-hydroxyls of partially O-methylated flavonols (Latchinian *et al.*, 1987). Other GTs mediate the stepwise transfer of sugars in the biosynthesis of flavonol 3-O-triglucosides (Jourdan and Mansell, 1982), 3-O-triglycosides (Kleinehollenhorst *et al.*, 1982), anthocyanidin 3-(*p*-coumaroyl) rutinoside-5-O-glucoside (Jonsson *et al.*, 1984) and flavone 7-di-O-glucuronosyl-4'-glucuronide (Schultz and Weissenböck, 1988), to mention a few. Recently, 6-/8-C-glucosylation has been shown to be catalysed by a 2-hydroxyflavanone-specific enzyme from *Fagopyrum esculentum*, where the resulting glucosides are believed to be intermediates in C-glucosylflavone biosynthesis (Kerscher and Franz, 1988).

1. Assay

The following standard assay applies to any flavonoid acceptor and its preferred nucleotide diphosphate (NDP) sugar donor. The assay mixture consists of 1–10 μM of the flavonoid substrate (dissolved in DMSO or 2-methoxyethanol), 2–20 μM of the labelled NDP-sugar, and 1–20 μg enzyme protein in the appropriate buffer, pH 7.5–8.0, containing 14 mM 2-mercaptoethanol in a total volume of 100 μl. The reaction mixture is incubated at 30°C for 10–30 min and terminated by the addition of 10 μl of 6 N HCl. The glycosylated products are extracted with 200 μl ethyl acetate and an aliquot (*c.* 50 μl) is counted to determine the glycosylating activity. The remaining organic layer is co-chromatographed with reference glycosides (if available) on cellulose TLC using one or more of the solvent systems recommended for different classes of flavonoid glycosides (Harborne, 1989).

A non-radioactive assay may be used after increasing the concentrations of substrate and co-substrate 5- to 10-fold. The chromatographed products are then visualised in UV light and the compounds of interest are eluted for identification and quantity determination using standard techniques.

2. Purification

Depending on the degree of abundance of the GT to be studied, as well as the presence of other contaminating GT activities in a given source, enzyme purification to apparent homogeneity may be achieved with as few as two chromatographic steps after ammonium sulphate precipitation, as in the case of buckwheat C-GT (Kerscher and Franz, 1988), or after several steps as with the flavonol 3-GT from *Hippeastrum* petals (Hrazdina, 1988), the flavanone 7-GT from *Citrus paradisi* (McIntosh *et al.*, 1990), and the partially methylated flavonol 2'-/5'-GT from *Chrysosplenium americanum*

(Latchinian-Sadek and Ibrahim, 1991). The latter enzyme has been purified using a combination of conventional and FPLC columns including Sephacryl S-200, UDP-glucuronic acid agarose, Mono P, Superose 12 and Mono Q columns. This protocol co-purified the 2′- and 5′-GTs to apparent homogeneity and resulted in >3500-fold increase in specific activity but with 2% yield. A similar protocol resulted in a 940- to 1470-fold purification of the flavanone 7-GT (McIntosh *et al.*, 1990). On the other hand, purification of the more abundant enzyme from *Hippeastrum* petals was achieved by gel filtration on AcA44, ion-exchange chromatography on DEAE-Bio-gel and chromatofocusing on PBE 94 with 750-fold purification and 13.4% recovery (Hrazdina, 1988).

3. Properties

Most GTs exhibit pH optima in the range 7.5 to 8.5, although exceptions are known with optima as high as 9.0 to 9.5 (Köster and Barz, 1981; Jonsson *et al.*, 1984; Kerscher and Franz, 1988), and as low as 5.8 to 6.2 (Sun and Hrazdina, 1991). Their molecular-weight vary between 40 and 59 kDa and their pI are in the range of 4.2 to 5.4. Except for the red cabbage 3-GT, which preferentially accepts flavonols (Sun and Hrazdina, 1991), other flavonol GTs have been reported to accept anthocyanidins as well (e.g. Jonsson *et al.*, 1984; Teusch *et al.*, 1986; Hrazdina, 1988). Other reported GTs are substrate- and position-specific enzymes (Hösel, 1981; Heller and Forkmann, 1988), especially those which catalyse stepwise glucosylation (Jourdan and Mansell, 1982), glycosylation (Kleinehollenhorst *et al.*, 1982), and glucuronosylation (Schultz and Weissenböck, 1988), as well as the 2′- and 5′-GTs which only accept partially *O*-methylated flavonols (Latchinian *et al.*, 1987). While a few GTs are stimulated by divalent cations (Mg^{2+}, Ca^{2+}, Mn^{2+}), others are inhibited by similar cations (Zn^{2+}, Cu^{2+}, Co^{2+}), as well as by EDTA and SH-group reagents (Hösel, 1981; Ibrahim *et al.*, 1986; Heller and Forkmann, 1988). The K_m values for the flavonoid substrates have been reported to be in the micromolar range, whereas those for the sugar donor range from 0.25 to several mM.

The kinetic mechanism of the flavonoid-ring B GT has been investigated in detail (Khouri and Ibrahim, 1984). Substrate interaction kinetics of the flavonol and UDPG gave converging lines consistent with a sequential binding mechanism. The enzyme reaction is inhibited by high concentration of the flavonoid substrate with a K_b of 10 μM. Product inhibition studies show competitive inhibition between UDPG and UDP (K_{iq} 20 μM) and non-competitive inhibition between the flavonol substrate and its glucoside (K_{ip} 1 mM). The kinetic patterns are consistent with an ordered bi-bi mechanism, where UDPG is the first substrate to bind to the enzyme and UDP is the final product released. The high K_{ip} value, when compared with that of the K_b, indicates that the reaction is not inhibited by the glucosylated product, and is consistent with their accumulation in this tissue (Khouri and Ibrahim, 1984).

C. Malonyltransferase

Flavonoid compounds are commonly acylated with malonyl residues, although acetyl or succinyl residues may also occur (Harborne, 1988). Malonyltransferases (MTs) catalyse the malonylation of the sugar moieties of flavone and flavonol glycosides

(Matern *et al.*, 1981), isoflavone glucosides (Köster *et al.*, 1984) and anthocyanins (Teusch and Forkmann, 1987) with the formation of their 6″-malonyl esters. Malonylation is considered the last reaction in the enzymatic synthesis of flavonoid conjugates (Heller and Forkmann, 1988).

1. Assay

If not available commercially, flavonoid glycosides may be isolated and purified from plant sources using standard phytochemical methods or may be synthesised enzymatically (Section IV.B). The standard assay mixture consists of 2–60 nmol flavonoid glycoside (dissolved in DMSO or 2-methoxyethanol), 20 nmol [1,3-^{14}C] malonyl-CoA dissolved in 10 μl of 0.1 N H_2SO_4 and 20–80 μg protein in 0.1 M Tris-HCl buffer, pH 8.0, containing 2 mg ml^{-1} BSA in a total volume of 150 μl (Matern *et al.*, 1981; Köster *et al.*, 1984). The enzyme reaction is incubated for 10 min at 30°C, and terminated by the addition of 100 μl MeOH, and the precipitated protein removed by centrifugation. The supernatant is concentrated and chromatographed by TLC or HPLC using standard techniques for the quantification of malonylated products (Harborne, 1989).

2. Purification

A few MTs have been partially purified from illuminated parsley cell cultures (Matern *et al.*, 1981) and chickpea roots (Köster *et al.*, 1984). The tissue is extracted with 0.2 M Tris-HCl buffer, pH 7.5 containing 10% (v/v) glycerol and filtered. The filtrate is treated sequentially with Dowex 1×2, 10% polyethyleneimine (0.3% final concentration) and solid ammonium sulphate (40–80% salt saturation). The desalted protein pellet is chromatographed on DEAE cellulose and the enzyme eluted with a linear gradient (40 to 200 mM) of phosphate buffer, pH 8.0. The protein is further chromatographed on a hydroxyapatite column and eluted with a linear gradient (5 to 80 mM) phosphate buffer, pH 6.8. This step separates the flavonoid 7- and 3-MTs with 350- and 1310-fold purification and specific activities of 5.9 and 8.5 pkat kg^{-1} protein, respectively (Matern *et al.*, 1981). Both MTs are further purified by successive chromatography on CM-Sephadex, octyl Sepharose CL-4B and Sephacryl S-200. Alternatively, the chickpea enzyme was partially purified, after the DEAE-Sephacel step, by chromatography twice on octyl-Sepharose, concanavalin A-Sepharose and finally by gel filtration on Ultrogel AcA34. The latter protocol resulted in 157-fold purification of the isoflavone 7-*O*-glucoside MT with a specific activity of 125 nkat mg^{-1} protein and a 13% recovery (Köster *et al.*, 1984).

3. Properties

Malonyltransferases have a pH optimum of 8.0, a pI of 5.3, and a mol. wt. of 55 kDa as determined by gel filtration and SDS-PAGE, although a mol. wt. of 110 kDa has been reported for the chickpea enzyme. It is not certain whether this difference is due to a dimeric form of the enzyme (Köster *et al.*, 1984) or to the presence of carbohydrate residues (Matern *et al.*, 1981) . Partially purified MTs are fairly stable for several months when stored at −70°C, whereas highly purified preparations have a half-life of only 2–3 days. They exhibit pronounced specificity for the flavonoid glycosides present in their parent tissues. The 7-MT from parsley accepts apiin (apigenin 7-apioglucoside), as well as apigenin- and diosmetin 7-O-glucosides for the 7-MT, despite the fact that the two latter compounds are not found in parsley cells. The 3-MT accepts the flavonol 3-glucosides found in parsley, but with little or no activity towards other phenolic gluco-/glycosides (Matern *et al.*, 1981). Likewise, the chickpea enzyme accepts the 7-O-glucosides of biochanin A and formononetin as preferred substrates. However, it will also accept genistein and daidzein, as well as the 3'-OH derivatives of biochanin A (pratensein) and of genistein (orobol) if they are glucosylated at position 7, but none of the flavone, flavanone or chalcone glucosides (Köster *et al.*, 1984). The anthocyanin MT accepts pelargonidin-, cyanidin- and delphinidin 3-O-glucosides equally well for malonylation, although this is not the case with cyanidin 3,5-diglucoside which is acylated at 10% compared to the 3-O-glucosides (Teusch and Forkmann, 1987). The K_m values for the flavonoid substrates and malonyl CoA are in the micromolar range and vary between 4–36 μM and 20–50 μM, respectively. The malonylation reaction of anthocyanins is variably inhibited by 20–50% in the presence of 1 mM of several divalent cations, and is strongly inhibited (by 64%) in presence of 0.2 mM PCMB.

D. *O*-Methyltransferase

O-Methyltransferase (OMT; EC 2.1.1.6.-) catalyses the transfer of the methyl group of an activated methyl donor, *S*-adenosyl-L-methionine (SAM), to the hydroxyl groups of various flavonoid acceptors with the formation of the corresponding methyl ether and *S*-adenosyl-L-homocysteine (SAH) as products. As with GTs, OMTs are substrate- and position-specific enzymes (Poulton, 1981; Ibrahim *et al.*, 1987; Heller and Forkmann, 1988). A number of flavonoid-specific OMTs have been shown to methylate positions 3' of flavones and flavonols (Poulton *et al.*, 1977; Sütfeld and Wiermann, 1978), 3' and 5' of anthocyanins (Jonsson *et al.*, 1982), 4' of isoflavones (Wengenmayer *et al.*, 1974) and flavones (Kuroki and Poulton, 1981), 5 of isoflavones (Khouri *et al.*, 1988a), 7 of flavonols (Tsang and Ibrahim, 1979; Khouri *et al.*, 1988b), *C*-glycoflavones (Knogge and Weissenböck, 1984) and isoflavones (Edwards and Dixon, 1991), 8 of flavonols (Jay *et al.*, 1985), and the 3-hydroxyl group of pterocarpans (Preisig *et al.*, 1989). Whereas these enzymes catalyse single methylation steps, *Chrysosplenium americanum* (Saxifragaceae) has been shown to contain five flavonol-specific, position-oriented OMTs which catalyse the sequential methylation of quercetin to its 3-mono-, 3,7-di-, 3,7,4'-tri-, 3,6,7,4'-/3,7,4',5'-tetra- and 3,6,7,2',4'-pentamethyl ether derivatives (De Luca and Ibrahim, 1985a; Khouri *et al.*, 1988a; Ibrahim *et al.*, 1987; Khouri *et al.*, 1988b). However, no enzyme has yet been reported to catalyse the *C*-methylation of flavonoids.

1. Assay

The assay mixture consists of 10–20 μM of the flavonoid substrate dissolved in 50% DMSO, 25 μM [^{14}CH$_3$]SAM, and 10–50 μg enzyme protein in 0.1 M phosphate buffer, pH 7.5, containing 1.4 mM 2-mercaptoethanol in a total volume of 100 μl. The mixture is incubated for 15–30 min at 30°C, and the reaction terminated by the addition of

$10\,\mu l$ 6 M HCl and $250\,\mu l$ ethyl acetate or ethyl acetate–benzene (1:1; v/v) depending on the methylation level of the product. An aliquot of the organic layer is counted for radioactivity and the remaining portion is concentrated and co-chromatographed on TLC or HPLC with reference compounds in the appropriate solvent systems. The chromatographed products are visualised in UV light and autoradiographed in order to locate the labelled compounds (De Luca and Ibrahim, 1985a).

In the case of anthocyanins, which are dissolved in 5 mM HCl, the enzyme reaction is terminated by the addition of a mixture of chloroform–1% methanolic HCl (2:1; v/v). The upper phase of this Folch partition (Jonsson et al., 1982) is used for activity determination and identification of the reaction products.

2. Purification

The tissue is frozen in liquid N_2, mixed with Polyclar AT (5% w/w), and homogenised with 0.1 M phosphate buffer, pH 7.5, containing 5 mM EDTA, 10 mM diethylammonium diethyldithiocarbamate, and 14 mM 2-mercaptoethanol. The slurry is filtered and the filtrate centrifuged for 15 min at $20\,000 \times g$. The supernatant is stirred with Dowex 1×2 (10% w/v), previously equilibrated in the same buffer, and filtered. The filtrate is then fractionated with solid ammonium sulphate (35–70% saturation) and the protein is collected by centrifugation. The protein pellet is first desalted by gel filtration on Sephacryl S-200 and the active fractions purified by successive chromatography on an anion-exchange column using a salt gradient (0 to 300 mM KCl in buffer) and a hydroxyapatite column using a gradient (c. 25 to 250 mM) of phosphate buffer (De Luca and Ibrahim, 1985a). Further purification may be achieved by affinity chromatography on a SAH-Agarose column pre-equilibrated with 25 mM imidazole-HCl, pH 7.5, containing 14 mM 2-mercaptoethanol and 10% glycerol, and elution with a linear gradient (0 to 1 M) NaCl (Jay et al., 1985). The enzyme protein may be subjected to chromatofocusing on PBE 94 or Mono P column. Despite the partial loss of enzyme activity, chromatofocusing is very efficient in the separation of the component OMT activities (Khouri and Ibrahim, 1987) including the abundant caffeic acid OMT. The fractions eluted in Polybuffer (c. 0.5–1.0 ml) are collected in tubes containing $250\,\mu l$ of 0.2 M phosphate buffer, pH 8.0, in order to prevent inactivation of the enzyme. In order to remove the Polybuffer, the active fractions are applied to a hydroxyapatite column and the enzyme protein eluted stepwise with phosphate buffer. Using a combination of these procedures, various OMTs have been partially purified with several hundred-fold increase in specific activity as compared with crude extracts (e.g. Knogge and Weissenbock, 1984; Jay et al., 1985; Khouri et al., 1988a,b; Ibrahim et al., 1989).

A recent protocol involves the purification of an isoflavone 7-OMT from alfalfa cell cultures challenged with an elicitor preparation of Colletotrichum lindemuthianum (Edwards and Dixon, 1991). The purification procedure consists of successive chromatography of the ammonium sulphate pellet on DEAE-Sepharose CL-6B, hydrophobic interaction FPLC column, Mono P, and Mono Q. This protocol yields a highly purified enzyme preparation as revealed by photoaffinity labelling with [^{3}H-methyl] SAM and SDS-PAGE (Preisig et al., 1989). It affords 362-fold purification with a specific activity of 20 pkat mg^{-1} protein and a 4.5% recovery (Edwards and Dixon, 1991).

3. Properties

Different flavonoid OMTs exhibit pH optima in the range of 7.0 to 9.7 and pI values between 4.3 and 6.0. Their molecular weights vary between 42 kDa (Edwards and Dixon, 1991) and 57 kDa (De Luca and Ibrahim, 1985a). Other OMTs have molecular weights as low as 35 kDa (Sütfeld and Wiermann, 1978) or as high as 110 kDa (Wengenmayer *et al.*, 1974). The substrate and position specificities of OMTs have already been mentioned in this section. However, most enzymes accept aglycones as substrates, although few gluco/glycosides act as methyl acceptors, such as 3,6,7,4'-trimethylquercetin 5'-*O*-glucoside (2'-OMT) and 3,7,4'-trimethylquercetin 2'-*O*-glucoside (5'-OMT) from *C. americanum* (Latchinian *et al.*, 1987), vitexin 2'-*O*-rhamnoside (7-OMT) from oat (Knogge and Weissenböck, 1984) and anthocyanidin 3(*p*-coumaroyl)-rutinosido-5-glucoside (3'/5'-OMT) from *Petunia* (Jonsson *et al.*, 1982). Whereas some OMTs exhibit an absolute requirement for Mg^{2+} as with the flavonol 6-OMT (De Luca and Ibrahim, 1985a), others show only a slight activation in the presence of this ion (Poulton, 1981), and still others are not affected by the presence of divalent cations or EDTA. On the other hand, OMT activity is severely inhibited by SH-group reagents, especially PCMB and NEM. Such inhibition is partially restored by the addition of 2-mercaptoethanol or dithiothreitol. Most OMTs exhibit high affinities for their flavonoid substrates with K_m values of 2–60 μM, whereas those for SAM vary between 50 and 150 μM. Almost all enzymes are competitively inhibited by SAH with K_i values of 4.5–35 μM (Ibrahim *et al.*, 1989).

The kinetic mechanism of the methylation reaction has been shown to be consistent with an ordered bi-bi mechanism (De Luca and Ibrahim, 1985b; Khouri *et al.*, 1988a,b), a mono-iso Theorell–Chance mechanism (Knogge and Weissenböck, 1984) or either of the two mechanisms (Jay *et al.*, 1985). In most cases, SAM is the first substrate to bind to the enzyme and SAH is the final product released. However, product inhibition patterns for the flavonol 8-OMT seem to indicate that the flavonol substrate binds to the enzyme before SAM, followed by the release of SAH and 8-methoxyflavonol (Jay *et al.*, 1985).

E. Sulphotransferase

In spite of the increasing reports on the natural occurrence of flavonoid sulphate esters (for review see Barron *et al.*, 1988), there has been no information on their enzymology. This recently discovered enzyme (EC 2.8.2.-) catalyses the transfer of sulphate groups of 3'-phosphoadenosine 5'-phosphosulphate (PAPS) to different hydroxyl groups of flavonols (Varin, 1988). Four distinct, flavonol-specific sulphotransferases (STs) have been highly purified from shoot tips of *Flaveria chloraefolia* and *F. bidentis* (Asteraceae). They exhibit strict specificity for positions 3 of various flavonol acceptors, 3' and 4' of flavonol 3-sulphate, and 7 of flavonol 3,3'- or 3,4'-disulphate (Varin and Ibrahim, 1989, 1991). These novel STs constitute the enzyme complement involved in the sequential sulphation (3 $\rightarrow$ 3'/4' $\rightarrow$ 7) of polysuphated flavonols in these tissues (Varin and Ibrahim, 1991).

1. Preparation of sulphated flavonols

Since these compounds are not available commercially, they may be isolated and purified from natural sources for use as substrates and reference compounds (Barron *et al.*, 1988). Alternatively, they can be synthesised by N,N'-dicyclohexylcarbodiimide-mediated esterification of flavones and flavonols with tetrabutylammonium hydrogen sulphate. Sulphation which occurs at positions 7 > 4' > 3 allows the preparation of a number of flavonoid mono- to trisulphates (Barron and Ibrahim, 1987). Flavonoid 3',4'-disulphates can be synthesised from the corresponding 4'-sulphate esters using sulphur trioxide–trimethylamine complex (Barron and Ibrahim, 1988a). On the other hand, flavonol 3-sulphates can be prepared from highly sulphated analogues by treatment with aryl sulphatase, where the rate of hydrolysis follows the order 7 or 4' ⋙ 3. The resistance of the 3-sulphate ester to enzyme hydrolysis allows the synthesis of a number of naturally occurring flavonol 3-sulphates (Barron and Ibrahim, 1988b).

2. Assay

The standard enzyme assay consists of 0.1 nmol flavonoid substrate, 0.1 nmol [35S]PAPS, and the enzyme protein in either 25 mM *bis*-Tris-HCl, pH 6.5, for the flavonol 3-ST, or 50 mM Tris-HCl, pH 7.5, for the 3'-, 4'- and 7-STs, in a total volume of 100 μl. The mixture is incubated at 30°C for 10 min, and the reaction terminated by the successive addition of 10 μl of 2.5% acetic acid and 20 μl of 0.1 M tetrabutylammonium dihydrogen phosphate. This results in the formation of an ion-pair with the flavonoid sulphate esters, rendering them soluble in non-polar solvents, with PAPS remaining in the aqueous layer (Varin *et al.*, 1987a). The sulphated products are extracted with 250 μl of ethyl acetate and an aliquot of the organic phase is counted for radioactivity. The remaining fraction is concentrated and chromatographed with reference compounds on cellulose TLC using H_2O or *n*-BuOH–HOAc–H_2O (3:1:1; v/v/v) as solvents (Varin *et al.*, 1987b).

2. Purification

The 3-, 3'- and 4'-STs have been partially purified from *F. chloraefolia* (Varin and Ibrahim, 1989) and the 7-ST from *F. bidentis* (Varin and Ibrahim, 1991). The protein extract is fractionated with solid ammonium sulphate (35–75% saturation) and the resulting pellet desalted on Sephacryl S-200 or Sephadex G-100. The active fractions are chromatographed on a 3'-phosphoadenosine 5'-phosphate (PAP)-agarose affinity support and eluted with a 0.0 to 0.7 M NaCl gradient. Individual ST actitivities are resolved by chromatofocusing on Mono P using a 7.0–4.0 pH gradient. This procedure results in a 500- to 600-fold enrichment of activity for the 3-, 3'- and 4'-STs and 200-fold for the 7-ST.

The 3-ST has recently been purified to apparent homogeneity by a five-step purification procedure which included chromatography on Sephadex G-100, DEAE-Sephacel, hydroxyapatite, PAP-agarose, and finally ion-exchange on Mono-Q. This procedure resulted in a 2000-fold increase in purification, with a specific activity of $1.4\,\text{nkat}\,\text{mg}^{-1}$ protein and a recovery of 9% (Varin and Ibrahim, 1992).

3. Properties

The flavonol STs are monomers with a mol. wt. of 35 kDa. Except for the 3-ST, which exhibits two pH optima at pH 6.5 in bis-Tris-HCl and 8.5 in Tris-HCl buffers, the 3'-, 4'- and 7-STs have an optimum pH of 7.5 in Tris-HCl buffer. The four STs have pI values in the range pH 5.1 to 6.0. The highly purified enzymes exhibit no requirement for divalent cations and are not inhibited by EDTA or SH-group reagents at concentrations up to 10 mM. Their K_m values for the flavonol and PAPS vary between 0.2 and 0.4 μM, and are competitively inhibited by PAP (Varin and Ibrahim, 1989, 1991).

F. Prenyltransferase

This particulate enzyme catalyses the prenylation of isoflavones at positions 6 and 8 (Schröder *et al.*, 1979), as well as pterocarpans (Scheme 4.2) at positions 2 and 4 (Zähringer *et al.*, 1981) and position 10 (Biggs *et al.*, 1987). Prenylated isoflavones and prenylated pterocarpan phytoalexins are believed to act as potent antimicrobial substances and feeding deterrents (Lane *et al.*, 1985). The enzyme utilises dimethylallyl pyrophosphate (DMAPP), but not isopentenyl pyrophosphate (IPP), as the prenyl donor.

1. Synthesis of DMAPP

[^{3}H]DMAPP is synthesised by an elaborate procedure (Davisson *et al.*, 1986) which consists of the reduction of dimethylallyl aldehyde using NaB[^{3}H$_4$]. The tritiated derivative is reacted with sodium pyrophosphate to form [^{3}H]DMAPP. Alternatively, [1-^{14}C]DMAPP may be prepared enzymatically from [1-^{14}C]IPP, which is available commercially, using a partially purified IPP isomerase from yeast cells (Reardon and Ables, 1986). In this way, the assay of the latter enzyme can be coupled to the prenyltransferase (PT) assay. Unlabelled DMAPP is synthesised from dimethylallyl alcohol as described by Conforth and Popjak (1969). If not available commercially, the isoflavonoid substrates may be isolated and purified from plant sources using standard phytochemical methods (Harborne, 1989).

2. Enzyme preparation

The pterocarpan PT has been prepared from bean (*Phaseolus vulgaris*) cell cultures that were challenged with yeast extract (0.3% w/v) for 12 h in the dark. The filtered cells are homogenised in buffer and a microsomal fraction is prepared as previously described (Sections III.D or IV.A). The microsomal pellet is suspended in 0.1 M Tris-HCl, pH 7.5, containing 14 mM 2-mercaptoethanol and 20% sucrose at 5–7 mg protein ml^{-1} of buffer (Biggs *et al.*, 1987).

3. Assay

The assay mixture consists of 10.6 μmol Tris-HCl, pH 7.5, 1.3 μmol MnCl$_2$, 3.2 μmol fluoride, 1.4 μmol glutathione, 8 nmol isoflavone (or pterocarpan), 0.4–1.0 nmol [^{3}H]DMAPP, and 5–10 μg microsomal protein in a total volume of 100 μl. The reaction is incubated for 15–30 min at 20°C and terminated by the addition of solid NaCl and 200 μl ethyl acetate. After centrifugation, an aliquot of the organic layer is counted for radioactivity and the remainder used for the identification of reaction products. Thin layer chromatography is performed on silica gel plates using either toluene–chloroform–acetone (45:25:35; v/v/v) or CHCl$_3$–MeOH (19:1; v/v) for pterocarpans, and either CHCl$_3$–acetone–NH$_4$OH (50:50:1; v/v/v) or CHCl$_3$–MeOH (9:1; v/v) for isoflavones (Biggs *et al.*, 1987).

4. Properties

The pH optimum for the prenylation of pterocarpans is in the range of 7 to 8, however, the enzyme activity decreases at lower pH values, reaching zero at pH 5.5. Prenyltransferase has a strict requirement for a divalent cation (preferably Mn^{2+}), and DMAPP (but not IPP) as the prenyl donor. The bean enzyme accepts 3,9-dihydroxypterocarpan for prenylation at position 10 and gives rise to phaseollidin, a precursor of the bean phytoalexin, phaseollin (Biggs *et al.*, 1987). On the other hand, the soybean enzyme prenylates positions 2 and 4 of 3,6a,9-trihydroxypterocarpan giving rise to glyceollidin (glycinol), the precursor of glyceollins (Zähringer *et al.*, 1981). The bean enzyme does not exhibit absolute specificity since medicarpin (3-methoxy-9-hydroxypterocarpan) and coumestrol (3,9-dihydroxycoumestan) are prenylated at 24% and 27%, respectively, of the rate of dihydroxypterocarpan. The K_m values for DMAPP and 3,9-dihydroxypterocarpan are 1.5 and 2.8 μM, respectively. Prenyltransferase activity is competitively inhibited by phaseollidin and phaseollin with K_i values of 8 and 24 μM, respectively (Biggs *et al.*, 1987). Recent work seems to indicate that prenylation of 6-, 8- and 3′-positions of the isoflavones, genistein and 2′-hydroxygenistein is catalysed by a number of distinct, position-specific prenyltransferases (P. Laflamme *et al.*, unpub. res.).

G. Prenylated Pterocarpan Cyclase

This microsomal NADPH- and O$_2$-dependent enzyme catalyses the cyclisation of 2- and 4-dimethylallylglycinols (3,6a,9-tri-hydroxypterocarpans) to glyceollin I, and glyceollins II and III, respectively (Scheme 4.2). It has recently been characterised from elicitor-challenged soybean cell cultures (Welle and Grisebach, 1988).

1. Synthesis of substrates

DMAPP and [^{3}H]DMAPP are synthesised as previously described (Section IV.F). The mixture of 2- and 4-dimethylallyl[^{3}H]glycinol is prepared using glycinol, DMAPP, and a soybean prenyltransferase preparation (Section IV.F).

2. Assay

The reaction mixture of the cyclase assay consists of 20 μmol Tris-HCl, pH 7.5, 100 nmol NADPH, 310 pmol of a mixture of [^{3}H]glycinol, and 5–30 μg microsomal protein (Section IV.F) in a total volume of 100 μl. The reaction mixture is incubated for 20 min at 20°C and then extracted twice with 100 μl portions of ethyl acetate. The organic layers are concentrated and chromatographed on silica gel plates in toluene–chloroform–acetone (45:25:35; v/v/v) where the substrate, glycinol, and the product, glyceollin, exhibit R_f values of 0.3 and 0.45, respectively. Chromatography of the reaction mixture on formamide-impregnated plates in n-hexane–diethyl ether (3:1; v/v) resolves the labelled glyceollin isomers I, II and III (Komives, 1983) in the ratio of about 65:5:27, respectively. It should be noted that since the substrate used consists of a mixture of 2- and 4-dimethylallylglycinol, glyceollin I results from the cyclisation of 4-dimethylallylglycinol, whereas glyceollin II and III result from the cyclisation of 2-dimethylallylglycinol (Scheme 4.2) (Welle and Grisebach, 1988).

3. Properties

The cyclisation reaction has a broad pH optimum between pH 7.5 and 9.5 with half-maximal activity at pH 7.0. The K_m values for glycinol and NADPH are 2 and 30 μM, respectively. The enzyme is inhibited by its reaction product with 50% inhibition at 58 μM glyceollin. Prenylated pterocarpan cyclase (PPC) is a cytochrome P-450-dependent monooxygenase as determined by CO inhibition and its partial reversal by illumination with white light, inhibition by cytochrome c (25–85% in presence of 10–50 μM cytochrome c), the effect of a number of known inhibitors for cyrochrome P-450 enzymes, and inhibition by NADP$^+$ (Welle and Grisebach, 1988).

Subcellular localisation of the cyclase on a Percoll gradient suggests its association with the endoplasmic reticulum. Maximum PPC activity coincides with those of NADPH-cytochrome reductase, cinnamate 4-monooxygenase, isoflavone synthase, and flavone synthase II. In contrast, highest PT activity appears at a lower density of the gradient which remains to be investigated (Welle and Grisebach, 1988).

Information is not yet available as to the substrate specificity of cyclases. However, the enzymatic synthesis of three glyceollin isomers suggests the existence of more than one cyclase in soybean. This must await the solubilisation and separation of individual cytochrome P-450 species involved, which may prove to be a challenging task!

REFERENCES

Ayabe, S., Udagawa, A. and Furuya, T. (1988). *Arch. Biochem. Biophys.* **261**, 458–462.
Barron, D. and Ibrahim, R. K. (1987). *Tetrahedron* **43**, 5197–5202.
Barron, D. and Ibrahim, R. K. (1988a). *Z. Naturforsch.* **43c**, 625–630.
Barron, D. and Ibrahim, R. K. (1988b). *Z. Naturforsch.* **43c**, 631–635.
Barron, D., Varin, L., Ibrahim, R. K., Harborne, J. B. and Williams, C. A. (1988). *Phytochemistry* **27**, 2375–2395.
Barton, G. M. (1968). *J. Chromatogr.* **34**, 562.
Bednar, R. A. and Hadcock, J. R. (1988). *J. Biol. Chem.* **263**, 9582–9588.
Beerhues, L. and Wiermann, R. (1985). *Z. Naturforsch.* **40c**, 160–165.
Beerhues, L. and Wiermann, R. (1988). *Planta* **173**, 532–543.
Beerhues, L., Robenek, H. and Wiermann, R. (1988). *Planta* **173**, 544–553.
Biggs, D. R., Welle, R., Visser, F. R. and Grisebach, H. (1987). *FEBS Lett.* **220**, 223–226.
Bless, W. and Barz, W. (1988). *FEBS Lett.* **235**, 47–50.
Boland, M. J. and Wong, E. (1975). *Eur. J. Biochem.* **50**, 383–389.

Bonhoff, A., Loyal, R., Ebel, J. and Grisebach, H. (1986). *Arch. Biochem. Biophys.* **246**, 149–154.

Britsch, L. (1990a). *Arch. Biochem. Biophys.* **276**, 348–354.

Britsch, L. (1990b). *Arch. Biochem. Biophys.* **282**, 152–160.

Britsch, L. and Grisebach, H. (1985). *Phytochemistry* **24**, 1975–1976.

Britsch, L. and Grisebach, H. (1986). *Eur. J. Biochem.* **156**, 569–577.

Britsch, L., Heller, W. and Grisebach, H. (1981). *Z. Naturforsch.* **36c**, 742–750.

Chmiel, E., Sutfeld, R. and Wiermann, R. (1983). *Biochem. Physiol. Pflanzen* **178**, 139–146.

Conforth, R. H. and Popjak, G. (1969). *Methods Enzymol.* **15**, 359–390.

Davisson, U. J., Woodside, A. B., and Poulter, C. D. (1986). *Methods Enzymol.* **110**, 130–144.

De Luca, V. and Ibrahim, R. K. (1985a). *Arch. Biochem. Biophys.* **238**, 596–605.

De Luca, V. and Ibrahim, R. K. (1985b). *Arch. Biochem. Biophys.* **238**, 606–618.

Deutscher, M. P. (ed.) (1990). *Methods Enzymol.* **182**, 9–780.

Dewick, P. M. (1977). *Phytochemistry* **16**, 93–97.

Dewick, P. M. and Steele, M. J. (1982). *Phytochemistry* **21**, 1599–1603.

Dixon, R. A., Dey, P. M. and Whitehead, I. M. (1982). *Biochim. Biophys. Acta* **715**, 25–33.

Dixon, R. A., Blyden, E. R., Robbins, M. P., Van Tunen, A. J. and Mol, J. N. (1988). *Phytochemistry* **27**, 2801–2808.

Ebel, J. (1986). *Annu. Rev. Phytopathol.* **24**, 235–264.

Edwards, R. and Dixon, R. A. (1991). *Arch. Biochem. Biophys.*, in press.

Fischer, D., Stich, K., Britsch, L. and Grisebach, H. (1988). *Arch. Biochem. Biophys.* **264**, 40–47.

Fischer, D., Ebenau-Jehle, C. and Grisebach, H. (1990a). *Arch. Biochem. Biophys.* **276**, 390–395.

Fischer, D., Ebenau-Jehle, C. and Grisebach, H. (1990b). *Phytochemistry* **29**, 2879–2882.

Forkmann, G., Heller, W. and Grisebach, H. (1980). *Z. Naturforsch.* **35c**, 691–695.

Forkmann, G., de Vlaming, P., Spribille, R., Wiering, H. and Schram, A. W. (1986). *Z. Naturforsch.* **41c**, 179–186.

Freifelder, D. (1982). "Physical Biochemistry: Applications to Biochemistry and Molecular Biology". W. H. Freeman, New York.

Fritsch, H. and Grisebach, H. (1975). *Phytochemistry* **14**, 2437–2442.

Geissman, T. A. and Clinton, R. O. (1946). *J. Am. Chem. Soc.* **40**, 697–700.

Hagmann, M. and Grisebach, H. (1984). *FEBS Lett.* **175**, 199–202.

Hagmann, M., Heller, W. and Grisebach, H. (1983). *Eur. J. Biochem.* **134**, 547–554.

Hagmann, M., Heller, W. and Grisebach, H. (1984). *Eur. J. Biochem.* **142**, 127–131.

Harborne, J. B. (ed.) (1989). "Methods in Plant Biochemistry", Vol. 1. Academic Press, London.

Harborne, J. B. (ed.) (1988). "The Flavonoids: Advances in Research since 1980". Chapman and Hall, London.

Harborne, J. B. and Mabry, T. J. (eds) (1982). "The Flavonoids: Advances in Research". Chapman and Hall, London.

Heller, W. and Forkmann, G. (1988). *In* "The Flavonoids: Advances in Research since 1980" (J. B. Harborne, ed.), pp. 399–425. Chapman and Hall, New York.

Heller, W., Forkmann, G., Britsch, L. and Grisebach, H. (1985). *Planta* **165**, 284–287.

Heller, W. and Grisebach, H. (1984). *Eur. J. Biochem.* **142**, 127–131.

Hinderer, W., Noe, W. and Seitz, H.U. (1983). *Phytochemistry* **22**, 2417–2420.

Hinderer, W., Flentje, U. and Barz, W. (1987). *FEBS Lett.* **214**, 101–106.

Hösel, W. (1981) *In* "The Biochemistry of Plants", Vol. 7 (E. E. Conn, ed.), pp. 725–753. Academic Press, New York.

Hrazdina, G. (1988). *Biochim. Biophys. Acta.* **955**, 301–309.

Hrazdina, G., and Wagner, G. J. (1985). *Arch. Biochem. Biophys.* **237**, 88–100.

Hrazdina, G., Lifson, E. and Weeden N. F. (1986). *Arch. Biochem. Biophys.* **247**, 414–419.

Hrazdina, G., Zobel, A. and Hoch, H. C. (1987). *Proc. Natl. Acad. Sci. USA* **84**, 8966–8970.

Hrazdina, G., Kreuzaler, F., Hahlbrock, K. and Grisebach, H. (1976). *Arch. Biochem. Biophys.* **175**, 392–399.

Ibrahim, R. K., Latchinian, L. and Brisson, L. (1989). *Am. Chem. Soc. Symp.* **399**, 122–136.

Ibrahim, R. K., Khouri, H., Brisson, L., Latchinian, L., Barron, D. and Varin, L. (1986). *Bull. Liaison Groupe Polyphénols* **13**, 3–14.

Ibrahim, R. K., DeLuca, V., Khouri, H., Latchinian, L., Brisson, L., Barron, D. and Charest, P. M. (1987). *Phytochemistry* **26**, 1237–1245.

Ishikura, N., Murhani, H. and Fujii, Y. (1988). *Plant Cell Physiol.* **29**, 795–799.

Jakoby, W. B. (1984). *Methods Enzymol.* **34**, 3–755.

Janson, J. C. and Rydén, L. (1989). *Protein Purification.* VCH Publishers, New York.

Jay, M., De Luca, V. and Ibrahim, R. K. (1985). *Eur. J. Biochem.* **153**, 321–325.

Jonsson, L. M. V., Aarsman, M. E. G., Schram, A. W. and Bennink, G. J. H. (1982). *Phytochemistry* **21**, 2457–2459.

Jonsson, L. M. V., Aarsman, M. E. G., van Diepen, J., de Vlaming, P., Smit, N. and Schram, A. W. (1984). *Planta* **160**, 341–347.

Jourdan, P. S. and Mansell, R. L. (1982). *Arch. Biochem. Biophys.* **213**, 434–443.

Kerscher, F. and Franz, G. (1988). *J. Plant Physiol.* **132**, 110–115.

Khouri, H. E. and Ibrahim, R. K. (1984). *Eur. J. Biochem.* **142**, 559–564.

Khouri, H. E. and Ibrahim, R. K. (1987). *J. Chromatogr.* **407**, 291–297.

Khouri, H. E., Tahara, S. and Ibrahim, R. K. (1988a). *Arch. Biochem. Biophys.* **262**, 592–598.

Khouri, H. E., De Luca V. and Ibrahim, R. K. (1988b). *Arch. Biochem. Biophys.* **265**, 1–7.

Kleinehollenhorst, G., Behrens, H., Pegels, G., Srunk, N. and Wiermann, R. (1982). *Z. Naturforsch.* **37c**, 587–599.

Knogge, W. and Weissenböck, G. (1984). *Eur. J. Biochem.* **140**, 113–118.

Kochs, G. and Grisebach, H. (1986). *Eur. J. Biochem.* **155**, 311–318.

Kochs, G. and Grisebach, H. (1987). *Z. Naturforsch.* **42c**, 343–348.

Komives, T. (1983) *J. Chromatogr.* **261**, 423–426.

Kondo, A., Blanchard, J. S. and Englard, S. (1981). *Arch. Biochem. Biophys.* **212**, 338–346.

Köster, J. and Barz, W. (1981). *Arch. Biochem. Biophys.* **212**, 98–104.

Köster, J., Strack, D. and Barz, W. (1983). *Planta Med.* **48**, 131–135.

Köster, J., Bussmann, R. and Barz, W. (1984). *Arch. Biochem. Biophys.* **234**, 513–521.

Kreuzaler, F. and Hahlbrock, K. (1975). *Eur. J. Biochem.* **56**, 205–213.

Kristiansen, K. N. (1986). *Carlsberg Res. Commun.* **51**, 51–60.

Kuroki, G. and Poulton, J. E. (1981). *Z. Naturforsch.* **36c**, 916–920.

Lane, G. A., Biggs, D. R., Russell, G. B., Sutherland, O. R. W., Williams, E. M., Maindonald, J. H. and Donnell, D. J. (1985). *J. Chem. Ecol.* **11**, 1713–1735.

Larson, R. L. and Bussard, J. B. (1986). *Plant Physiol.* **80**, 483–486.

Latchinian-Sadek, L. and Ibrahim, R. K. (1991). *Arch. Biochem. Biophys.*, **289**, 230–236.

Latchinian, L., Khouri, H. E. and Ibrahim, R. K. (1987). *J. Chromatogr.* **388**, 235–242.

Matern, U., Potts, J. R. M. and Hahlbrock, K. (1981). *Arch. Biochem. Biophys.* **208**, 233–241.

McIntosh, C. A., Latchinian, L. and Mansell, R. L. (1990). *Arch. Biochem. Biophys.* **282**, 50–57.

Meyer, P., Heidmann, I., Forkmann, G. and Saedler, H. (1987). *Nature* **330**, 677–678.

Mol, J. N. M., Robbins, M. P., Dixon, R. A. and Veltkamp, E. (1985). *Phytochemistry* **24**, 2267–2269.

Ozeki, Y., Sakano, K., Komamine, A., Tanaka, Y., Noguchi, H., Sankawa, U. and Suzuki, T. (1985). *J. Biochem. Tokyo* **98**, 9–17.

Peters, A., Schwartz, H., Schneider-Poetsch, H. A. W. and Weissenböck, G. (1988). *J. Plant Physiol.* **133**, 178–182.

Poulton, J. E. (1981). *In* "The Biochemistry of Plants", Vol. 7 (E. E. Conn, ed.), pp. 667–723. Academic Press, New York.

Poulton, J. E., Hahlbrock, K., and Grisebach, H. (1977). *Arch. Biochem. Biophys.* **180**, 543–549.

Preisig, C., Matthews, D. E. and VanEtten, H. D. (1989). *Plant Physiol.* **91**, 559–566.

Preisig, C. L., Bell, J. N., Sun, Y., Hrazdina, G., Matthews, D. E. and VanEtten, H. D. (1990). *Plant Physiol.* **94**, 1444–1448.

Raymond, W. R. and Maier, V. P. (1977). *Phytochemistry* **16**, 1535–1539.

Reardon, J. E. and Abeles, R. H. (1986). *Biochemistry* **25**, 5609–5616.

Robbins, M. P. and Dixon, R. A. (1984). *Eur. J. Biochem.* **145**, 195–202.

Rolf, C. and Kindl, H. (1984). *Plant Physiol.* **75**, 489–492.

Schröder, G., Zähringer, U., Heller, W., Ebel, J. and Grisebach, H. (1979). *Arch. Biochem. Biophys.* **194**, 635–636.

Schulz, M. and Weissenböck, G. (1988). *Phytochemistry* **27**, 1261–1267.

Scopes, H. A. (1987). "Protein Purification Principles and Practice". Springer-Verlag, New York.

Spribille, R. and Forkmann, G. (1982a). *Phytochemistry* **21**, 2231–2234.

Spribille, R. and Forkmann, G. (1982b). *Planta* **155**, 176–182.

Spribille, R. and Forkmann, G. (1984). *Z. Naturforsch.* **39c**, 714–719.

Stafford, H. A. (1990). "Flavonoid Metabolism". CRC Press, Boca Raton, FL.

Stafford, H. A. and Lester, H. H. (1982). *Plant Physiol.* **70**, 695–698.

Stafford, H. A. and Lester, H. H. (1985). *Plant Physiol.* **78**, 791–794.

Stich, K. and Forkmann, G. (1988). *Phytochemistry* **27**, 785–789.

Stöckigt, J. and Zenk, M. H. (1975). *Z. Naturforsch.* **30c**, 332–338.

Stotz, G. and Forkmann, G. (1981). *Z. Naturforsch.* **36c**, 737–741.

Stotz, G. and Forkmann, G. (1982). *Z. Naturforsch.* **37c**, 19–23

Stotz, G., Spribille, R. and Forkmann, G. (1984). *J. Plant Physiol.* **116**, 173–183.

Stotz, G., de Vlaming, P., Wiering, H., Schram, A. W. and Forkmann, G. (1985). *Theor. Appl. Genet.* **70**, 300–305.

Sun, Y. and Hrazdina, G. (1991). *Plant Physiol.* **95**, 570–576.

Sütfeld, R. and Wiermann, R. (1978). *Biochem. Physiol. Pflanzen* **172**, 111–123.

Sütfeld, R. and Wiermann, R. (1980). *Arch. Biochem. Biophys.* **201**, 64–72.

Sütfeld, R. and Wiermann, R. (1981). *Z. Naturforsch.* **36C**, 30–34.

Sütfeld, R., Kehrel, B. and Wiermann, R. (1978). *Z. Naturforsch.* **33c**, 841–846.

Sutter, A., Poulton, J. E. and Grisebach, H. (1975). *Arch. Biochem. Biophys.* **170**, 547–556.

Teusch, M. and Forkmann, G. (1987). *Phytochemistry* **26**, 2181–2183.

Teusch, M., Forkmann, G. and Seyffert, W. (1986). *Z. Naturforsch.* **41c**, 699–706.

Tiemann, K., Hinderer, W. and Barz, W. (1987). *FEBS Lett.* **213**, 324–328.

Tsang, Y. F. and Ibrahim, R. K. (1979). *Phytochemistry* **18**, 1131–1136.

Van Tunen, H. J. and Mol, J. (1987). *Arch. Biochem. Biophys.* **257**, 85–91.

Varin, L. (1988). *Bull. Liaison Groupe Polyphénols* **14**, 248–257.

Varin, L. and Ibrahim, R. K. (1989). *Plant Physiol.* **90**, 977–981.

Varin, L. and Ibrahim, R. K. (1991). *Plant Physiol.,* **95**, 1254–1258.

Varin, L. and Ibrahim, R. K. (1992). *J. Biol. Chem.,* **267**, 1858–1863.

Varin, L., Barron, D. and Ibrahim, R. K. (1987a). *Anal. Biochem.* **161**, 176–180.

Varin, L., Barron, D. and Ibrahim, R. K. (1987b). *Phytochemistry* **26**, 135–138.

Welle, R. and Grisebach, H. (1988). *Arch. Biochem. Biophys.* **263**, 191–198.

Wengenmayer, H., Ebel, J. and Grisebach, H. (1974). *Eur. J. Biochem.* **50**, 135–143.

Wilson, J. M. and Wong, E. (1976). *Phytochemistry* **15**, 1333–1341.

Woodward, M. D. (1980). *Phytochemistry* **19**, 921–927.

Zähringer, U., Schaller, E. and Grisebach, H. (1981). *Z. Naturforsch.* **36c**, 234–241.

5 Heterocyclic β-Substituted Alanines

FUMIO IKEGAMI and ISAMU MURAKOSHI

Faculty of Pharmaceutical Sciences, Chiba University, Yayoi-cho 1-33, Inage-ku, Chiba 263, Japan

I. INTRODUCTION

During the last decades a large number of non-protein amino acids have been isolated from plants as secondary metabolites. While some of those compounds have a well-defined biological activity, many have been regarded as biochemical oddities. Earlier results by several research groups indicated the ecological importance of some non-protein amino acids. Several recent reports indicate a renewed interest in the biochemistry and the ecology of these secondary metabolites. In particular, in the field of neurochemistry, some non-protein amino acids of plant origin have become important research tools. The small group of heterocyclic β-substituted alanines contains compounds of special interest to biochemistry, ecology and neurochemistry. Our research interest has recently focused on the enzymes involved in their biosyn-

METHODS IN PLANT BIOCHEMISTRY Vol. 9
ISBN 0–12–461019–6

thesis. These enzymes represent a link between the metabolisms of protein and non-protein amino acids.

Heterocyclic β-substituted alanines such as L-quisqualic acid (Takemoto *et al.*, 1975), L-mimosine (Hegarty and Court, 1964), β-(pyrazol-1-yl)-L-alanine (Noe and Fowden, 1960), L-willardiine (Gmelin, 1959), L-isowillardiine (Lambein and Van Parijs, 1968) and β-(3-isoxazolin-5-on-2-yl)-L-alanine (Lambein *et al.*, 1976) have been isolated from selected plants (Scheme 5.1). Some of these amino acids are toxic to, or physiologically active in, organisms in which they do not normally occur. For example, L-quisqualic acid present in *Quisqualis* spp. (Combretaceae) is a neuro-excitatory amino acid and is utilised as a vermicide in Chinese medicine (Ishizaki *et al.*, 1973). L-Mimosine present in *Mimosa* and *Leucaena* spp. (Leguminosae) is a thyro-toxic amino acid and causes loss of hair in growing animals (Hegarty *et al.*, 1964). Furthermore, β-(3-isoxazolin-5-on-2-yl)-L-alanine from *Pisum* and *Lathyrus* spp. shows antimycotic activity towards *Saccharomyces cerevisiae* (Schenk and Werner, 1991). The biosynthesis of these heterocyclic β-substituted alanines is interesting in relation to secondary product metabolism in plants.

Fowden *et al.* (1979) and Bell (1976) proposed that the biosynthesis of non-protein amino acids in plants may be due to the action of enzymes normally involved in the biosynthesis of protein amino acids that, during the course of evolution, have acquired a different specificity. In the case of the heterocyclic β-substituted alanines, it was first suggested that these natural compounds may arise by a non-specific action of tryptophan synthase (Dunnill and Fowden, 1963; Hadwiger *et al.*, 1965; Tiwari *et al.*, 1967).

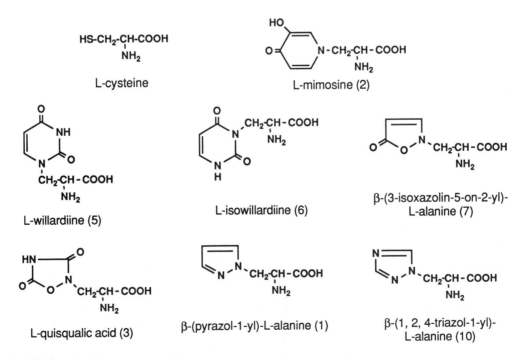

SCHEME 5.1. Some heterocyclic β-substituted alanines in higher plants.

Since 1972, Murakoshi and associates (1972, 1973, 1974a, b, 1975, 1977, 1978a, b, 1983) have demonstrated that these naturally occurring heterocyclic β-substituted alanines are enzymatically synthesised by condensation of the appropriate *N*-heterocyclic compounds with *O*-acetyl-L-serine (OAS) as an activated form of L-serine (Scheme 5.2). When OAS was recognised as the donor for the alanyl moiety, the enzymatic formation of heterocyclic β-substituted alanines was compared to the biosynthesis of *S*-substituted L-cysteines. A number of *S*-substituted L-cysteines occurring in plants are formed by the cysteine synthase (*O*-acetylserine(thiol)lyase [EC4.2.99.8]) catalysed reaction of OAS and thiol compounds (Giovanelli and Mudd, 1968; Thompson and Moore, 1968; Smith and Thompson, 1971; Masada *et al.*, 1975; Bertagnolli and Wedding, 1977). This enzyme also catalyses the formation of seleno-cysteine in higher plants (Hock and Anderson, 1978b). The group of enzymes producing heterocyclic β-substituted alanines generally establishes an N—C bond between the β-carbon of alanine and a ring-nitrogen.

Occasionally a C–C bond is formed in a natural product (Murakoshi *et al.*, 1975) or an O—C bond is formed with a synthetic precursor *in vitro* (Murakoshi *et al.*, 1978b). *O*-Acetyl-L-serine (OAS) is an intermediate in the L-cysteine biosynthesis, and is formed by an acetyl-CoA-dependent activation of L-serine. *O*-Acetyl-L-serine has been shown to be present in plants (Smith, 1977; Murakoshi *et al.*, 1979).

In 1984, Murakoshi *et al.* (1984a, b) purified two such β-substituted alanine synthases forming heterocyclic β-substituted alanines to homogeneity: β-(pyrazol-1-yl)-L-alanine synthase from *Citrullus vulgaris* and L-mimosine synthase from *Leucaena leucocephala*. Both enzymes are specific for OAS as a donor for the alanyl moiety but show differences in specificity for the heterocyclic substrates as an acceptor. The similarities between these two enzymes, including the amino acid composition, suggested a common evolutionary origin of the cucurbit and the legume enzyme. Both enzymes also could synthesise *S*-methyl-L-cysteine.

Recently, Ikegami and associates (Murakoshi *et al.*, 1985a, b, 1986b; Ikegami *et al.*, 1987, 1988a, c, 1990a, b, 1991), during an attempt to purify cysteine synthases from higher plants, found evidence of additional catalytic activities of cysteine synthases. In some plants these enzymes can catalyse the formation of heterocyclic β-substituted alanines, such as L-quisqualic acid (Murakoshi *et al.*, 1986b) and L-mimosine (Ikegami *et al.*, 1990b) as secondary metabolites, without jeopardising the biosynthesis of L-cysteine as a primary metabolite. However, the formation of L-willardiine and L-isowillardiine could not be catalysed by these enzymes (Ikegami *et al.*, 1987). Moreover, a study on the intracellular localization of cysteine synthases indicated that non-protein amino acid biosynthesis occurred in specific sites of the plant and plant cells (Ikegami *et al.*, 1992).

II. HETEROCYCLIC β-SUBSTITUTED ALANINE SYNTHASE

A. Extraction

The seedlings of a number of plants (*Pisum sativum*, *Leucaena leucocephala*, etc.) were grown in moistened vermiculite in the dark for 5–6 days at 26–28°C. After harvesting and washing, the cotyledons were removed and the seedlings were cooled for 1 h at

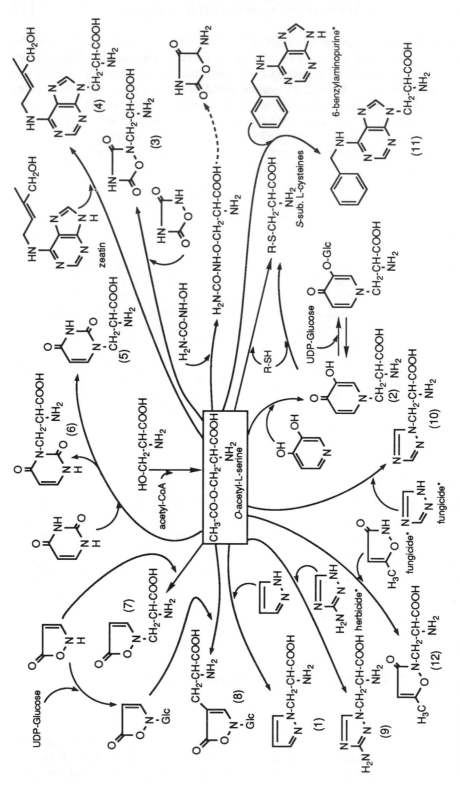

SCHEME 5.2. Biosynthetic pathways for heterocyclic β-substituted alanines derived from *O*-acetyl-L-serine in higher plants.
*Chemical substances used as herbicide, fungicide or cytokinin, and their alanyl-substituted metabolites, were isolated from plants.

0–4°C before enzyme extraction. Some plants such as spinach (*Spinacia oleracea*) and *Quisqualis indica* var. *villosa* were grown in the medicinal plant garden of our faculty and fresh leaves were collected for the enzyme extraction. In our laboratory we have established the following method to isolate β-substituted alanine synthases from several plant seedlings or leaves: plant material was homogenised in 3 to 5 volumes of 200 mM potassium phosphate (K-Pi) buffer (pH 8.0) containing 10 mM 2-mercaptoethanol, 0.5 mM EDTA and 250 mM sucrose with polyvinylpolypyrrolidone (10% of the fresh weight of plant material) and centrifuged at 15 000 × g for 30 min. The supernatant was desalted on a Sephadex G-25 column pre-equilibrated with 50 mM K-Pi buffer (pH 8.0) and then assayed for the formation of heterocyclic β-substituted alanines and L-cysteine.

B. Assays

1. Cysteine synthase activity

The enzyme preparations obtained were dissolved in 50 mM K-Pi buffer (pH 8.0). Substrate concentrations were 4 6 mM for sulphide and 12.5 mM for OAS. The incubation was carried out at 30°C for 10 min; the total reaction volume was 0.5 ml, utilising up to 0.2 ml of enzyme (corresponding to 2–400 μg of protein). Reactions were terminated by the addition of 0.1 ml of 7.5% trichloroacetic acid and the formation of L-cysteine was spectrophotometrically measured at 560 nm using an acidic ninhydrin reagent (Gaitonde, 1967). The unit of enzyme activity used is equivalent to 1 μmol of L-cysteine or β-substituted alanines produced per min. Protein was determined by a dye-binding method (Bradford, 1976).

2. Identification of heterocyclic β-substituted alanines and S-substituted L-cysteines as reaction products

The normal reaction mixtures contained 12.5 mM of OAS, 70 mM of thiol compounds or *N*-heterocyclic compounds and 0.2 ml of enzyme preparation (corresponding to 2–400 μg of protein), in a final volume of 0.5 ml of 50 mM K-Pi buffer (pH 8.0). Incubation was at 30°C for 10–30 min, and the reactions were terminated by the addition of 30 μl of 1 N KOH followed by heating in hot water for 1 h. The resulting solution was acidified with 15 μl of 6 N HCl and examined for the formation of heterocyclic β-substituted alanines and S-substituted L-cysteines using an automatic amino acid analyser (Hitachi 835-10) under standard operating conditions (2.6 × 25 cm column, 33–68°C, Li-citrate buffer system pH 3.0–7.0, flow rate 0.275 ml min^{-1}) (Murakoshi *et al.*, 1984c). The formation of L-mimosine was determined by using 2% ammonium dihydrogen phosphate buffer (pH 2.4) on HPLC (Wakosil 5C18 column, Hitachi 655) (Lowry *et al.*, 1985).

3. Spectrophotometric measurements of cysteine synthases

The absorption spectra of purified enzymes were measured at 200–600 nm by using an automatic spectrophotometer (Hitachi 557). The protein concentrations were 0.25 to 1.65 mg per ml of 50 mM K-Pi buffer (pH 8.0) for cysteine synthase isoenzymes A and B. The identification of bound pyridoxal 5'-phosphate (PLP) in the purified enzymes was determined by measuring the absorption at 410 nm in comparison with that of standard PLP solution (Kumagai *et al.*, 1970). The bound PLP concentration was shown to be 1.9–2.1 molecules per molecule of native enzyme.

4. Molecular weight determination

The apparent native molecular weight of the purified enzymes was determined by gel filtration chromatography using Sephadex G-100 (1.5 × 115 cm) according to the method of Andrews (1965). The column was equilibrated with 50 mM K-Pi buffer (pH 8.0), which contained 0.1 M NaCl. The enzyme was detected by assaying fractions for activity using a flow rate of 0.04 ml min^{-1}. The following standard marker proteins (Pharmacia) were used: chymotrypsinogen A (mol. wt. = 25 kDa), ovalbumin (mol. wt. = 43 kDa) and bovine serum albumin (mol. wt. = 67 kDa).

The column void volume was determined in a separate run by using Blue Dextran 2000.

5. Sodium dodecylsulphate polyacrylamide gel electrophoresis

The purified enzymes were subjected to sodium dodecylsulphate (SDS) polyacrylamide gel electrophoresis (PAGE) on 4–20% gradient gels (Daiichi Pure Chemicals) at pH 8.8 (Tris-glycine buffer) following the method of King and Laemmli (1971) and the running time was 1 h at 60 mA per gel. Gels were stained with Coomassie Brilliant Blue R-250, followed by destaining in HOAc–MeOH–H$_2$O (1:5:5; v/v). The enzyme concentrations were 0.5–2 μg per well and protein standards were α-lactalbumin (mol. wt. = 14.4 kDa), trypsin inhibitor (mol. wt. = 20.1 kDa), carbonic anhydrase (mol. wt. = 30 kDa), ovalbumin (mol. wt. = 43 kDa) and bovine serum albumin (mol. wt. = 67 kDa). Bromophenol blue was used as a marker.

6. Determination of amino acid compositions

The purified enzyme (0.2–1.0 mg) was hydrolysed in 2 ml of 6 N HCl in a sealed tube at 110°C for 24, 48 and 72 h using Reacti-Therm Heating Module (Pierce). After hydrolysis and evaporation *in vacuo*, 2.0 ml of 0.02 N HCl was added and then the sample was analysed on a Hitachi amino acid analyser. Determination of tryptophan was made by the general method described (Gaitonde and Dovey, 1970) and also by alkaline hydrolysis: the enzyme was hydrolysed in 1 ml of 2.5 N NaOH containing 0.375% starch in a sealed tube at 110°C for 24 h. After hydrolysis and evaporation *in vacuo*, 1.0 ml of 0.02 N HCl was added and the sample was analysed. Duplicate analyses were carried out and the mean of these results was used to determine the amino acid composition of cysteine synthases purified from plants.

7. Kinetics and other properties

The pH optima and the curves relating enzyme activities to OAS and sulphide concentrations were determined under standard assay conditions by using 100 μl samples of purified enzymes (equivalent to 2–7 μg of protein).

TABLE 5.1. Summary of the purification of cysteine synthases from *Leucaena leucocephala*.

Purification step	Total activity (units)[d]	Total protein (mg)	Specific activity (units mg^{-1} protein)	Yield (%)	Purification (-fold)
1. Crude extract[a]	1030	5700	0.18	100	1
2. Heated supernatant[b]	920	4600	0.2	89	1.1
3. Ammonium sulphate precipitate[c]	890	1560	0.57	86	3.2
4. DEAE-Sephadex A-50					
Isoenzyme A (110–150 mM)	156	46.2	3.4	15.1	18.9
Isoenzyme B (150–210 mM)	143	31.8	4.5	13.9	25
5. Sephadex G-100 (peak fractions)					
Isoenzyme A	115	18.5	6.2	11.2	34.4
Isoenzyme B	107	12.7	8.4	10.4	46.7
6. Polyacrylamide-gel electrophoresis					
Isoenzyme A	38	1.1	34.5	3.7	192
Isoenzyme B	35	0.7	50.0	3.4	278
7. AH-Sepharose 4B					
Isoenzyme A (75–85 mM)	33	0.4	82.5	3.2	458
Isoenzyme B (80–100 mM)	24	0.25	95.7	2.3	532

[a] Starting from 420 g of the fresh seedlings of *Leucaena leucocephala*.
[b] 60°C, 1 min.
[c] 30–70% saturation and desalted on Sephadex G-25.
[d] A unit of enzyme activity represents 1 μmol of product formed per min at 30°C in 50 mM K-Pi buffer, pH 8.0.
Reprinted with permission from Ikegami *et al.* (1990b), Copyright 1990, Pergamon Press Ltd.

C. Purification of Cysteine Synthases from Plants

All operations were carried out at 0–4°C. Extraction and purification of cysteine synthase from fresh seedlings or leaves followed our own methods as described above. The enzyme was prepared simultaneously with the β-substituted alanine synthase activities by a procedure including heat treatment, ammonium sulphate fractionation, gel filtration on Sephadex G-100, ion-exchange chromatography on DEAE-Sephadex A-50, preparative PAGE and hydrophobic chromatography on EAH-Sepharose 4B (Table 5.1).

Step 1. Fresh 5- to 6-day-old seedlings (cotyledons removed) or fresh leaves were homogenised in 200 mM K-Pi buffer, pH 8.0, containing 10 mM 2-mercaptoethanol, 0.5 mM EDTA and 250 mM sucrose with polyvinylpolypyrrolidone (10% of the fresh weight plant material). Homogenisation was performed by Waring blender at top speed for 2 min. The homogenate was pressed through nylon cloth and the filtrate centrifuged at 15 000 × g for 30 min. This supernatant desalted on Sephadex G-25 column (8.2 × 20 cm) was designated as crude extract.

Step 2. The crude extract was placed in a boiling water bath and stirred vigorously until the temperature of the preparation reached 60°C; it was then kept in the water bath for an additional 1 min and gently swirled by hand. At the end of this it was immediately chilled to 4°C in ice water, and then centrifuged at 10 000 × g for 20 min.

Step 3. The Step 2 supernatant was subjected to $(NH_4)_2SO_4$ fractionation and the protein precipitating between 30 and 70% saturation was collected and dissolved in 30 mM K-Pi buffer (pH 8.0) containing 10 mM 2-mercaptoethanol and 0.5 mM EDTA (buffer A). After desalting on Sephadex G-25 column pre-equilibrated with buffer A, enzyme activity was stable for at least 7 days at 4°C.

Step 4. The resulting solutions were then applied to the DEAE-Sephadex A-50 column (2.1 × 15 cm) pre-equilibrated with buffer A. The column was washed extensively with buffer A and the enzymes eluted with a linear gradient of K-Pi (30–350 mM) in the same buffer. Two peaks (A and B) exhibiting cysteine synthase activity were separated completely after the DEAE-Sephadex A-50 column was eluted with a concentration gradient of K-Pi buffer. Cysteine synthase activities normally eluted at 100–150 mM and 150–200 mM K-Pi buffer, respectively, and enzyme activity for β-substituted alanine synthase (except L-willardiine and L-isowillardiine synthases) correlated with the two peaks of cysteine synthase (Fig. 5.1). These enzyme fractions were concentrated by immersible membrane filter CX-10 (Millipore).

Step 5. The first and second active fractions were individually applied to a column of Sephadex G-100 (5.0 × 95 cm) pre-equilibrated with buffer A. The eluates were collected in 5 ml fractions, and two series of active fractions were pooled and concentrated by immersible membrane filter CX-10.

Step 6. The resulting solutions were individually subjected to preparative PAGE on 7.5% gels (10 mm) at pH 8.3 (Tris-glycine buffer).

Step 7. Cysteine synthase fractions obtained from gel slices were finally applied to a column (1.0 × 3.5 cm) of EAH-Sepharose 4B pre-equilibrated in buffer A and then eluted with a linear gradient of K-Pi (30–200 mM) in buffer A. The highly purified enzyme fractions (normally isoenzyme A: 65–85 mM, isoenzyme B: 80–100 mM) were concentrated by immersible membrane filter CX-10 and these enzyme

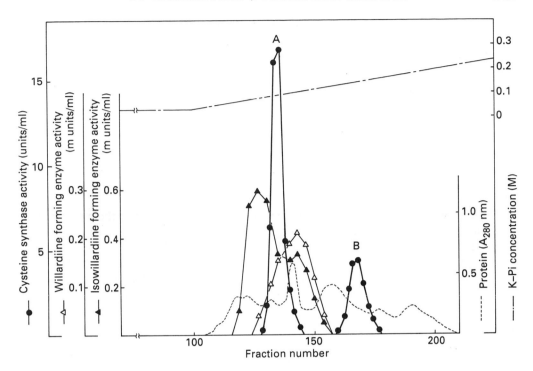

FIG. 5.1. Elution patterns of cysteine synthase and uracilylalanine synthase activities after the DEAE-Sephadex A-50 column chromatography. Cysteine synthase activity (●–●), willardiine synthase activity (△–△), isowillardiine synthase activity (▲–▲) and protein (A₂₈₀, ---) were monitored as described in Section II. Reprinted with permission from Ikegami *et al.* (1987), Copyright 1987, Pergamon Press Ltd.

preparations in 50 mM K-Pi buffer (pH 8.0) were used as isoenzymes A and B in all further experiments. The purified enzyme fraction was a yellow solution and was stable for at least one month at 0–2°C. The typical purification steps of cysteine synthases from *L. leucocephala* seedlings are shown in Table 5.1. In this study the complete procedure afforded apparent purifications of *c*. 500-fold for isoenzymes A and B with specific activity of 82.5 U mg^{-1} protein for A, 95.7 U mg^{-1} protein for B, and yields of 3.2 and 2.3%, respectively, as compared to the total cysteine synthase activity of the crude extract.

D. Purification of Cysteine Synthase from Etioplasts, Mitochondria and Chloroplasts

In order to discover the site of non-protein amino acid biosynthesis in plants, the intracellular localisation of cysteine synthases was studied in pea and spinach. Etioplasts and mitochondria were isolated from etiolated pea shoots following the methods of Sandelius and Selstam (1984) and Jackson *et al.* (1979). Chloroplasts were isolated mechanically from green pea shoots or spinach leaves following the method of Mills and Joy (1980). Cysteine synthases were prepared by a procedure including heat treatment, ammonium sulphate fractionation, gel filtration on Sephadex G-100, ion-exchange chromatography on DEAE-Sephadex A-50 and hydrophobic chromatography on EAH-Sepharose 4B as our initial purification.

E. Properties and Structure

Several different types of cysteine synthase have been isolated to apparent homogeneity from various higher plants, and the existence of isoenzymes of cysteine synthases catalysing the formation of heterocyclic β-substituted alanines has been demonstrated. A comparison of their properties and substrate specificities has been made to clarify the biosynthetic mechanism of heterocyclic β-substituted alanine formation in higher plants.

The cysteine synthase isoenzymes purified have very similar physicochemical properties (Table 5.2): the molecular weights of plant cysteine synthases have been determined to be in the range 52–72 kDa, estimated by analytical gel filtration using Sephadex G-100. They can be dissociated into two identical subunits of approximately half that of the intact enzyme by SDS-PAGE on 4–20% gradient gels. The purified enzymes have absorbance peaks of 280 and 410 nm, typical for a PLP-enzyme. Direct spectrophotometric measurements indicated that cysteine synthases have one molecule of PLP bound to each subunit. The enzymes exhibited almost the same pH optimum of 8.0, although there was a rapid acetyl shift from O to N atoms in OAS above c. pH 8.

The K_m values for OAS are normally within the range 1.5–7.1 mM. The enzymes are not inhibited by OAS at concentrations up to 20 mM. The response of cysteine synthase isoenzymes to sulphide concentrations below 0.2 mM was also examined, and the K_m values for sulphide are normally within the range of 22–59 μM. Remarkably, kinetic parameters obtained for isoenzymes of L. leucocephala indicate Hill numbers of 2.3 for each enzyme.

The addition of PLP had an inhibitory effect of about 5% on the activities of enzymes at a concentration of 0.1 mM, and a higher concentration of 1 mM caused 15–30% inhibition. Cysteine synthases were also sensitive to PLP-enzyme inhibitors, 10 mM sodium borohydride caused about 95% inhibition, while 10 mM hydroxylamine had a less inhibitory effect.

The isoenzymes of cysteine synthase from plants display somewhat different relative activities, but their responses to OAS are essentially the same. Under standard assay conditions, the cysteine synthase isoenzymes are specific for OAS as a donor of the alanyl moiety, even though some activities were found in the presence of β-chloro-L-alanine and O-sulpho-L-serine under identical conditions. No detectable activity was found with O-phospho-L-serine or with L-serine.

Cysteine synthases also showed a distinct substrate specificity when a variety of thiol compounds or N-heterocyclic compounds were used as an acceptor for the alanyl moiety. The relative enzyme activities for these different substrates compared to L-cysteine formation are presented in Table 5.3. Among the substrates studied thus far, plant cysteine synthase isoenzymes could catalyse the formation of S-substituted L-cysteines and also heterocyclic β-substituted alanines such as β-(pyrazol-1-yl)-L-alanine, β-(1,2,4-triazol-1-yl)-L-alanine and β-(3-amino-1,2,4-triazol-1-yl)-L-alanine in low yields, when suitable substrates were provided. On the other hand, the formation of L-quisqualic acid, L-mimosine, L-willardiine or L-isowillardiine could not be

TABLE 5.2. Some properties of cysteine synthases in higher plants.

| Enzyme source | Enzyme | Molecular weight | | Pyridoxal 5′-phosphate (mol enzyme^{-1}) | K_m value | |
		Native enzyme (kDa)	Subunit (kDa)		O-Acetyl-L-serine (mM)	H$_2$S (mM)
Leucaena leucocephala	Cysteine synthase					
	Isoenzyme A	64	32	2	16.7	–[a]
	Isoenzyme B	64	32	2	6.7	–
Quisqualis indica	Cysteine synthase					
var. *villosa*	Isoenzyme A	58	29	n.d.[b]	1.9	0.059
	Isoenzyme B	58	29	n.d.	7.1	n.d.
Pisum sativum	Cysteine synthase					
	Isoenzyme A	52	26	2	2.1	0.036
	Isoenzyme B	52	26	2	2.3	0.038
Citrullus vulgaris	Cysteine synthase					
	Isoenzyme A	58	29	2	2.6	0.036
	Isoenzyme B	58	29	2	1.5	0.033
Spinacia oleracea	Cysteine synthase					
	Isoenzyme A	66	33	2	2.9[c]	0.022[c]
	Isoenzyme B	72	36	2		
Brassica juncea	Cysteine synthase	52	26	2	2.5	0.043
Brassica chinensis var. *Komatsuna*[d]	Cysteine synthase	62	31	2	6.1	n.d.
Raphanus sativus[d]	Cysteine synthase	66	33	n.d.	2.2	n.d.
Phaseolus vulgaris[e]	Cysteine synthase					
	Isoenzyme A	65	n.d.	n.d.	3.8	0.28
	Isoenzyme B	70	n.d.	n.d.	2.3	0.33

[a] Hill numbers are 2.3 for each enzyme.
[b] Not determined.
[c] Murakoshi *et al.* (1985b).
[d] Masada *et al.* (1975); Tamura *et al.* (1976).
[e] Bertagnolli and Wedding (1977).

TABLE 5.3. Relative synthetic rates of S-substituted L-cysteines and β-substituted alanines by cysteine synthases in higher plants.

Amino acid synthesised	Leucaena leucocephala Cysteine synthase		Quisqualis indica var. villosa Cysteine synthase		Pisum sativum Cysteine synthase		Citrullus vulgaris Cysteine synthase		Spinacia oleracea Cysteine synthase	
	Isoenzyme A	Isoenzyme B	Isoenzyme A	Isoenzyme B	Isoenzyme A	Isoenzyme B	Isoenzyme A	Isoenzyme B	Isoenzyme A	Isoenzyme B
L-Cysteine	100[a]	100	100	100	100	100	100	100	100	100
S-Methyl-L-cysteine	29.6	12.7	7.7	n.d.[b]	4.0	1.75	21.5	71.9	5.63	9.64
S-Allyl-L-cysteine	21.6	73.2	4.0	n.d.	2.6	7.08	0.46	5.76	3.05	2.45
S-Carboxymethyl-L-cysteine	0.35	0.97	0	n.d.	2.31	1.62	3.46	2.81	4.37	5.18
L-Quisqualic acid (3)	0	0	0	0.33	0	0	0	0	0	0
L-Mimosine (2)	0.03	15.7	n.d.	n.d.	0	0	0	0	0.01	0.02
L-Willardiine (5)	0	0	0	0	0	0	0	0	0	0
L-Isowillardiine (6)	0	0	0	0	0	0	0	0	0	0
β-(Pyrazol-1-yl)-L-alanine (1)	0.23	0.19	1.2	0.44	1.33	0.54	1.88	1.99	3.37	5.28
β-(3-Amino-1,2,4-triazol-1-yl)-L-alanine (9)	0.74	0.33	0.59	0.48	0.74	0.65	0.98	1.31	1.04	1.50
β-(1,2,4-Triazol-1-yl)-L-alanine (10)	n.d.	n.d.	n.d.	n.d.	5.81	3.13	n.d.	n.d.	17.8	11.9
β-Cyano-L-alanine	0.13	0	1.58	0	22.1	0	35.1	0	2.34	3.06

[a] The relative rates of synthesis (%) were compared with that of L-cysteine formed by each enzyme, respectively.
[b] Not determined.

demonstrated. These observations are in line with all our findings. In relation to the biosynthesis of L-willardiine and L-isowillardiine, Ahmmad *et al.* (1984) partially purified an uracilylalanine synthase from pea seedlings. This enzyme could be resolved into a L-willardiine synthase and a L-isowillardiine synthase, which were also separated from cysteine synthases on DEAE-Sephadex A-50 column chromatography (Fig. 5.1) (Ikegami *et al.*, 1987). Both uracilylalanine synthases were localised in the etioplasts in pea seedlings (Ikegami *et al.*, 1992). Moreover, the cysteine synthase isoenzyme B from *L. leucocephala* could catalyse the formation of L-mimosine in relatively high yields (Ikegami *et al.*, 1990b). The physicochemical properties of this enzyme are very similar to those of a L-mimosine synthase purified previously (Murakoshi *et al.*, 1984b). Therefore, we consider that the cysteine synthase isoenzyme B from *L. leucocephala* seedlings also functions as a L-mimosine synthase in the same manner as the formation of L-quisqualic acid in *Q. indica* var. *villosa* (Murakoshi *et al.*, 1986a) and β-(pyrazol-1-yl)-L-alanine in *Citrullus vulgaris* (Ikegami *et al.*, 1988a) are catalysed by a cysteine synthase isoenzyme.

The purified cysteine synthase also catalysed the formation of β-cyano-L-alanine from OAS and CN⁻ as an additional activity. This activity is different from the biosynthesis of β-cyano-L-alanine from L-cysteine and CN⁻ as catalysed by the β-cyanoalanine synthases (EC4.4.1.9) from microorganisms (Macadam and Knowles, 1984) and some higher plants (Manning, 1986; Ikegami *et al.*, 1988b,d, 1989).

The amino acid compositions of the purified cysteine synthases from plants are given in Table 5.4. In comparison with those from microorganisms (Kredish *et al.*, 1969; Yamagata, 1976) we found that only the enzyme from yeast contains tryptophan. The number of cysteine and methionine residues is invariably 1 and 5 respectively for the enzymes from microorganisms, while the plant enzymes contain 18–22 S-containing amino acids except for isoenzyme B from pea seedlings which contains 48. The plant enzymes and the yeast enzyme normally have 26–32 structurally important proline residues, while the *Salmonella typhimurium* isoenzymes contain 16–17 proline residues. Comparison of the amino acid compositions determined for purified cysteine synthases by a mathematical method (Chernoff, 1973) indicates that cysteine synthases from higher plants are closely related to each other. This makes it likely that a phylogenetic relationship exists among those enzymes.

The two cysteine synthase isoenzymes catalysing the formation of some heterocyclic β-substituted alanines have a different intracellular localisation (Ikegami *et al.*, 1992).

Among two cysteine synthase isoenzymes present in etiolated and green pea shoots, isoenzyme B was mainly located in the chloroplasts and etioplasts. Isoenzyme A, on the other hand, was located in mitochondria and possibly in the cytosol. This was also observed by Hock and Anderson (1978a) who reported large amounts of cysteine synthase activity in the chloroplast stroma of plants.

Our findings represent ample proof that the enzymes responsible for the biosynthesis of this group of non-protein amino acids are at the same time also responsible for the biosynthesis of the protein amino acid L-cysteine. Although these plant enzymes do not have significant differences in substrate specificity towards sulphide and OAS, they can have very different specificities for the heterocyclic substrate. As the amino acid sequences of these enzymes are not yet available, any conclusion on their evolutionary relatedness would be speculative. It has however been suggested (Murakoshi *et al.*, 1985b) that the isoenzymes of cysteine synthase may have arisen by gene duplication,

TABLE 5.4. Amino acid compositions of cysteine synthases purified from some higher plants.

Amino acids	Pisum sativum Cysteine synthase		Citrullus vulgaris Cysteine synthase		Spinacia oleracea Cysteine synthase	Brassica juncea Cysteine synthase
	Isoenzyme A	Isoenzyme B	Isoenzyme A	Isoenzyme B		
	Residues per 52 000 g		Residues per 58 000 g		Residues per 60 000 g[a]	Residues per 52 000 g
Asp	40	28	42	42	36	26
The	30	12	32	28	28	28
Ser	36	42	44	52	38	30
Glu	50	38	56	54	66	62
Pro	30	68	30	28	32	26
Gly	52	72	50	56	60	58
Ala	48	34	46	44	48	54
Val	36	30	40	36	42	32
Cys	16	44	4	14	6	6
Met	6	4	14	4	14	14
Ile	26	24	34	26	40	42
Leu	44	34	46	44	46	42
Tyr	2	2	18	18	14	10
Phe	22	14	22	24	26	14
Trp	0	0	0	0	0	0
Lys	28	18	28	24	42	32
His	6	6	10	12	4	4
Arg	22	36	22	32	18	16
Total	494	506	538	538	560	496

Results are expressed as residues per mole and based on the mol. wt. of each enzyme. Values for Thr and Ser are extrapolated to zero-time hydrolysis. The numbers of residues of amino acids were calculated based on the results of analyses after 24, 48 and 72 h acid hydrolysis of native enzyme. Means of duplicate analyses are given. Determination of tryptophan was made by alkaline hydrolysis.
[a] This result was recalculated based on the results of Murakoshi et al. (1985b).

after which one of the isoenzymes could more freely adapt to an evolving biochemical environment, eventually detoxifying internal toxins like cyanide, pyrazole, 3,4-dihydroxypyridine with the formation of β-cyano-L-alanine, β-(pyrazol-1-yl)-L-alanine and L-mimosine, respectively. These compounds may be less toxic for the plant species producing them and at the same time may also give an added evolutionary advantage as potential allelochemicals. Meanwhile at least one of the isoenzymes would ensure the biosynthesis of the primary metabolite L-cysteine. In some cases it was demonstrated that external heterocyclic toxins, e.g. the herbicides 1,2,4-triazole and Tachigaren (3-hydroxy-5-methylisoxazole), can be alanylated in a similar way by isoenzymes of cysteine synthase (Ikegami *et al.*, 1990a; Murakoshi *et al.*, 1985a).

III. BIOMIMETIC SYNTHESIS OF HETEROCYCLIC β-SUBSTITUTED ALANINES

In the course of our ongoing study on the β-substituted alanine synthases in higher plants forming heterocyclic β-substituted alanines, we have developed a very simple biomimetic synthesis of these β-substituted alanines catalysed by pyridoxal 5′-phosphate (PLP) and metal ions, in particular gallium (Scheme 5.3) (Murakoshi *et al.*, 1986a). The formation of tryptophan and β-(pyrazol-1-yl)alanine in low yield by the non-enzymatic reaction of indole or pyrazole and L-serine with pyridoxal or PLP and aluminium ion has already been demonstrated by Metzler *et al.* (1954) and by Dunnill and Fowden (1963).

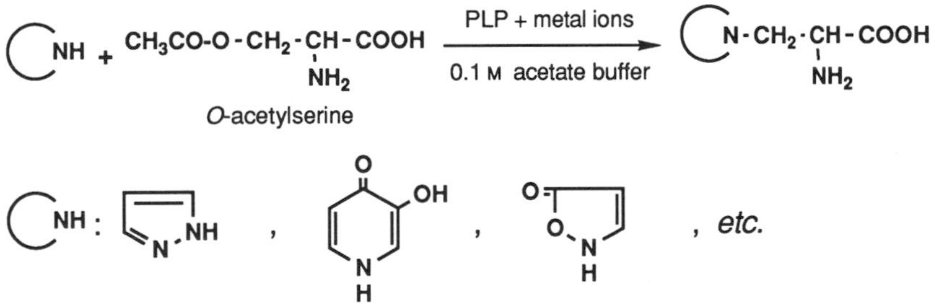

SCHEME 5.3. Biomimetic synthesis of heterocyclic β-substituted alanines by the PLP-catalysed reaction.

The results demonstrated that PLP and metal ions could catalyse the biomimetic synthesis of heterocyclic β-substituted alanines in the absence of protein, using the same substrates as are used in the enzymatic synthesis. In this biomimetic synthesis, several heterocyclic β-substituted alanines were synthesised by incubating 0.1 M acetate buffer solution (pH 3.5–5.5) containing the appropriate heterocyclic compounds and O-acetylserine or serine in the presence of PLP and metal ions. This PLP-catalysed chemical reaction depends upon pH, metal ions and temperature (Fig. 5.2, Table 5.5). The addition of Ga^{3+} ($Ga(NO_3)_3$), Fe^{3+} (iron alum) or Al^{3+} (potassium alum) enhanced the rate of synthesis about 15- to 20-fold compared to the controls. The chemical reaction shows no specificity for the optical form of the substrate and takes place at a lower pH than the biochemical reaction.

TABLE 5.5. Summary of the optimal reaction conditions for the formation of heterocyclic β-substituted alanines in the presence of PLP and Ga^{3+} by the PLP-catalysed chemical reaction.

Reaction products	pH	Temperature (°C)	Reaction time (h)	Yield (%)
β-(Pyrazol-1-yl)alanine (**1**)	3.7	62–65	2	40–45
Mimosine (**2**)	4.5	62–65	0.5	1–2
Quisqualic acid (**3**)	5.0	62–65	2	8–10
Lupinic acid (**4**)	4.0	30–33	2	1.5–1.7
Willardiine (**5**)	5.5	33–35	2	0.8–1.0
Isowillardiine (**6**)	5.5	83–85	2	0.1
β-(3-Isoxazolin-5-on-2-yl)alanine (**7**)	4.0	30	2	0.15
β-(3-Amino-1,2,4-triazol-1-yl)alanine (**9**)	4.7	62–65	2	30–35
β-(6-Benzylaminopurin-9-yl)alanine (**11**)	4.5	62–65	2	1.5–2
β-(5-Methylisoxazolin-3-on-2-yl)alanine (**12**)	4.7	62–65	2	35–40
β-(2-Furoyl)alanine[a]	4.0	62–65	1	5–7
Histidinoalanine[b]	3.5	62–65	2	0.7–0.8

[a] This compound was obtained by hydrolytic cleavage with 6 N HCl, providing indirect evidence of the formation of ascorbalamic acid (Couchman *et al.*, 1973).
[b] A cross-linking amino acid found in proteins (Fujimoto *et al.*, 1982).
From Murakoshi *et al.* (1986a) with permission from the Pharmaceutical Society of Japan.

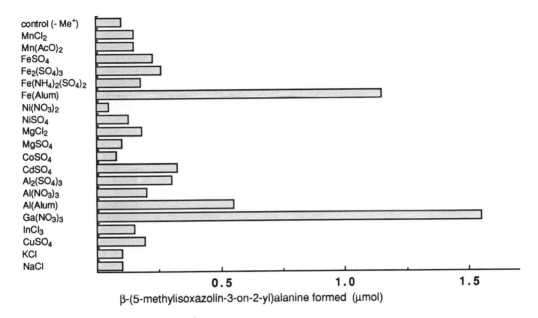

FIG. 5.2. Comparative effects of various metal ions on the biomimetric synthesis of β-(5-methylisoxazo-lin-3-on-2-yl)alanine in the presence of PLP under the optimal reaction conditions. The concentration of metal ions was fixed at 1.25 mM per 12.5 mM O-acetylserine. From Murakoshi *et al.* (1986a) with permission from the Pharmaceutical Society of Japan.

Our results suggest that this biomimetic method can be applied to the synthesis of heterocyclic β-substituted alanines which are relatively difficult to synthesise, or which have not been synthesised before, from OAS and the corresponding heterocyclic compounds, using simple reaction conditions. This procedure might be used as a more economic way for the synthesis of some amino acids.

ACKNOWLEDGEMENTS

We are grateful to Sir Leslie Fowden, Rothamsted Experimental Station, Harpenden, UK, for his encouragement during this work, and to Drs F. Lambein, Y.-H. Kuo and G. Ongena, State University of Ghent, Belgium, for their helpful advice and criticism of the manuscript. We also are indebted to Professor E. A. Bell (King's College, London, UK), Dr L. Fellows (Kew Gardens, UK), Professor E. G. Brown (University College of Swansea, UK) and Professor C. H. Stammer (University of Georgia, Athens, USA) for cordially providing authentic samples.

REFERENCES

Ahmmad, M. A. S., Maskall, C. S. and Brown, E. G. (1984). *Phytochemistry* **23**, 265–270.
Andrews, P. (1965). *Biochem. J.* **96**, 595–606.
Bell, E. A. (1976). *FEBS Lett.* **64**, 29–35.

Bertagnolli, B. L. and Wedding, R. T. (1977). *Plant Physiol.* **60**, 115–121.
Bradford, M. M. (1976). *Anal. Biochem.* **72**, 248–254.
Chernoff, H. (1973). *J. Am. Stat. Assoc.* **68**, 361–368.
Couchman, R., Eagles, J., Hegarty, M. P., Laid, W. M., Self, R. and Synge, R. L. M. (1973). *Phytochemistry* **12**, 707–718.
Dunnill, P. M. and Fowden, L. (1963). *J. Exp. Bot.* **14**, 237–248.
Fowden, L., Lea, P. J. and Bell, E. A. (1979). *Adv. Enzymol.* **50**, 117–175.
Fujimoto, D., Hirama, M. and Iwashita, T. (1982). *Biochem. Biophys. Res. Commun.* **104**, 1102–1106.
Gaitonde, M. K. (1967). *Biochem. J.* **104**, 627–633.
Gaitonde, M. K. and Dovey, T. (1970). *Biochem. J.* **117**, 907–911.
Giovanelli, J. and Mudd, S. H. (1968). *Biochem. Biophys. Res. Commun.* **31**, 275–280.
Gmelin, R. (1959), *Hoppe-Seyler's Z. Physiol. Chem.* **316**, 164–169.
Hadwiger, L. A., Floss, H. G., Stoker, J. R. and Conn, E. E. (1965). *Phytochemistry* **4**, 825–830.
Hegarty, M. P. and Court, R. D. (1964). *Austral. J. Agric. Res.* **15**, 165–167.
Hegarty, M. P., Schinckel, P. G. and Court, R. D. (1964). *Austral. J. Agric. Res.* **15**, 153–165.
Hock Ng, B. and Anderson, J. W. (1978a). *Phytochemistry* **17**, 879–885.
Hock Ng, B. and Anderson, J. W. (1978b). *Phytochemistry* **17**, 2069–2074.
Ikegami, F., Kaneko, M., Lambein, F., Kuo, Y.-H. and Murakoshi, I. (1987). *Phytochemistry* **26**, 2699–2704.
Ikegami, F., Kaneko, M., Kamiyama, H. and Murakoshi, I. (1988a). *Phytochemistry* **27**, 697–701.
Ikegami, F., Takayama, K., Tajima, C. and Murakoshi, I. (1988b). *Phytochemistry* **27**, 2011–2016.
Ikegami, F., Kaneko, M., Kobori, M. and Murakoshi, I. (1988c). *Phytochemistry* **27**, 3379–3383.
Ikegami, F., Takayama, K. and Murakoshi, I. (1988d). *Phytochemistry* **27**, 3385–3389.
Ikegami, F., Takayama, K., Kurihara, T., Horiuchi, H., Tajima, C., Shirai, R. and Murakoshi, I. (1989). *Phytochemistry* **28**, 2285–2291.
Ikegami, F., Komada, Y., Kobori, M., Hawkins, D. R. and Murakoshi, I. (1990a) *Phytochemistry* **29**, 2507–2508.
Ikegami, F., Mizuno, M., Kihara, M. and Murakoshi, I. (1990b). *Phytochemistry* **29**, 3461–3465.
Ikegami, F., Itagaki, S., Ishikawa, T., Ongena, G., Kuo, Y.-H., Lambein, F. and Murakoshi, I. (1991). *Chem. Pharm. Bull.* **39**, 3376–3377.
Ikegami, F., Horiuchi, H., Kobori, M., Morishige, I. and Murakoshi, I. (1992). *Phytochemistry* **31**, 1991–1996.
Ishizaki, T., Kato, K., Kumada, M., Takemoto, T., Nakajima, T., Takagi, N. and Koike, K. (1973). *Jap. J. Parasit.* **22**, 181–186.
Jackson, C., Dench, J. E., Hall, D. O. and Moore, A. L. (1979). *Plant Physiol.* **64**, 150–153.
King, J. and Laemmli, U. K. (1971). *J. Mol. Biol.* **62**, 465–477.
Kredish, N. M., Becker, M. A. and Tomkins, G. M. (1969). *J. Biol. Chem.* **244**, 2428–2439.
Kumagai, H., Yamada, H., Matsui, H., Ohkishi, H. and Ogata, K. (1970). *J. Biol. Chem.* **245**, 1773–1777.
Lambein, F. and Van Parijs, R. (1968). *Biochem. Biophys. Res. Commun.* **32**, 474–479.
Lambein, F., Kuo, Y.-H. and Van Parijs, R. (1976). *Heterocycles* **4**, 567–593.
Lowry, J. B., Tangendjaja, B. and Cook N. W. (1985). *J. Sci. Food Agric.* **36**, 799–807.
Macadam, A. M. and Knowles, C. J. (1984). *Biochim. Bioshys. Acta* **786**, 123–132.
Manning, K. (1986). *Planta* **168**, 61–66.
Masada, M., Fukushima, K. and Tamura, G. (1975). *J. Biochem.* **77**, 1107–1115.
Metzler, D. E., Ikawa, M. and Snell, E. E. (1954). *J. Am. Chem. Soc.* **76**, 648–652.
Mills, W. R. and Joy, K. W. (1980). *Planta* **148**, 75–83.
Murakoshi, I., Kuramoto, H., Haginiwa, J. and Fowden, L. (1972). *Phytochemistry* **11**, 177–182.
Murakoshi, I., Kato, F., Haginiwa, J. and Fowden, L. (1973) *Chem. Pharm. Bull.* **21**, 918–920.

Murakoshi, I., Kato, F., Haginiwa, J. and Takemoto, T. (1974a). *Chem. Pharm. Bull.* **22**, 473–475.

Murakoshi, I., Kato, F. and Haginiwa, J. (1974b). *Chem. Pharm. Bull.* **22**, 480–481.

Murakoshi, I., Ikegami, F., Kato, F., Haginiwa, J., Lambein, F., Van Rompuy, L. and Van Parijs, R. (1975). *Phytochemistry* **14**, 1515–1517.

Murakoshi, I., Ikegami, F., Ookawa, N., Haginiwa, J. and Letham, D. S. (1977). *Chem. Pharm. Bull.* **25**, 520–522.

Murakoshi, I., Ikegami, F., Ookawa, N., Ariki, T., Haginiwa, J., Kuo, Y.-H. and Lambein, F. (1978a). *Phytochemistry* **17**, 1571–1576.

Murakoshi, I., Ikegami, F., Harada, K. and Haginiwa, J. (1978b). *Chem. Pharm. Bull.* **26**, 1942–1945.

Murakoshi, I., Ikegami, F., Ariki, T., Harada, K. and Haginiwa, J. (1979). *Chem. Pharm. Bull.* **27**, 2484–2487.

Murakoshi, I., Koide, C., Ikegami, F. and Nasu, K. (1983). *Chem. Pharm. Bull.* **31**, 1777–1779.

Murakoshi, I., Ikegami, F., Hinuma, Y. and Hanma, Y. (1984a). *Phytochemistry* **23**, 973–977.

Murakoshi, I., Ikegami, F., Hinuma, Y. and Hanma, Y. (1984b). *Phytochemistry* **23**, 1905–1908.

Murakoshi, I., Ikegami, F., Hama, T. and Nishino, K. (1984c). *Shoyakugaku Zasshi* **38**, 355–358.

Murakoshi, I., Ikegami, F., Nishimura, T. and Tomita, K. (1985a). *Phytochemistry* **24**, 1693–1695.

Murakoshi, I., Ikegami, F. and Kaneko, M. (1985b). *Phytochemistry* **24**, 1907–1911.

Murakoshi, I., Ikegami, F., Yoneda, Y., Ihara, H., Sakata, K. and Koide C. (1986a). *Chem. Pharm. Bull.* **34**, 1473–1478.

Murakoshi, I., Kaneko, M., Koide, C. and Ikegami, F. (1986b). *Phytochemistry* **25**, 2759–2763.

Noe, F. F and Fowden, L. (1960). *Biochem. J.* **77**, 543–546.

Sandelius, A. S. and Selstam, E. (1984). *Plant Physiol.* **76**, 1041–1046.

Schenk, S. U. and Werner, D. (1991). *Phytochemistry* **30**, 467–470.

Smith, I. K. (1977). *Phytochemistry* **16**, 1293–1294.

Smith, I. K. and Thompson, J. F. (1971). *Biochim. Biophys. Acta* **227**, 288–295.

Takemoto, T., Takagi, N., Nakajima, T. and Koike, K. (1975). *Yakugaku Zasshi* **95**, 176–179.

Tamura, G., Iwasawa, T., Masada, M. and Fukushima, K. (1976). *Agric. Biol. Chem.* **40**, 637–638.

Thompson, J. F. and Moore, D. P. (1968). *Biochem. Biophys. Res. Commun.* **31**, 281–286.

Tiwari, H. P., Penrose, W. R. and Spenser, I. D. (1967). *Phytochemistry* **6**, 1245–1248.

Yamagata, S. (1976). *J. Biochem.* **80**, 787–797.

6 Enzymes of N-Heterocyclic Ring Synthesis

E. G. BROWN

Biochemistry Research Group, School of Biological Sciences, University College of Swansea, Swansea SA2 8PP, UK

I. INTRODUCTION

Although the enzymes of N-heterocyclic ring biosynthesis are appropriately included in this volume on secondary metabolism, it should be remembered that there are few,

METHODS IN PLANT BIOCHEMISTRY Vol. 9
ISBN 0–12–461019–6

if any, primary metabolic pathways that do not depend, directly or indirectly, on the participation of N-heterocyclic compounds. For example, most coenzymes have within their structures N-heterocyclic ring components. The ubiquitous redox coenzymes NAD and NADP are pyridines, so too is the coenzyme of most amino acid-metabolising enzymes, i.e. pyridoxal 5-monophosphate. The riboflavin-derived coenzymes FMN and FAD are isoalloxazines, and S-adenosylmethionine and 3'-phosphoadenosine 5'-phosphosulphate (PAPS) are purines. Since they each contain an AMP moiety within their respective molecular structures, NAD, NADP, FAD and coenzyme A could also be regarded as purines. Thiamine pyrophosphate has a pyrimidine moiety, and biotin, the carboxylation coenzyme, is a fused imidazole. The one-carbon unit carrier tetrahydrofolic acid is a pteridine; another carrier coenzyme system, that of the 'sugar nucleotides' (e.g. UDP-glucose and ADP-glucose) uses pyrimidine and purine ribotides; and alcohols such as ethanolamine and ribitol are transported by the pyrimidine ribotide CDP.

In man and most animals, the majority of coenzymes represent the functional form, at a cellular level, of vitamins. By definition, man cannot synthesise vitamins and they are essential dietary constituents for the maintenance of health. Higher plants, being autotrophic, can meet all their own coenzyme requirements and thus they possess the necessary enzyme-catalysed mechanisms for biosynthesising all these N-heterocyclic rings. In addition to these rings and those of the more complex alkaloids, plants synthesise a variety of other unusual N-heterocycles, such as pyrazole and azetidine. Despite their importance to the plants concerned and their potential importance to man, many of these biosynthetic processes in plants remain obscure and little or nothing is known of many of the enzymes involved in them. This is an area of plant biochemistry ripe for development; some of these enzymes catalyse novel chemical transformations of high potential in biotechnology and industrial biotransformation processes. Many of the products have potential as industrial feedstocks and are, or eventually will be, targets of genetic manipulation. However, until more is known of the enzymes concerned, little progress can be made in that direction. It is hoped, therefore, that this Chapter will not only provide the basic information for further studies of the known enzymes but will stimulate further research in this neglected but nevertheless important field of plant biochemistry.

In juxtaposing the topics of N-heterocyclic ring formation and plant secondary metabolism, it is inevitable that alkaloid biosynthesis will spring to mind. The more complex polycyclic ring systems of most alkaloids are, however, beyond the scope of this present Chapter which, instead, concerns itself primarily with the enzymic production of simple N-heterocycles. Table 6.1 shows the structures considered and their occurrence in Nature.

II. ENZYMES CATALYSING-FORMATION OF SINGLE RINGS

A. Aziridine

Little is known of the biosynthetic origin of this ring system (Table 6.1) which is represented in Nature by the fused aziridine ring of the mitomycins produced by various species of *Streptomyces*.

B. Azetidine

This higher homologue of aziridine is found within the molecule of the non-protein amino acid azetidine 2-carboxylic acid which is produced by several species of the Liliaceae and by *Delonix regia*. Work with *Convallaria majalis* (lily-of-the-valley) and employing radioactive precursors (Leete *et al.*, 1974) is consistent with the biosynthetic scheme shown below (Scheme 6.1).

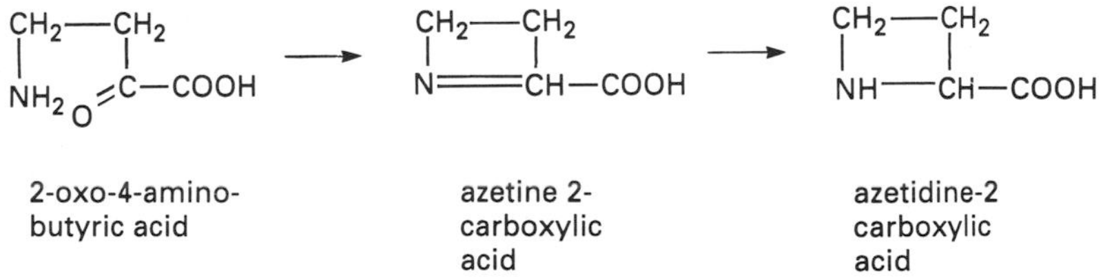

| 2-oxo-4-amino-
butyric acid | azetine 2-
carboxylic
acid | azetidine-2
carboxylic
acid |

SCHEME 6.1.

There are no reports of the preparation of active enzymic extracts capable of catalysing these steps.

C. Pyrrole

Pyrrole synthesis occurs in all plants, animals and microorganisms in the form of porphobilinogen production during porphyrin biosynthesis. in this process, two molecules of 5-aminolaevulinic acid are condensed by the enzyme porphobilinogen synthase (ALA-dehydratase; EC 1.5.1.2) as shown in Scheme 6.2.

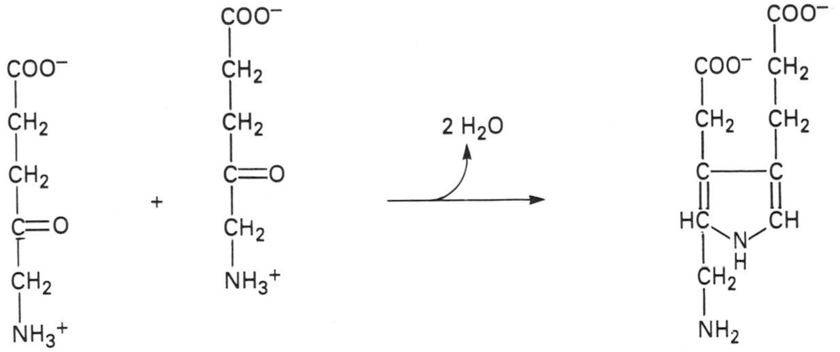

SCHEME 6.2.

1. Porphobilinogen synthase (ALA-dehydratase)

(a) From soybean callus cultures. The following procedure for extracting and partially purifying porphobilinogen synthase from soybean callus culture cells was described by Tigier *et al.* (1968). All the steps are carried out at 4°C.

TABLE 6.1. *N*-heterocyclic ring synthesis by plants.

Name	Ring system	Examples	Plant source
Aziridine	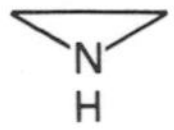	Only fused derivatives, e.g. the mytomycins, are known in nature	*Streptomyces* spp.
Azetidine	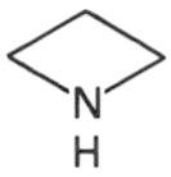	Azetidine 2-carboxylic acid	Many species of Liliaceae. Seedlings of *Delonix regina*
Pyrrole	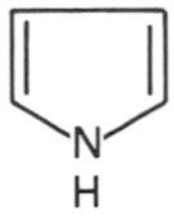	Porphobilinogen	All plants (precursor of chlorophyll)
Pyrrolidine	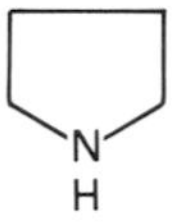	Proline Nicotine Stachydrine	Proline is ubiquitous Nicotine in *Nicotiana tabacum* Stachydrine in *Stachys tuberifera*
Δ^1-Pyrroline	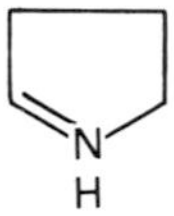	Free base; Δ^1-pyrroline-2-carboxylic acid	Intermediate in proline synthesis; ubiquitous
Pyrazole	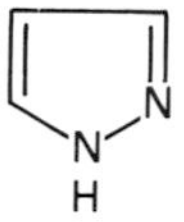	Pyrazol-1-yl-L-alanine	Various species of Cucurbitaceae
Imidazole	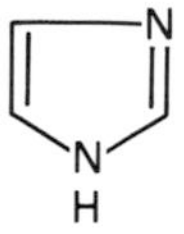	Histidine; AICR	Ubiquitous
Pyridine	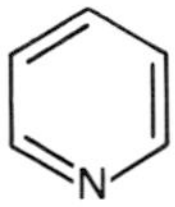	Pyridoxal; NAD; NADP; Nicotine; anabasine; picolinic acid; quinolinic acid; mimosine; ricinine; trigonelline	The coenzymes are ubiquitous. Tobacco, *Mimosa* spp. *Ricinus communis*, *Trigonella* spp.
Piperidine	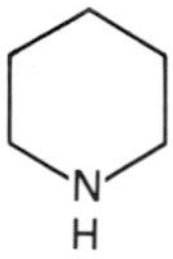	Pipecolic acid; Baikiain; *trans*-4-hydroxy pipeidic acid; *trans*-5-hydroxy pipecolic acid	*Lobelia* spp. *Conium maculatum*

TABLE 6.1. Cont.

Name	Ring system	Examples	Plant source
Pyrazine		Aspergillic acid; pulcherriminic acid	*Aspergillus* spp.; *Candida pulcherrima*
Pyrimidine		Uracil; cytosine; pyrimidine amino acids; vicine; convicine	The common bases are ubiquitous. Various legumes
Indole		Trytophan; indole alkaloids	Tryptophan is ubiquitous; indole alkaloids in Apocynaceae, Loganiaceae, Rubiaceae
Benzimidazole		5,6-Dimethyl-benzimidazole	*Streptomyces* spp., *Bacillus* spp., *Propionibacterium* spp.
Purine		Adenine; guanine	Ubiquitous as ribotides
Pyrrolopyrimidine		Tubericidin; toyocamycin; sangivamycin	*Streptomyces* spp.
Pteridine		Folic acid; THFA	Ubiquitous
Triazolopyrimidines		Toxoflavin; fervenulin	*Pseudomonas cocovenenans; Streptomyces fervens*
Isoalloxazine (benzo-pteridine)		Riboflavin; FAD; FMN	Ubiquitous

The callus culture cells are harvested and a 60% (wet volume) suspension in 0.1 M glycine-NaOH buffer (pH 9.0) is homogenised using a Potter–Elvehjem homogeniser. The homogenate is centrifuged at 800 × g for 10 min and the sediment washed by resuspension and recentrifuging a further three times. All the supernatants are pooled and centrifuged at 70 000 × g. The enzymic activity remains in the supernatant, which is then taken and made 50% saturated with respect to ammonium sulphate. The pH is maintained at 9.0 by adding NaOH as necessary. The precipitated protein is collected by centrifuging and redissolved in 0.1 M glycine-NaOH buffer (pH 9.0). This is then loaded onto a column (1.8 cm × 30 cm) of Sephadex G-100 previously equilibrated with the glycine-NaOH buffer. The same buffer is used to elute the column and the eluate monitored for porphobilinogen synthase activity. The pooled, active fractions are again 50% saturated with ammonium sulphate and the sediment stored at −15°C. This fraction is the final enzymic preparation. Tigier et al. (1968) reported that starch-gel electrophoresis of it at pH 7.4 and pH 8.8 revealed only a single band; the properties are similar to those of the enzyme isolated from other sources.

(b)   From yeast (Saccharomyces cerevisiae).   An early report (De Barreiro, 1967) also described the preparation of porphobilinogen synthase from yeast. In this procedure, cells are washed and freeze-dried, ground to a dry powder, and the powder suspended in acetone at −15°C using an homogeniser (Virtis; 40 000 rpm for 20 min). The homogenate is kept in a solid CO_2–ethanol bath during the homogenisation step. The powder is removed by filtration and washed with cold acetone. This yields a stable powder which, as required, can be extracted with 20 mM phosphate buffer at pH 7.6 (20 g per 100 ml of buffer).

The extract is further purified by ammonium sulphate fractionation, the 30–50% saturation fraction, which contains the activity, is dialysed against 20 mM phosphate buffer (pH 7.6) and the non-diffusible fraction treated with four portions, each of 1 ml, of $Ca_3(PO_4)_2$ gel (32 mg ml^{-1}). The gel is eluted with 0.1 M glycine-NaOH buffer (pH 9.6) containing 5% (w/v) ammonium sulphate. This is done by centrifuging. The supernatant is dialysed against 10 mM glycine-NaOH buffer at pH 7.7 and passed through a DEAE-cellulose column previously equilibrated with the same buffer. Protein is eluted by increasing the molarity of the glycine buffer, stepwise, from 10 to 200 mM. Four peaks of increasing activity are obtained at glycine concentrations of 20, 50, 100 and 200 mM. The most active is the latter. This can be further purified by polyacrylamide gel electrophoresis.

(c)   Immobilised porphobilinogen synthase from Rhodopseudomonas spheroides. An interesting preparation of this enzyme from a biotechnology viewpoint, is that described by Gurne and Shemin (1973). They have developed an immobilised porpho-bilinogen synthase from cells of the photosynthetic bacterium *Rhodopseudomonas spheroides*. This preparation takes the form of a column which has been operated for 27 days at 36°C with only 30% loss of catalytic activity.

(d)   Assay.   Porphobilinogen synthase is assayed by measuring the rate of production of porphobilinogen. The concentration of the pyrrole is determined colorimetrically following addition of a modified Ehrlich's reagent (Urata and Granick, 1963). The reagent is prepared by adding to 168 ml of glacial acetic acid, 40 ml of 70%

(v/v) perchloric acid, 4.0 g of *p*-dimethylaminobenzaldehyde, and 0.7 g of $HgCl_2$; the solution is made up to a final volume of 220 ml with water. To 2 ml of the porphobilinogen-containing sample is added 2 ml of reagent and after 15 min the increase in A_{555} is measured (ε 68 000 litre mol cm^{-1}).

D. Pyrroline and pyrrolidine

1. Δ¹-Pyrroline 5-carboxylate reductase

The pyrrolidine ring of the amino acid proline arises from the spontaneous cyclisation of glutamic 5-semialdehyde to Δ¹-pyrroline 5-carboxylic acid and subsequent enzymic reduction of the pyrroline ring as shown below in Scheme 6.3.

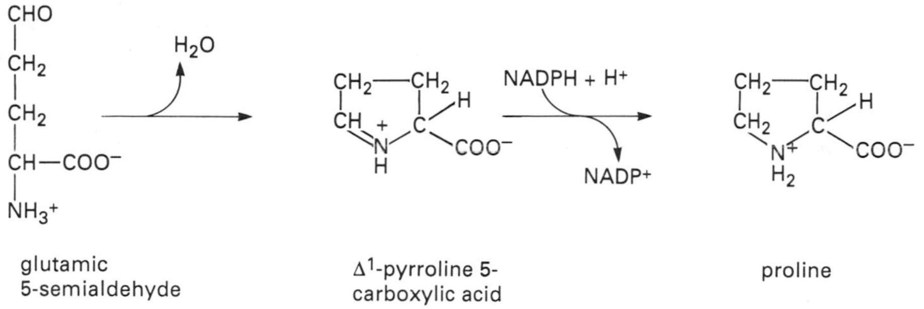

| glutamic 5-semialdehyde | Δ¹-pyrroline 5-carboxylic acid | proline |

SCHEME 6.3.

The enzyme catalysing the process is Δ¹-pyrroline 5-carboxylate reductase (EC 1.5.1.2). It can be partially purified from tobacco leaves as described below (Noguchi *et al.*, 1966).

(a) Extraction and purification of Δ¹-pyrroline 5-carboxylate reductase. Young, expanded leaves of *Nicotiana tabacum* (1 kg) are freed from midribs and rapidly frozen in a deep-freezer. Using a Waring blender they are then homogenised for 5 min in 3 l of 0.1 M phosphate buffer (pH 7.5) containing 0.01 M sodium ascorbate and 2 mM EDTA. The homogenate is filtered through four layers of cheesecloth and the filtrate clarified by centrifuging at 2500 × g for 15 min. The supernatant is 25% saturated with ammonium sulphate; the pH is kept at 7.0–7.5 by adding aq. ammonia as necessary. The precipitate is collected by centrifuging at 16 000 × g for 5 min and discarded. A further addition of ammonium sulphate is made to bring the supernatant to 50% saturation and the precipitate, which contains the bulk of the enzymic activity, is collected by centrifuging and then redissolved in 300 ml of 0.05 M phosphate buffer (pH 7.0) containing 5 mM 2-mercaptoethanol and 1 mM EDTA. Noguchi *et al.* (1966) designated this 'Fraction A'.

Fraction A is again fractionated using ammonium sulphate, and the precipitate obtained at 30–45% saturation dissolved in 150 ml of phosphate buffer (Fraction B). Fraction B is again fractionated and the 35–45% saturation fraction collected and dissolved in 10 ml of phosphate buffer (Fraction C). This latter fraction is then loaded onto a column (3 × 25 cm) of Sephadex G-200 equilibrated and eluted with the same buffer. The active fractions are pooled (Fraction D) and the 45–50% saturation ammonium sulphate fraction obtained. This is redissolved in 10 ml of the phosphate buffer and subjected to the same Sephadex G-200 chromatographic step as before. The active fractions are again pooled and the 45–50% saturation ammonium sulphate fraction collected and redissolved in phosphate buffer. Using this procedure, Noguchi *et al.* (1966) obtained a 158-fold purification.

The final preparation so obtained, has a pH optimum of 7.0 with either NADH or NADPH, and has a preference for the latter as coenzyme. The activity is inhibited by Ag$^+$ and by *p*-chloromercuribenzoate.

2. N-*methylputrescine oxidase*

The *N*-methylpyrrole moiety of nicotine originates as the *N*-methyl Δ^1-pyrrolinium cation and is formed from putrescine via *N*-methylputrescine and 4-methylamino-butanal as shown in Scheme 6.4. The oxidative deamination of *N*-methylputrescine to 4-methylaminobutanal is catalysed by *N*-methylputrescine oxidase and cyclisation of 4-methylaminobutanal to *N*-methyl Δ^1-pyrrolinium cation then follows spontaneously. The enzyme has been extracted from tobacco roots and partially purified by Mizusaki *et al.* (1972).

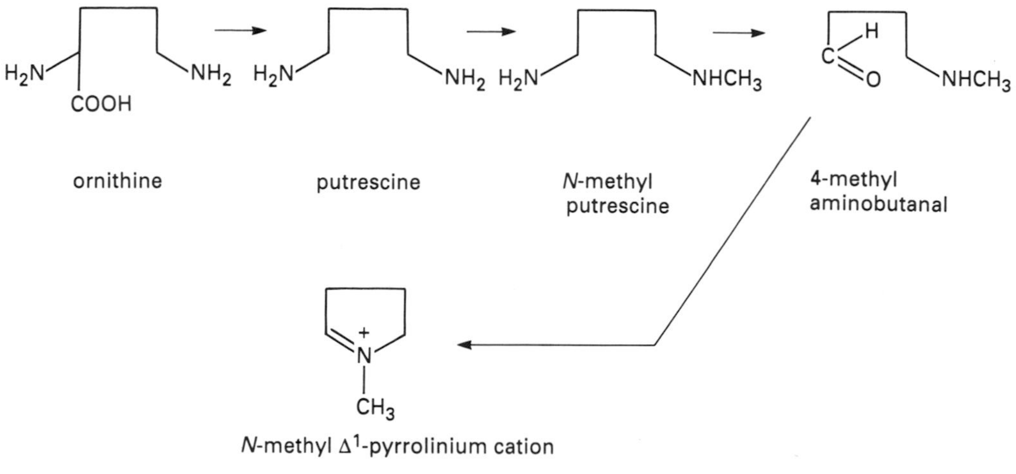

ornithine   putrescine   *N*-methyl putrescine   4-methyl aminobutanal

N-methyl Δ^1-pyrrolinium cation

SCHEME 6.4.

(a) Extraction and purification of N-*methylputrescine oxidase.* In the original method, tobacco plants were grown hydroponically for 1 week then decapitated and left for 24 h to increase the activity of *N*-methylputrescine oxidase in the roots (Mizusaki *et al.*, 1971). The basic extraction and purification procedure is as follows. Roots (600 g sample) are homogenised with 1–2 l of a 50 mM Tris-HCl buffer (pH 7.4) containing 10 mM 2-mercaptoethanol, 5 mM EDTA, 0.5% (w/v) sodium ascorbate and 1% (v/v) polyethyleneglycol 400. After removing crude debris by filtration

through nylon cloth, the filtrate is clarified by centrifuging at $10\,000 \times g$ for 15 min. Ammonium sulphate is added to the supernatant to give 65% saturation and after standing for 1 h the precipitate is resuspended in 200 ml of a buffer essentially similar to that used for extraction except that the polyethylene glycol is omitted. This preparation is dialysed overnight against 10 mM Tris-HCl buffer (pH 7.4) containing 10 mM mercaptoethanol and 1 mM EDTA (buffer A). The non-diffusible fraction is loaded onto a DEAE-cellulose column (2.5×15 cm) previously equilibrated with the buffer used for dialysis. The column is washed with 300 ml of the same buffer and the enzyme eluted in buffer A containing 0.2 M NaCl. Fractions comprising the enzymic activity are pooled and the protein concentrated by adding solid ammonium sulphate to 60% saturation. The precipitate is collected by centrifuging and dissolved in 3 ml of 25 mM Tris-HCl buffer (pH 7.4) containing 1 mM EDTA, and 0.2 M NaCl. This same buffer is then used to equilibrate a column (2.8×90 cm) of Sephadex G-200 and the sample loaded. The protein is allowed to flow into the bottom of the column and then eluted by reversing the flow of buffer; the recommended flow rate is 15 ml h^{-1} and 10 ml fractions are collected. In Mizusaki's paper, fractions 18–24 were pooled and brought to 60% saturation with ammonium sulphate. The precipitated enzymic activity is dissolved in 10 ml of buffer A and dialysed overnight against the same buffer. The non-diffusible fraction is applied to a DEAE-cellulose column (1×12 cm) previously equilibrated with buffer A. The enzyme is eluted in a linear gradient of NaCl (0–0.4 M) in buffer A in a total of 200 ml; 2 ml fractions are collected. In Mizusaki's work, fractions 36–42 were found to exhibit high activity and were accordingly pooled. Ammonium sulphate is added to the pooled, active fractions to give 60% saturation. This suspension can be stored at 0°C for up to a week without loss of activity. When required, portions of this suspension are dialysed against 10 mM Tris-HCl buffer (pH 7.4) containing 10 mM mercaptoethanol, or desalted on a Sephadex G-25 column.

(b) Assay. *N*-Methylputrescine oxidase activity is determined by measuring the radioactivity of 1-[^{14}C]methyl-2-cyanopyrrolidine converted from *N*-[^{14}C] methyl-pyrrolinium salt, a reaction product of the enzymic incubation, in the presence of excess KCN. Aqueous KCN reacts quickly and completely with *N*-methylpyrrolinium salt to give 1-methyl-2-cyanopyrrolidine (Leonard and Hauck, 1957). The enzymic reaction mixture consists of 20 μmol of Tris-HCl (pH 8.0), 0.4 μmol of *N*-[^{14}C]methyl-putrescine (2.35×10^5 dpm μM^{-1}), 50 μg of catalase and the enzyme preparation, all in a final volume of 0.25 ml. After 30 min incubation at 30°C, the reaction is stopped by adding 0.2 ml of 1% (w/v) KCN solution and 5 ml of toluene. The tube is stoppered and shaken vigorously. One ml of the toluene-based phase is counted for radioactivity in 10 ml of an appropriate scintillation fluid. Mizusaki *et al.* (1972) used a toluene-based scintillant (0.01% 1,4-bis[2(4-methyl-5-phenyloxazolyl)]-benzene and 0.4% 2,5-diphenyloxazole).

3. Polyamine oxidase

Spermine is oxidised by polyamine oxidase from barley leaves to yield 1,3-diamino-propane and 1-(3-aminopropyl)pyrroline (see Scheme 6.5). The position of the double bond in the pyrroline is not yet confirmed.

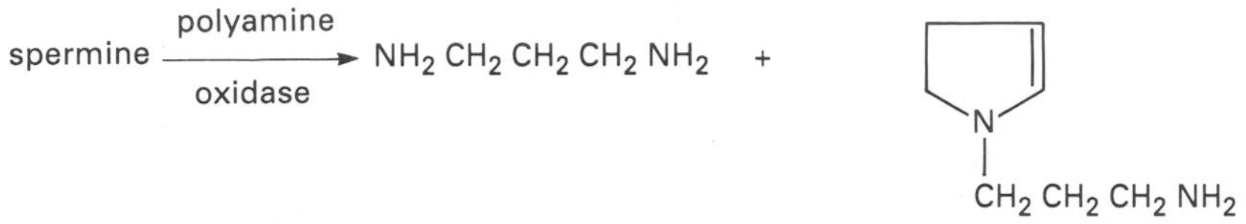

spermine $\xrightarrow[\text{oxidase}]{\text{polyamine}}$ $NH_2\ CH_2\ CH_2\ CH_2\ NH_2$ +

(pyrroline ring structure with N)

$CH_2\ CH_2\ CH_2\ NH_2$

SCHEME 6.5.

This enzyme is associated with nucleic acid-containing particles which sediment in low centrifugal fields. The enzyme can be released by washing the particles with 0.5 M NaCl and has been extracted and purified from etiolated oat leaves (Smith 1974, 1976, 1977). The same enzyme also catalyses the oxidation of spermidine to diaminopropane and pyrroline (Smith 1970).

(a) Extraction and purification of polyamine oxidase from barley. The procedure described by Smith (1976) for the extraction and purification of the enzyme from barley leaves is as follows.

The particulate fraction of barley leaves is extracted with 0.5 M NaCl and the extract used undialysed. After cooling the extract to 0°C, an equal volume of acetone at -15°C is added and the precipitate is collected by centrifuging at $3000 \times g$ for 5 min at 0°C. The precipitate is redissolved in 0.5 M NaCl (pH 4.5) and to this solution cellulose phosphate (Whatman P11) is added in the proportion of $2\ mg\ ml^{-1}$. The suspension is dialysed for 18 h against NaCl-free citrate buffer (pH 4.5) and the cellulose collected by centrifuging. After further washing with NaCl-free buffer, the enzyme is recovered by eluting the cellulose phosphate in 0.5 M NaCl in acetate buffer (pH 4.5). The enzyme (2 ml) is chromatographed on a column ($3.5 \times 87\ cm$) of Sephadex G-100 equilibrated with 0.5 M NaCl in the pH 4.5 acetate buffer. Smith (1976) used a flow rate of $3\ ml\ h^{-1}$ and 8 ml fractions were collected. His column was calibrated with blue dextran (monomer and dimer) and cytochrome c, and he reported that the enzyme elutes as a sharp band with a retention volume corresponding to a molecular mass of 85 kDa. The procedure gives a 60- to 70-fold purification with 50–100% recovery.

(b) Extraction from oat leaves. To extract polyamine oxidase from oat, the following procedure is described by Smith (1977). Oat leaves are macerated in 4 volumes of cold water and the macerate squeezed through muslin. The residue is re-extracted with 2 volumes of cold water. The activity is located in the combined filtrate from which it can be recovered by acidification to pH 4 with citric acid. The precipitate is collected by centrifuging at $2500 \times g$ for 15 min and extracted into pH 6 citrate buffer containing 1 M NaCl. After centrifuging the NaCl extract at $3000 \times g$ for 15 min, the supernatant is cooled to 0°C and 1 volume of acetone (-15°C) added. The precipitate, collected by centrifugation, is extracted into 1 M NaCl in pH 6 citrate buffer and chromatographed on the Sephadex G100 column as before. After elution, the enzyme is concentrated by ultrafiltration (Amicon YM10 membrane). In the original work, Smith (1977) reported that after the G-100 step, the final fraction (60 ml) contained $20\ nkat\ ml^{-1}$ at a specific activity of $6020\ nkat\ mg^{-1}$.

(c) Further purification of polyamine oxidase.   Smith (1977) described the further purification of the enzyme by electrophoresis. The procedure employs small pore gels (7.5% acrylamide at pH 4.3) with a β-alanine buffer. Sucrose (50 mg ml^{-1}) is added to samples containing *c.* 3000 nkat ml^{-1}. Samples (10–100 μl) are added to the anodic end of each tube without stacking gel, together with 10 μl 0.1% (w/v) methyl green in 5% (w/v) sucrose as marker. After stacking for 1 h, at 2 mA per tube, the voltage is increased to 100 V (10 mA per tube). When, after about 30 min, the methyl green has migrated to within 3 mm of cathode end, the tubes are removed. In the experiment described, the final voltage was 120 V and the current 8 mA per tube. Proteins were stained for protein with Coomassie Blue.

To detect the enzyme, the unstained gel is immersed in 10 ml of pH 6.5 phosphate buffer containing spermidine (50 μM) guaiacol (50 μM) and peroxidase (200 purpurogallin units). The enzyme gives an intense brown band within 5 min of immersion.

Assay is by the method of Smith (1974).

4. Diamine oxidase

Diamine oxidase will also catalyse the formation of a pyrroline ring. For example, putrescine is oxidised by this enzyme to pyrroline as shown in Scheme 6.6. Diamine oxidase has been purified to homogeneity from pea seedlings (Hill, 1971), and more recently from *Lathyrus sativus* (Suresh *et al.*, 1976; Suresh and Adiga, 1979).

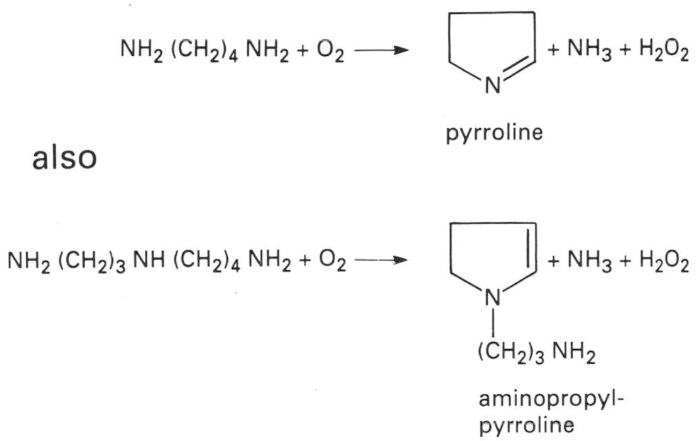

$$NH_2 (CH_2)_4 NH_2 + O_2 \longrightarrow \text{pyrroline} + NH_3 + H_2O_2$$

also

$$NH_2 (CH_2)_3 NH (CH_2)_4 NH_2 + O_2 \longrightarrow \text{aminopropyl-pyrroline} + NH_3 + H_2O_2$$

aminopropyl-
pyrroline

SCHEME 6.6.

E. Pyrazole

1. Pyrazole synthase

Recent studies (Brown *et al.*, 1982; Brown and Diffin, 1990) have shown that the pyrazole ring of the non-protein amino acid β-pyrazol-1-ylalanine originates from the cyclisation of 1,3-diaminopropane. This is a particularly interesting biochemical

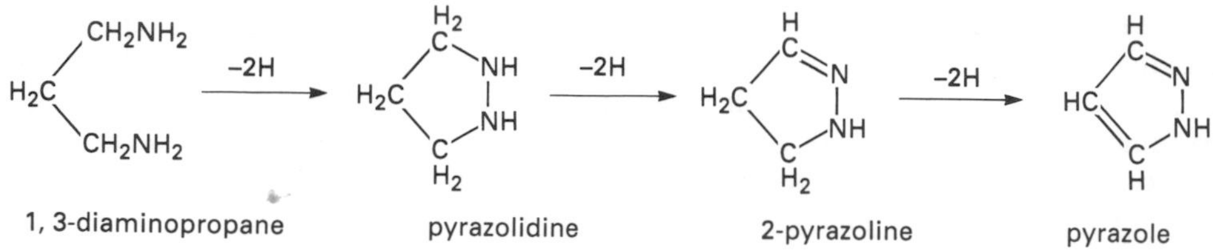

SCHEME 6.7.

process since it involves formation of a N–N covalent bond; the enzyme catalysing the reaction is the only one of its kind known at this time (see Scheme 6.7).

Enzymic cyclisation of 1,3-diaminopropane is catalysed by a pyrazole synthase preparation obtained from cucumber seedlings. Formation of pyrazole by the preparation not only involves cyclisation but also sequential dehydrogenation of the initial saturated ring product, pyrazolidine, via 2-pyrazoline. The enzyme preparation requires FAD for its activity (Brown and Diffin, 1990). The substrate, 1,3-diaminopropane, arises from the oxidative degradation of spermine catalysed by diamine oxidase (see Section II.D.3).

(a) Assay of pyrazole synthase. Under appropriate conditions, pyrazole reacts with trisodium pentacyanoaminoferrate ($Na_3[Fe(CN)_5NH_3]$) to yield a yellow colour which can be determined spectrophotometrically at 458 nm to form the basis of a colorimetric assay (LaRue, 1965). This method is, however, unsuitable for assaying pyrazole biosynthesised from 1,3-diaminopropane since the latter compound also forms a coloured complex with the reagent. Instead, Brown *et al.* (1982) and Brown and Diffin (1990) have made use of the enzyme β-pyrazol-1-yl-L-alanine synthase, also found in cucumber seedlings, and which effectively scavenges for pyrazole, rapidly reacting it with *O*-acetylserine to form β-pyrazol-1-ylalanine. This is a stoichiometric reaction rapidly going to completion. The product is chemically and physiologically stable and easily isolated.

For extraction of the β-pyrazol-1-yl-L-alanine synthase activity required for this pyrazole assay, the procedure of Murakoshi *et al.* (1984) is recommended. In this procedure, 23 day-old cucumber seedlings (*Cucumis sativus*) are washed, cooled for 30 min at 0°C, and homogenised in a Waring blender at top speed for 2 min in a 0.1 M potassium phosphate buffer (pH 7.5) containing sucrose (0.25 M) and 2-mercaptoethanol (10 mM); buffer is used in the proportion of 1.25 ml g^{-1} fresh weight of tissue. For each 1 ml of buffer, 50 mg of insoluble polyvinyl-*N*-pyrrolidone (Polyclar AT) is added before homogenising. The homogenate is passed through nylon cloth and the filtrate centrifuged at 15 000 × g for 30 min. The supernatant represents a crude β-pyrazol-1-ylalanine synthase preparation. To purify it further, it is placed in a water bath at 80°C and stirred vigorously until the temperature of the preparation reaches 55°C. It is then gently swirled, by hand, in the water bath for a further 1 min and chilled to 4°C in an ice bath. The precipitate is removed by centrifuging at 10 000 × g for 20 min and the supernatant taken for fractionation with ammonium sulphate.

Sufficient ammonium sulphate is added to the supernatant to give 40% saturation. After stirring at 4°C for 60 min, the protein precipitate is removed by centrifuging at

$10\,000 \times g$ for 10 min. The supernatant is brought to 60% saturation with ammonium sulphate and again stirred for 60 min at 4°C. This precipitate, which contains the enzymic activity, is removed by centrifuging at $10\,000 \times g$ for 10 min and then redissolved in potassium phosphate buffer (30 mM; pH 7.4). After dialysis overnight against a large volume of the same buffer, the non-diffusible fraction is used as the β-pyrazol-1-yl-L-alanine preparation for the pyrazole synthase assay.

For assay of pyrazole synthase activity, the incubation mixture (final volume 0.45 ml) consists of 0.1 ml of the sample containing 150–200 μg of protein in potassium phosphate buffer (30 mM; pH 7.4), 1,3-diaminopropane (45 μmol), 0.1 ml of pyrazol-1-yl-L-alanine synthase (100 μg protein) in potassium phosphate buffer (30 mM; pH 7.4) and *O*-acetyl-L-[3-¹⁴C]serine (1.25 μmol; 0.6 μCi). Appropriate controls are advisable to guard against effects of extracts on the coupling enzyme; incubation is at 28°C for 3 h.

After the reaction has been terminated by adding 20 μl of 10% (v/v) aq. NH_3 to each tube, precipitated protein is removed by centrifugation. Aliquots (200 μl) of the aqueous supernatants are spotted onto Whatman No. 1 chromatography paper and the chromatograms developed for 3 days at room temperature allowing the solvent (butan-1-ol saturated with 3 M aq. NH_3) to run off the end of the paper. The radioactive β-pyrazol-1-yl-L-alanine is eluted in 0.01 M HCl and its radioactivity determined by scintillation spectrometry.

O-Acetyl-L-[3-¹⁴C]serine can be synthesised for use in the above assay by appropriate modification of the method of Sheehan *et at.* (1956). L-[3-¹⁴C]Serine (50 μCi; specific radioactivity 55 mCi mmol⁻¹) in 2% (v/v) ethanol is evaporated to dryness under N_2. Glacial acetic acid (0.5 ml) is added and the resulting solution is saturated with anhydrous HCl at 0°C. It is then left to stand at room temperture for 15 h. The reaction mixture is subjected to high voltage paper electrophoresis and the *O*-acetyl-L-[3-¹⁴C]serine separated from the residuum of unreacted L-[3-¹⁴C]serine. For this purpose, Whatman 3 MM paper is used in a formic acid–acetic acid buffer (1.5 M; pH 2.0) prepared by mixing 50 ml of glacial acetic acid with 40 ml of formic acid (98–100%) and making up to 1 l with distilled water. A current of 50 mA (35 V cm⁻¹) is applied for 1 h. The *O*-acetylserine band is located by simultaneously running reference samples and using ninhydrin reagent (0.1 w/v in acetone) to detect them.

(b)   Preparation of pyrazole synthase.   The enzyme system responsible for cyclising 1,3-diaminopropane to pyrazolidine and then dehydrogenating this, first to 2-pyrazoline and then to pyrazole, can be extracted from *Cucumis sativus* seedlings as follows.

Shoots from 23-day-old seedlings are cooled for 30 min at 0°C and homogenised at top speed in a Waring blender for 2 min in a potassium phosphate buffer (0.1 M; pH 7.5) containing sucrose (0.25 M), 2-mercaptoethanol (10 mM), and insoluble poly-*N*-vinylpyrrolidone (Polyclar AT; 50 mg ml⁻¹). For each 1 g of tissue, 1.25 ml of buffer solution is used. After squeezing the homogenate through nylon cloth to remove coarse debris, the filtrate is centrifuged at $15\,000 \times g$ for 30 min. The supernatant is dialysed overnight against a large excess of the same phosphate buffer.

Further purification of this crude dialysed preparation is achieved by fractionation with ammonium sulphate. All the activity resides in the 0–40% saturation fraction. A further increase in specific activity is attained by gel-filtration chromatography on a column (1 × 20 cm) of Sephadex G-100 kept at 2–4°C. Before use, the column is washed with 4 bed-volumes of potassium phosphate buffer (30 mM; pH 7.4) at a flow rate of 2.5 ml cm² h⁻¹. Samples (0.5 ml) are loaded and elution commenced with the same buffer at the same flow rate. Fractions (0.5 ml) are collected. In the work described by Brown and Diffin (1990) the major peak of activity eluted in fractions 13–19. The pooled active fractions can be concentrated by ultrafiltration or by ammonium sulphate precipitation followed by dialysis.

The enzyme requires FAD for its activity; 50 μM FAD gives a doubling of activity. NAD$^+$ and NADP$^+$ are without effect. Except for Ca^{2+}, none of a variety of metal ions have any significant effect; Ca^{2+} (50 μM) gives a 30% inhibition. If the enzyme is previously dialysed against EDTA, this is increased to 40% inhibition. Neither FAD nor Ca^{2+} had any detectable effect on the coupling enzyme (β-pyrazol-1-yl-L-alanine synthase) used in the enzyme assay. The pH optimum of the cyclising and dehydrogenating enzyme is 6.2 and that for β-pyrazol-1-ylalanine synthase is 8.0. With the coupled assay, described above, two pH optima are seen, corresponding to these two values. The apparent molecular mass of pyrazole synthase is 66 kDa (Brown and Diffin, 1990).

F.   Imidazole

The imidazole ring of the amino acid histidine originates in a pathway which involves ring-opening of the pyrimidine moiety of an adenine derivative, namely N^1-5'-phosphoribosyl-AMP. The essential step is shown below (Scheme 6.8). Although this process in animals and microorganisms has been studied in some detail, nothing appears to be known of the enzymes involved in plants.

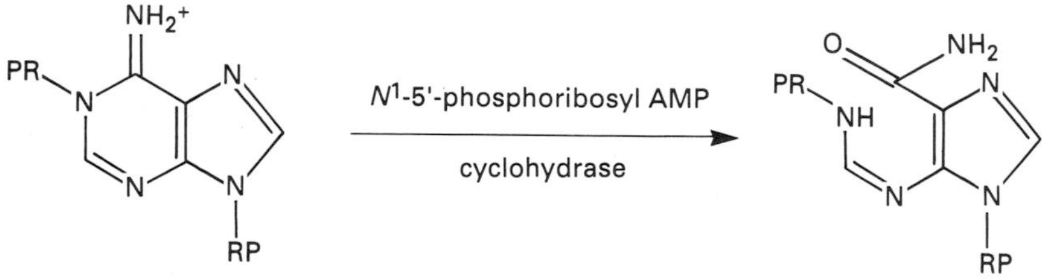

SCHEME 6.8.

Another imidazole derivative, 5-aminoimidazole ribotide (**IV**), is an intermediate in purine biosynthesis and for this purpose it is biosynthesised from glycine, ribose, and ATP via glycinamide ribotide (**I**), N-formylglycinamide ribotide (**II**), and N-formylglycinamidine ribotide (**III**) (see Scheme 6.9).

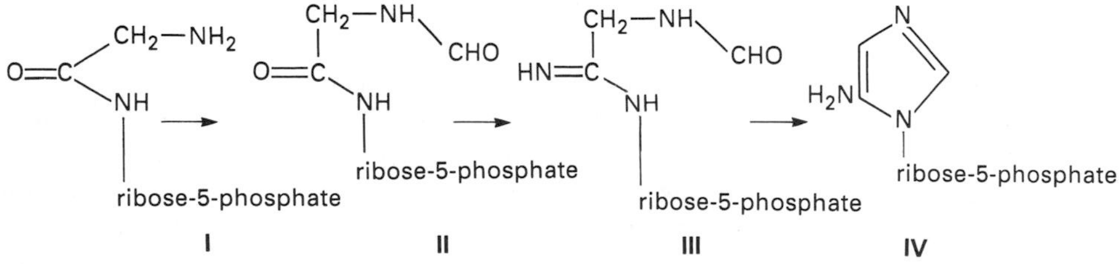

SCHEME 6.9.

Once again, however, although this pathway has been studied in plant tissues, using radioactive precursors, and it has been confirmed to be essentially similar in plants to that in animals and microorganisms, nothing is known of the plant enzymes involved.

G. Pyridine

In addition to the pyridine nucleotides NAD and NADP, which are of ubiquitous occurrence in Nature, there are a number of pyridine alkaloids and other pyridine-derived secondary products found in plants; examples include nicotine, anabasine, picolinic acid, quinolinic acid, mimosine, and trigonelline.

The pyridine nucleotides NAD and NADP arise in higher plants, via nicotinic acid from aspartate (V) and a three-carbon unit derived from glycerol, and believed to be D-glyceraldehyde 3-phosphate (VI). A key intermediate in this process is quinolinic acid (VIII) (see Scheme 6.10). Nothing is known of the enzymes involved.

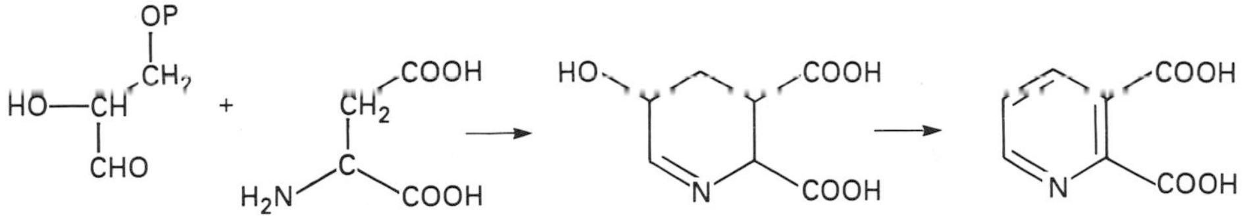

SCHEME 6.10.

Ricinine (3-cyano-4-methoxy-*N*-methylpyridin-2-one) is synthesised from nicotinic acid via *N*-methyl-3-cyanopyridine. Nicotine- and anabasine-related alkaloids are formed from nicotinic acid in a number of plant families. In *Nicotiana tabacum*, nicotine is produced by condensation of 3,6-dihydronicotinic acid with the *N*-methyl-Δ^1-pyrrolinium cation (see Section I.D.2).

Quinolinic acid is also formed in plants from 3-hydroxyanthranilic acid which is oxidised by 3-hydroxyanthranilate 3,4-dioxygenase to 2-amino-3-carboxymuconic acid 6-semialdehyde; the latter spontaneously cyclises to quinolinic acid as shown in Scheme 6.11. The oxygenase however has only been extracted and purified from beef liver (Decker *et al.*, 1961) and there are no reports of studies of the enzyme in plants.

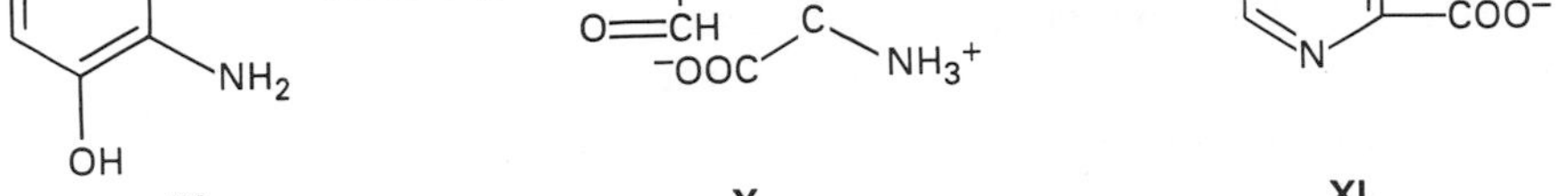

SCHEME 6.11.

H. Piperidine

Pipecolic acid (piperidine 2-carboxylic acid) is formed in higher plants from lysine (**XII**) via 2-aminoadipic semialdehyde (**XIII**) which results from the action of lysine aminotransferase. The semialdehyde spontaneously cyclises to piperideine 6-carboxylic acid (**XIV**) and this is reduced to pipecolic acid (**XV**) by pipecolate dehydrogenase (Scheme 6.12).

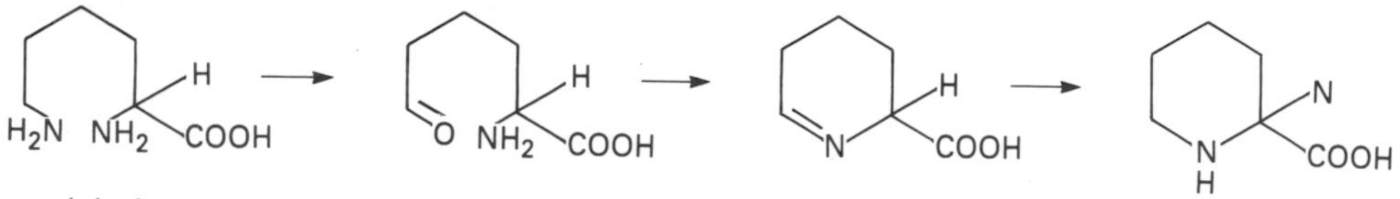

L-lysine   L-α-aminoadipic   L-Δ'-piperideine-6-   L-pipecolic acid
                        semialdehyde   carboxylic acid

XII   **XIII**   **XIV**   **XV**

SCHEME 6.12.

During the formation of γ-coniceine (Roberts, 1975), a piperideine ring is synthesised from 5-oxooctanal (**XVI**). This is essentially a transamination to give the corresponding amine (**XVII**) since the latter spontaneously cyclises to γ-coniceine (**XVIII**).

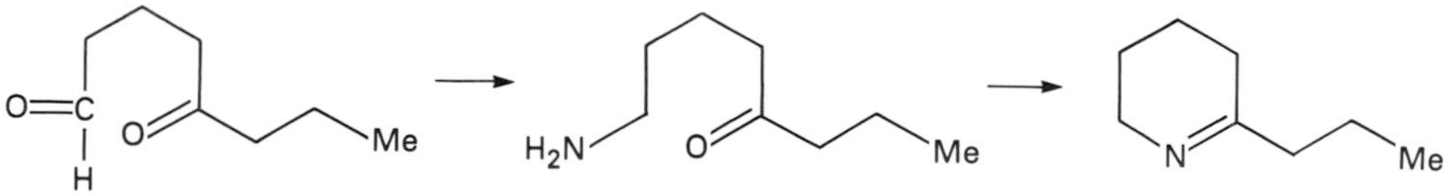

5-oxooctanal   1-amino-5-oxooctane   γ-coniceine

XVI   **XVII**   **XVIII**

SCHEME 6.13.

(a) Preparation of 5-oxooctanal aminotransferase. The enzyme catalysing the above transamination and resulting in the formation of a piperideine ring has been purified 369-fold from leaves of *Conium maculatum* (Roberts, 1977).

To obtain the enzyme by the method described by Roberts (1977) an acetone powder is first prepared. Fresh young leaves, with petioles removed, are homogenised with acetone at −20°C at top speed in a blender; five successive volumes of acetone are used. After filtration, the powder is vacuum-dried at room temperature. Extraction of this powder and all subsequent procedures are effected at 0–5°C.

The acetone powder is extracted by gentle stirring for 1 h with 10 volumes of 50 mM Tris buffer (pH 7.5) containing dithiothreitol (1 M) and EDTA (1 mM). After centrifuging at 38 000 × g for 10 min, the supernatant is collected and 0.5 volume of protamine sulphate solution is added to it slowly with stirring. The resulting dark brown precipitate is sedimented by centrifuging at 38 000 × g for 10 min and discarded. Sufficient ammonium sulphate is added to the supernatant to bring it to 65% saturation and the precipitate obtained is collected by centrifuging (30 000 × g for 10 min). This is redissolved in a minimal volume of 50 mM Tris buffer (pH 7.5) containing dithiothreitol (1 mM), and dialysed against the same standard buffer. Calcium phosphate gel (BDH; 3% w/v suspension in water) was added to the dialysed preparation at 1 ml per 120 mg of protein. After stirring for 5 min, the gel was removed by low speed centrifuging. The remaining solution was dialysed for 18 h against the standard buffer.

The dialysed preparation (non-diffusible fraction) is loaded onto a column (1.6 × 30 cm) of DEAE-cellulose previously equilibrated with the same buffer as before. Elution is commenced with this buffer and after 3 volumes have passed through, it is replaced with a similar one containing 0.1 M KCl. The active protein elutes in the KCl buffer solution and is dialysed against the original buffer to remove the KCl. The protein fraction is then loaded onto a second similar DEAE-cellulose column and eluted using a linear gradient of 0–0.3 M KCl in the standard buffer. Fractions (3 ml) are collected and the activity elutes at 0.04 M KCl. The active fractions are collected and pooled.

Roberts (1977) reported that at all stages of purification the preparation showed aminotransferase activity without the need for addition of pyridoxal phosphate. Even after dialysis for 2 days little alteration in activity was seen. Nevertheless, as addition of pyridoxal phosphate (1 mM) doubled the activity at all stages of purification, this coenzyme was included, at 1 mM concentration, in all assays. Pre-incubation of the enzyme with pyridoxal phosphate for 5 min was enough to give maximum activation. The enzyme is inhibited by carbonyl reagents, and by reagents that react with $-SH$ groups. Hydroxylamine, KCN and isonicotinic acid hydrazide were the most effective inhibitors.

I. Pyrazine

There are a number of pyrazine derivatives produced by fungi, e.g. pulcherriminic acid (*Candida pulcherrima*), aspergillic acid, neoaspergillic acid and flavacol (*Aspergillus* spp.). The pyrazine ring is produced biosynthetically by the condensation of two amino acid molecules. These can be two identical amino acid molecules, as illustrated in Scheme 6.14, or two different amino acid molecules, as illustrated in Scheme 6.15. Nothing is known of the enzymes involved in pyrazine formation.

J. Pyrimidine

The pyrimidine ring mainly arises via the orotic acid pathway. Carbamoyl phosphate and aspartic acid are condensed to yield carbamoyl aspartate which ring-closes to a dihydropyrimidine. This is subsequently dehydrogenated (Scheme 6.16). The enzyme forming 5,6-dihydroorotate (i.e. a dihydropyrimidine) from carbamoyl aspartate is dihydroorotase.

1. Dihydrooratase

The enzyme catalysing ring closure of carbamoyl aspartate to form the 5,6-dihydro-pyrimidine dihydroorotic acid has been purified 320-fold from pea seedlings by Mazuś and Buchowicz (1968).

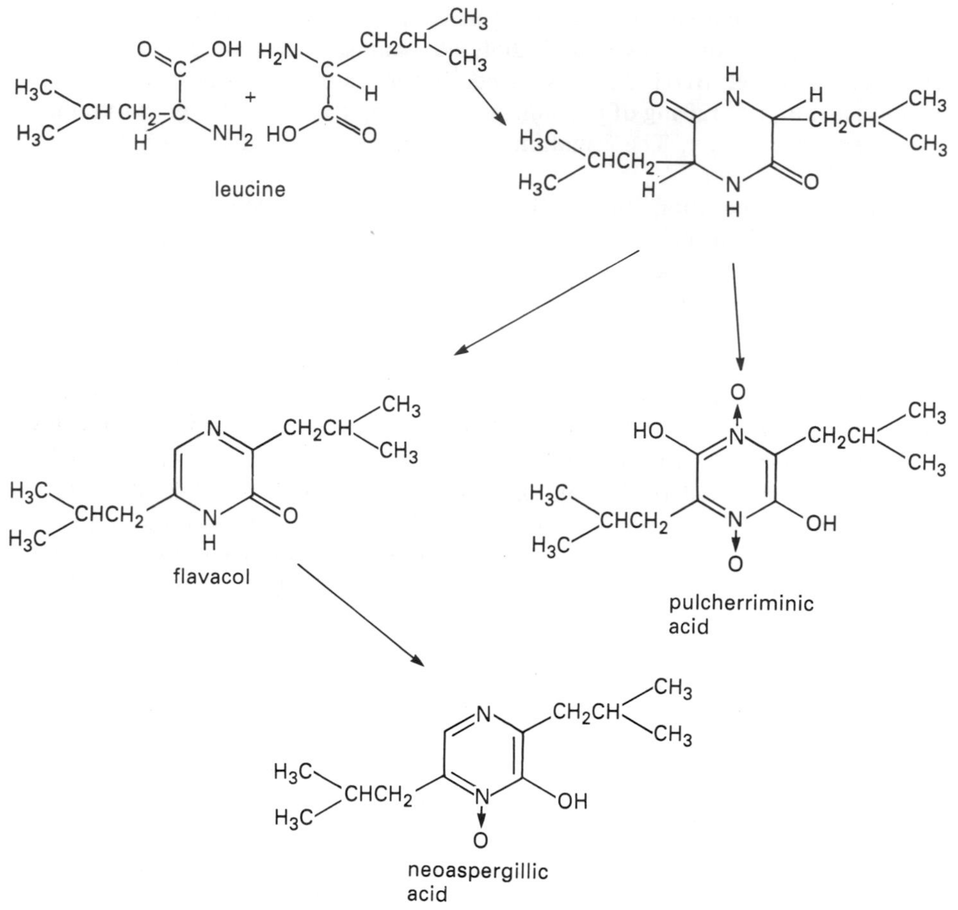

SCHEME 6.14.

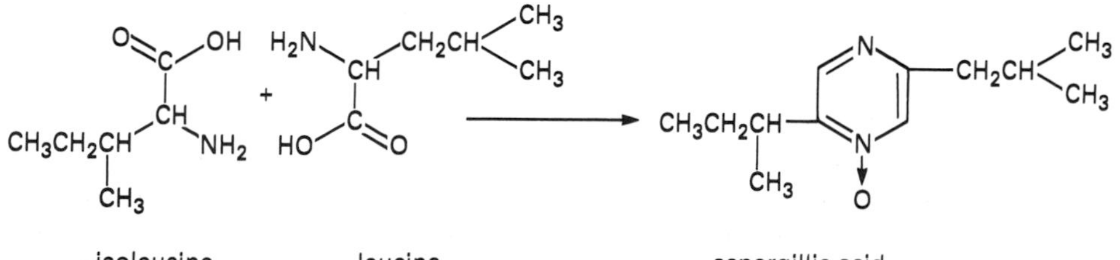

SCHEME 6.15.

(a) Assay of dihydroorotase (synthesis of dihydroorotate). For the standard incubation mixture, the enzymic activity is prepared in 0.1 M citrate-potassium phosphate buffer (pH 6.0) containing 1 mM 2-mercaptoethanol and 0.1 mM EDTA. To 0.2 ml portions, containing about 1 mg of protein, 1 μmol of the substrate dissolved

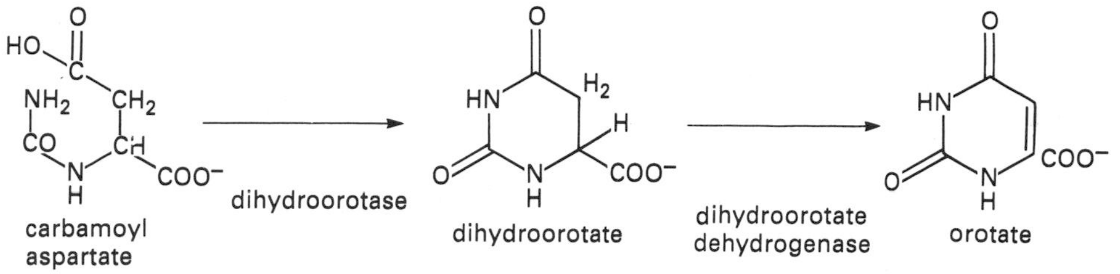

dihydroorotase

carbamoyl
aspartate

dihydroorotate

dihydroorotate
dehydrogenase

orotate

SCHEME 6.16.

in 0.02 ml of the same buffer is added and the mixture incubated at 37°C. Omission of EDTA and 2-mercaptoethanol decreases activity. It is not affected by dialysis nor by addition of divalent cations. After incubating for 1 h at 37°C, the reaction is stopped by adding 5 ml of 0.3 M perchloric acid. The precipitated protein is removed by centrifuging and the supernatant chromatographed on paper using propan-1-ol–water–HCl (30:10:1; v/v) and butan-1-ol–water–acetic acid (2:1:1; v/v). Dihydroorotate is detected with Ehrlich's reagent (*p*-dimethylaminobenzaldehyde) by the method of Fink *et al.* (1956) and after elution, is determined by the procedure of Janion and Shugar (1960).

(b) Enzyme extraction and purification (Mazuś and Buchowicz, 1968). Top leaves (500 g) of 2-week-old pea seedlings are homogenised with 100–200 ml of 0.1 M potassium phosphate buffer (pH 7.4) containing 1 mM 2-mercaptoethanol and 0.1 mM EDTA. Homogenisation and all subsequent steps are carried out at 0–4°C. After filtering the homogenate through cheesecloth, the filtrate is clarified by centrifuging at 6000 × g for 10 min. The supernatant is taken and made 40% saturated with ammonium sulphate. After 1 h, the precipitate is removed by centrifuging and discarded. The supernatant is made 65% saturated with ammonium sulphate and after standing, the protein precipitate is collected by centrifuging. From 500 g of fresh plant material about 1 g of protein is obtained. This crude preparation contains both dihydroorotase and dihydropyrimidinase activities of similar magnitude but the dihydropyrimidinase can be selectively destroyed by heat. For this purpose, 1 g of the crude preparation is dissolved in 36 ml of 0.1 M potassium phosphate buffer (pH 6.0) and heated at 70°C for 3 min.

After heat treatment, the preparation is again subjected to fractional precipitation with ammonium sulphate. The fraction precipitating at 20–40% saturation is collected by centrifuging at 10 000 × g for 10 min. This is dissolved in 2 ml of 0.01 M potassium phosphate buffer (pH 7.2) and loaded onto a Sephadex G-100 column (1 × 17 cm) that is previously equilibrated with the same buffer. The effluent is collected in 1 ml fractions at a flow rate of 0.5 ml min^{-1}. Dihydroorotase elutes in fractions 19–25. These fractions are pooled and layered onto a column of DEAE-cellulose (1 × 15 cm) previously equilibrated with 0.01 M potassium phosphate buffer (pH 7.2). The column is washed successively with 50 ml portions of the same buffer and then with 50 ml of 0.03 M phosphate buffer at the same pH. The effluents are discarded. To elute the enzyme, 0.05 M phosphate buffer is used at a flow rate of 0.5 ml min^{-1}. The enzyme elutes between 6 and 18 ml. These fractions are pooled and concentrated by placing in dialysis tubing and immersing in dry Aquacide I for 3 h at 4°C. The latter procedure reduces the volume tenfold without significant loss of enzymic activity. A 321-fold purification of dihydroorotase was obtained by Mazuś and Buchowicz (1968) using this overall procedure.

2. Dihydroorotate dehydrogenase

The first true pyrimidine in the biosynthetic sequence is orotate. This is produced by the dehydrogenation of 5,6-dihydroorotate, catalysed by dihydroorotate dehydrogenase. One of the few reports of studies of this enzyme in higher plants is that of Kapoor and Waygood (1965) in which they show that wheat embryos contain an NAD-dependent enzyme reducing orotate to dihydroorotate. Their method for the isolation and partial purification of this enzyme is described below (Kapoor and Waygood, 1965).

(a) Isolation and purification of dihydroorotate dehydrogenase from wheat. Viable wheat embryos (10 g) are ground in a mortar with 100 ml of 0.05 M K_2HPO_4–0.01 M cysteine hydrochloride (pH 7.4). The extract is passed through four layers of cheesecloth at 20 000 × g for 10 min. The supernatant is 33% saturated with ammonium sulphate and the protein precipitate sedimented by centrifuging and discarded. Ammonium sulphate is added to the supernatant to give 65% saturation and the precipitated protein collected by centrifuging. The protein pellet is redissolved in 10 ml of 0.02 M phosphate buffer, pH 6.5. This is the enzyme preparation; it shows a broad pH optimum curve with a peak at pH 6.

    Other pathways do exist in plants for the biosynthesis of individual pyrimidines but these are specific processes for the pyrimidine concerned and not general routes for the assembly of pyrimidine molecules. For example, a 5,6-diaminopyrimidine intermediate in riboflavin biosynthesis is produced by loss of C-8 from a purine (see Sections III.A(a); III.F). Also, the pyrimidine moiety of thiamine appears to be biosynthesised by a process involving ring expansion of 5-aminoimidazole ribotide (AIR), an intermediate in purine biosynthesis (see Scheme 6.17).

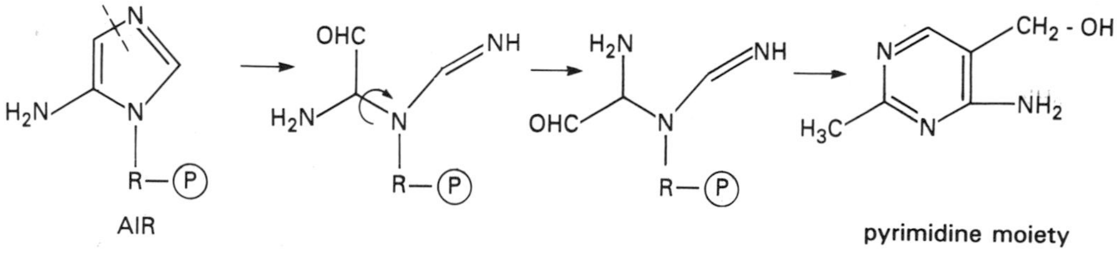

SCHEME 6.17.

III. ENZYMES CATALYSING FORMATION OF FUSED RINGS

A. Pyrrolopyrimidine

The pyrrolopyrimidine ring system is represented in Nature by the antibiotics tubericidin (**XIX**), toyocamycin (**XX**), and sangivamycin (**XXI**). All are microbial products.

XIX   XX   XXI

They are produced by ring-opening of the imidazole moiety of the purine ring system, loss of N-7 together with C-8, and reclosure with insertion of a two-carbon unit. It is now well established that C-8 of guanine, guanosine or GTP is lost as formate in the biosynthesis of riboflavin, pteridines, the azapteridine ring and the benzimidazole ring. The enzyme that catalyses conversion of GTP to formate and 6-*N*-[(5'-triphospho)-1'-ribosylamino]-2,5-diamino-4-hydroxypyrimidine by *Streptomyces rimosus* has been described. This enzyme, GTP-8-formylhydrolase, also appears to be involved in the biosynthesis of the pyrrolopyrimidines.

(a)   Assay of GTP-8-formylhydrolase.   The assay (Elstner and Suhadolnik, 1975) is based on two procedures: (1) release of ^{14}C-labelled formate from C-8 of the imidazole ring of [8-^{14}C]ATP, the formate being oxidised to $^{14}CO_2$; and (2) addition of the enzyme reaction mixture after a 2 h incubation, to a 1 ml column packed with a mixture of Norit A and Celite (1:1; w/w). The column is washed with 5 ml of 1 M formic acid to elute the [^{14}C]formate. Recovery for procedure (2) is 75% of that achieved by procedure (1).

The assay incubation mixture (total volume 1 ml) consists of [8-^{14}C]GTP (12.5 nmol; 80 000 cpm), Tris buffer 25 mM (pH 8.0) and enzyme (0.4 unit* ml^{-1}). This is incubated for 2 h at 38°C and formic acid released by (1) or (2) described above. If oxidation of formic acid is used, $^{14}CO_2$ is removed by bubbling N_2 through the solution for 10 min at 100°C. The N_2 stream is passed into 1 ml of 1 M Hyamine; $^{14}CO_2$ is counted in 0.5 ml of Hyamine.

(b)   Extraction and purification of GTP-8-formylhydrolase (Elstner and Suhadolnik, 1975).   Cells of *S. rimosus* are harvested 48 h after inoculation (maximum enzymic activity) and centrifuged at 5000 × g for 5 min. They are washed by resuspension in 50 mM Tris buffer (pH 8.0) centrifuged and freeze-dried. Freeze-dried cells (5 g dry wt.) are suspended in cold Tris buffer (50 mM; pH 8) and put through a French pressure cell (Aminco No. 4-3396) at 16 000 psi (~110 MPa). The slurry is centrifuged at 48 000 × g for 30 min and the supernatant is taken as the crude enzyme. This is dialysed overnight against 5 mM Tris buffer (pH 8.0).

Dialysed crude extract (70 ml; 1200 mg protein) is loaded onto a column (2 × 30 cm) of DE 52 cellulose after equilibration with 10 mM phosphate buffer at pH 7.4. A linear gradient is set up to elute (10 mM–0.3 M phosphate buffer; pH 7.4) using 200 ml

---

* 1 unit = 1 nmol of HCOOH h^{-1} at 38°C.

of each buffer. Fractions (10 ml) are collected at 4°C and at a flow rate of 0.5 ml min^{-1}.

The active fractions are pooled and 30% saturated with ammonium sulphate. After 15 min, the precipitate is removed by centrifuging at 15 000 × g for 10 min. The supernatant is brought to 60% saturation with ammonium sulphate, stirred for 15 min, and again centrifuged at 15 000 × g for 10 min. The pellet is dissolved in 10 ml Tris buffer (50 mM; pH 8.0) and dialysed overnight against 2 l of the same buffer. Assays of the crude undialysed extract are unreliable because of presence of inhibitors but these are removed by dialysis.

Dialysed extract (2 ml; 26 mg protein) is loaded onto a Sephadex G-200 column (1.5 × 100 cm) previously equilibrated with 10 mM phosphate buffer (pH 7.4) containing 1 mM EDTA. The protein is eluted at a flow rate of 0.5 ml min^{-1}. In the description by Elstner and Suhadolnik (1975) peak activity was found in tube 30.

B. Triazolopyrimidine

The triazolopyrimidine ring system, like that of pyrrolopyrimidine, is found in antibiotic products of molds. Toxoflavin (**XXII**) and fervenulin (**XXIII**) are examples of triazolopyrimidines. Their biosynthetic origin is also similar to that of the pyrrolo-pyrimidines in that the imidazole moiety of a purine molecule is enzymically ring-opened at C-8 but in this case a C–N unit is inserted in place of C-8 and ring-closed, resulting in ring expansion. GTP-8-formylhydrase (Section III. A(a)) may be involved in the initial removal of C-8.

XXII   XXIII

C. Indole

The indole ring is ubiquitous in Nature and is represented in the structures of a variety of indole alkaloids arising from the preformed indole ring system of the amino acid tryptophan. This is produced by the shikimic acid pathway during which 1-(*o*-carboxyphenylamino)-1′-deoxyribulose (**XXIV**) 5′-phosphate is cyclised to yield indole 3-glycerol phosphate (**XXV**). The reaction (see Scheme 6.18) is catalysed by indole 3-glycerol phosphate synthase.

(a) Assay of indole 3-glycerol phosphate synthase. During the conversion of 1-(*o*-carboxyphenylamino)-1′-deoxyribulose 5′-phosphate (**XXIV**) to indole 3-glycerol phos-

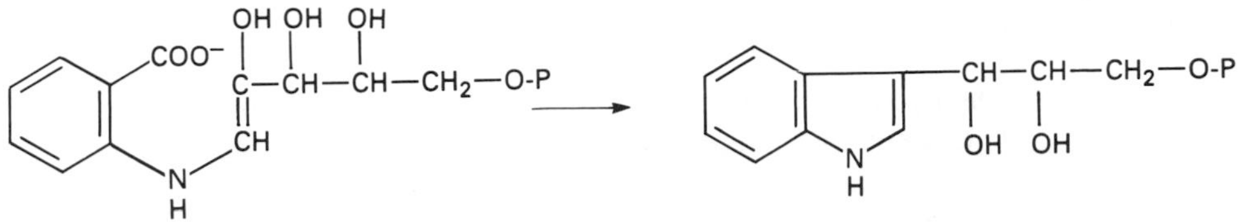

SCHEME 6.18.

phate (**XXV**) there is an increase in A_{280} which can be used to monitor the rate of reaction. To do this, a 0.1 mM solution of the substrate (**XXIV**) in 0.1 M Tris-HCl buffer at pH 7.8 is used. After adding the enzyme preparation, the mixture is incubated at 37°C. The complete conversion of the 0.1 mM substrate solution to indole 3-glycerol phosphate gives an increase in A_{280} of 0.448 under these conditions.

(b)   Extraction and purification of the enzyme.   The method described below is that of Wegman and DeMoss (1965), who used it to obtain an 83-fold purification of the enzyme from a mutant strain (*trypt*-4) of *Neurospora crassa*.

The mycelium is ground with 16 volumes of 0.05 M potassium phosphate buffer and centrifuged at 13 000 × g for 40 min. The supernatant is taken and treated with 1.5% (w/v) protamine sulphate solution allowing 14 ml per 100 ml of crude extract. After stirring for 20 min, this is centrifuged at 13 000 × g for 30 min and the pellet is discarded. Sufficient ammonium sulphate is added to the supernatant to give 40% saturation and after stirring for 15 min, it is centrifuged at 13 000 × g for 20 min. The pellet is resuspended in 0.05 M potassium phosphate buffer (pH 7.0) and loaded onto a column of Sephadex G-25 previously equilibrated with the same buffer. This buffer is also used to elute the activity. The active fractions are pooled and 45% saturated with ammonium sulphate; the pH (7.0) is maintained by adding aq. ammonia as required. After stirring for 10 min, the suspension is centrifuged at 13 000 × g for 10 min and the pellet discarded. The supernatant is brought to pH 4.9 with acetic acid, stirred for 10 min and centrifuged at 13 000 × g for 10 min. The pellet is resuspended in 0.05 M potassium phosphate buffer (pH 7.0) and again put through the Sephadex G-25 column step.

The active fractions, which elute in 0.05 M potassium phosphate buffer (pH 7.0), are diluted to 2 mg of protein ml^{-1} with the same buffer, and mixed with calcium phosphate gel (3.1 mg ml^{-1}), allowing 1.25 ml of gel for each ml of the diluted fraction. The suspension is stirred for 5 min, centrifuged at 13 000 × g for 5 min and the supernatant discarded. Using 1 ml of 0.05 M potassium phosphate buffer (pH 7.0) for each 1.5 ml of discarded supernatant, the pellet is resuspended and stirred for 5 min. This gel is centrifuged at 13 000 × g for 5 min and ammonium sulphate added to the decanted supernatant allowing 105 ml of a saturated solution of ammonium sulphate for each 100 ml of the gel. The solution is stirred for 10 min, centrifuged at 13 000 × g for 10 min and the pellet resuspended in 0.05 M potassium phosphate buffer at pH 7.0.

The activity in the final preparation is stable to freezing and thawing and has no cofactor or coenzyme requirements.

D. Benzimidazole

The benzimidazole ring system is represented within the molecular structures of vitamin B_{12} and its derivatives (cobalamins). There is however, to date, no well authenticated report of the occurrence of these compounds in higher plants. Nevertheless, the cobalamins are synthesised by a variety of microorganisms (e.g. species of Streptomyces and Bacillus). Many of the biosynthetic studies have used *Propionibacterium shermanii* as the synthetic organism. The benzimidazole ring was reported to arise in these organisms from riboflavin (Renz, 1970): subsequently Alworth *et al.* (1971) found that 6,7-dimethyl[^{14}C]-8-ribityllumazine, an intermediate in riboflavin biosynthesis (see Isoalloxazine, Section III.F), is converted into 5,6-dimethylbenzimidazole (Scheme 6.19). Whereas the work described above shows a connection between riboflavin and 5,6-dimethylbenzimidazole synthesis, it does not eliminate the possibility that both arise from a comon precursor rather than one giving rise to the other. No associated enzymology has been described.

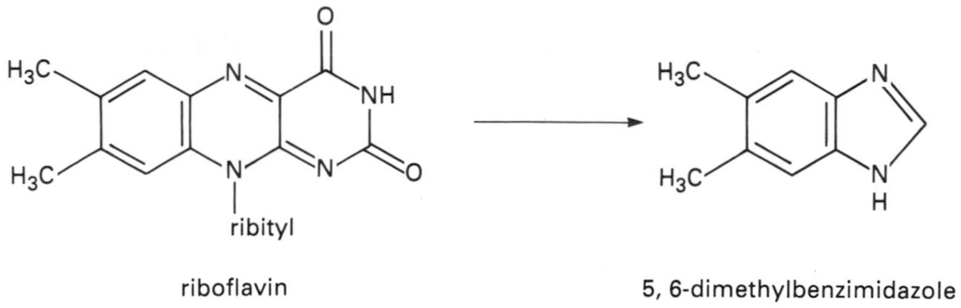

riboflavin   5, 6-dimethylbenzimidazole

SCHEME 6.19.

E. Purine

The pathway of purine biosynthesis in plants has been confirmed to be essentially similar to that in animal tissues and microorganisms by a number of workers using ^{14}C-labelled precursors and intermediates. The essential biosynthetic strategy is construction of a pyrimidine ring on an imidazole ribotide; the final steps in the process i.e. production of the first formed purine, are shown in Scheme 6.20.

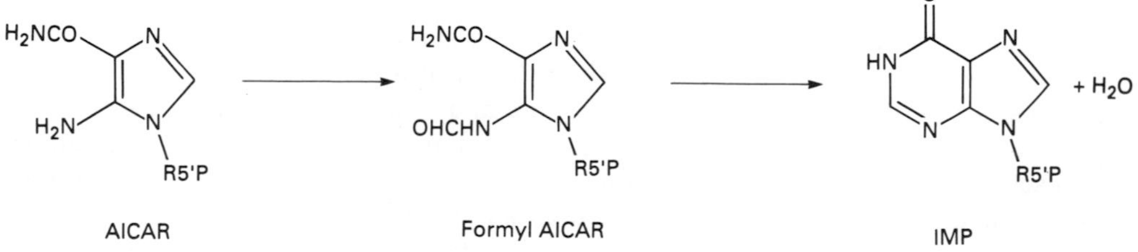

AICAR   Formyl AICAR   IMP

SCHEME 6.20.

Relatively few of the enzymes of purine biosynthesis have been demonstrated directly in plants. The enzyme catalysing the formylation of 5-aminoimidazole carboxamide ribotide (AICAR formyl transferase) has been purified 40-fold from pea seedlings by Iwai *et al.* (1972). However, since the product obtained with this enzyme is IMP it is concluded that IMP cyclohydrase, which catalyses ring closure, is also present in this partially purified extract. The preparation can therefore be said to catalyse formation of a purine ring.

1.   5-Aminoimidazolecarboxamide ribotide (AICAR) formyl transferase

(a)   Assay.   This enzyme is assayed by measuring the disappearance of the substrate AICAR, using the procedure of Flaks *et al.* (1957). The assay incubation mixture, final volume 0.5 ml, consists of the following: AICA ribotide (0.2 mM), N^{10}-formyltetra-hydrofolic acid (0.4 mM), KCl (20 mM), Tris-HCl buffer pH 7.4 (60 mM), and the enzyme preparation. After incubating at 37°C for 20 min, the reaction is terminated by adding 0.4 ml of 10% (v/v) perchloric acid. A zero-time (unincubated) control is included. Acetic anhydride (0.1 ml) is added immediately to each incubate and after *exactly* 10 min further incubation, residual non-acetylated arylamine is determined. This is done by first adding 3 ml of 0.2 M H_2SO_4 and 0.2 ml of 0.1% (w/v) $NaNO_2$ to each incubate at 0°C, followed 5 min later by addition of 0.2 ml of 0.5% ammonium sulphamate. This is followed after a further 3 min by 0.2 ml of 0.1% (w/v) N-1-naphthylethylenediamine dihydrochloride. After standing for 30 min at room temperature, the incubates are clarified by centrifuging and the A_{540} of the supernatants determined.

(b)   Extraction and purification of the enzyme.   Pea seedlings are germinated and grown in the dark for 10 days, then their cotyledons are removed. These seedlings (5 kg) are homogenised with 5 l of 0.1 M Tris-HCl buffer, pH 7.4, and the homogenate squeezed through muslin. To the extract, ammonium sulphate is added to give 30% saturation and the pH maintained at 7.4 by adding 10% aq. NH_3 as necessary. The mixture is kept stirred for 30 min, centrifuged to remove the precipitate, and then brought to 70% saturation with ammonium sulphate. The precipitate is collected and dissolved in cold water, then dialysed against 10 l of 5 mM Tris-HCl buffer (pH 7.4) for 24 h; the buffer is changed every 6 h. The precipitate formed during dialysis is removed by centrifugation and the supernatant is again fractionated by ammonium sulphate precipitation. The 35–65% saturation fraction is dissolved in water and dialysed as before. To it is added 1% neutralised protamine sulphate solution, allowing 16 ml for each 230 ml of the enzyme preparation. After removing the precipitate by centrifugation, the supernatant is brought to 35–65% saturation with ammonium sulphate. The precipitate is dissolved in a small volume of cold water and loaded onto a Sephadex G-150 column (3 × 60 cm) previously equilibrated with 0.05 M Tris-HCl buffer (pH 7.4). Elution is with the same buffer at a flow rate of 1 ml per 3 min; 10 ml fractions are collected. In the original method of Iwai *et al.* (1972) the enzymic activity was concentrated in fractions 40–50. The pooled active fractions are combined and brought to 65% saturation with ammonium sulphate. The precipitate is collected and dialysed as before. The non-diffusible fraction is used as the final enzymic preparation and represents a 39-fold purification.

As mentioned above, since the product of the incubation of this preparation with 5-aminoimidazolecarboxamide ribotide and N^{10}-formyltetrahydrofolic acid has been identified (Iwai *et al.* 1972) as IMP, it is concluded that this enzyme preparation also contains IMP cyclohydrase.

F. Isoalloxazine

In the form of riboflavin and its derivatives FMN and FAD, the isoalloxazine ring system is of ubiquitous occurrence in plants, animals and microorganisms. This ring system, which can also be regarded as a benzopteridine, arises biosynthetically from a purine precursor by ring expansion to a pteridine followed by assembly of the benzene ring. The first part of this process involves an opening of the imidazole ring of the purine precursor GTP (**XXVI**) with loss of C-8 (see Section III.A(a)) which involves GTP-8-formylhydrolase. This is followed by an Amadori rearrangement of the resulting pyrimidine ribotide (**XXVII**) and ring closure to form the new pteridine ring of 6,7-dimethyl-8-ribityllumazine (**XXVIII**) (see Scheme 6.21).

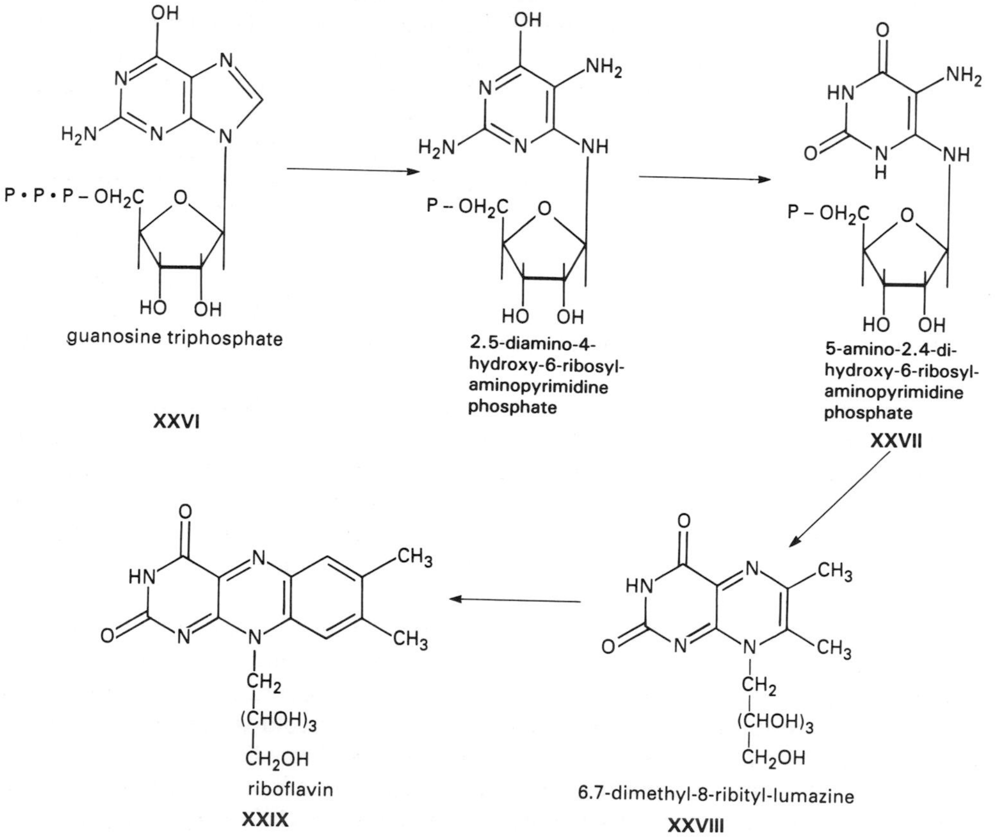

guanosine triphosphate

XXVI

2.5-diamino-4-hydroxy-6-ribosyl-aminopyrimidine phosphate

5-amino-2.4-di-hydroxy-6-ribosyl-aminopyrimidine phosphate

XXVII

riboflavin

XXIX

6.7-dimethyl-8-ribityl-lumazine

XXVIII

SCHEME 6.21.

The enzyme, riboflavin synthase, which converts 6,7-dimethyl-8-ribityllumazine into riboflavin (**XXIX**) and thereby makes an isoalloxazine ring, is particularly interesting.

Even crude preparations of this enzyme from the mycelium of *Eremothecium ashbyii* rapidly convert the colourless lumazine substrate into the bright yellow of riboflavin, such that under appropriate conditions the progress of the reaction can be followed by eye. The reaction catalysed is unusual in that two molecules of the substrate are consumed for each molecule of product (riboflavin) produced. This is because 6,7-dimethyl-8-ribityllumazine acts both as donor and recipient of the four-carbon unit needed to complete the riboflavin molecule. Riboflavin synthase has been extracted and purified from bakers' yeast (Harvey and Plaut, 1966) and from spinach (*Spinacea oleraceae*) by Mitsuda *et al.* (1971). The procedure for spinach is described below.

(a) Assay of riboflavin synthase. The incubation mixture consists of 0.2 µmol of 6,7-dimethyl-8-ribityllumazine, 20 µmol of cysteine, 20 µmol of ascorbic acid, enzyme sample, and 0.1 mM phosphate buffer (pH 7.4) to give a final volume of 2 ml. Incubation is at 37°C for 1–3 h and the reaction is terminated by placing the tubes in a boiling water bath for 2 min. After rapidly cooling the incubates, precipitated protein is removed by centrifugation and a 0.1 ml portion of the supernatant is chromatographed on paper in butanol–acetic acid–water (12:3:5; v/v). The yellow riboflavin band, which also fluoresces bright yellow in UV light, is eluted with distilled water and riboflavin determined spectrophotometrically. The initial concentration of the lumazine substrate can be determined spectrophotometrically at 405 nm ($\varepsilon = 10\,300$).

(b) Extraction and purification of riboflavin synthase. Using a pre-cooled mortar and pestle, fresh green leaves of spinach (5 kg) are macerated in 200 g portions, each with 50 ml of 0.1 M phosphate buffer (pH 7.0) as extracting medium. This homogenate is squeezed through muslin and the combined extracts (4.5 l) are adjusted to pH 5.4 with 10% aq. acetic acid. After standing overnight, the precipitate is removed by centrifuging at $3500 \times g$ for 20 min and the supernatant brought to pH 6.5 with NaOH.

To the supernatant, ammonium sulphate is added to give 38% saturation; the pH is maintained at 6.5 by adding NaOH. The precipitate is collected by centrifuging and resuspended in a small volume of 0.02 M phosphate buffer (pH 7.0). This is dialysed against the same buffer for 2 h. At this stage, the enzyme is stable if kept frozen. To the non-diffusible fraction after dialysis is added 26.4 ml of protamine sulphate solution ($20\,\text{mg ml}^{-1}$). After stirring for 20 min, the precipitate is removed by centrifuging at $3500 \times g$ for 20 min.

The supernatant (320 ml) is brought to 23% saturation with ammonium sulphate and the precipitate removed by centrifuging. More ammonium sulphate is added to the supernatant to give 40% saturation. This time, the precipitate is removed by centrifuging and retained. After redissolving it in a small volume of 0.025 M phosphate buffer (pH 6.5), it is dialysed against the same buffer. The non-diffusible fraction (75 ml) is divided into three portions of 25 ml. Each is loaded onto a column (2.2 cm diameter) made up of 8 g of CM-cellulose previously washed with 0.025 M phosphate buffer (pH 6.5). The column is then eluted with 150 ml of the same buffer, and the eluate collected and brought to 55% saturation with ammonium sulphate. The precipitate is recovered by centrifuging at $9700 \times g$ for 20 min, suspended in 0.025 M phosphate buffer (pH 6.5), and dialysed against the same buffer for 5 h.

Samples (12.5 ml) of the dialysate (non-diffusible fraction) are applied to a column (2.1 cm diameter) prepared from 5 g DEAE-cellulose previously washed and equilibrated with 0.01 M Tris–maleate–NaOH buffer (pH 6.2). The column is washed with 60 ml of this Tris buffer, followed by 80 ml of the same buffer containing 0.2 M NaCl. Most of the enzyme is eluted by increasing the NaCl concentration to 0.5 M. The eluate is brought to 60% saturation with ammonium sulphate and the precipitate collected by centrifuging. Following resuspension in 0.01 M Tris–maleate–NaOH buffer (pH 6.2) it is dialysed against the same buffer for 5 h.

After dialysis, the preparation is loaded onto a column (2.1 cm diameter) containing 4.2 g of DEAE-cellulose previously equilibrated with 0.01 M Tris–maleate–NaOH buffer (pH 6.0). Stepwise elution is then effected with NaCl solutions buffered by 0.01 M Tris–maleate–NaOH (pH 6.0): (1) 60 ml of the buffer without NaCl; (2) 120 ml with 0.2 M NaCl; (3) 80 ml with 0.3 M NaCl; (4) 90 ml with 0.4 M NaCl; and (5) 120 ml with 0.7 M NaCl. Fractions (10 ml) are collected; the major activity elutes with 0.3 M salt. Active fractions are combined. According to Mitsuda et al. (1971) the final product represented a 648-fold purification.

The purified enzyme remains active for 24 h if kept at 0–4°C, and is stable for 1 week at 0–4°C in the presence of ammonium sulphate. Freezing the preparation causes inactivation. The enzyme is also inhibited by thiol reagents such as p-chloromercuribenzoate.

G. Pteridine

Folic acid and its coenzyme form tetrahydrofolic acid (THFA) are ubiquitous in Nature. In addition, there are a variety of other pteridine pigments in plants, animals and microorganisms. They are biosynthesised by a pathway similar to that used to produce the pteridine intermediate (6,7-dimethyl-8-ribityllumazine) in riboflavin biosynthesis (Section III.F) and involving the enzyme GTP-8-formylhydrolase (Section III.A(a)).

REFERENCES

Alworth, W. L., Lu, S.-H. and Winkler, M. F. (1971). *Biochemistry* **10**, 1421–1424.
Brown, E. G. and Diffin, F. M. (1990). *Phytochemistry* **29**, 469–478.
Brown, E. G., Flayeh, K. A. M. and Gallon, J. R. (1982). *Phytochemistry* **21**, 863–867.
De Barreiro, O. L. C. (1967). *Biochim. Biophys. Acta* **139**, 479–486.
Decker, R. H., Kang, H. H., Leach, F. R. and Henderson, L. M. (1961). *J. Biol. Chem.* **236**, 3076–3082.
Elstner, E. F. and Suhadolnik, R. J. (1975). *Methods Enzymol.* **43**, 515–520.
Fink, R. M., Cline, R. E., McGaughey, C. and Fink, K. (1956). *Anal. Chem.* **28**, 4–6.
Flaks, J. G., Erwin, M. J. and Buchanan, J. M. (1957). *J. Biol. Chem.* **229**, 603–612.
Gurne, D. and Shemin, D. (1973). *Science* **180**, 1188.
Harvey, R. A. and Plaut, G. W. E. (1966). *J. Biol. Chem.* **241**, 2120–2136.
Hill, J. M. (1971). *Methods Enzymol.* **17B**, 730–735.
Iwai, K., Fujisawa, Y. and Suzuki, N. (1972). *Agric. Biol. Chem.* **36**, 398–408.
Janion, C. and Shugar, D. (1960). *Acta Biochim. Polon.* **7**, 309–329.
Kapoor, M. and Waygood, E. R. (1965). *Can. J. Biochem.* **43**, 143–151.
LaRue, T. (1965). *Anal. Chem.* **37**, 246–248.

Leete, E., Davies, G. E., Hutchinson, C. R., Woo, K. W. and Chedekel, M. R. (1974). *Phytochemistry* **13**, 427–433.

Leonard, N J. and Hauck, F. P. (1957). *J. Am. Chem. Soc.* **79**, 5279–5292.

Mazuś, B. and Buchowicz, J. (1968). *Acta Biochimica Polonica* **15**, 317–325.

Mitsuda, H., Kawai, F. and Suzuki, Y. (1971). *Methods Enzymol.* **18B**, 539–543.

Mizusaki, S., Tanabe, Y., Noguchi, M. and Tamaki, E. (1971). *Plant Cell Physiol.* **12**, 633–640.

Mizusaki, S., Tanabe, Y., Noguchi, M. and Tamaki, E. (1972). *Phytochemistry* **11**, 2757–2762.

Murakoshi, I., Ikegami, F., Hinuma, Y. and Hanma, Y. (1984). *Phytochemistry* **23**, 973–977.

Noguchi, M., Koiwai, A. and Tamaki, E. (1966). *Agric. Biol. Chem.* **30**, 452–456.

Renz, P. (1970). *FEBS Lett.* **6**, 187–189.

Roberts, M. F. (1977). *Phytochemistry* **14**, 2393–2397.

Sheehan, J. C., Goodman, M. and Hess, G. P. (1956). *J. Am. Chem. Soc.* **78**, 1367–1369.

Smith, T. A. (1970). *Biochem. Biophys. Res. Commun.* **41**, 1452–1456.

Smith, T. A. (1974). *Phytochemistry* **13**, 1075–1081.

Smith, T. A. (1976). *Phytochemistry* **15**, 633–636.

Smith, T. A. (1977). *Phytochemistry* **16**, 1647–1649.

Suresh, M. R. and Adiga, P. R. (1979). *J. Biosci.* **1**, 109–124.

Suresh, M. R., Ramakrishna, S. and Adiga, P. R. (1976). *Phytochemistry* **15**, 483–485.

Tigier, H. A., Batlle, D. C. and Locascio, G. (1968). *Biochim. Biophys. Acta* **151**, 300–302.

Urata, G. and Granick, S. (1963). *J. Biol. Chem.* **238**, 811–820.

Wegman, J. and DeMoss, J. A. (1965). *J. Biol. Chem.* **240**, 3781–3788.

7 Cyanogenic Glycosides

BIRGER LINDBERG MØLLER[1] and JONATHAN E. POULTON[2]

[1]*Plant Biochemistry Laboratory, Royal Veterinary and Agricultural University, DK-1871 Frederiksberg C, Copenhagen, Denmark*

[2]*Department of Biological Sciences, University of Iowa, Iowa City, IA 52242, USA*

I. INTRODUCTION

Several thousand species of higher plants, including some used as food sources by humans and as feed for domestic animals, release hydrogen cyanide upon tissue disruption, infection or food processing (Seigler, 1991). This phenomenon of cyanogenesis has been responsible for numerous cases of acute cyanide poisoning. In areas of the world where cyanogenic plants such as cassava (*Manihot esculenta*) are major components of the human diet, chronic cyanide poisoning and associated pathological conditions still exist (Poulton, 1983). Most frequently, HCN arises during catabolism of cyanogenic glycosides. The approximately 60 documented cyanogenic glycosides are *O*-β-glycosidic derivatives of α-hydroxynitriles (cyanohydrins). Depending on their precursor amino acid, they may be aliphatic, aromatic or cyclopentenoid in nature (Seigler, 1991). Most are monoglycosides in which the unstable cyanohydrin moiety

METHODS IN PLANT BIOCHEMISTRY Vol. 9
ISBN 0–12–461019–6

is stabilised by glycosidic linkage to a single sugar residue. Alternatively, in the cyanogenic di-, tri- and tetraglycosides, two, three or four sugar moieties, respectively, are involved in such stabilisation. Sulphated, malonylated, and acylated derivatives have also been isolated but little is known about their biochemistry. In the Sapindaceae and Hippocastanaceae, a limited number of cyanolipids constitute an alternative source of HCN (Selmar *et al.*, 1990; Seigler, 1991).

Scientific interest in cyanogenic glycosides focused initially on structural elucidation of these unusual compounds. A subsequent major task has been to clarify the pathways leading to their synthesis and degradation. The biosynthetic pathway is particularly fascinating because it contains intermediates not previously encountered in amino acid metabolism. Additional intermediates are still being identified. All biosynthetic steps, except the final glycosylation of the α-hydroxynitrile, are catalysed by membrane-bound enzymes. Of these, only an unspecific NADPH-cytochrome P450 oxidoreductase has been isolated. In contrast, the soluble enzymes involved in cyanogenic glycoside degradation have been purified from many sources and are well characterised.

This Chapter focuses upon the enzymology of cyanogenic glycoside metabolism. The purification and assay methods described here may require modification when applied to other experimental systems. Since many biosynthetic intermediates are not commercially available, procedures for their chemical synthesis are listed. The interested reader is directed toward several recent reviews on cyanogenic glycosides which describe their detection, isolation and characterisation (Nahrstedt, 1981; Brinker and Seigler, 1989, 1991; Seigler, 1991), taxonomic distribution (Saupe, 1981; Hegnauer, 1986; Nahrstedt, 1987; Seigler *et al.*, 1989), biochemistry and molecular biology (Halkier *et al.*, 1988; Poulton, 1988, 1990; Hughes *et al.*, 1988), toxicology (Poulton, 1989), and likely physiological roles (Nahrstedt, 1985, 1987, 1988; Jones, 1988; Hruska, 1988).

II. CYANOGENIC GLYCOSIDE BIOSYNTHESIS

A. The Membrane-bound Enzymes Catalysing the Conversion of Amino Acids to α-Hydroxynitriles

The biosynthesis of the different cyanogenic glycosides is thought to follow a common pathway (Conn, 1973, 1980). Several additional intermediates in this pathway have recently been identified as illustrated for the biosynthesis of the tyrosine-derived cyanogenic glucoside dhurrin in Fig. 7.1 (Halkier *et al.*, 1989, 1991b; Halkier and Møller, 1990). Initially, feeding experiments using excised seedlings demonstrated that amino acids serve as effective precursors whereas none of the expected intermediates was found to accumulate. Identification of intermediates was accomplished when a microsomal fraction from etiolated seedlings of *Sorghum bicolor* was shown to catalyse the *in vitro* conversion of the parent amino acid tyrosine to *p*-hydroxymandelonitrile, the aglycone of the cyanogenic glucoside dhurrin (McFarlane *et al.*, 1975; Møller and Conn, 1979). Subsequently, active microsomal preparations have been obtained from a number of other cyanogenic plants including flax (*Linum usitatissimum*) (Cutler *et al.*, 1981), white clover (*Trifolium repens*) (Collinge and Hughes, 1982b), *Triglochin maritima* (Hösel and Nahrstedt, 1980; Cutler *et al.*, 1981b), California poppy (*Esch-*

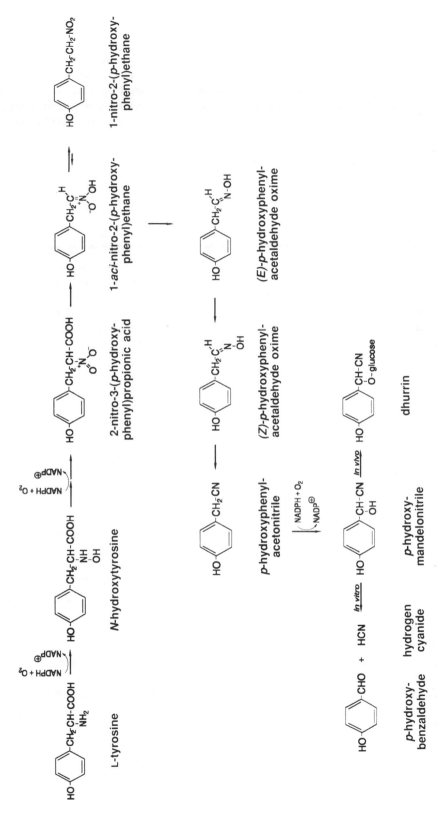

FIG. 7.1. Biosynthesis of the cyanogenic glucoside dhurrin.

scholtzia californica) (Hösel *et al.*, 1985) and cassava (*Manihot esculenta*) (Koch *et al.*, 1991). These studies demonstrate that all steps in the biosynthesis of cyanogenic glycosides except the last are catalysed by membrane-bound enzymes and that the intermediates are channelled (Møller and Conn, 1980). The hydroxylation reactions are photoreversibly inhibited by carbon monoxide, demonstrating the involvement of cytochrome P450 (Halkier and Møller, 1991). The soluble UDP-glucose:glucosyl-transferases catalysing the conversion of α-hydroxynitriles to cyanogenic glycosides are discussed in Section II.B.

1. Comments on the isolation of active microsomal systems

The microsomal preparations have been obtained almost exclusively from etiolated plant material. Those preparations obtained from young folded shoot tips of white clover grown under a light regime (Collinge and Hughes, 1984) are unstable and exhibit low biosynthetic activity. Any seed coats or endosperm residues which remain associated with the seedlings should be completely removed before homogenisation of the plant material. Without exception, these tissues have been found to contain a strong inhibitor which abolishes the enzymatic activity. To permit easy harvest of seedlings of *S. bicolor* and at the same time avoid contamination by seed coats, seeds are germinated between two sheets of gauze stretched over a metal screen placed in a germination tray (Halkier and Møller, 1989). For seeds of other species, metal screens may be designed. It is important that the relationship between cyanogenic glycoside content and germination time be determined to select the optimal harvesting time (Collinge and Hughes, 1982a; Halkier and Møller, 1989; Koch *et al.*, 1991).

To stabilise enzymatic activity, dithiothreitol (DTT), phenylmethylsulphonyl fluoride (PMSF) and polyvinylpolypyrrolidone (PVPP) should be added to the homogenisation medium. Microsomal preparations obtained by homogenising fresh plant material using a mortar and pestle are more active compared to those obtained by grinding plant material frozen in liquid nitrogen.

Generally, the *N*-hydroxylation reaction converting the primary amino acid to the *N*-hydroxyamino acid is the rate limiting step in the *in vitro* production of the aglycone (Halkier and Møller, 1991; Halkier *et al.*, 1991b). The enzyme system catalysing this conversion has a very high substrate specificity (Koch *et al.*, 1991). The specificity of the enzymes catalysing the subsequent steps appears to be less pronounced as experimentally demonstrated in flax by the metabolism of a number of different oximes (Cutler *et al.*, 1985) and in cassava by the metabolism of a range of different oximes as well as nitriles (Koch *et al.*, 1991). The metabolic rates observed using aldoximes as substrates are higher than those obtained with amino acids. In initial experiments to isolate an active microsomal preparation from a new plant species, the metabolic activity may be too low to be detected. In this situation, a rational approach may be to test a set of different oximes as substrates. The higher rates observed with oximes compared to amino acids may render the metabolic activity detectable and thus provide a platform from which improvements can be made. The experimental procedure listed below for the preparation of the microsomal system from *S. bicolor* is generally applicable for other species.

2. Isolation of the microsomal system from S. bicolor

Etiolated *S. bicolor* seedlings (2–3 cm tall) are harvested (approx. 150 g) and homogenised using a mortar and pestle in 2 volumes (v/w) of 250 mM sucrose, 100 mM Tricine (pH 7.9), 50 mM NaCl, 2 mM ethylenediaminetetraacetic acid (EDTA), 1 mM PMSF, 2 mM DTT. PVPP is added (0.1 g per g fresh weight) prior to homogenisation. The homogenate is filtered through 22 μm nylon cloth and centrifuged for 15 min at 15 000 × g. The supernatant is centrifuged for 1 h at 165 000 × g. The microsomal pellet is resuspended and homogenised in isolation buffer using a Potter–Elvehjem homogeniser fitted with a Teflon pestle. After recentrifugation and rehomogenisation, the homogenate is dialysed overnight against 50 mM Tricine (pH 7.9), 2 mM DTT under a nitrogen atmosphere. Additional purification of the microsomal system is achieved by sucrose gradient centrifugation (Halkier and Møller, 1989). Detergent solubilisation of the microsomal system and subsequent fractionation by column chromatography have resulted in the isolation of the unspecific NADPH-cytochrome P450 oxidoreductase which mediates electron transfer to the cytochrome P450 dependent hydroxylation steps (Halkier and Møller, 1991).

3. Chemical synthesis of intermediates

Most known cyanogenic glycosides are derived from one of the six amino acids valine, isoleucine, leucine, phenylalanine, tyrosine and 2-cyclopentenylglycine (Conn, 1980). The different sets of intermediates derived from each of these parent amino acids are therefore of potential interest for biosynthetic studies. Since none of the intermediates derived from 2-cyclopentenylglycine (David *et al.*, 1944) and only a few of the others are available commercially, procedures used for their synthesis are briefly summarised.

The *N*-hydroxyamino acids (Møller, 1981) are generally synthesised by chemical reduction of the corresponding ketoximes with sodium cyanoborohydride. The ketoximes are obtained by reacting the ketoacid with hydroxylamine (Møller, 1977, 1978; Møller *et al.*, 1977). *N*-Hydroxyamino acids should be stored as solids or in acidic solutions. *N*-Hydroxyamino acids are oxidatively decarboxylated to aldoximes in alkaline solutions (Møller *et al.*, 1977; Møller, 1978).

The α-nitroso- and α-nitrocarboxylic acids are labile. At physiological pH, they decarboxylate forming oximes and *aci*-nitro compounds, respectively (Pedersen 1934, 1949; Pritzkow and Rösler, 1967). Since the decomposition products are subsequent intermediates in the pathway, no attempts have been made to directly test the parent compounds as substrates for the microsomal system (Halkier and Møller, 1990). Procedures are available for the chemical synthesis of α-nitroso- and α-nitrocarboxylic acids (Finkbeiner and Stiles, 1963; Pritzkow and Rösler, 1967). The α-nitrocarboxylic acids are stabilised by decreasing the concentration of the free anion. This may be done by chelation or by lowering the pH (Pedersen, 1949). The esters of α-nitrocarboxylic acids are stable and have been used in biosynthetic experiments with fungi (Baxter *et al.*, 1988), the reasoning being that their hydrolysis *in vivo* would release the active α-nitrocarboxylic acid within the cells. However, no data on the relative rates of hydrolysis, decarboxylation and biosynthetic conversion are available.

aci-Nitro compounds are in equilibrium with the tautomeric nitro compound and their common nitronate anions (Nielsen, 1969). At physiologically relevant pH values around 7, the equilibrium favours the stable nitro compound. The nitro alkane intermediates are prepared by condensation of an aldehyde and a lower primary nitroalkane (Bachman *et al.*, 1972) or by modifications thereof (Halkier *et al.*, 1991a; Hösel *et al.*, 1985; Schiefer and Kindl, 1971). Aldoximes are generally prepared (Bousquet, 1943) by reacting the corresponding aldehydes with hydroxylamine (Cutler *et al.*, 1981; Cutler and Conn, 1981; Shimada and Conn, 1977). Aldoximes may also be obtained by oxidative decarboxylation of the corresponding *N*-hydroxyamino acid (Møller, 1978; Halkier *et al.*, 1989). *p*-Hydroxyphenylacetaldehyde is not commercially available but is conveniently synthesised by a pinacol–pinacolone-type rearrangement of synephrine (Robbins, 1966). Generally, a mixture of the (*E*)- and (*Z*)-isomers are obtained. The isomers may be separated by high performance liquid chromatography (HPLC) (Halkier *et al.*, 1989) and identified by nuclear magnetic resonance (NMR) spectrometry (Karabatsos *et al.*, 1963). Both isomers are true intermediates in the pathway (Halkier *et al.*, 1989).

Nitrile intermediates, except that derived from 2-cyclopentenylglycine, are commercially available. 2-Cyclopentenylcarbonitrile is synthesized by condensation of 3-chlorocyclopentene and cyanide (David *et al.*, 1944; 'Olafsdóttir, 1990).

For procedures regarding the chemical synthesis of radioisotopically labelled intermediates, the following references should be consulted: Shimada and Conn (1977), Møller (1978), Cutler *et al.* (1981), Jaroszewski *et al.* (1981), 'Olafsdóttir (1990), Halkier *et al.* (1991a). Alternatively, radioactively labelled intermediates accumulated in *in vitro* experiments may be isolated from the incubation mixtures and used as substrates. This approach has proven of special value for the generation of isotopically labelled *p*-hydroxyphenylacetaldehyde oxime using the *S. bicolor* system prepared in the absence of dithiothreitol (Møller and Conn, 1979; Shimada and Conn, 1977).

4. Biosynthetic experiments using microsomal preparations

The biosynthetic reactions are carried out in septum-covered glass tubes (3 ml). Typically, the reaction mixture contains 12.5 μmol Tricine (pH 7.9), 100–250 nmol substrate (saturating concentration), and 0.3 μmol NADPH in a total volume of 250 μl. The enzyme reaction is started by addition of 50 μl of the microsomal preparation (1.5 mg protein ml^{-1}) and the tube closed with a silicone septum. The reaction mixture is incubated for 30 min in a shaking water-bath at 30°C after which the composition of products is analysed using the methods listed below.

5. Cyanide determination (Halkier *et al.*, 1988)

The aglycone (cyanohydrin) accumulates as the end product in *in vitro* experiments with microsomal preparations. The cyanohydrin is quantitatively dissociated into the corresponding carbonyl compound and cyanide at alkaline pH. The overall biosynthetic activity of the microsomal preparation can therefore be monitored by a spectrophotometric cyanide assay. At the end of the incubation period, 40 μl 6 N NaOH is injected through the septum to (1) stop the enzyme reaction, (2) ensure complete dissociation of the accumulated cyanohydrin, and (3) trap free HCN from the gas phase. After 10 min, the septum is removed and the cyanide in the reaction mixture quantified by a modified Lambert procedure (Lambert *et al.*, 1975). The assay involves sequential addition of $50\,\mu l$ 100% HOAc, $200\,\mu l$ N-chlorosuccinimide $(1\,g\,l^{-1})$/succinimide $(2.5\,g\,l^{-1})$ and $200\,\mu l$ pyridine/barbituric acid (60 g barbituric acid in 300 ml pyridine with H_2O added to a final volume of 1000 ml) to the reaction mixture. To compensate for the turbidity of some samples, the coloured complex formed is quantified by dual wavelength spectroscopy with wavelength settings of 585 and 650 nm using an Aminco DW2000 spectrophotometer. To compensate for absorbances resulting from endogenous cyanide, reaction mixtures without added NADPH or substrate are also prepared and measured. This procedure permits the detection of as little as 0.2 nmol cyanide.

The N-chlorosuccinimide/succinimide reagent should be prepared by initial solubilisation of succinimide and subsequent addition of solid N-chlorosuccinimide. The N-chlorosuccinimide/succinimide and the pyridine/barbiturate reagents are prepared directly in air-tight dispensers and are stored at 4°C. These precautions prolong the stability of the reagents from a few days to more than a month.

Cyanide assays used in earlier studies required an overnight incubation of septum-covered reaction vials to permit diffusion of HCN from an acidified reaction mixture to a suspended centre well containing NaOH (McFarlane *et al.*, 1975; Møller and Conn, 1979).

6. Analysis of intermediates

Intermediates which accumulate in the *in vitro* reaction mixtures may be quantitated after fractionation by thin layer chromatography (TLC), gas–liquid chromatography (GLC) or HPLC. These procedures are more informative than the cyanide assay since they permit the monitoring of individual steps in the pathway.

Some of the intermediates derived from valine, isoleucine and leucine are volatile. Analysis by TLC is therefore applicable only after initial derivatisation with 2,4-DNPH (Collinge and Hughes, 1982b). The generation of derivatisation artefacts and the lack of UV absorbance of the underivatisable intermediates render separation based on GLC more favourable. Gas–liquid chromatographic procedures have been developed using Porapak S (Cutler and Conn, 1981) and Porapak Q (Koch *et al.*, 1991).

Three main procedures based on TLC, GLC and HPLC are available for the analysis of tyrosine-derived intermediates. Separation by TLC is achieved on silica gel (Merck DC-Alufolien Kieselgel $60F_{254}$, 0.2 mm) using toluene–ethyl acetate (5:1; v/v) as the mobile phase (Møller and Conn, 1979). The GLC method requires initial silylation of the reaction mixture and is performed using 3% SP-2250 coated on 80/100 mesh Supelcoport (Møller, 1977). This is the only procedure which permits specific analysis of the N-hydroxyamino acid (Møller, 1977; Møller and Conn, 1979). The HPLC procedure is the method of choice in cases where specific analyses of the (E)- and (Z)-isomers of the oxime are required and is carried out using Nucleosil $100\text{-}10C_{18}$ and isocratic elution with 1.5% 2-propanol in 25 mM Hepes (pH 7.9) (Fig. 7.2.) (Halkier *et al.*, 1989). When coupled to a gas proportional counter (Møller, 1977) and a radioactivity monitor equipped with a solid-state scintillator (Halkier *et al.*, 1989), respectively, GLC and HPLC procedures provide simultaneous and accurate determination of the radioactivity distribution within the intermediates.

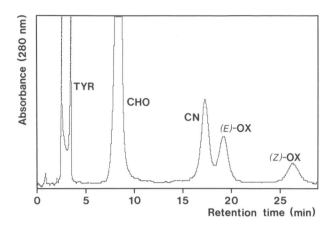

FIG. 7.2. Analysis of tyrosine-derived intermediates by HPLC.

B. Cyanohydrin β-Glycosyltransferases

Glycosylation of α-hydroxynitriles constitutes the final step in cyanogenic glycoside biosynthesis. Cyanogenic monoglycosides usually contain D-glucose as stabilising sugar. Their synthesis is catalysed by soluble β-glucosyltransferases (EC 2.4.1.-) which are discussed in detail below. Enzymes which glycosylate cyanogenic monoglycosides to their corresponding di-, tri- and tetraglycosides have not yet been demonstrated *in vitro* and are not considered further.

1. Properties

β-Glucosyltransferases have been partially purified and characterised from a few cyanogenic species (Table 7.1). All have pH optima in the range 6.5–9.0 and they usually lack a requirement for metal ions or cofactors. Most are unstable but may be partially stabilised by sulphhydryl compounds or glycerol (Reay and Conn, 1974; Poulton and Shin, 1983; Hösel and Schiel, 1984). Where tested, these enzymes showed absolute specificity for UDPG (K_m 0.029–1.0 mM). Broader specificity is displayed towards their acceptor substrates. While cyanohydrins acted as the best substrates, chemically related alcohols, acids and phenols are also glucosylated by some preparations. This breadth of activity may in part reflect contamination by other glucosyltransferases, as was demonstrated for the sorghum enzyme (Reay and Conn, 1974). The dissociation of cyanohydrins under assay conditions prevents calculation of exact K_m values, but optimal substrate concentrations range from 5 to 40 mM.

The most highly purified glucosyltransferases available show intriguing differences in their specificities towards cyanohydrins (Table 7.1). The flax enzyme accepts a limited number of aliphatic ketone cyanohydrins but utilises benzaldehyde cyanohydrin and p-hydroxybenzaldehyde cyanohydrin only poorly. By contrast, the sorghum enzyme rapidly glucosylates these aromatic cyanohydrins but is inactive toward acetone cyanohydrin and acetaldehyde cyanohydrin. While the flax enzyme glucosylates both enantiomers of 2-hydroxy-2-methylbutyronitrile, absolute stereospecificity is displayed

TABLE 7.1. Properties of cyanohydrin β-glucosyltransferases from cyanogenic plants.

Enzyme	Purification (-fold)	Purification methods[e]	pH optimum	Major substrates	Stereospecificity	Requirements
Linum usitatissimum[a]	120	Acet, MnCl$_2$, Dow AS, DC, G-100	8–9	Acetone cyanohydrin Butanone cyanohydrin 3-Pentanone cyanohydrin 2-Butanol	No	–
Sorghum bicolor[b]	77	Acet, AS, CP, P-60, DC	8.2–8.5	*p*-Hydroxybenzaldehyde cyanohydrin Benzaldehyde cyanohydrin Hydroquinone	Yes	DTT (5mM)
Triglochin maritima[c]	<200	AS, DS, MRA, G-200	6.5–8.5	*p*-Hydroxybenzaldehyde cyanohydrin 3,4-Dihydroxybenzaldehyde cyanohydrin	Yes	–
Prunus serotina[d]	–	–	7–8	Benzaldehyde cyanohydrin Benzoic acid	Yes	–

Authors: [a]Hahlbrock and Conn, 1970; [b]Reay and Conn, 1974; [c]Hösel and Schiel, 1984; [d]Poulton and Shin, 1983.
[e]Purification methods: AB, antibody affinity; ACA, Ultrogel ACA 34; Acet, acetone powder; AS, ammonium sulphate; CF, chromatofocusing; CL, Sepharose CL 6B; CM, CM-cellulose; CMB, CM Bio-Gel; CMS, CM-Sephadex; Con A, Concanavalin A–Sepharose (agarose); CP, calcium phosphate gel; DB, DEAE Bio-Gel; DC, DEAE-cellulose; DCX, DEAE-Cellex; Dow, Dowex 1 treatment; DS, DEAE-Sephadex; $\Delta\theta$, heat treatment; ΔpH, pH adjustment; EC, Ecteola-cellulose; EtOH, ethanol precipitation; G-100, Sephadex G-100; G-150, Sephadex G-150; G-200, Sephadex G-200; HA, hydroxyapatite; MnCl$_2$, MnCl$_2$ precipitation; MnSO$_4$, MnSO$_4$ treatment; MRA, Matrex Red A; P-60, Bio-Gel P-60; P-200, Bio-Gel P-200; PAGE, polyacrylamide gel electrophoresis; PEI, polyethyleneimine; Red, Reactive Red 120-Agarose; S6B, Sepharose 6B; S-200, Sephacryl S-200.

by the sorghum, *Prunus serotina* and *Triglochin maritima* enzymes, leading to the synthesis of (*S*)-dhurrin, (*R*)-prunasin and (*S*)-taxiphyllin, respectively.

2. *Extraction and purification*

Cyanohydrin β-glucosyltransferases are routinely extracted from fresh tissue or acetone powders in Tris-HCl or potassium phosphate buffers (pH 7.6–8.0). Where necessary, Polyclar AT (0.1–2 × tissue weight), DIECA (25 mM), EDTA (1 mM), DTT (5 mM) and β-mercaptoethanol (14 mM) have been included in extraction media. Significant enzyme purification was achieved using the conventional methods listed in Table 7.1, but in no case was homogeneity reached. With few exceptions (e.g. Poulton and Shin, 1983), crude extracts of cyanophoric plants exhibit high β-glycosidase activity towards endogenous cyanogenic glycosides (see Section III.A). It is highly desirable that these hydrolases be removed during early purification stages to prevent their interference in glucosyltransferase assays by degrading glycosidic products. Since many cyanogenic glycosidases, but not cyanohydrin glucosyltransferases, are glycoproteins, lectin affinity chromatography may be a promising solution for this problem. Indeed, the *Prunus serotina* mandelonitrile glucosyltransferase was successfully separated from amygdalin hydrolase and prunasin hydrolase by Concanavalin A–Sepharose 4B chromatography (Poulton, unpubl. data).

3. *Assay*

Most authors assay glucosyltransferase activity by measuring the rate of formation of labelled glucoside when enzyme is incubated with unlabelled cyanohydrin and UDP-[^{14}C]glucose. The product is separated from unreacted substrates by paper chromatography (PC) (Reay and Conn, 1974; Poulton and Shin, 1983) or TLC (Hösel and Schiel, 1984). A typical assay (Reay and Conn, 1974) contains 10 μmol Tris-HCl buffer (pH 8.0), 2 μmol aromatic or aliphatic cyanohydrin, 0.3 μmol UDP-[^{14}C]glucose, 0.1 μmol DTT, 0.1 mg BSA and 0.05–5 μg protein in a total volume of 100 μl. The mixture is incubated at 30°C for 10–60 min before stopping the reaction with 25 μl 10% acetic acid. The assay mixture is then applied to a 4-cm wide strip of Whatman No. 1 chromatography paper and developed with butanone–acetone–H$_2$O (15:5:3; v/v). After drying the papers, radioactive product zones are detected by strip scanner or hand-held monitor, cut up into a vial, and their radioactivity determined by scintillation counting. While somewhat time-consuming, this method is widely applicable since (1) solvent systems for chromatographic separation of most common cyanogenic glycosides are well documented (Nahrstedt, 1981; Brimer, 1988; Brinker and Seigler, 1991), and (2) UDP-[^{14}C]glucose and unlabelled cyanohydrins are commercially available.

Alternatively, glucosylation reactions may be followed using radiolabelled cyanohydrins with product separation by PC (Reay and Conn, 1974) or other means. As described by Hahlbrock and Conn (1970), the enzyme preparation (30–500 μg protein) is incubated for 60 min at 30°C with 50 μmol Tris-HCl (pH 7.6), 0.5 μmol UDPG, and 2 μmol 1-[^{14}C]cyanohydrin in a total volume of 200 μl. The reaction is stopped by boiling the mixture in a water bath for 2 min, and aliquots (100 μl) are transferred into scintillation vials. The solvent and volatile cyanohydrin are evaporated under reduced pressure in a desiccator over NaOH, and the radioactivity of the remaining non-volatile products is measured by scintillation counting using Bray's solution. Chemical synthesis of 1-[^{14}C]cyanohydrins of acetaldehyde and aliphatic ketones and determination of their specific activity are described by Hahlbrock *et al.* (1968).

Due to the instability of cyanohydrins and some cyanogenic glycosides in alkaline media, glucosyltransferase assays are usually performed at lower pH values than the optima given in Table 7.1. When assaying glucosyltransferase activity of crude extracts, inhibitors such as β-D-glucosylamine (10 mM) (Wurtele *et al.*, 1982) or castanospermine (1–5 mM) may be included to reduce interference by endogenous β-glycosidases. Whether these inhibitors also affect cyanohydrin glucosyltransferase activity has not been rigorously tested.

4. Localisation

Few localisation studies have been reported. In young, light-grown sorghum seedlings, the epidermal and mesophyll tissues exhibited 70% and 30% respectively of the total *p*-hydroxybenzaldehyde cyanohydrin glucosyltransferase activity of the leaf blade. The bundle sheath strands lacked activity (Kojima *et al.*, 1979). Fractionation of epidermal-enriched protoplast preparations showed that at least 41% of the recoverable glucosyltransferase activity was associated with plastids (Wurtele *et al.*, 1982).

III. CATABOLISM OF CYANOGENIC GLYCOSIDES

Upon tissue disruption, cyanogenic glycoside degradation is initiated by cleavage of the glycosidic linkage(s) by one or more soluble β-glycosidases (EC 3.2.1.-), yielding the corresponding α-hydroxynitrile. This intermediate may decompose either spontaneously or enzymatically in the presence of an α-hydroxynitrile lyase (EC 4.1.2.-) to yield HCN and an aldehyde or ketone. While non-enzymatic decomposition of cyanohydrins occurs rapidly at alkaline pH, it is negligible below pH 5.0. Plant macerates are commonly slightly acidic (pH 5.0–6.5). In such macerates, α-hydroxynitrile lyases greatly accelerate the release of HCN and carbonyl compounds (Gross *et al.*, 1982;

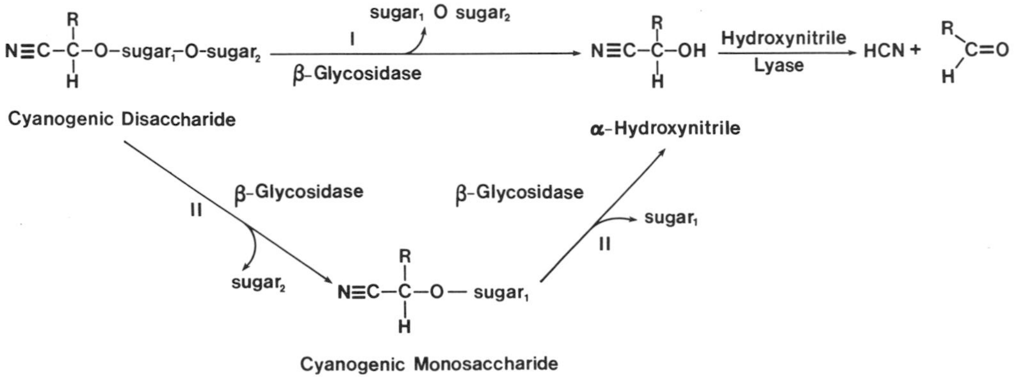

FIG. 7.3. Hydrolysis of cyanogenic diglycosides.

Selmar *et al.*, 1989b). Most likely, these enzymes serve the same function *in vivo*, although the pH of their cellular environment is not known.

Cyanogenic diglycosides may undergo hydrolysis by either the 'sequential' (I) or 'simultaneous' (II) pathways depending on whether the two sugar residues are removed stepwise or as a disaccharide unit (Fig. 7.3) (Kuroki *et al.*, 1984). In the former case, the two hydrolytic steps are generally catalysed by distinct β-glycosidases. Well-documented examples of the sequential mechanism are provided by the catabolism of amygdalin in *Prunus serotina* and of linustatin and neolinustatin in *Linum usitatissimum* (for review, see Poulton, 1988). The hydrolyses of vicianin in *Davallia trichomanoides* and *Vicia angustifolia* and of linustatin in *Hevea brasiliensis* occur via the simultaneous mechanism.

A. β-Glycosidases Involved in Cyanogenesis

1. Properties

β-Glycosidases involved in cyanogenesis are remarkably similar in their kinetic and molecular properties (Tables 7.2 and 7.3). All possess acidic pH optima (pH 4.0–6.2) and, with few exceptions, are glycoproteins having isoelectric points in the range pH 4–5.5. Despite wide variation in native molecular masses (40–600 kDa), most have subunit molecular masses of 55–65 kDa, suggesting a common structural relationship among these and other plant β-glycosidases. Multiple forms of cyanogenic β-glycosidases are common, but the nature of this apparent microheterogeneity is largely unexplained. Unlike the cyanohydrin glucosyltransferases, they usually show excellent stability. Many can be stored for several months at 4°C (with sodium azide) or at −20°C (with glycerol) with little activity loss. They have no requirement for metal ions and are generally unaffected by thiol reagents, sulphhydryl compounds and metal chelators. Where tested (e.g. Eksittikul and Chulavatnatol, 1988), cyanogenic β-glycosidases were inhibited by common β-glycosidase inhibitors such as δ-gluconolactone, nojirimycin, l-amino-β-D-glucoside, glucosylpiperidine, and galactosylpiperidine. The pyrrolizidine alkaloid castanospermine potently inhibited β-glycosidases from *Prunus serotina* and *Davallia trichomanoides*.

In past years, the use of synthetic rather than natural substrates during the purification and characterisation of relatively crude but highly active enzyme preparations contributed to the widely accepted view that plant β-glycosidases lack aglycone specificity (see discussion by Hösel, 1981; Hösel and Conn, 1982). Since 1975, this viewpoint has been challenged by the isolation of β-glycosidases exhibiting pronounced specificity for their endogenous cyanogenic glycosides. Examples include β-glucosidases showing specificity towards taxiphyllin (*Triglochin maritima*), triglochinin (*Alocasia macrorrhiza, Triglochin maritima*), vicianin (*Davallia trichomanoides, Vicia angustifolia*), dhurrin (*Sorghum bicolor*), amygdalin (*Prunus serotina*), and prunasin (*P. serotina*) (for review, see Poulton, 1988). In comparison, linamarases, such as those from *Linum usitatissimum, Hevea brasiliensis* and *Phaseolus lunatus*, often show broader substrate specificities.

TABLE 7.2. Properties of purified β-glycosidases hydrolysing aromatic cyanogenic glycosides.

Enzyme	Purification (-fold)	Purification methods[f]	Native molecular mass (kDa)	Subunit molecular mass (kDa)	pI	Carbohydrate	pH optimum
Prunus amygdalus[a] (Isozyme A)	170	ΔpH, MnSO₄, AS HA, DS	140 ± 20	–	–	–	5.0
Prunus serotina amygdalin hydrolases (I and II)[b]	209 (I) 216 (II)	Con A, CM, HA, CF	55–60	62	6.6 (I) 6.5 (II)	Yes	4.5–5.0
Prunus serotina prunasin hydrolases (I, IIa and IIb)[c]	122 (I) 85 (IIb)	Con A, HA, S-200	68 (I) 140 (IIa) 68 (IIb)	69.5 (I) 69.5 (IIa) 69.5 (IIb)	4.0–4.5	Yes	5.0
Sorghum bicolor dhurrinases (1A, 1B and 2A)[d]	56 (1A) 54 (1B) 100 (2A)	CP, AS, DC, MRA, Con A, ACA	200–240 (1A) 100–110 (1B) 250–300 (2A)	57 (1A) 57 (1B) 61 (2A)	5.5 (1A, 1B)	No	6.0–6.2 (1B, 2A)
Davallia trichomanoides vicianin hydrolase[e]	7	DC, HA	340	32.5, 49, 56	4.6–4.7	No	5.0–6.0

Authors: [a]Lalégerie, 1974; [b]Kuroki and Poulton, 1986; [c]Kuroki and Poulton, 1987; [d]Hösel et al., 1987; [e]Lizotte and Poulton, 1988. [f]Purification methods: see legend for Table 7.1 for explanation of abbreviations.

TABLE 7.3. Properties of β-glycosidases hydrolysing aliphatic cyanogenic glycosides.

Enzyme	Purification (-fold)	Purification methods[l]	Native molecular mass (kDa)	Subunit molecular mass (kDa)	pI	Carbohydrate	pH optimum
Linum usitatissimum linustatinase[a]	100	Acet, PEI, AS, MRA, Con A, CF	56–61	19, 39	7–8	Yes	4–6
Hevea brasiliensis linustatinase[b]	7	AS, G-150	40	–	–	–	4.5
Linum usitatissimum linamarase[c]	283	Acet, AS, Con A, DC, CL, CF	570–670	62.5, 65	4–5	Yes	5.5–6
Trifolium repens linamarase[d,e]	16	Acet, G-200, CM, (Con A)	105	62	–	Yes	4–5
Manihot esculenta linamarase[f,g]	10	Acet, Con A, P-200	–	62	4.3–4.4	Yes	–
	350	AS, DC, S6B	600	–	–	–	6.0
	2–17	AS, S6B, CF	~600	63	2.9, 3.3, 4.3	–	6.0
Hevea brasiliensis linamarase[h]	4–15	AS, DB, CMB, G-150, PAGE (AB)	64 and higher oligomers	64	–	Yes	5.6
Phaseolus lunatus linamarase[i]	11,700	AS, Acet, CMS, DS, G-200	124 ± 9	59 ± 2.4	–	–	5.1–6.2
Alocasia macrorrhiza triglochininase[j]	9.1	Acet, AS, DCX, CMS, G-200	310	55–60	4.5–5.0	–	5.5
Triglochin maritima triglochininase[k]	50–100	Acet, AS, DCX, CMS, G-200	120–125	–	4.5–5.0	–	5.2
Triglochin maritima glucosidase C[k]	100	Acet, AS, DCX, CMS, G-200	220–290	–	4–4.5	–	5.0

Authors: [a]Fan and Conn, 1985; [b]Selmar, 1986; [c]Fan and Conn, 1985; [d]Hughes and Dunn, 1982; [e]Pócsi et al., 1989; [f]Cooke et al., 1978; [g]Eksittikul and Chulavatnatol, 1988; [h]Itoh-Nashida et al., 1987b; [i]Selmar et al., 1987b; [j]Hösel and Nahrstedt, 1975; [k]Nahrstedt et al., 1979.
[l]Purification methods: see legend for Table 7.1 for explanation of abbreviations.

2. *Extraction and purification*

Extraction of β-glycosidases from acetone powders or fresh tissues is accomplished using standard buffers in the pH range 5.0–7.6. Protease inhibitors, including PMSF, pepstatin A, TLCK and Na₂EDTA, may be added to inhibit proteolysis during enzyme isolation. Enzyme browning caused by high tissue phenolic contents may be suppressed by PVPP, DIECA, cysteine and DTT. As summarised in Tables 7.2 and 7.3, conventional methods have allowed extensive purification of cyanogenic β-glycosidases, and, in some cases, apparent homogeneity was achieved. Taking advantage of the glycoproteinous nature of many cyanogenic glycosidases, Concanavalin A–Sepharose 4B chromatography has proven outstanding, often recording purification factors of 20- to 40-fold (Fan and Conn, 1985; Kuroki and Poulton, 1986, 1987).

3. *β-Glycosidase assays employing synthetic substrates*

Popular methods for assaying β-glycosidase activities in plant extracts utilise chromogenic and fluorogenic substrates such as nitrophenyl and 4-methylumbelliferyl glycosides. Such assays are relatively rapid, facile, inexpensive, and accommodate large numbers of samples. Furthermore, given the commercial availability of a wide range of PNP- and ONP-sugars, they facilitate analysis of the sugar specificity and stereospecificity of purified glycosidases. The major drawback of these assays is their lack of specificity with respect to the aglycone moiety. During procedure development to isolate β-glycosidases, specific assay methods should be included (see Section III.A.4).

To measure PNP-glycosidase activity, the reaction mixture typically contains 3.75 μmol PNP-sugar, 50 μmol citrate-phosphate buffer (pH 5.0), and up to 100 μg of protein in a total volume of 250 μl (Kuroki and Poulton, 1987). Bovine serum albumin (BSA; 500 μg) may be added when assaying preparations with low protein concentration. Assay mixtures are pre-incubated without enzyme at 30°C for 5 min. After incubation at 30°C with enzyme for 5–60 min, reactions are terminated by adding 3 ml of 0.2 M borate–NaOH buffer (pH 9.8) or 5% (w/v) Na₂CO₃. *p*-Nitrophenol production is quantitated by absorbance at 400 nm using a standard curve recorded under assay conditions.

4-Methylumbelliferyl-β-D-glucosidase activity may be determined by replacing the PNP-sugar by 4-MUG in the above assay. After terminating the reaction with borate buffer, 4-methylumbelliferone levels are quantitated by absorbance measurement at 360 nm using suitable standard curves. Alternatively, product formation may be monitored fluorometrically ($\lambda_{exc.}$ 360 nm; $\lambda_{flu.}$ 450 nm) (Boersma *et al.*, 1983).

4. *β-Glycosidase assays employing cyanogenic glycosides as substrates*

It is highly recommended that endogenous cyanogenic glycosides be used as substrates wherever possible, especially during enzyme purification (Hösel, 1981). Prunasin, amygdalin and linamarin may be purchased (Sigma, Calbiochem). Procedures for isolation of several other cyanoglycosides are well documented (Nahrstedt, 1981; 'Olafsdóttir, 1990; Brinker and Seigler, 1991). The rate of cyanogenic glycoside hydrolysis may be determined by assaying either the liberated sugars (usually glucose) or the HCN or carbonyl compound which arise from dissociation of the cyanohydrin aglycone.

(a) Assay of liberated glucose by chemical or enzymatic methods. Cyanide interferes with the determination of glucose by the usual copper methods for reducing sugars (Haisman and Knight, 1967). Glucose can be determined by the glucosazone method of Wahba *et al.* (1956). This method is convenient and accurate but requires correction for concurrent formation of benzaldehyde phenylhydrazone (Haisman and Knight, 1967).

Many laboratories prefer enzymatic methods to quantitate the glucose released during hydrolysis of cyanogenic glycosides. A typical reaction mixture contains 3.75 μmol cyanogenic glycoside, 50 μmol citrate–phosphate buffer (pH 5.0), 0.5 mg BSA, and up to 100 μg of protein in a total volume of 250 μl (Kuroki and Poulton, 1987). Assay mixtures are pre-incubated for 5 min at 30°C before addition of enzyme. The reaction is terminated after 5–60 min by adding 20 μl 9.4% (w/v) HCl. Aliquots (150 μl) are then removed, and the glucose produced is measured by the glucose oxidase–peroxidase method as described by Sigma (Sigma Procedure No. 510). Alternatively, glycoside hydrolysis is terminated by boiling the assay mixture for 30 s and assaying liberated glucose by the hexokinase–glucose-6-phosphate dehydrogenase method as modified by Boehringer (Fan and Conn, 1985). In either case, appropriate control reactions, in which active enzyme is replaced by buffer or by boiled enzyme preparation, should be routinely included. Extreme caution should be exercised when using these coupled assays to gauge the effect on β-glucosidase activity of metal ions, chelators, thiol reagents or inhibitors, as these components may interfere with the enzymatic method for glucose determination. In that event, assays employing synthetic substrates (e.g. PNPGlc) are commonly used.

(b) Assay of carbonyl compounds released. β-Glucosidase activity towards triglochinin and certain aromatic cyanoglucosides may be assayed by measuring the carbonyl compound formed upon dissociation of their respective cyanohydrin aglycones:

(i) Assay for triglochininase activity. The standard assay contains 40 μM triglochinin, 50 mM citrate–sodium phosphate buffer (pH 5.0), and enzyme in a total volume of 0.5 ml (Nahrstedt *et al.*, 1979). After pre-incubating the substrate at 30°C for 5 min, the reaction is initiated by addition of β-glucosidase. The rate of triglochinin hydrolysis is measured spectrophotometrically by following the decrease in A_{276}. Initial velocities are determined from linear portions of graphs ($\varepsilon = 12\ 000\ \text{M}^{-1}\text{cm}^{-1}$).

(ii) Assay for dhurrinase and taxiphyllinase activities. Standard assays contain 1 mM cyanogenic glycoside, 0.1 M citrate–phosphate buffer (pH 6.2) and enzyme in a total volume of 0.5 ml (Hösel *et al.*, 1987). After 5–60 min incubation at 30°C, the reaction is terminated by 0.5 ml 0.2 M Tris-HCl buffer (pH 8.8) which shifts the pH to 8.3. Control reactions lack substrate or active enzyme. The A_{330}, due to liberated 4-hydroxybenzaldehyde ($\varepsilon = 21\ 000\ \text{M}^{-1}\text{cm}^{-1}$), is immediately determined. This procedure, which is highly useful for routine assay of column fractions, can be adapted to follow prunasin and sambunigrin hydrolysis by terminating reactions with 1 M potassium carbonate and measuring at 255 nm to quantitate levels of liberated benzaldehyde ($\varepsilon = 11\ 000\ \text{M}^{-1}\text{cm}^{-1}$).

(iii) Continuous assay for dhurrinase activity. To obtain initial rates, the enzyme is incubated at 37°C in 1.5 ml quartz cuvettes with 1.8 μmol dhurrin, 80 μmol sodium acetate buffer (pH 5.5), and a large excess (55 mU) of highly purified sorghum oxynitrilase (Bové and Conn, 1961) in a total volume of 1.2 ml (Kojima *et al.*, 1979). The latter enzyme catalyses immediate dissociation of the liberated cyanohydrin to 4-hydroxybenzaldehyde, the levels of which are followed spectrophotometrically at 285 nm ($\varepsilon = 14\,800\,\text{M}^{-1}\text{cm}^{-1}$) for at least 3 min.

(iv) Continuous assay for prunasin hydrolase activity. Prunasin hydrolase activity is assayed by monitoring the catabolism of prunasin to benzaldehyde in the presence of excess purified mandelonitrile lyase (Gross *et al.*, 1982). The hydrolase preparation (up to 0.2 nkat activity) is incubated at 30°C with 5 μmol prunasin, 50 μmol sodium acetate buffer (pH 6.0), and 2.35 nkat of mandelonitrile lyase in a total volume of 1.01 ml. The rate of prunasin hydrolysis is calculated from the increase in A_{249} observed during the first few minutes of the reaction ($\varepsilon = 15\,250\,\text{M}^{-1}\text{cm}^{-1}$). A single-stage ion-exchange chromatographic procedure yields nearly homogeneous mandelonitrile lyase from commercial almond emulsin (Sigma) (Gross *et al.*, 1982).

(c) Assay of HCN released. Many laboratories determine β-glucosidase activity towards linamarin (and lotaustralin) by assaying the HCN that is released upon alkalinisation of terminated reaction mixtures. As exemplified by the method of Kojima *et al.* (1983), the cassava linamarase is incubated at 30°C with 0.75 μmol linamarin and 5 μmol phosphate buffer (pH 6.8) in a total volume of 650 μl. Control reactions which lack either substrate or enzyme are also included. After 10 min incubation, reactions are terminated by addition of 350 μl 0.3 N NaOH. Under these conditions, the product acetone cyanohydrin decomposes spontaneously to HCN, which is quantitated by any of the following methods.

(i) Modified Lambert method (Lambert *et al.*, 1975). This popular method is relatively specific for cyanide ion and detects as little as 5 μg l^{-1}. The NaOH concentration of terminated reaction mixtures is adjusted to 0.1 M with distilled water. To 1 ml aliquots are then added the following reagents in order: 0.5 ml 1 M acetic acid, 5 ml N-chlorosuccinimide reagent (containing 0.25 g N-chlorosuccinimide and 2.5 g succinimide in 1 litre of H$_2$O), and 1 ml barbituric acid/pyridine reagent (prepared by mixing 12 g barbituric acid with a small amount of water to make a paste, then 120 ml pyridine is added with enough water to reach a final volume of 400 ml). The tubes are shaken or vortexed vigorously. After 10 min, the colours are read at 580 nm and quantitated using a standard curve prepared using NaCN solutions (0–1 μg CN$^-$ ml^{-1}). The colours are stable for at least 30 min. Detailed experimental procedures have been described by Brinker and Seigler (1989). For a micro-scale version of this procedure, see Section II.A.5.

(ii) Spectroquant method. After neutralising alkaline samples with HCl, cyanide may be determined using the Merck Spectroquant kit as described in data sheet 130 259 8Do dt/5r by Merck, D-6100 Darmstadt, Pb 4119, FRG (Selmar *et al.*, 1988).

(iii) Pyrazolone method (Epstein, 1947). As described by Mao *et al.* (1965), aliquots of the terminated reaction mixture containing 0.05–0.6 μg HCN are pipetted into test tubes in an ice-bath. The following additions are then made with thorough mixing: 0.1 N NaOH to bring the volume to 1 ml; 1 ml 0.15 M NaH$_2$PO$_4$; and 0.5 ml

0.14% chloramine-T. Two to five min after addition of chloramine-T, 1 ml pyridine–pyrazolone reagent (85 mg of 3-methyl-1-phenyl-2-pyrazoline-5-one : 3,3′-dimethyl-1,1′-diphenyl-4,4′-bi-2-pyrazoline-5,5′-dione, 5:1 (w/w), in 25 ml pyridine AR, freshly made up) is added, and the A_{620} is read after 90 min at room temperature.

(d)   Assay of diglycosidases working by the simultaneous mode. Hydrolysis of cyanogenic diglycosides via the simultaneous mechanism yields disaccharides (e.g. gentiobiose, vicianose) which are not detectable by the enzymatic methods described above (Section III.A.4). To obviate this problem, Lizotte and Poulton (1988) utilised prunasin as an alternative substrate when purifying vicianin hydrolase from *Davallia trichomanoides*. This seemed justifiable after initial experiments confirmed that the elution profiles for vicianin hydrolase activity (monitored by the semi-quantitative test of Feigl and Anger (1966)) exactly matched those of prunasin hydrolase activity after each stage. Vicianin hydrolysis by the purified enzyme was quantitated using an HPLC method which measured the decline in substrate concentration (rather than product formation). More accurate is the method of Selmar *et al.* (1988) who incubated the *Hevea brasiliensis* diglucosidase at 30°C with 10 mM cyanogenic diglycoside (linustatin or amygdalin) and McIlvaine buffer (pH 4.5) in a total volume of 2.5 ml. After 10–60 min, the reaction was terminated by adding 1 ml 0.5 N NaOH. The liberated HCN was estimated by the Merck Spectroquant method (see above) immediately after acidification by HCl.

5.   Localisation

In 6-day-old, light-grown sorghum seedlings, dhurrin β-glucosidase resides almost exclusively within the mesophyll tissue where it is largely associated with chloroplasts (Kojima *et al.*, 1979; Thayer and Conn, 1981). In contrast, an apoplastic location appears likely for linamarases in leaves of *Trifolium repens* (Kakes, 1985), *Phaseolus lunatus* (Frehner and Conn, 1987) and cassava (Mkpong *et al.*, 1990) and in *Hevea brasiliensis* endosperm (Selmar *et al.*, 1989b). β-Glucosidase (linamarase) activity was detected histochemically in the phloem, epidermis and mesophyll of healthy *Lotus corniculatus* leaves (Rissler and Millar, 1977).

B.   α-Hydroxynitrile Lyases

1.   Properties

α-Hydroxynitriles arising from the catabolism of cyanogenic glycosides are further degraded to HCN and an aldehyde or ketone by α-hydroxynitrile lyases. Since the first isolation of mandelonitrile lyase from almond emulsin in 1908, hydroxynitrile lyases have been extensively purified and characterised from plants accumulating aromatic and aliphatic cyanogenic glycosides (for review, see Poulton, 1988). These enzymes generally show greatest, but not always exclusive, activity toward endogenous cyanohydrins. While all known lyases possess slightly acidic pH optima and isoelectric points between pH 3.9 and 4.8, they appear to fall into at least three fundamentally distinct groups based on their FAD and carbohydrate contents (Table 7.4). The first group contains (*R*)-(+)-mandelonitrile lyases (EC 4.1.2.10) from seeds of the Prunoideae and

TABLE 7.4. Properties of purified plant hydroxynitrile lyases.

Enzymes	Purification (-fold)	Purification methods[l]	Native molecular mass (kDa)	Subunit molecular mass (kDa)	PI	FAD	Carbo-hydrate	pH optimum	Substrates
Prunus amygdalus[a]	—	ΔpH, EtOH, EC, G-100	60.4–62.1	—	4.42–4.46	Yes	Yes	5.5–6.0	(R)-Mandelonitrile
Prunus laurocerasus[b]	—	EtOH, DC, G-100	60	—	4.20–4.37	Yes	Yes	—	(R)-Mandelonitrile
Prunus serotina[c,d]	10 / 34	Con A, CF / Con A, DC, Red, PAGE	55.6	57–59	4.58–4.63	Yes	Yes	6.0–7.0	(R)-Mandelonitrile
Prunus lyonii[e]	4.3	DC, Con A	50	59	4.75	Yes	Yes	5.5	(R)-Mandelonitrile
Sorghum vulgare[f,g]	253	Δθ, ΔpH, AS, CP, AS, DC, G-75	180	—	4.5	No	—	5.0–6.0	p-Hydroxymandelonitrile, vanillin cyanohydrin
Linum usitatissimum[h]	136	AS, CF, DC, HA, Con A, S-200, MRA	82	42	4.5–4.8	No	No	5.5	Acetone cyanohydrin, 2-butanone cyanohydrin
Manihot esculenta[i]	163	Acet, AS, Δθ, MnCl$_2$, G-150, DC, HA	91.2	16.5	4.7	No	No	5.4	Acetone cyanohydrin, 2-butanone cyanohydrin, 2-pentanone cyanohydrin
Hevea brasiliensis[j]	7	AS, G-150	46	—	—	—	—	5.5	Acetone cyanohydrin, 2-butanone cyanohydrin, mandelonitrile
Ximenia americana[k]	122	Acet, CM, CF	38	36.5	3.9	No	Yes	5.5	(S)-Mandelonitrile, p-hydroxymandelonitrile

Authors: [a]Aschhoff and Pfeil, 1970, and additional references herein; [b]Gerstner and Kiel, 1975; [c]Yemm and Poulton, 1986; [d]Wu, 1990; [e]Xu et al., 1986; Bové and Conn, 1961; [f]Seely et al., 1966; [h]Xu et al., 1988; [i]Carvalho, 1981; [j]Selmar, 1986; Kuroki and Conn, 1989.
[l]Purification methods: see legend for Table 7.1 for explanation of abbreviations.

Maloideae. Requiring only 5- to 35-fold purification to reach homogeneity, these glycoproteins are clearly major seed constituents and quite possibly serve an additional role as storage proteins. Curiously, they possess FAD bound close to their catalytic sites although these enzymes do not catalyse a net oxidation–reduction reaction. The second group of hydroxynitrile lyases includes those isolated from vegetative tissues of sorghum (Seely *et al.*, 1966), cassava (Carvalho, 1981), and flax (Xu *et al.*, 1988). These proteins required 100- to 300-fold purification to attain homogeneity, and, where tested, they lacked FAD and carbohydrate. They usually possessed higher native molecular weights and, in the case of the cassava and flax enzymes, showed a tendency towards aggregation. Recently, a unique mandelonitrile lyase was characterised from leaves of *Ximenia americana* (Olacaceae) (Kuroki and Conn, 1989). Like lyases from rosaceous species, it is glycoproteinous but it lacks FAD and is stereospecific for (*S*)-(−)-mandelonitrile. Whether FAD plays a direct role in determining the stereo-specificity of mandelonitrile lyases remains unclear.

2. Extraction and purification

(a) Purification of rosaceous mandelonitrile lyases. In early isolations of rosaceous mandelonitrile lyases, seed material was freed of sclerenchyma and testa, ground, and defatted with petroleum ether. The meal was extracted for 12–15 h with dilute NH_3 (pH 7.5) and centrifuged. The milky supernatant was brought to pH 5.4 by adding 6 N acetic acid, cooled to 3–4°C, and recentrifuged to remove inactive contaminants. This crude fraction, which exhibited lyase and β-glycosidase activities, was subjected to ethanol precipitation, ion-exchange chromatography and gel filtration to yield highly purified and, in one case even crystalline, mandelonitrile lyase preparations. Recent improvements to this procedure include: (1) omitting defatting, (2) replacing the time-consuming steps of NH_3 extraction, pH adjustment, and ethanol precipitation by more rapid procedures, (3) taking advantage of Concanavalin A–Sepharose chromato-graphy to remove non-glycoprotein contaminants, and (4) including protease inhibitors and PVPP to reduce potential artefacts from proteolysis and phenolic compounds, respectively. Thus, Yemm and Poulton (1986) described the extensive purification of *Prunus serotina* mandelonitrile lyase by Concanavalin A–Sepharose 4B chromato-graphy and chromatofocusing. A modified procedure employing Concanavalin A–Sepharose, DEAE–cellulose and Reactive Red 120–Agarose chromatography followed by preparative non-denaturing polyacrylamide gel electrophoresis (PAGE) allowed homogeneity to be reached after 34-fold purification (Wu and Poulton, 1991). Tissues were homogenised in the presence of PVPP, PMSF (1 mM), Na_2EDTA (5 mM), pepstatin A (10^{-7} M), TLCK (10^{-6} M), and iodoacetamide (1 mM). Only 4.3-fold purification was necessary to attain homogeneity of the *Prunus lyonii* mandelonitrile lyase (Xu *et al.*, 1986). After extraction in buffer containing PVPP, the enzyme was recovered in 60% yield by DEAE–cellulose and Concanavalin A–Sepharose chromatography. Hochuli (1983) described the rapid purification of mandelonitrile lyase from 1.2 kg almond meal by affinity chromatography using methyl *p*-(3-aminopropoxy) benzoate as ligand.

In most species of Prunoideae and Maloideae, mandelonitrile lyase exists as mixtures of isozymes. These may be partially resolved by chromatofocusing, isoelectric focusing and disc electrophoresis and obtained preparatively by DEAE–cellulose chromato-graphy (Aschhoff and Pfeil, 1970; Gerstner and Pfeil, 1972; Gerstner and Kiel, 1975; Gross *et al.*, 1982; Yemm and Poulton, 1986). Within a particular species, the isozymes displayed slight differences in molecular weight, electrophoretic mobility, isoelectric point, and specific activity but had identical antigenic properties. The nature of their structural heterogeneity is poorly understood. Small but reproducible quantitative differences were revealed by total amino acid analysis of two *Malus communis* isozymes (Gerstner and Pfeil, 1972). Furthermore, Concanavalin A–Sepharose chro-matography suggested that *Prunus serotina* mandelonitrile lyases possess at least two distinct carbohydrate side-chain patterns (Yemm and Poulton, 1986).

(b) Purification of other hydroxynitrile lyases. Hydroxynitrile lyases from other sources have been extracted from acetone powders or fresh tissue in neutral or mildly acidic media (pH 5.4–7.4). Where required, ascorbate (2–100 mM), β-mercaptoethanol (10 mM) and Amberlite XAD-4 are added for protection against endogenous phenolic compounds. Glycerol (25%) is required to stabilise the *Ximenia americana* lyase. Conventional techniques used in purifying non-rosaceous hydroxynitrile lyases are listed in Table 7.4.

3. Assays

(a) Lyase activity towards aromatic cyanohydrins. A facile spectrophotometric assay monitors the decomposition of mandelonitrile at 249 nm, a wavelength at which the product benzaldehyde absorbs strongly ($\varepsilon = 13\,200\,\text{M}^{-1}\text{cm}^{-1}$) (Jorns, 1979, 1980). Reaction rates are measured immediately after the simultaneous addition of small amounts of enzyme and (R, S)-mandelonitrile (final concentration $2.1 \times 10^{-4}\,\text{M}$ to 1 ml 0.1 M sodium citrate buffer (pH 5.5) at 25°C and are corrected for the rate of non-enzymic cyanohydrin decomposition. While commercial preparations (e.g. Sigma) may be used, Jorns (1980) reports the synthesis and crystallisation of pure (R,S)-mandelonitrile. A stock solution may be prepared freshly each day since it exhibits negligible decomposition when kept at 0°C. The enzymatic decomposition of other aromatic cyanohydrins may be assayed by analogous spectrophotometric methods which detect production of their corresponding aldehydes (Bové and Conn, 1961). With *p*-hydroxybenzaldehyde cyanohydrin as substrate, the conditions are as follows: 2 mM aqueous stock solution of substrate (prepared daily and kept on ice); assay buffer, 0.1 M acetate buffer (pH 5.4); initial substrate concentration, 0.2 mM; detection wavelength, 285 nm. After adding the cyanohydrin to the acetate buffer in a quartz cuvette, the non-enzymic decomposition is assessed at room temperature by measuring the change in A_{285} at 1 min intervals for 5 min. Enzyme is then introduced and measurement continued. After correction for non-enzymic decomposition, the rate of the enzymatic reaction is constant for several minutes and is proportional to the amount of enzyme added. Activity towards the cyanohydrins of vanillin and isovanillin may be monitored at 308 nm and 278 nm, respectively, by assays in which the acetate buffer is replaced by Na_2HPO_4 (22 mM)–citric acid (56 mM), pH 5.5 (Seely *et al.*, 1966). The foregoing racemic cyanohydrins may be synthesised according to Laden-burg *et al.* (1936).

(b)   Lyase activity towards acetone cyanohydrin.   Acetone cyanohydrin lyase activity may be assayed by determining levels of enzymatically released HCN by the modified Lambert procedure (1975) as described by Selmar *et al.* (1987a). The substrate must previously be freed of HCN by distillation (use within 4 weeks!) and by evacuation for several minutes under a water pump immediately before use. A stock solution of 10% (v/v) (approx. 1.08 M) substrate is then prepared in 0.1 M citric acid (final pH approx. 2.2). Aliquots (0.1 ml) are added at 30°C to 9.9 ml McIlvaine buffer (pH 5.6) or 0.1 M acetate buffer (pH 5.6) as a blank or to 9.9 ml of the same buffer containing the enzyme to be assayed. Depending on enzyme activity, aliquots (10 or 25 μl) of the reaction mixture are removed at 5 and 10 min and added to 5 ml *N*-chlorosuccinimide reagent (for composition, see Section III.A.4). This terminates the enzymic reaction and lowers the pH below 4.0 at which further substrate dissociation is negligible. Barbituric acid-pyridine reagent (1 ml; see Section III.A. 4) is added with thorough mixing. After 10 min, the A_{585} is read and quantitated using an appropriate standard curve. Corrections are made for non-enzymic dissociation of acetone cyanohydrin by subtraction of values obtained with the blanks. This method may be extended for use with other cyanohydrins. Released HCN may also be quantitated by the Merck-Spectroquant cyanide test (see Section III.A.4).

4. Localisation

Hydroxynitrile lyases have been isolated from blossoms, seeds, leaves, and both etiolated and light-grown shoots, but to large extent their tissue and subcellular localisations have not been unequivocally established. In leaf blades of light-grown sorghum seedlings, virtually all of the lyase activity was associated with the mesophyll tissue (Kojima *et al.*, 1979) where it probably has a cytoplasmic location (Thayer and Conn, 1981). Protoplast studies also suggest intracellular localisations for the lima bean and *Hevea* lyases (Frehner and Conn, 1987; Selmar *et al.*, 1989a). In almond and peach seeds, mandelonitrile lyase activity was detectable in cotyledon cells but not in the growing axis (Gerstner, 1974). The subcellular localisation of *P. serotina* mandelonitrile lyase was determined by immunogold labelling studies at the electron microscope level using monospecific polyclonal antibodies raised against deglycosylated enzyme (Swain *et al.*, 1992). Highest levels of mandelonitrile lyase were observed in the protein bodies of the cotyledonary parenchyma cells, with lesser amounts in the procambial cell protein bodies.

REFERENCES

Aschhoff, H. J. and Pfeil, E. (1970). *Hoppe-Seyler's Z. Physiol. Chem.* **351**, 818–826.
Bachman, G. B. and Maleski, R. J. (1972). *J. Org. Chem.* **37**, 2810–2814.
Baxter, R. L., Hanley, A. B. and Chan, H. W.-S. (1988). *J. Chem. Soc. Chem. Commun.*, 757–758.
Boersma, P., Kakes, P. and Schram, A. W. (1983). *Acta Bot. Neerl.* **32**, 39–47.
Bousquet, E. W. (1943). *Org. Syn. Coll.* **2**, 313–315.
Bové, C. and Conn, E. E. (1961). *J. Biol. Chem.* **236**, 207–210.
Brimer, L. (1988). *In* "Cyanide Compounds in Biology" (D. Evered and S. Harnett, eds.), pp. 177–200. John Wiley and Sons, Chichester.

Brinker, A. M. and Seigler, D. S. (1989). *Phytochemical Bull.* **21**, 24–31.

Brinker, A. M. and Seigler, D. S. (1991). *In* "Modern Methods of Plant Analysis" (L. F. Linskens and J. F. Jackson, eds.), in press.

Carvalho, F. J. P. C. (1981). Ph.D Dissertation, University of California, Davis, CA.

Collinge, D. B. and Hughes, M. A. (1982a). *J. Exp. Bot.* **33**, 154–161.

Collinge, D. B. and Hughes, M. A. (1982b). *Arch. Biochem. Biophys.* **218**, 38–45.

Collinge, D. B. and Hughes, M. A. (1984). *Plant. Sci. Lett.* **34**, 119–125.

Conn, E. E. (1973). *Biochem. Soc. Symp.* **38**, 277–302.

Conn, E. E. (1980). *Ann. Rev. Plant Physiol.* **31**, 433–451.

Cooke, R. D., Blake, G. G. and Battershill, J. M. (1978). *Phytochemistry* **17**, 381–383.

Cutler, A. J. and Conn, E. E. (1981). *Arch. Biochem. Biophys.* **212**, 468–474.

Cutler, A. J., Hösel, W., Sternberg, M. and Conn, E. E. (1981). *J. Biol. Chem.* **256**, 4253–4258.

Cutler, A. J., Sternberg, M. and Conn, E. E. (1985). *Arch. Biochem. Biophys.* **238**, 272–279.

David, S., Dupont, G. and Paquot, C. (1944). *Bull. Soc. Chim. Fr.* **11**, 561.

Eksittikul, T. and Chulavatnatol, M. (1988). *Arch. Biochem. Biophys.* **266**, 263–269.

Epstein, J. (1947). *Anal. Chem.* **19**, 272–274.

Fan, T. W.-M. and Conn, E. E. (1985). *Arch. Biochem. Biophys.* **243**, 361–373.

Feigl, F. and Anger, V. (1966). *Analyst* **91**, 282–284.

Finkbeiner, H. L. and Stiles, M. (1963). *J. Am. Chem. Soc.* **85**, 616–622.

Frehner, M. and Conn, E E. (1987). *Plant Physiol.* **84**, 1296–1300.

Gerstner, E. (1974). *Naturwissenschaften* **61**, 687.

Gerstner, E. and Kiel, U. (1975). *Hoppe-Seyler's Z. Physiol. Chem.* **356**, 1853–1857.

Gerstner, E. and Pfeil, E. (1972). *Hoppe-Seyler's Z. Physiol. Chem.* **353**, 271–286.

Gross, M., Jacobs, G. H. and Poulton, J. E. (1982) *Anal. Biochem.* **119**, 25–30.

Hahlbrock, K. and Conn. E. E. (1970). *J. Biol. Chem.* **245**, 917–922.

Hahlbrock, K., Tapper, B. A., Butler, G. W. and Conn, E. E. (1968). *Arch. Biochem. Biophys.* **125**, 1013–1016.

Halkier, B. A and Møller, B. L. (1989). *Plant Physiol.* **90**, 1552–1559.

Halkier, B. A. and Møller, B. L. (1990). *J. Biol. Chem.* **265**, 21114–21121.

Halkier, B. A. and Møller, B. L. (1991). *Plant Physiol,.* in press.

Halkier, B. A., Scheller, H. V. and Møller, B. L. (1988). *In* "Cyanide Compounds in Biology" (D. Evered and S. Harnett, eds.), pp. 49–66. John Wiley and Sons, Chichester.

Halkier, B. A., Olsen, C. E. and Møller, B. L. (1989). *J. Biol. Chem.* **264**, 19487–19494.

Halkier, B. A., Olsen, C. E. and Marcussen, J. (1991a). *J. Lab. Compd. Radiopharm.* **29**, 1–7.

Halkier, B. A., Lykkesfeldt, J., and Møller, B. L. (1991b). *Proc. Natl. Acad. Sci. USA* **88**, 487–491.

Haisman, D. R. and Knight, D. J. (1967) *Biochem. J.* **103**, 528–534.

Hegnauer, R. (1986). *In* "Chemotaxonomie der Pflanzen", Vol. 7. Birkhauser Verlag, Basel.

Hochuli, E. (1983). *Helvetica Chimica Acta* **66**, 489–493.

Hösel, W. (1981). *In* "Cyanide in Biology" (B. Vennesland, E. E. Conn, C. J. Knowles, J. Westley and F. Wissing, eds.), pp. 217–232. Academic Press, London.

Hösel, W. and Conn, E. E. (1982). *Trends Biochem. Sci.* **7**, 219–221.

Hösel, W. and Nahrstedt, A. (1975). *Hoppe-Seyler's Z. Physiol. Chem.* **356**, 1265–1275.

Hösel, W. and Nahrstedt, A. (1980). *Arch. Biochem. Biophys.* **203**, 753–757.

Hösel, W. and Schiel, O. (1984). *Arch. Biochem. Biophys.* **229**, 177–186.

Hösel, W., Berlin, J., Hanzlik, T. N. and Conn, E. E. (1985). *Planta* (*Berl.*) **166**, 176–181.

Hösel, W., Tober, I., Eklund, S. H. and Conn, E. E. (1987). *Arch. Biochem. Biophys.* **252**, 152–162.

Hruska, A. J. (1988). *J. Chem. Ecol.* **14**, 2213–2217.

Hughes, M. A. and Dunn, M. A. (1982). *Plant Mol. Biol.* **1**, 169–181.

Hughes, M. A., Sharif, A. L., Dunn, M. A. and Oxtoby, E. (1988). *In* "Cyanide Compounds in Biology" (D. Evered and S. Harnett, eds.), pp. 111–130. John Wiley, Chichester.

Itoh-Nashida, T., Hiraiwa, M. and Uda, Y. (1987). *J. Biochem.* (*Tokyo*) **101**, 847–854.

Jaroszewski, J., Szancer, J. and Ettlinger, M.G. (1981) *J. Lab. Compd. Radiopharm.* **18**, 703–706.

Jones, D. A. (1988). *In* "Cyanide Compounds in Biology" (D. Evered and S. Harnett, eds.), pp. 151–176. John Wiley and Sons, Chichester.

Jorns, M. S. (1979). *J. Biol. Chem.* **254**, 12145–12152.

Jorns, M. S. (1980). *Biochim. Biophys. Acta* **613**, 203–209.

Kakes, P. (1985). *Planta (Berl).* **166**, 156–160.

Karabatsos, G. J., Taller, R. A. and Vane, F. M. (1963). *J. Am. Chem. Soc.* **85**, 2326–2327.

Koch, B., Nielsen, V. S., Halkier, B. A. and Møller, B. L. (1991). *Arch. Biochem. Biophys.* submitted.

Kojima, M., Poulton, J. E., Thayer, S. S. and Conn, E. E. (1979). *Plant Physiol.* **63**, 1022–1028.

Kojima, M., Iwatsuki, N., Data, E. S., Villegas, C. D. V. and Uritani, I. (1983). *Plant Physiol.* **72**, 186–189.

Kuroki, G. W. and Conn. E. E. (1989). *Proc. Natl. Acad. Sci. USA* **86**, 6978–6981.

Kuroki, G. W. and Poulton, J. E. (1986). *Arch. Biochem. Biophys.* 247, 433–439.

Kuroki, G. W. and Poulton, J. E. (1987). *Arch. Biochem. Biophys.* **255**, 19–26.

Kuroki, G. W., Lizotte, P. A. and Poulton, J. E. (1984). *Z. Naturforsch.* **39**c, 232–239.

Lambert, J. L., Ramasamy, J. and Paukstelis, J. V. (1975). *Anal. Chem.* **47**, 916–918.

Ladenburg, K., Folkers, K. and Major, R. T. (1936). *J. Am. Chem. Soc.* **58**, 1292–1294.

Lalégerie, P. (1974). *Biochemie* **56**, 1297–1303.

Lizotte, P. A. and Poulton, J. E. (1988). *Plant Physiol.* **86**, 322–325.

Mao, C.-H., Blocher, J. P., Anderson, L. and Smith, D. C. (1965). *Phytochemistry* **4**, 297–303.

McFarlane, I. J., Lees, E. M., and Conn, E. E. (1975). *J. Biol. Chem.* **250**, 4708–4713.

Mkpong, O. E., Yan, H., Chism, G. and Sayre, R. T. (1990). *Plant Physiol.* **93**, 176–181.

Møller, B. L. (1977). *Anal. Biochem.* **81**, 292–304.

Møller, B. L. (1978). *J. Lab. Compd. Radiopharm.* **14**, 663–671.

Møller, B. L. (1981). *In* "Cyanide in Biology" (B. Vennesland, E. E. Conn, C. J. Knowles, J. Westley, and F. Wissing, eds.), pp. 197–215. Academic Press, London.

Møller, B. L., and Conn, E. E. (1979). *J. Biol. Chem.* **254**, 8575–8583.

Møller, B. L., and Conn, E. E. (1980). *J. Biol. Chem.* **255**, 3049–3056.

Møller, B. L., McFarlane, I. J., and Conn, E. E. (1977). *Acta Chem. Scand. B.* **31**, 343–344.

Nahrstedt, A. (1981). *In* "Cyanide in Biology" (B. Vennesland, E. E. Conn, C. J. Knowles, J. Westley and F. Wissing, eds.), pp. 145–181. Academic Press, London.

Nahrstedt, A. (1985). *Plant Syst. Evol.* **150**, 35–47.

Nahrstedt, A. (1987). *In* "Biologically Active Natural Products" (K. Hostettman and P. J. Lea, eds.), pp. 213–234. Clarendon Press, Oxford.

Nahrstedt, A. (1988). *In* "Cyanide Compounds in Biology" (D. Evered and S. Harnett, eds.) pp. 131–150. John Wiley and Sons, Chichester.

Nahrstedt, A., Hösel, W. and Walther, A. (1979). *Phytochemistry* **18**, 1137–1141.

Nielsen, A. T. (1969). *In*: "The Chemistry of the Nitro and Nitroso Groups" (H. Feuer, ed.), pp. 349–486. John Wiley and Sons, New York.

'Olafsdóttir, E. S. (1990). Ph.D. Thesis, Royal Danish School of Pharmacy, Copenhagen, 95 pp.

Pedersen, K. J. (1934). *J. Phys. Chem.* **38**, 559–571.

Pedersen, K. J. (1949). *Acta Chem. Scand.* **3**, 676–696.

Pócsi, I., Kiss, L., Hughes, M. A. and Nánási, P. (1989). *Arch. Biochem. Biophys.* **272**, 496–506.

Poulton, J. E. (1983). *In* "Handbook of Natural Toxins", Vol. I (R. F. Keeler and A. T. Tu, eds.), pp. 117–157. Marcel Dekker, New York.

Poulton, J. E. (1988). *In* "Cyanide Compounds in Biology" (D. Evered and S. Harnett, eds.), pp. 67–91. John Wiley and Sons, Chichester.

Poulton, J. E. (1989). *In* "Food Proteins" (J. E Kinsella and W. G. Soucie, eds.), pp. 381–401. American Oil Chemists' Society, Champaign, IL.

Poulton, J. E. (1990). *Plant Physiol.* **94**, 401–405.

Poulton, J. E. and Shin, S-I. (1983). *Z. Naturforsch.* **38**c, 369–374.

Pritzkow, W. and Rösler, W. (1967). *Liebigs Ann. Chem.* **703**, 66–76.

Reay, P. F. and Conn, E. E. (1974). *J. Biol. Chem.* **249**, 5826–5830.

Rissler, J. F. and Millar, R. L. (1977). *Protoplasma* **92**, 57–70.

Robbins, J. H. (1966). *Arch. Biochem. Biophys.* **114**, 576–584.

Saupe, S. G. (1981). *In* "Phytochemistry and Angiosperm Phylogeny" (D. A. Young and D. S. Seigler, eds.), pp. 80–116. Praeger, New York.

Schiefer, S. and Kindl, H. (1971). *J. Lab. Compd. Radiopharm.* **7**, 291–297.

Seely, M. K., Criddle, R. S. and Conn, E. E. (1966). *J. Biol. Chem.* **241**, 4457–4462.

Seigler, D. S. (1991). *In* "Herbivores: Their Interactions with Secondary Plant Metabolites" (G. A. Rosenthal and M. R. Berenbaum, eds.), Vol. I, pp. 35–77. Academic Press, San Diego.

Seigler, D. S., Maslin, B. R. and Conn, E. E. (1989). *In* "Advances in Legume Biology", *Monogr. Syst. Bot. Missouri Bot. Gard.* **29**, 645–672.

Selmar, D. (1986). Ph. D. Dissertation. University of Braunschweig, Germany.

Selmar, D., Carvalho, F. J. P. and Conn, E. E. (1987a). *Anal. Biochem.* **166**, 208–211.

Selmar, D., Lieberei, R., Biehl, B. and Voigt, J. (1987b). *Plant Physiol.* **83**, 557–563.

Selmar, D., Lieberei, R. and Biehl, B. (1988). *Plant Physiol.* **86**, 711–716.

Selmar, D., Frehner, M. and Conn, E. E. (1989a). *J. Plant Physiol.* **135**, 105–109.

Selmar, D., Lieberei, R., Biehl, B. and Conn, E. E. (1989b). *Physiologia Pl.* **75**, 97–101.

Selmar, D., Grocholewski, S. and Seigler, D. S. (1990). *Plant Physiol.* **93**, 631–636.

Shimada, M. and Conn, E. E. (1977). *Arch. Biochem. Biophys.* **180**, 199–207.

Swain, E., Li, C. P. and Poulton, J. E. (1992). *Plant Physiol.* **100**, in press.

Thayer, S. S. and Conn, E. E. (1981). *Plant Physiol.* **67**, 617–622.

Wahba, N., Hanna, S. and El-Sadr, M. M. (1956). *Analyst* **81**, 430–432.

Wu, H.-C. (1990). Ph.D. Dissertation. University of Iowa, Iowa City, IA.

Wu, H.-C. and Poulton, J. E. (1991). *Plant Physiol.*, in press.

Wurtele, E. S., Thayer, S. S. and Conn, E. E. (1982). *Plant Physiol.* **70**, 1732–1737.

Xu, L. L., Singh, B. K. and Conn, E. E. (1986). *Arch. Biochem. Biophys.* **250**, 322–328.

Xu, L. L., Singh, B. K. and Conn, E. E. (1988). *Arch. Biochem. Biophys.* **263**, 256–264.

Yemm, R. S. and Poulton, J. E. (1986). *Arch. Biochem. Biophys.* **247**, 440–445.

8 Glucosinolates

JONATHAN E. POULTON[1] and BIRGER LINDBERG MØLLER[2]

[1]*Department of Biological Sciences, University of Iowa, Iowa City, IA 52242, USA*

[2]*Plant Biochemistry Laboratory, Royal Veterinary and Agricultural University, DK-1871 Frederiksberg C, Copenhagen, Denmark*

I. INTRODUCTION

Originally known as mustard oil glucosides, the glucosinolates are hydrophilic, non-volatile thioglucosides found within several orders of dicotyledonous angiosperms (Cronquist, 1988). Of greatest economic significance is their presence in all members of the Brassicaceae (order Capparales), whose many cultivars have for centuries provided mankind with a source of condiments, relishes, salad crops and vegetables as well as fodders and forage crops. More recently, rapeseed (principally *Brassica napus* and *campestris*) has emerged as a major oilseed of commerce.

Almost 100 glucosinolates are known possessing the same general structure (Fig. 8.1) but differing mainly in the nature of their side-chain (R). The nomenclature system for glucosinolates has been designed so that the prefix used for each glucosinolate corresponds to the side-chain of the parent amino acid (Ettlinger and Dateo, 1961).

METHODS IN PLANT BIOCHEMISTRY Vol. 9
ISBN 0-12-461019-6

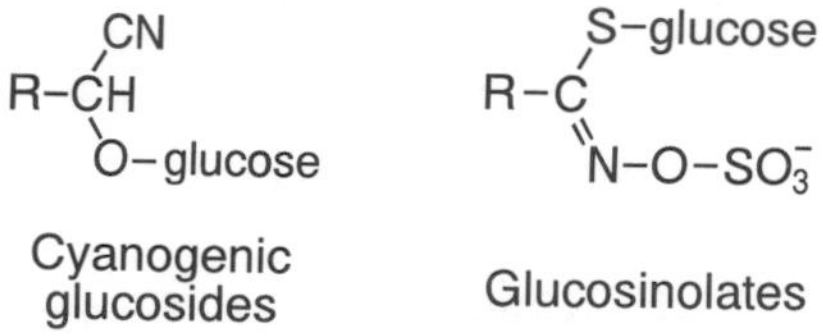

$$
\begin{array}{cc}
\overset{\displaystyle CN}{\underset{\displaystyle O-glucose}{R-CH}} & \overset{\displaystyle S-glucose}{\underset{\displaystyle N-O-SO_3^-}{R-C}} \\[2em]
\text{Cyanogenic} & \\
\text{glucosides} & \text{Glucosinolates}
\end{array}
$$

FIG. 8.1.   General structural formulae for cyanogenic glucosides and glucosinolates.

The generally accepted structure for glucosinolates, including the Z configuration around the $C=N$ bond (Ettlinger and Lundeen 1956; Ettlinger *et al.*, 1961), was proven for allylglucosinolate by X-ray analysis (Marsh and Waser, 1970). Isotopic tracer studies indicate that glucosinolates are formed from protein amino acids either directly or after single or multiple chain extension (Underhill *et al.*, 1973). Our knowledge of the first part of the biosynthetic pathway is still far from complete. In contrast, the last two biosynthetic steps involving the conversion of thiohydroximates to glucosinolates are well understood and are catalysed by two soluble enzymes, a glucosyltransferase (EC 2.4.1.-) and a sulphotransferase (EC 2.8.2.-).

Glucosinolate-containing plants generally possess a thioglucoside glucohydrolase (EC 3.2.3.1) commonly referred to as myrosinase. Upon plant injury or during food processing, this enzyme catalyses the rapid hydrolysis of glucosinolates to D-glucose, HSO_4^- and a multitude of physiologically active products (Fig. 8.2). The latter, which include isothiocyanates, thiocyanates, organic nitriles and oxazolidine-2-thiones, not only contribute to the distinctive flavour and aroma characteristic of crucifers, but also may have undesirable effects in animal feedstuffs due to their pungency and goitrogenic activity (Van Etten and Tookey, 1983; Langer, 1983). Additionally, they exhibit a wide range of physiological effects in plants, affecting such interactions as insect choice of host plants and attack by microbial pathogens (Chew, 1988a, b).

This chapter reviews our present knowledge about the pathways of glucosinolate metabolism and provides methods for the extraction, purification and assay of thiohydroximate glucosyltransferase, desulphoglucosinolate sulphotransferase and myrosinase. The procedures described should be considered as starting points for future research and will most likely require optimisation when applied to other experimental systems. Excellent reviews are available which focus upon glucosinolate chemistry and biochemistry (Underhill, 1980; Larsen, 1981; Fenwick *et al.*, 1983), taxonomic distribution (Rodman, 1981), and biology (Van Etten and Tookey, 1979; Chew, 1988a, b).

II.   GLUCOSINOLATE BIOSYNTHESIS

A.   The Conversion of Amino Acids to Thiohydroximates

1.   Introduction

The conversion of amino acids to thiohydroximic acids constitutes the first part of the biosynthesis of glucosinolates. The individual processes responsible for this conversion

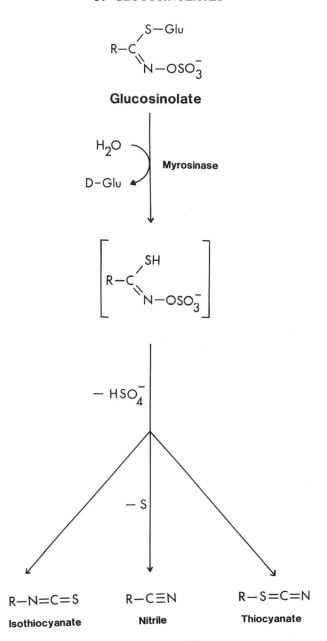

FIG. 8.2. Degradation of glucosinolates initiated by myrosinase.

are poorly understood. Therefore, standard experimental protocols for analysis of the intermediates cannot be outlined. Instead, present knowledge about thiohydroximate biosynthesis is discussed in the context of future approaches which might be undertaken to elucidate the pathway.

2. Biosynthetic experiments

Based on the structural similarity between naturally occurring amino acids and glucosinolates, Kjaer (1954) originally suggested that amino acids are direct precursors for
the aglycone moiety of glucosinolates. This relationship was subsequently confirmed
by numerous biosynthetic experiments where isotopically labelled amino acids were
administered to excised plant parts (Benn 1962; Kutacek et al., 1962; Underhill et al.,
1962). Conversion percentages as high as 38% were reported (Underhill, 1967). The
use of selectively labelled amino acids demonstrated that the α-carbon atom of the
amino acid is recovered as the thiohydroximate carbon and that the complete carbon
skeleton of the parent amino acid, with the exception of the carboxyl group, is
incorporated as a unit into the equivalent positions of the glucosinolate (Underhill
et al., 1962; Underhill and Chisholm, 1964). Feeding double-labelled $^{14}C/^{15}N$ amino
acids resulted in the production of glucosinolates with essentially unaltered $^{14}C/^{15}N$
ratios (Underhill and Chisholm, 1964; Matsuo and Yamazaki, 1966). The biosynthetic
pathway thus proceeds without cleavage of the bond between the α-carbon atom and
the amino nitrogen atom of the parent amino acid. Consequently, all pathway
intermediates must be nitrogenous.

As stated earlier, our knowledge of the intermediates involved in the conversion of
amino acids to thiohydroximates is regrettably limited. None of the intermediates has
been properly established, and participating enzymes have neither been isolated nor
characterised as components of crude preparations. It is generally assumed, but has
never been proven, that these biosynthetic steps are catalyzed by microsomal systems
analogous to those shown to catalyse the synthesis of aglycones of cyanogenic glycosides (Fig. 8.3). This assumption is partly based on the structural similarity between
glucosinolates and cyanogenic glycosides which suggests that the conversion of an
amino acid to these two groups of natural products could proceed via common
intermediates (Ettlinger and Kjaer, 1968; Dewick, 1984). Due to the lack of direct
information about the early intermediates in glucosinolate biosynthesis, the pathway
outlined for cyanogenic glycoside biosynthesis is briefly summarised here. The first
step involves N-hydroxylation of the parent amino acid (Møller and Conn, 1979). This
reaction is dependent on the presence of molecular oxygen and is most likely catalyzed
by a cytochrome P450 dependent monooxygenase (Halkier and Møller, 1991). The N-
hydroxyamino acid is subsequently converted to the α-nitrocarboxylic acid by an N-
oxidation reaction and an additional N-hydroxylation reaction (Halkier et al., 1991).
Again, the latter reaction appears dependent on cytochrome P450 (Halkier and Møller,
1991). An α-nitrosocarboxylic acid may function as an intermediate (Halkier et al.,
1991). The α-nitrocarboxylic acid is then converted to the aci-nitro compound (Halkier
et al., 1991). In a subsequent reduction step, the aci-nitro compound yields the E-oxime
(Halkier et al., 1989). The biosynthetic incorporation studies on which this pathway
is based are strongly supported by the experimental demonstration of two obligatory
hydroxylation steps in the conversion of the parent amino acid to the oxime as
determined by simultaneous measurements of oxygen consumption and biosynthetic
activity (Halkier and Møller, 1990). Finally, mass spectrometric analyses combined
with the use of ^{18}O-labelled molecular oxygen have shown that the enzyme system is
able to discriminate between the oxygen atoms introduced by the two hydroxylation
reactions such that the oxygen atom introduced by the first hydroxylation reaction is

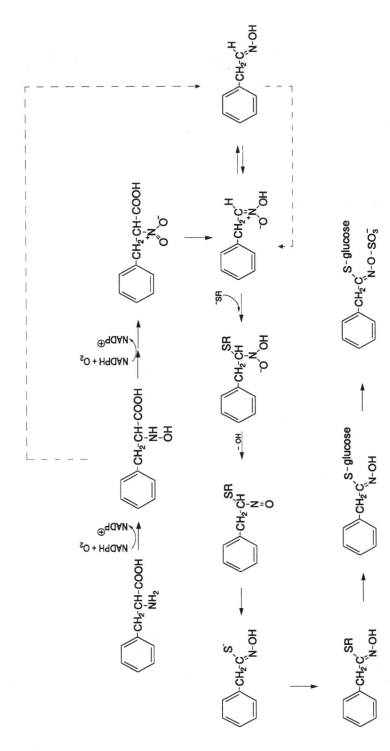

FIG. 8.3. Putative biosynthetic pathway for benzylglucosinolate derived from L-phenylalanine.

preferentially lost in the subsequent conversion of the *aci*-nitro compound to the *E*-oxime (Halkier *et al.*, 1991). The occurrence of cyanogenic glycosides and gluco- sinolates within the same plant species is apparently mutually exclusive (Kjaer, 1980–1981; Larsen, 1981). While the putative intermediate constituting the branch- point between the synthesis of cyanogenic glycosides and glucosinolates has not been unequivocally identified, it was previously thought to be the oxime (Conn, 1988). However, after the recent demonstration of the involvement of an *aci*-nitro compound as an intermediate in cyanogenic glycoside biosynthesis, the *aci*-nitro compound has now been postulated to serve as the branch-point (Halkier *et al.*, 1991) (Fig. 8.4). Nucleophilic attack on the α-carbon atom of the *aci*-nitro compound, most likely by the mercapto group of cysteine, provides a plausible mechanism for the formation of an *S*-alkylthiohydroximic acid. This may then be converted to the thiohydroximic acid by the action of a C–S lyase.

Some of the intermediates known from the biosynthesis of cyanogenic glycosides have in fact been isolated from glucosinolate-producing plants. These include the *aci*- nitro compound (isolated as the tautomeric nitro compound) (Matsuo *et al.*, 1972) and aldoximes (Underhill, 1967). In a biosynthetic experiment, the nitro compound was shown to be formed from the oxime (Matsuo *et al.*, 1972). This observation and the numerous occasions (Tapper and Butler, 1967; Underhill, 1967; Matsuo, 1968b; Kindl and Schiefer, 1969) on which oximes have been demonstrated to serve as good precursors for glucosinolates point to the oxime as an intermediate in glucosinolate biosynthesis (Dewick, 1984). *N*-Hydroxyamino acids have also been reported as precursors of glucosinolates (Kindl and Underhill, 1968). *N*-Hydroxyamino acids are easily converted to the corresponding oximes by a non-enzymatic oxidative decar- boxylation reaction (Møller, 1978; Halkier *et al.*, 1988). Conclusions made from biosynthetic experiments become ambiguous if the compounds tested or isolated are in an enzyme-catalysed or chemical equilibrium with a true intermediate (Tapper and Butler, 1972; Underhill *et al.*, 1973). According to the suggested pathway for glucosinolate biosynthesis (Fig. 8.4), the oxime is not a true intermediate (Halkier *et al.*, 1991). It is interesting to note that enzymatic conversion of the *N*-hydroxyamino acid to the oxime has been reported to progress with a stoichiometric consumption of molecular oxygen in cell-free extracts (Kindl and Underhill, 1968). The reaction was stimulated by flavin mononucleotide (FMN). The possibility of an additional requirement for NADPH, which would have made the reaction analogous to the *N*-oxidation and *N*-hydroxylation reactions demonstrated in cyanogenic glycoside biosynthesis (Halkier *et al.*, 1991), was not tested. Unfortunately, the proper control experiments to support these pioneering studies are missing and the interpretation of the results obtained are therefore ambiguous (Halkier *et al.*, 1988).

The introduction of the thioglucosidic sulphur atom of glucosinolates may take place by addition of a thiol to the *aci*-nitro compound. A subsequent step in the pathway would be the C–S lyase-induced severance of the C–S linkage leading to a thiohydro- ximic acid. Cysteine and methionine serve as good donors of the thioglucosidic sulphur atom in biosynthetic experiments (Wetter, 1964; Kindl, 1965; Matsuo, 1968a; Wetter and Chisholm, 1968). Direct incorporation of a thioglucose unit as a means of introducing the thioglucosidic sulphur atom into the glucosinolate has not gained experimental support (Matsuo, 1968a; Wetter and Chisholm, 1968; Underhill and Wetter, 1969). Several of the putative intermediates contain a C=N double bond,

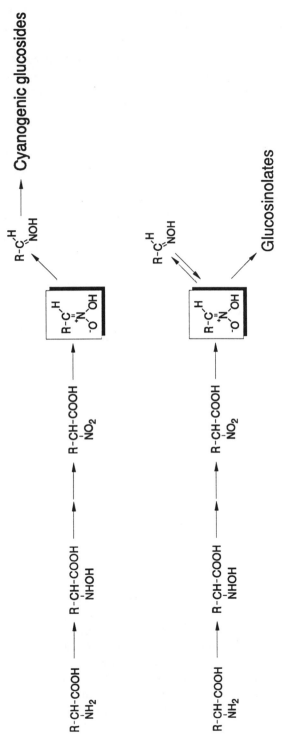

FIG. 8.4. The putative involvement of an *aci*-nitro compound as the branch point between the biosynthetic pathways for cyanogenic glucosides and glucosinolates.

giving the possibility of geometrical isomers. The actual configuration of these inter-mediates has not been examined but is assumed to be the same as observed in the final glucosinolate.

Higher plants contain several selenium isologues of sulphur amino acids. The isolation of a selenium-containing glucosinolate (selenosinigrin) from horseradish (*Armoracia lapathifolia*) plants after administration of $^{75}SeO_4$ (Stewart *et al.*, 1974) might indicate the possibility of achieving a more specific labelling of glucosinolates by feeding selenium rather than sulphur isotopes. However, a careful reinvestigation did not substantiate this report, and additional attempts to establish selenogluco-sinolates as natural constituents of other glucosinolate-producing plants, including the selenium tolerant species *Stanleya pinnata*, were negative (Bertelsen *et al.*, 1988). Use of selenium isotopes for the labelling of glucosinolates is therefore not recommendable.

Only a few amino acids serve as precursors for cyanogenic glycosides (Conn, 1973). This greatly limits the possibility for structural diversity among this group of natural products. During their biosynthesis, chirality of the cyanohydrin carbon atom may result in the formation of epimers (Conn, 1973). Secondary modification of the aglycone moiety, as observed in the formation of triglochinin (Jaroszewski and Ettlinger, 1981), seldom occurs. Variation in the sugar moiety thus constitutes the major possibility for structural diversity among the cyanogenic glycosides (Seigler, 1975). In contrast, all glucosinolates hitherto isolated contain a β-D-1-glucopyranosyl residue. A large number of protein amino acids as well as non-protein amino acids serve as precursors for glucosinolates (Underhill *et al.*, 1973; Kjaer, 1980–1981; Larsen, 1981), and the diversity is significantly extended by the presence of series of chain-elongated parent amino acids (Chisholm and Wetter, 1964; Matsuo and Yamazaki, 1966; Dörnemann *et al.*, 1974; Löffelhardt and Kindl, 1975; Glover *et al.*, 1988). Under normal 'flow conditions', these homologues do not tend to accumulate. Secondary modification processes (e.g. hydroxylation (Kjaer and Christensen, 1959; Kindl, 1965; Kindl and Schiefer, 1969; Josefsson, 1971b; Lee and Serif, 1971; Underhill and Kirkland, 1972), oxidation (Chisholm and Wetter, 1966; Ettlinger and Kjaer, 1968; Chisholm and Matsuo, 1972), methylation (Elliott and Stowe, 1971; Chisholm, 1973), and elimination (double bond formation) (Dewick, 1984)), which may occur at different levels during biosynthesis (Underhill *et al.*, 1973; Larsen, 1981), further serve to increase the number of naturally occurring glucosinolates.

The major breakthrough in elucidating the biosynthetic pathway for cyanogenic glycosides has been the availability of a microsomal system which catalyses the *in vitro* synthesis of the aglycone. It has not yet been possible to obtain such an *in vitro* system from a glucosinolate-producing plant. The plant species hitherto investigated are mustard (*Sinapis alba* L.), cress (*Lepidium sativum* L.), and garden nasturtium (*Tropaeolum majus* L.) (J. Lykkesfeldt and B. L. Møller, unpubl. data; J. E. Poulton, unpubl. data; E. W. Underhill, unpubl. data). While such lack of success may indicate that the enzyme system has different properties with respect to stability and cofactor requirements or that it more easily loses essential components, it more likely may reflect the choice of plant material. The *in vitro* systems obtained from cyanogenic plants were isolated from etiolated seedlings of plant species (e.g. *Sorghum bicolor*: McFarlane *et al.*, 1975) where the seed itself is very low in or devoid of cyanogenic glycosides (Erb *et al.*, 1981) but where their rapid synthesis is induced upon germination (Halkier and Møller, 1989). The use of etiolated seedlings provides a good opportunity to obtain the enzyme system from young actively growing plant material devoid of high concentrations of plant phenolics. In contrast, glucosinolates are often produced at a late stage in the life-cycle of the plant. Seeds of glucosinolate-producing plants typically have a glucosinolate content similar to that of young seedlings, and the seedling content decreases with germination time (Larsen, 1981; Palmer *et al.*, 1987; Macfarlane Smith and Griffiths, 1988). Although biosynthetic experiments with excised seedlings from such plants reveal *de novo* synthesis (Benn, 1962; Tapper and Butler, 1967), it may be quantitatively insignificant compared to the mobilisation of glucosinolates from the seed. Some plants contain one major glucosinolate (e.g. benzylglucosinolate in *Tropaeolum majus* (Benn, 1962)) whereas others contain numerous glucosinolates (e.g. *Arabidopsis thaliana*: Hogge *et al.*, 1988). In the latter case, the content of an individual glucosinolate may actually increase during seedling development, although the total glucosinolate content remains constant or decreases. This applies to the indolylglucosinolates in seedlings of *Brassica* spp. (Uppstrom, 1983; Sang *et al.*, 1984) and *Arabidopsis thaliana* L. (Hogge *et al.*, 1988) as well as to the ω-methylsulphinylalkyl glucosinolates in *Arabidopsis thaliana* L. (Hogge *et al.*, 1988). Based on these observations, an attempt was made to obtain an active microsomal system from *Brassica napus* using tryptophan as the parent amino acid but it too was unsuccessful (B. L. Møller, unpubl. res.). In studying the biosynthesis of individual glucosinolates, the operation of sequential secondary modification processes needs to be considered when precursors are selected. Thus, hydroxyl groups positioned in an aliphatic side-chain of a glucosinolate are apparently introduced as the very last biosynthetic step (Underhill and Kirkland, 1972; Rossiter *et al.*, 1990). In contrast, phenolic hydroxyl groups appear to be introduced already at the amino acid stage of biosynthesis (Kindl, 1965; Kindl and Schiefer, 1969).

Infection of crucifers by parasitic fungi may induce increased synthesis of gluco-sinolates. This phenomenon has been extensively studied in *Brassica napus* infected with *Plasmodiophora brassicae* (Rausch *et al.*, 1983). The synthesis of indol-3-ylmethylglucosinolate from tryptophan is significantly increased in clubroot diseased plants as compared to control plants. Such experimental systems should be exploited in future attempts to obtain active *in vitro* systems for glucosinolate synthesis.

Traditional plant breeding has resulted in varieties having low glucosinolate levels. In *B. napus*, the cultivar Bronowski has a low glucosinolate content in the seeds. The vegetative parts of this cultivar contain normal levels of indolylglucosinolates compared to other cultivars but the levels of 3-butenylglucosinolate and 2-hydroxy-3-butenylglucosinolate are low (Josefsson, 1971a). Biosynthetic experiments showed that the amino acids methionine and 2-amino-6-(methylthio)-caproic acid are poor precursors for the latter two glucosinolates in Bronowski (Josefsson, 1971b). 5-Methyl-thiopentenal oxime serves as an efficient precursor for 3-butenylglucosinolate but poorly as a precursor of 2-hydroxy-3-butenylglucosinolate (Josefsson, 1973). The cultivar Bronowski is therefore considered to have at least two metabolic blocks. One block is situated before the site where the oxime (or an intermediate derived from it) enters the pathway (Josefsson, 1973). The hydroxylation process converting

3-butenylglucosinolate to 2-hydroxy-3-butenylglucosinolate constitutes the second block (Josefsson, 1973; Rossiter et al., 1990). Accumulation of intermediates as a result of the metabolic blocks in Bronowski has not been reported.

In the absence of in vitro systems, the possibility to accumulate intermediates in mutants generated by treatment with mutagens or by transformation may constitute a valuable alternative experimental approach for elucidating the glucosinolate biosynthetic pathway. In vitro cultures of Brassica spp. generally exhibit very low overall levels of glucosinolates. Consequently, determination of the endogenous glucosinolate content cannot be used as a screening system in the selection of glucosinolate-free cultures (GrootWassink et al., 1988). Biosynthetic experiments indicate that, except for the enzymes which are thought to catalyse the initial hydroxylation reactions of the parent amino acid, the subsequent enzymes appear to be fully expressed (GrootWassink et al., 1988). As in cyanogenic glycoside biosynthesis (Cutler et al., 1985; Koch et al., 1991), the enzymes catalysing the later steps of glucosinolate biosynthesis appear to exhibit a broader substrate specificity (Groot-Wassink et al., 1988). Administration of xenobiotic oximes to a number of gluco-sinolate-producing plants resulted in the production of artificial glucosinolates as well as different sulphatoglucopyranosides (GrootWassink et al., 1990). The broad substrate specificity of the latter enzymes was also exploited in the use of the artificial precursor 2-nitrobenzaldoxime for the production of 2-nitrophenylglucosinolate. Myrosinase treatment of this artificial glucosinolate yields 2-nitroaniline, a yellow compound, which may be converted to a red azo dye with a low detection limit and thereby used in visual screening for glucosinolate-free mutants (GrootWassink et al., 1988).

Mutants which over-produce certain amino acids are known. Since cysteine and methionine serve as direct donors of the thioglucosidic sulphur atom of glucosinolates, it might be interesting to select among glucosinolate-producing plants for mutants which over-produce these amino acids. The subsequent sulphation of the desulpho-glucosinolate requires an additional sulphur donor which has been identified as PAPS (Glendening and Poulton, 1988). Because APS is required in cysteine synthesis, the selected mutants may also have increased ability to synthesise activated sulphate. In wild-type plants, competition between the use of sulphur amino acids for glucosinolate and protein synthesis appears to be a limiting factor for glucosinolate synthesis. This has been elegantly demonstrated using transfected plants of Brassica campestris harbouring a mammalian metallothionein cDNA (Lefebvre, 1990). The cysteine content of the metallothionein is 30 mol %, and the protein is efficiently expressed constituting about 5% of the soluble protein. The production of the metallothionein severely limits the availability of cysteine for glucosinolate synthesis, thereby reducing the glucosinolate content of the transfected plants to approximately 50% of wild-type levels. According to these results, mutants which over-produce cysteine or methionine may be expected to contain higher glucosinolate levels and possibly also to have induced higher levels of the enzymes responsible for glucosinolate synthesis. This would prevent the excessive release of toxic hydrogen sulphide from an increased cysteine or methionine pool (Harrington and Smith, 1980). Higher levels of the enzyme system catalysing the initial steps in glucosinolate biosynthesis would make it easier to obtain in vitro systems from such mutants compared to wild-type.

3. Chemical synthesis of intermediates

General methods for the chemical synthesis of *N*-hydroxyamino acids, α-nitro-carboxylic acids, nitro compounds and oximes are compiled in Chapter 7 of this volume. Several methods are available for chemical synthesis of thiohydroxamic acids and thiohydroximates (Matsuo and Underhill, 1971; Walter and Schaumann, 1971). A recent 'one-flask' procedure describes their chemical synthesis from nitro compounds using thiosilanes as counter-attack reagents (Hwu and Tsay, 1990). Thiohydroxamic acids are labile compounds which in the presence of light decompose to produce nitriles, sulphur and water (Ettlinger and Lundeen, 1956; Holm *et al.*, 1975). Thiohydroximates are stable in the light (Hwu and Tsay, 1990) and can exist in either the *E*- or *Z*-form. These isomers may be separated on the basis of different solubility (Davies *et al.*, 1968). When chemically synthesized from nitro compounds, thiohydroximates which have a small substituent at the C=N carbon atom are formed as an *E/Z* mixture. Thiohydroximates with large substituents are predominantly produced as the *Z*-isomer, the isomer most likely involved in the biosynthetic pathway (Hwu and Tsay, 1990). Procedures for the chemical synthesis of intermediates related to the chain elongation pathway are also available (Löffelhardt and Kindl, 1975; Chapple *et al.*, 1988).

For procedures regarding the chemical synthesis of radioisotopically labelled intermediates, the following references should be consulted: chain-elongated amino acids and their corresponding hydroxy- or ketoacids (Kjaer and Wagner, 1955; Chisholm and Wetter, 1966; Dörnemann *et al.*, 1974); nitro compounds (Matsuo *et al.*, 1972); thiohydroxamic acids (Underhill and Wetter, 1969; Rossiter *et al.*, 1990).

B.   UDPG:Thiohydroximate Glucosyltransferase

S-Glucosylation of thiohydroximates catalyzed by UDPG:thiohydroximate glucosyl-transferases (EC 2.4.1.-) is thought to be the penultimate stage in glucosinolate biosynthesis.

1. Extraction and purification

Thiohydroximate glucosyltransferase (GT) was initially extracted from *Tropaeolum majus* by homogenising 50 g leaves in a chilled mortar with washed sand, 1 g Polyclar AT, and 50 ml Tris-HCl buffer (pH 7.4) containing 50 mM β-mercaptoethanol (Matsuo and Underhill, 1969, 1971). After filtration and centrifugation, the enzyme was purified 20-fold in 30% yield by ammonium sulphate fractionation and gel filtration on Sephadex G-50 and G-100 columns. The final preparation was unstable, losing 68% of its activity when frozen for 24 h and becoming totally inactive upon freeze-drying. Upon removal of β-mercaptoethanol, it underwent irreversible inactivation. Cell-free extracts exhibiting thiohydroximate GT activity were obtained by similar means from leaves of *Sinapis alba*, *Nasturtium officinale* and *Armoracia lapathifolia* but were not further purified.

2. Assay

Thiohydroximate GT assays follow the production of radiolabelled desulphoglucosinolates from thiohydroximates and UDP-[^{14}C]glucose. The reaction mixture contains 25 μl enzyme, 0.5 μmol thiohydroximate, 0.5 μmol UDP-[^{14}C]glucose (0.2 μCi)
and buffer (0.1 M Tris-HCl (pH 7.4) containing 50 mM β-mercaptoethanol) in a total
volume of 0.1 ml (Matsuo and Underhill, 1971). After incubation at 30 °C for 10 min,
the reaction is terminated by adding 2 vol methanol. The supernatant is passed through
a small column containing 300 mg anionotropic alumina (grade 1, Woelm, Eschwege,
Germany) which absorbs reaction substrates but allows passage of desulphoglucosinolates. The column is washed with 1.5 ml methanol, and the radioactivity of the
combined eluate and washings is determined by scintillation spectrometry. The identity
of desulphoglucosinolate products may be confirmed by PC (Whatman No. 1 paper,
n-butanol–acetic acid–water, 4:1:1.8, v/v), TLC (silica gel, n-butanol–n-propanol–
acetic acid–water, 3:1:1:1;v/v), or GLC (Matsuo and Underhill, 1969) using authentic
standards. Sodium phenylacetothiohydroximate, synthesised according to Underhill
and Wetter (1969), is commonly employed as acceptor substrate, but activity towards
several other thiohydroximates was successfully demonstrated by this assay.

An alternative procedure utilises ion-exchange chromatography on DEAE-cellulose
discs for rapid analysis of reaction rates (Jain et al., 1989b). The assay mixture contains
1 mM sodium phenylacetothiohydroximate, 1 mM UDP-[^{14}C]glucose (0.05 μCi) and
enzyme (0.4 mg protein) in a total volume of 0.1 ml buffer (0.2 M HEPES, 14 mM
β-mercaptoethanol, 5 mM EDTA, pH 7.5). In control reactions, active enzyme is
replaced by heat-inactivated (10 min at 100 °C) preparations. Incubations are undertaken at 30 °C for 30 min. A pad of two DE-81 filter discs (Whatman) is stacked within
a filtration manifold and prewashed with 15 ml water. After incubation, each assay
mixture is diluted two-fold with water to lower its ionic strength and then transferred,
either directly or after heat inactivation (2 min at 100°C), to a pad of discs which is
then washed with approx. 15 ml water. The radioactivity of the discs, due to residual
UDP-[^{14}C]glucose, is determined by scintillation spectrometry using 5 ml Aquasol-2
cocktail (NEN). Assay linearity is observed with incubation times up to 50 min and
with protein levels up to 0.5 mg. Data obtained with this procedure are in good
agreement with those derived using TLC methods (Matsuo and Underhill, 1971).
However, since this assay measures disappearance of radiolabelled substrate rather
than product formation, control reactions lacking thiohydroximate must be included
to correct for UDPG catabolism by endogenous UDPG hydrolases. With Brassica
extracts, the latter activity accounted for approx. 25% of observed reaction rates but
could be eliminated by Mono Q ion-exchange chromatography.

3. Properties

The Tropaeolum majus GT has a native molecular mass of approx. 50 kDa, as estimated by Sephadex G-100 chromatography, and a pH optimum of 6.5–7.5 in phosphate or Tris-HCl buffers. It shows high specificity for thiohydroximates as glucose
acceptors. The endogenous compound phenylacetothiohydroximate (K_m, 0.65 mM) is
most rapidly glucosylated, but the enzyme is also active toward propiothiohydro-
ximate, butyrothiohydroximate, isobutyrothiohydroximate, 4-methylthiobutyrothio-

hydroximate (K_m, 1.05 mM) and benzothiohydroximate. The oxygen analogue phenylacetohydroximic acid and several aliphatic alcohols and thiols are not substrates. Of several sugar nucleotides tested, only UDPG (K_m, 1.54 mM) and TDPG (one-tenth as active as UDPG) served as sugar donors. The enzyme is inhibited strongly by mercuric chloride and PCMB, moderately by NEM, but is insensitive to KCN and iodoacetic acid. Its activity is stimulated 39–69% by 10 mM sulphhydryl compounds (e.g. β-mercaptoethanol, DTT) and 21–36% by chelators. The *Brassica juncea* GT, which has a native molecular mass of 44 kDa, co-purifies with the PAPS:desulpho-glucosinolate sulphotransferase (see Section II.C) (Jain *et al.*, 1990b). It is most stable at acidic pHs and at temperatures below 30°C.

4. Localisation

The distribution of GT activity in 5-day-old, dark-grown *Brassica juncea* seedlings was investigated using partially purified preparations from cotyledons, hypocotyls and roots (Jain *et al.*, 1989a). Although cotyledons show greatest activity per g fresh weight, hypocotyls have approximately seven-fold higher specific activity than the other organs. Recovered mostly in the 12 000 × *g* supernatant fraction of protoplast lysates, GT appears cytosolic although an association with vacuoles and/or the endoplasmic reticulum (ER) was not unequivocally ruled out.

C. Desulphoglucosinolate Sulfphotransferase

The final step of glucosinolate biosynthesis, the sulphation of desulphoglucosinolates, is catalysed by 3'-phosphoadenosine 5'-phosphosulphate (PAPS):desulphogluco-sinolate sulphotransferases (EC 2.8.2.-).

1. Extraction and purification

Although desulphoglucosinolate sulphotransferase (ST) activity has been detected in cell-free extracts from several crucifers (Underhill *et al.*, 1973; Glendening and Poulton, 1990), this enzyme has been extensively purified from only two species. It was extracted from light-grown cress (*Lepidium sativum*) seedlings in 0.1 M Tris-HCl buffer (pH 8.5) containing 20 mM MgCl$_2$ and 10 mM β-mercaptoethanol (Glendening and Poulton, 1988). After removing low molecular weight compounds by Sephadex G-25 filtration, the enzyme was subsequently purified 40-fold with 62% recovery of activity by Con A–Sepharose 4B, Matrex Gel Green A, and Mono Q FPLC chromatography (Glendening and Poulton, 1990). At this stage, the cress ST was extremely unstable in dilute protein concentrations and could only be stabilised by addition of 2.5 mg ml^{-1} BSA, a procedure which precluded further purification. With 0.25 M sucrose, 10 mM β-mercaptoethanol and BSA, however, its activity remained constant during storage at −20 °C for 6 months. When BSA was omitted, enzyme activity was lost at the rate of 1% per min. Attempts to stabilise the enzyme by adding including additional 1–10 mM MgCl$_2$, β-mercaptoethanol, DTT or reduced gluta-thione proved unsuccessful. To exclude the possibility that activity losses observed in the absence of BSA resulted from proteolysis during enzyme isolation, all purification procedures were repeated with protease inhibitors present. Such action did not reduce these losses or alter elution profiles of the ST during purification.

Jain *et al.* (1990a) described the 230-fold purification of ST in 58% yield from *Brassica juncea* (cv Cutlass) cell cultures. Cells were disrupted in 0.2 M Hepes buffer (pH 7.5) containing 14 mM β-mercaptoethanol and 5 mM EDTA using a bead beater homogeniser (Biospec Products, Bartlesville, OK). After filtration and centrifugation, the enzyme was concentrated by ammonium sulphate precipitation (40–70% fraction) and further purified by Mono Q FPLC and Sephadex G-100 chromatography. Like the cress enzyme, the partially purified *Brassica* ST required SH-group protectors for stability. With 14 mM β-mercaptoethanol, the half-life was approx. 2 weeks, while, in its absence, more than 50% activity was lost within 48 h at 40 °C. When stored at −20 °C in 20 mM Tris-HCl buffer (pH 7.5) containing 14 mM β-mercaptoethanol and 10% glycerol, ST activity remained stable for at least 2 months.

Interestingly, the *Brassica juncea* STase co-purified with thiohydroximate GT upon ammonium sulphate fractionation, Mono Q ion-exchange chromatography, and gel filtration (Jain *et al.*, 1990b). Furthermore, these activities showed similar electro-phoretic mobility and isoelectric point and were not resolvable by hydrophobic interaction, metal chelation, dye ligand or affinity chromatography. This behaviour raises the following intriguing possibilities: (1) that the two proteins have very similar physicochemical properties leading to their persistent co-elution; (2) that these enzymes constitute a multienzyme complex; or (3) that both active sites reside on a single polypeptide. SDS-PAGE data, which might help distinguish between these possibili-ties, were not presented. However, ST and GT activities displayed different pH and temperature stabilities, a finding which lends support to the existence of a multienzyme complex. It remains to be seen whether similar linkage exists between ST and GT activities in other crucifers including cress.

2. Assay

Desulphoglucosinolate ST activity may be assayed by measuring the rate of formation of radiolabelled benzylglucosinolate from [^{35}S]PAPS and desulphobenzylgluco-sinolate (Glendening and Poulton, 1990). The standard assay mixture contains 10 μmol Tris-HCl buffer (pH 9.0), 8 nmol desulphobenzylglucosinolate, 7 nmol [^{35}S]PAPS (containing 100 000 dpm), 1.1 μmol MgCl$_2$, and 50 μl of enzyme preparation in a total volume of 120 μl. After incubation at 30 °C for 60 min, the reaction is terminated by adding 5 μl 50% (v/v) acetic acid. Precipitated protein is removed by centrifugation, and the reaction mixture is kept at 4 °C until product and substrates are resolved by PC or extraction into *n*-butanol as described below. Control incubations, in which active enzyme is replaced by boiled enzyme or the substrate desulphobenzylgluco-sinolate by buffer, are routinely included.

During ST purification, when large numbers of fractions require rapid analysis due to the instability of the enzyme, benzylglucosinolate and PAPS may be resolved as follows. An aliquot (100 μl) of the centrifuged reaction mixture is placed in an Eppendorf tube sitting on ice. After adding 70 μl distilled water and 1 ml ice-cold *n*-butanol, the components are briefly vortexed before placing the tube on an Eppendorf shaker for 15 min at 4 °C. The phases are separated by centrifugation in a Beckman model B centrifuge for 2 min. An aliquot (0.8 ml) of the butanol supernatant, which contains benzylglucosinolate, is quantitated by scintillation spectrometry in 5 ml

Andersons cocktail (0.3% (w/v) PPO and 0.02% (w/v) POPOP in xylene–Triton X-114 (3:1; v/v). While this method allows rapid assay of column chromatography fractions, it suffers from high background counts (approx. 400 dpm). The following PC method is therefore recommended for enzyme characterisation studies. After centrifugation, aliquots (100 μl) of the terminated reaction mixture are applied to 4 cm wide strips of Whatman 3MM and chromatographed (descending) using n–butanol–pyridine–water (6:4:3; v/v) as solvent system. Product zones (R_f, 0.66) are located by chromatographing authentic benzylglucosinolate on parallel strips and spraying with alkaline AgNO$_3$ (Gmelin and Kjaer, 1970). Radioactive zones are cut out and quantitated by scintillation spectrometry using 10 ml Andersons cocktail. Reproducibility with this method is excellent, and background counts are approximately 50 dpm.

Jain *et al.* (1989b) have devised a rapid ion-exchange system for the separation of radiolabelled glucosinolate after the ST preparation is incubated with [^{14}C]desulpho-glucosinolate and unlabelled PAPS. The assay mixture contains 0.5 μM desulphobenzyl-glucosinolate (0.05 μCi), 0.5 mM PAPS, and 50 μl enzyme preparation in a total volume of 0.1 ml buffer (0.2 M Hepes (pH 7.5) containing 14 mM β-mercaptoethanol and 5 mM EDTA). In control reactions, active enzyme is replaced by boiled enzyme (10 min at 100 °C). Three DE-81 filter discs (Whatman) are stacked in a filtration manifold and pre-washed with 15 ml water. After 30 min incubation at 30 °C, the assay mixtures are diluted two-fold with water and transferred to the filter discs, either directly or after heat inactivation (2 min at 100 °C). The pad of discs is washed with approx. 15 ml water, and the residual radioactivity, due to [^{14}C]benzylglucosinolate, is determined by scintillation spectrometry using 5 ml Aquasol-2 fluid (NEN). Assay linearity is observed with incubation times up to at least 60 min and with protein amounts not exceeding 0.5 mg. The synthesis of [^{14}C]desulphobenzylglucosinolate is described by Jain *et al.* (1989b).

Myrosinase (β-thioglucoside glucohydrolase, EC 3.2.3.1), which initiates gluco-sinolate degradation, is commonly found in cell-free extracts of crucifers at far higher levels than those of desulphoglucosinolate STs. To prevent serious interference with ST assays, it is imperative that myrosinase be removed as early as possible during ST purification by Con A-Sepharose 4B chromatography or Mono Q FPLC (Glendening and Poulton, 1988; Jain *et al.*, 1989a, b).

3. Properties

The *Brassica* and *Lepidium* STs have many features in common (Table 8.1). Both enzymes are present in low abundance and exhibit similar pH optima, isoelectric points and molecular masses. Neither shows an absolute requirement for metal ions, although 10 mM MgCl$_2$ stimulates activity by 25–50%. Chelators have insignificant effect on activity. The sensitivity of the enzymes towards SH-group reagents and their protection from covalent modification by β-mercaptoethanol suggest the importance of one or more thiol groups for activity. Whether these groups are directly involved in catalysis or maintaining critical molecular conformation remains unknown. With the exception of the reaction product PAP, which is a potent competitive inhibitor with respect to PAPS, other nucleotides (e.g. 5′-AMP, 5′-ADP, 5′-ATP, and APS) do not significantly affect ST activity. Regarding substrate specificity, neither enzyme utilises APS as sulphate donor. The cress ST shows an absolute requirement for the

TABLE 8.1. Properties of purified desulphoglucosinolate sulphotransferases from *Lepidium sativum* and *Brassica juncea*.

Enzyme	Purification (-fold)	Purification Methods	Specific activity (nmol BGSL $h^{-1} mg^{-1}$)[c]	Native molecular mass (kDa)	pH optimum	pI	Substrates (K_m)	Inhibitors (K_i)	Activators
Lepidium sativum[a]	40	Con A-Sepharose Matrex Gel Green A Mono Q	935	31 ± 5	9.0	5.2	PAPS (60 μM) DSBG (82 μM) DSP (670 μM) DSA (6.5 mM)	PAP (36 μM) NEM DTNB PCMS	Mg^{2+} Thiol compounds
Brassica juncea[b]	120	Ammonium sulphate Mono Q	2	44	8.5–9.0	4.84	PAPS (0.78 μM) DSBG (2.3 μM)	PAP (0.4 μM) PCMB Iodoacetamide	Mg^{2+} DTT

Authors: [a] Glendening and Poulton (1990); [b] Jain *et al.* (1990a).
[c] Abbreviations: BGSL, benzylglucosinolate; PAPS, 3′-phosphoadenosine-5′-phosphosulphate; DSBG, desulphobenzylglucosinolate; DSP, desulpho-*p*-hydroxybenzylglucosinolate; DSA, desulphoallylglucosinolate; PAP, 3′-phosphoadenosine-5′-phosphate; NEM, *N*-ethylmaleimide; DTNB, 5,5-dithio-*bis*-(2-nitrobenzoic acid); PCMS, *p*-chloromercuriphenyl sulphonic acid; PCMB, *p*-chloromercuribenzoate.

desulphoglucosinolate structure. All other potential substrates tested, including flavonoids, flavonoid glycosides, cinnamic acids and phenylacetaldoxime, are not sulphated. The kinetics of desulphobenzylglucosinolate sulphation are consistent with a rapid equilibrium ordered mechanism with desulphobenzylglucosinolate binding first and PAPS second. This binding order may explain the successful binding of the *Brassica* ST to a desulphobenzylglucosinolate-agarose affinity column, while neither enzyme binds to a PAP-agarose matrix.

4. Localisation

The distribution of ST activity in dark-grown *Brassica juncea* seedlings mirrors that of the thiohydroximate GT (Jain *et al.*, 1989a). Cotyledons showed highest levels per gram fresh weight but hypocotyls had the greatest specific activity. ST was recovered in the 12 000 × *g* supernatant fraction of protoplast lysates, suggesting a cytosolic localization.

III. GLUCOSINOLATE DEGRADATION

A. Thioglucoside Glucohydrolase (Myrosinase)

Glucosinolate catabolism is initiated by a thioglucoside glucohydrolase (EC 3.2.3.1) known variously in the literature as myrosinase, glucosinolase, sinigrinase, and β-thioglucosidase. First reported in plants by Bussy (1840), similar activity has since been described in mammals, fungi, bacteria, and insects (Tani *et al.*, 1974; Björkman, 1976; MacGibbon and Beuzenberg, 1978; Goldner *et al.*, 1987; Nugon-Baudon *et al.*, 1990). An early contention that myrosinase was composed of two entities, a sulphatase and a thioglucosidase, was not subsequently confirmed. Myrosinase is now regarded as a single enzyme which cleaves only the thioglucosidic linkage of glucosinolates yielding D-glucose and an unstable aglycone (thiohydroxamate-O-sulphonate; Fig. 8.2). The fate of the latter depends upon the nature of the glucosinolate aglycone, pH, presence of metal ions and epithiospecifier protein (ESP), and perhaps also upon hitherto uncharacterised proteins or factors (Hasapis and MacLeod, 1982; Chew, 1988a). Under neutral conditions, the aglycone spontaneously undergoes a Lossen rearrangement with concerted loss of bisulphate to yield an isothiocyanate. At low pH or in the presence of Fe^{2+} ions, a nitrile, inorganic sulphate and elemental sulphur are formed. Certain glucosinolates also give rise to organic thiocyanates, the SCN^- ion, oxazolidine-2-thiones, and epithionitriles (Tookey *et al.*, 1980; Petroski and Tookey, 1982; Petroski, 1986; Chew, 1988a).

1. Extraction and purification

All crucifer organs contain myrosinase. White mustard (*Sinapis alba*) seeds constitute a particularly rich source from which homogeneous enzyme can be obtained in good yield after limited purification (Björkman and Janson, 1972; Palmieri *et al.*, 1986; Pessina *et al.*, 1990). Generally regarded as a soluble protein, myrosinase is usually extracted from vegetative organs by homogenising fresh tissue in appropriate buffers.

Its isolation from oilseeds traditionally started from hexane- or acetone-defatted seed meal but direct homogenization of seeds in water or aqueous extraction of crushed seeds is now popular (Palmieri *et al.*, 1986; Bones and Slupphaug, 1989). The crambe seed and Wasabi root enzymes appear exceptional in requiring sonication for solubilisation. Enzyme extraction is routinely undertaken with 0.01–0.2 M acetate, imidazole-HCl, phosphate, citrate-phosphate or Tris-HCl buffers in the pH range 5.9–9.0. Of these, imidazole-HCl buffer is highly recommended since it apparently stabilises the enzyme (Lönnerdal and Janson, 1973). Purified myrosinases are sensitive to freezing and are best stored in imidazole-HCl or phosphate buffer at 4 °C (Björkman and Lönnerdal, 1973; Durham and Poulton, 1989; Pessina *et al.*, 1990).

Using the conventional protein purification techniques summarised in Table 8.2, myrosinase has been extensively purified from many higher plants including *Sinapis alba* (Vose, 1972; Björkman and Janson, 1972; Lein, 1972; Pihakaski and Iversen, 1976; Palmieri *et al.*, 1986), *Brassica juncea* (Tsuruo *et al.*, 1967; Ohtsuru and Hata, 1972), *Brassica napus* (Björkman and Lönnerdal, 1973; Kozlowska *et al.*, 1983; Bones and Slupphaug, 1989), *Brassica oleracea* var. *gemmifera* (Springett and Adams, 1989), *Crambe abyssinica* (Tookey, 1973), *Wasabia japonica* (Ohtsuru and Kawatani, 1979), and *Lepidium sativum* (Durham and Poulton, 1989). Extracellular and intracellular myrosinases have also been characterised from *Aspergillus sydowi* and *A. niger*, respectively (Ohtsuru *et al.*, 1969a; Ohtsuru and Hata, 1973c). Among recently developed purification methods, FPLC chromatofocusing and ion-exchange chromatography appear especially promising, allowing rapid purification of myrosinase with excellent recoveries (Buchwaldt *et al.*, 1986). In view of the glycoproteinaceous nature of myrosinase, many authors have taken advantage of Con A-Sepharose 4B affinity chromatography. Starting with aqueous crude extracts of white mustard seeds, Palmieri *et al.* (1986) obtained nearly homogeneous myrosinase in a single step with recoveries exceeding 90%. Bones and Slupphaug (1989) reported 13-fold purification of the *B. napus* enzyme by this method. The cress and *Arabidopsis thaliana* myrosinases also bind tightly to this matrix, but recoveries are lower (15–60%) and inconsistent (Durham and Poulton, 1989; J. E. Poulton, unpubl. data).

Multiple forms of myrosinase have been demonstrated in several crucifers by starch gel electrophoresis (Vaughan *et al.*, 1968), native PAGE (MacGibbon and Allison, 1970; Vose, 1972; Pihakaski and Iversen, 1976) and IEF (Phelan and Vaughan, 1980), all used in conjunction with the activity stain of MacGibbon and Allison (1970). More recently, Buchwaldt *et al.* (1986) demonstrated the power of FPLC chromatofocusing and analytical IEF in isoenzyme screening. Isoenzyme patterns vary according to species, organ, degree of tissue differentiation, and presence of ascorbate (e.g. MacGibbon and Allison, 1970; Henderson and McEwen, 1972; Vose, 1972; Pihakaski and Pihakaski, 1978a, b; Phelan and Vaughan, 1980; Phelan *et al.*, 1984). Reports of isoenzymes isolated from the same source but by different authors are not always in agreement (e.g. Pihakaski and Pihakaski, 1978a; Phelan and Vaughan, 1980); this suggests that differences in cultivar or plant growth conditions may also be significant. In general, band patterns should be interpreted with extreme caution since they are not always stable and seem dependent on previous treatment of the enzyme. Both the molecular basis and the physiological significance of the apparent microheterogeneity remain poorly understood and beg further examination. In some cases, clear differences in carbohydrate and amino acid content have been established between

TABLE 8.2. Properties of purified plant myrosinases.

Enzymes	Purification methods[n]	Isozymes	Purification (-fold)	Native molecular mass (kDa)	Subunit molecular mass (kDa)	pI	Carbohydrate	pH optimum
Sinapis alba seed[a,b,c]	Hexane, ΔpH, G-50, DC, G-200, IEF	A,B,C (major)	13.6 (C)	135 (A), 145 (B), 151 (C)	62 (C) (peptide chain)	5.90(A), 5.45 (B), 5.08 (C)	18% (C)	4–5.5
Sinapis alba seed[d,e]	ConA, S-200, CF, G-75	A,B,C (major; C_1, C_2, C_3)	80–115	135.1	71.7	5.05–5.15 (C_2)	Yes	–
Brassica juncea seed[f,g]	AS, TEAE, DS, G-200, CMS	F-IA, F-IB, FIIA F-IIB	>100	153 (F-IA, F-IB, F-11A), 125 (F-IIB)	40 (F-IA, F-IB, F-IIA), 30 (F-IIB)	4.6 (F-IA, F-IB, F-IIA), 4.8 (F-IIB)	15.8% (F-IA), 17.8% (F-IB), 22.5% (FIIA), 8.6% (F-IIB)	5.5–7.0
Brassica napus cv. Panter seed[b,c,h]	Hexane, G-50, DC, S6B, IEF	A, B, C (major; C_1, C_2, C_3), D	188	135 (C)	65 (C) (peptide chain)	6.2 (A), 5.6 (B), 4.96 (C_1), 4.99 (C_2), 5.06 (C_3), 4.9 (D)	9.3% (C_1), 15.2% (C_2), 17.4% (C_3)	4.5–5.5
Brassica napus cv. Niklas seed[i]	Con A, DC, Mono Q	A,B,C (major; C_a, C_b, C_c) D	243–308	154 (C_a – C_c)	77 (C_a – C_c)	5.00 (C_a), 4.96(C_b), 4.94 (C_c)	18.9% (C_a), 11.7% (C_b), 9.6% (C_c)	5.2–5.5
Lepidium sativum seedlings[j,k]	Red, DC, IEF, G-75, Mono Q	1	14.5	130	62, 65	4.7–4.9	Yes	5.5
Wasabia japonica roots[l]	Son, AS, S-200, DS, S6B	1	100	580	45–47	–	–	6.5–7.0
Crambe abyssinica seed[m]	Son, AS, G-200	2	80	110 >200	–	–	–	8–9

Authors: [a]Björkman and Janson, (1972); [b]Björkman and Lönnerdal (1973); [c]Björkman (1976); [d]Palmieri *et al.* (1986); [e]Pessina *et al* (1990); [f]Tsuruo *et al.* (1967a); [g]Ohtsuru and Hata (1972); [h]Lönnerdal and Janson (1973); [i]Bones and Slupphaug (1989); [j]Durham and Poulton, (1990); [k]Durham and Poulton (1989); [l]Ohtsuru and Kawatani (1979); [m]Tookey (1973).

[n]Purification methods: AS, ammonium sulphate; CF, chromatofocusing; CMS, CM-Sephadex; Con A, concanaval in A-Sepharose; DC, DEAE-cellulose; DS, DEAE-Sephadex; ΔpH, pH adjustment; G-50, Sephadex G-50; G-75, Sephadex G-75; G-200, Sephadex G-200; Hexane, hexane extraction; IEF, isoelectric focusing; Mono Q, FPLC Mono Q; Red, Reactive Red 120-Agarose; S6B, Sepharose 6B; S-200, Sephacryl S-200; Son, sonication; TEAE, TEAE-cellulose.

isoenzymes from a single species (Ohtsuru and Hata, 1972; Lonnerdal and Janson, 1973; Bones and Slupphaug, 1989). However, few if any authors have rigorously excluded partial proteolysis, interaction with phenolics, or association–dissociation phenomena as alternative origins of the observed enzyme multiplicity. Furthermore, the interaction of isothiocyanates with *Brassica napus* seed proteins observed by Björkman (1973) raises the question whether some myrosinase 'isozymes' may not arise by reaction with glucosinolate breakdown products. Pertinent here is the demonstration that horseradish peroxidase, an enzyme which similarly exhibits various isozyme patterns, was modified by horseradish oil (Loomis *et al.*, 1981). Such modifications could be eliminated by extraction procedures using liquid N_2 homogenisation and adsorbent polystyrene.

Myrosinase isozymes may be resolved preparatively by IEF (Björkman and Janson, 1972; Vose, 1972; Lönnerdal and Jansen, 1973; Pihakaski and Iversen, 1976), chromatofocusing (Pessina *et al.*, 1990), ion-exchange chromatography (Tsuruo *et al.*, 1967; Björkman and Janson, 1972; Ohtsuru and Hata, 1972; Lönnerdal and Janson, 1973; Kozlowska *et al.*, 1983; Bones and Slupphaug, 1989), and gel filtration (Pihakaski and Iversen, 1976).

2. Assays employing glucosinolates as substrates

Myrosinase activity is usually assayed by monitoring glucosinolate utilisation or the production of glucose or sulphate. Less frequently, isothiocyanate or nitrile production is followed (Gil and MacLeod, 1980; Grob and Mathile, 1980; Uda *et al.*, 1986). Allylglucosinolate (sinigrin), easily obtainable from commercial sources (e.g. Sigma, Aldrich Chemical Company), is routinely used as substrate. Efforts to compare directly the most common methods have been made, at times marked by considerable differences in opinion (Pihakaski and Pihakaski, 1978a; Wilkinson *et al.*, 1984a; Palmieri *et al.*, 1987; Palmieri *et al.*, 1988; Wilkinson *et al.*, 1988). As indicated below, most methods have drawbacks ranging from suboptimal substrate or activator concentrations to lack of sensitivity. Final choice of assay may largely depend upon the number and nature of samples to be analysed, whether the presence of ascorbate is desired, and overall cost.

(a) The pH-stat assay.   Myrosinase activity may be assayed directly by titration of the released acid with alkali using a pH-stat apparatus. Regarded by many as a reference method against which other assays are compared, a typical assay contains 5 mM sinigrin, 1 mM ascorbate and 200 μl appropriately diluted myrosinase preparation in an initial volume of 2 ml (Wilkinson *et al.*, 1984a). All reagents including enzyme are prepared in double glass distilled, deionised water and the pH is adjusted to 6.5. After addition of sinigrin, enzyme activity is determined at 30 °C by measuring the acid release rate by titration with 1 mM NaOH using a pH-stat apparatus which maintains the assay pH at 6.5. Similar procedures are described by Tsuruo *et al.* (1967), Bjorkman and Lönnerdal (1973), and Palmieri *et al.* (1987), but, at pH 7.0 or above, reactions are undertaken under N_2. This procedure is reliable, has good sensitivity (thus suitable for analysing low activity fractions), and can also monitor reactions lacking ascorbate. Additional advantages are its low cost (after initial purchase of pH-stat!) and its applicability to very low assay pH, which rules out most direct coupled

assays (e.g. HK-G6PDH assay). However, since all buffered myrosinase extracts must be extensively desalted by dialysis or gel filtration before analysis, it is time-consuming and therefore not recommended for routine operations. Unlike other myrosinase assays, the assay volume changes during the reaction. An alkali concentration must therefore be carefully chosen which allows facile measurement of reaction rates but does not cause excessive dilution of assay components.

(b) Direct spectrophotometric assay for sinigrin utilisation. The myrosinase-catalysed decomposition of sinigrin may be followed spectrophotometrically using a continuous assay first proposed by Schwimmer (1961). As modified by Palmieri *et al.* (1987), the reaction mixture contains 0.5 mM sinigrin, 33 mM phosphate buffer (pH 6.5), 100 μl of appropriately diluted myrosinase, and 0.5 mM or 1 mM ascorbate when required, in a total volume of 1.5 ml. The mixture minus enzyme is pre-incubated at $30 \pm 1\,°C$ for several min in 0.5 cm path length quartz cells before initiating the reaction by adding myrosinase. The rate of substrate utilisation is calculated from the decrease in A_{227} using $\varepsilon = 6784\,M^{-1}\,cm^{-1}$. Linearity is observed with protein amounts up to approx. 2 μg enzyme. Despite its simplicity, this method records significantly lower reaction rates than the pH-stat assay since it must be run at suboptimal substrate concentrations due to the high absorbance of sinigrin at 227 nm (Palmieri *et al.*, 1987). This discrepancy is especially dramatic when using 1 cm cells, where the maximum permissible sinigrin concentration is only 0.2 mM (Wilkinson *et al.*, 1984a). Under these assay conditions, the optimal ascorbate concentration falls to 0.25 mM. The enzyme shows only 28% of this maximal activity at 1 mM ascorbate.

(c) Assay of liberated glucose by enzymatic methods. Several enzymatic methods have been utilised to quantitate glucose released during glucosinolate hydrolysis.

(i) *Glucose oxidase–peroxidase assays.* In these assays, liberated glucose is specifically oxidised by glucose oxidase to gluconic acid and H_2O_2. In the presence of peroxidase, H_2O_2 reacts with suitable chromogens to give coloured products which are quantitated spectrophotometrically. Alternatively, H_2O_2 may be decomposed by catalase to O_2 which is measured polarographically.

As described by Durham and Poulton (1989), thioglucosidase assays contain 0.5 μmol sinigrin, 50 μmol sodium citrate buffer (pH 5.5) and up to 100 μl enzyme preparation in a total volume of 0.5 ml. Mixtures are preincubated at 30 °C for 5 min before starting the reaction by adding enzyme. Reactions are terminated after 60 min by addition of 40 μl 2.9 M HCl. After centrifugation, aliquots (0.5 ml) are removed, and glucose production is measured by the glucose oxidase procedure as detailed by Sigma (Sigma Procedure No. 510). The assay is linear with amount of added purified enzyme up to 3.6 μg protein and in the range of 0–60 min incubation times. A similar assay by Björkman and Lönnerdal (1973) utilises the 'Glox' reagent (AB Kabi, Stockholm, Sweden) after terminating hydrolysis by heat-inactivation. Merck (Darmstadt, FRG) also markets a glucose oxidase reagent (Mercotest Glucose). Among drawbacks of this method, colour development is inhibited by ascorbate and by unidentified components in crude *Sinapis alba* extracts, especially from seeds and cotyledons (Pihakaski and Pihakaski, 1978a). Furthermore, it cannot be used with deeply coloured solutions.

The 'GOD-Perid' assay of Pihakaski and Pihakaski (1978a) utilises the ammonium salt of 2,2′-azido-di-(3-ethylbenzothiazoline sulphonic acid) as chromogen. The myrosinase preparation (0.5 ml) is incubated at 37 °C for 15–30 min with 0.3 ml sinigrin (15 mg ml^{-1}) and 1 ml 50 mM citrate buffer (pH 5.5) before terminating hydrolysis by boiling for 5 min. In control reactions, heat-inactivated preparations replace active enzyme. A sample of 10-fold diluted hydrolysate (50–200 μl) is made up to 1 ml with quartz-distilled water. Five ml 'GOD-Perid' reagent (Boehringer Mannheim; containing phosphate buffer (pH 7.0), peroxidase, glucose oxidase and chromogen) is added and thoroughly mixed. After incubating the mixture at 37 °C for 15 min, the A_{420} is immediately read against a blank solution. Glucose solutions (2–32 μg ml^{-1} water), receiving similar treatment, serve as standards. Slight modifications to this assay were made by Bones and Slupphaug (1989).

In the absence of ascorbate, myrosinase activity may also be assayed polarographically (Palmieri et al., 1987). Like the pH-stat assay, this method shows good linearity over a wide range of enzyme concentration. However, since it exhibits poor sensitivity and reproducibility when assaying extracts with low activities, it is not recommended for routine screening of crucifer extracts.

(ii) *Hexokinase/glucose-6-phosphate dehydrogenase assay.* The initial rate of myrosinase activity may be measured directly by monitoring the release of glucose using the hexokinase/glucose-6-phosphate dehydrogenase system (Wilkinson et al., 1984a, b, 1988). Enzyme assays contain 5 mM sinigrin, 3 mM MgCl$_2$, 1 mM ascorbate, 0.55 mM ATP, 0.72 mM NADP, 0.56 U hexokinase (EC 2.7.1.1), 0.35 U glucose-6-phosphate dehydrogenase (EC 1.1.1.49), 30 mM MES buffer (pH 6.5), 30 μl double glass-distilled deionised water, and 100 μl enzyme in a total volume of 1 ml. All reagents except the coupling enzymes, which are diluted with double glass-distilled deionised water, are prepared in 50 mM MES buffer (pH 6.5). Assay mixtures are preequilibrated in 1 cm path length quartz cells at 30 °C before initiating the reaction by adding sinigrin. Myrosinase activity is calculated from the observed increase in A_{340} immediately following substrate addition (i.e. within the first 20 s) using an extinction coefficient for NADPH of 6220 M^{-1} cm^{-1}. According to Wilkinson et al. (1984a), this method yields specific activities insignificantly different from those obtained by the pH-stat assay. Furthermore, the assay is sensitive, reproducible, and linear with respect to protein concentration in the range 0 to 30 μg purified *S. alba* myrosinase. However, such linearity was not observed with several other partially purified myrosinase preparations (Wilkinson et al., 1984b), a situation which might be remedied by raising the levels of Mg^{2+} and the coupling enzymes by at least 10-fold (Palmieri et al., 1987, 1988). A single-point glucose assay system employing HK-GPDH is described by Pihakaski and Pihakaski (1978a). Linearity extends to at least 120 μg glucose which gave an A_{340} of 1.35.

(d) *Assay of liberated glucose by chemical methods.* Among chemical methods for glucose determination, the o-toluidine method of Hultman (1959) shows good linearity but is relatively insensitive (Pihakaski and Pihakaski, 1978a). Tsuruo et al. (1967) modified the dinitrosalicylic acid method (Sumner, 1925).

(e) *Myrosinase activity stain.* MacGibbon and Allison (1970) introduced an activity stain to detect myrosinase isozymes after their separation by native PAGE. Electrophoresed gels are placed in an aqueous solution containing 5 mg ml^{-1} sinigrin, 10 mg ml^{-1} BaCl$_2$, 3 mM ascorbic acid and 1.7 N acetic acid. The enzyme's location is revealed by white bands of precipitated BaSO$_4$ (and perhaps also sulphur) measurable by densitometry. Band development, which requires several minutes to hours depending on enzyme activity and temperature, may be enhanced by Fe^{2+} ions (MacGibbon, 1975). The ascorbate dependency of individual isozymes may be revealed by exposing identical gels to stain solution lacking ascorbate (e.g. Vose, 1972). In a similar spot test for myrosinase activity, enzyme fraction (50–100 μl) is mixed with an equal volume of sinigrin solution (containing 1 mg sinigrin, 5 mg BaCl$_2$, 0.5 mg ascorbic acid and 0.1 ml acetic acid per ml solution) (Buchwaldt *et al.*, 1986).

3. *Assays employing synthetic substrates*

Since several myrosinases also hydrolyse PNPG (*p*-nitrophenyl-β-D-glucoside) (Tsuruo and Hata, 1968b; Ohtsuru *et al.*, 1969c; Durham and Poulton, 1989), their purification and characterisation may be greatly facilitated by spectrophotometric assay. A typical assay for PNPGase activity contains 7 μmol PNPG, 100 μmol potassium phosphate buffer (pH 6.5), and enzyme in a total volume of 1 ml. After 5 min pre-incubation at 30 °C, the reaction is initiated by adding enzyme and terminated 15–60 min later by addition of 2 ml 5% (w/v) sodium carbonate. The developed colours, due to liberated aglycone, are measured at 400 nm. Control incubations, in which active enzyme is replaced by buffer or by heat-inactivated enzyme, are routinely included. While this assay is rapid, inexpensive and suitable for large sample numbers, it lacks specificity and should be used in parallel with more specific methods when purifying myrosinase from novel sources.

4. *Properties*

Most plant myrosinases are dimeric or tetrameric glycoproteins (9–23% carbohydrate) with native molecular masses of 125–155 kDa and isoelectric points between pH 4.6 and 6.2 (Table 8.2). A notable exception is the *Wasabia japonica* root enzyme which has approx. 12 subunits and a molecular mass of 580 kDa. As discussed earlier, isozymes of myrosinase are common but their physiological significance remains unclear. Relatively few studies have focused upon substrate specificity (Gaines and Goering, 1962; Lein, 1972; Björkman, 1976; Durham and Poulton, 1990). Plant myrosinases show pronounced specificity towards glucosinolates. In some cases, they also hydrolyse the synthetic glucoside PNPG, allowing their purification to be monitored by the facile PNPGase assay (see Section III.A.3). Where examined in depth though, highly purified plant myrosinases showed little or no activity towards any other *O*- or *S*-glycosides (Lein, 1972; Durham and Poulton, 1990), thus indicating a high degree of aglycone specificity. Myrosinases from different species are not specifically adapted to a particular set of glucosinolates but instead hydrolyse all naturally occurring glucosinolates tested, although at different rates (Björkman, 1976; MacLeod and Rossiter, 1986). Regarding glycone specificity, the *Brassica juncea* myrosinase much preferred the natural thioglucoside benzylglucosinolate over the corresponding synthetic glycosides containing mannose, galactose and xylose moieties

(Gaines and Goering, 1962). With few exceptions, pH optima are often quite broad with maxima lying between pH 4.0 and 7.0.

While ascorbate-insensitive isozymes are also known (Ettlinger *et al.*, 1961; Vose, 1972), most myrosinases are specifically activated by ascorbic acid, an observation first made by Nagashima and Uchiyama (1959). Depending on species, degree of enzyme purity, assay pH and ascorbate concentration, up to 1000-fold stimulation has been recorded (Grob and Mathile, 1980). Purified myrosinases are activated maximally by 0.7–1 mM ascorbate while crude preparations may require up to 5 mM (Wilkinson *et al.*, 1984a). Higher concentrations are inhibitory with complete inhibition occurring between 10 and 100 mM (Lein, 1972; Björkman and Lönnerdal, 1973; Ohtsuru and Hata, 1973b). The mechanism of ascorbate activation is still not fully known but appears to involve a slight conformational change leading to an increase in both K_m and V_{max} (Tsuruo *et al.*, 1967; Ohtsuru and Hata, 1973b). Typical K_m ranges with sinigrin as substrate are: unactivated enzyme, 30–200 μM; activated enzyme, 0.2–1.0 mM.

In general, plant myrosinases lack a metal ion requirement and are unaffected by metal chelators. However, the Wasabi and *B. juncea* enzymes were stimulated up to two-fold by inorganic salts, and the former was significantly inhibited by EDTA and *o*-phenanthroline (Tsuruo and Hata, 1968a; Ohtsuru and Kawatani, 1979). Sulphhydryl groups appear essential for catalytic activity as judged by their sensitivity to SH-reagents (Nagashima and Uchiyama, 1959; Tookey, 1973; Ohtsuru and Hata, 1973a, 1979; Ohtsuru and Kawatani, 1979; Durham and Poulton, 1990). The action of glycosidase inhibitors on myrosinase activity has received little attention (Ohtsuru *et al.*, 1969b; Ohtsuru and Kawatani, 1979). Both thioglucosidase and PNPGase activities of the cress myrosinase were competitively inhibited by the pyrrolizidine alkaloid castanospermine with K_i values of 5 μM and 6 μM, respectively (Durham and Poulton, 1989). Several polyhydroxyalkaloids including castanospermine inhibited mustard and cabbage aphid myrosinases (Scofield *et al.*, 1990).

Fungal myrosinases display several distinctive properties (Ohtsuru *et al.*, 1969b, c; Ohtsuru and Hata, 1973c). They are ascorbate-insensitive and show higher affinities for PNPG (K_m, 0.1–1.5 mM) than sinigrin (K_m, 3.3–3.6 mM). The purified *A. sydowi* enzyme is less specific than plant myrosinases. It also hydrolyzes salicin, arbutin, phenyl-β-glucoside and cellobiose, but whether this behaviour reflects a major difference between fungal and plant myrosinases or contamination by other glycosidases remains in question.

5. Localisation

Since large-scale hydrolysis of glucosinolates by myrosinase occurs only after tissue disruption, these two components of the so-called 'mustard oil bomb' (Luthy and Mathile, 1984) must normally be kept separate in intact plants. Much evidence favours compartmentation at the subcellular level. In horseradish roots, storage parenchyma cells apparently contain both glucosinolates and myrosinase. Glucosinolates co-occur with ascorbate in vacuoles, while myrosinase is probably cytosolic but has a marked tendency to adhere to cellular membranes (Grob and Mathile, 1980; Luthy and Mathile, 1984). Such binding might in part explain earlier reports of myrosinase being associated with smooth ER, dictyosomes, mitochondria and the plasma membrane

(Iversen, 1970; Pihakaski and Iversen, 1976; Maheshwari *et al.*, 1981). The integrity of the tonoplast therefore determines the safety of glucosinolate-containing cells in this species.

Based on histological approaches with unspecific protein stains, it has long been assumed that myrosinase occurs in many crucifers in specialised protein-accumulating idioblasts termed 'myrosin cells' (Guignard, 1890). In some species, this assumption is supported by correlative studies of the occurrence and distribution of myrosin cells and myrosinase activity (e.g. Bones and Iversen, 1985). Using monoclonal antibodies against myrosinase, Thangstad *et al.* (1990) recently confirmed that this hydrolase is indeed restricted to myrosin cells in *B. napus* cotyledons and radicles. The enzyme was associated with the single tonoplast-like membrane surrounding the so-called myrosin grains. Similar results were obtained by Höglund *et al.* (1991) who immunocyto-chemically determined the distribution of rapeseed myrosinase in the embryo during seed development and germination as well as in adult plants. Myrosinase was localised in the cytoplasm, although occasional association with the myrosin grain membrane was noted. Whether glucosinolates are similarly restricted to myrosin cells, or, as originally believed (Guignard, 1890), are accumulated in all parenchymatous cells, remains to be unequivocally demonstrated. If the positive reaction shown by myrosin grains with Millon's reagent reflects reaction with methylindolylglucosinolates (Höglund *et al.*, 1991), it appears likely that, in rapeseed too, subcellular compart-mentation prevents premature glucosinolate hydrolysis.

In mature papaya seeds, compartmentation occurs at both tissue and subcellular levels (Tang, 1973). Benzylglucosinolate is present in the endosperm tissue but not in the sarcotestae, while the reverse is true for myrosinase. However, the embryo contains both enzyme and substrate.

ACKNOWLEDGEMENTS

B. L. Møller acknowledges support from the Rockefeller Foundation, the Commission of the European Communities Science and Technology for Development (contract TS2-0265-DK) and from the Center for Plant Biotechnology. The authors wish to thank Linda Donohoe for manuscript preparation.

REFERENCES

Benn, M. H. (1962). *Chem. Ind. (Lond.)*, 308–309.
Bertelsen, F., Gissel-Nielsen, G., Kjaer, A. and Skrydstrup, T. (1988). *Phytochemistry* **27**, 3743–3749.
Björkman, R. (1973). *Phytochemistry* **12**, 1585–1590.
Björkman, R. (1976). *In* "The Biology and Chemistry of the Cruciferae" (A. J. Vaughan, A. J. MacLeod and B. M. G. Jones, eds), pp. 191–205. Academic Press, London.
Björkman, R. and Janson, J.-C. (1972). *Biochim. Biophys. Acta* **276**, 508–518.
Björkman, R. and Lonnerdal, B. (1973). *Biochim. Biophys. Acta* **327**, 121–131.
Bones, A. and Iversen, T.-H. (1985). *Isr. J. Bot.* **34**, 351–375.
Bones, A. M. and Slupphaug, G. (1989). *J. Plant Physiol.* **134**, 722–729.
Buchwaldt, L., Larsen, L. M., Plöger, A. and Sørensen, H. (1986). *J. Chromatogr.* **363**, 71–80.
Bussy, A. (1840). *Ann. Chem. Pharm.* **34**, 223–230.

Chapple, C. C. S., Decicco, C. and Ellis, B. E. (1988). *Phytochemistry* **27**, 3461–3463.

Chew, F. S. (1988a). *In* "Biologically Active Natural Products" (H. G. Cutler, ed.), pp. 155–181. American Chemical Society Symposium, Washington, DC.

Chew, F. S. (1988b). *In* "Chemical Mediation of Coevolution" (K. A. Spencer, ed.), pp. 271–303. Academic Press, New York.

Chisholm, M. D. (1973). *Phytochemistry* **12**, 605–608.

Chisholm, M. D. and Matsuo, M. (1972). *Phytochemistry* **11**, 203–207.

Chisholm, M. D. and Wetter, L. R. (1964). *Can. J. Biochem.* **42**, 1033–1040.

Chisholm, M. D. and Wetter, L. R. (1966). *Can. J. Biochem.* **44**, 1625–1632.

Conn, E. E. (1973). *Biochem. Soc. Symp.* **38**, 277–302.

Conn E. E. (1988). *In* "Biologically Active Natural Products" (H. G. Cutler, ed.), pp. 143–154. American Chemical Society Symposium, Washington, DC.

Cronquist, A. (1988). "The Evolution and Classification of Flowering Plants". New York Botanical Garden, Bronx, NY.

Cutler, A. J., Sternberg, M. and Conn, E. E. (1985). *Arch. Biochem. Biophys.* **238**, 272–279.

Davies, J. H., Davis, R. H. and Kirby, P. (1968). *J. Chem. Soc. C.*, 431–435.

Dewick. P. M. (1984). *Nat. Prod. Rep.* **1**, 545–549.

Dörnemann, D., Löffelhardt, W. and Kindl, H. (1974). *Can. J. Biochem.* **52**, 916–921.

Durham, P. L. and Poulton, J. E. (1989). *Plant Physiol.* **90**, 48–52.

Durham, P. L. and Poulton, J. E. (1990). *Z. Naturforsch.* **45c**, 173–178.

Elliot, M. C. and Stowe, B. B. (1971). *Plant Physiol.* **47**, 366–372.

Erb, N., Zinsmeister, H. D., and Nahrstedt, A. (1981). *Plant Med.* **41**, 84–89.

Ettlinger, M. G. and Dateo, G. P., Jr. (1961). *Studies of Mustard Oil Glucosides* **12**, Final Report Contract DA19-129-QM-1059, US Army Natick Laboratories, Natick, MA.

Ettlinger, M. G. and Kjaer, A. (1968). *In* "Recent Advances in Phytochemistry" Vol. **1**, pp. 59–144. Appleton Century Crofts, New York.

Ettlinger, M. G. and Lundeen, A. J. (1956). *J. Am. Chem. Soc.* **78**, 4172–4173.

Ettlinger, M. G., Dateo, G. P., Jr., Harrison, B. W., Mabry, T. J. and Thompson, C. P. (1961). *Proc. Natl. Acad. Sci. USA* **47**, 1875–1880.

Fenwick, G. R., Heaney, R. K. and Mullin, W. J. (1983). *CRC Crit. Rev. Food Sci. Nutri.* **18**, 123–201.

Gaines, R. D. and Goering, K. J. (1962). *Arch. Biochem, Biophys.* **96**, 13–19.

Gil, V. and Macleod, A. J. (1980). *Phytochemistry* **19**, 2547–2551.

Glendening, T. M. and Poulton, J. E. (1988). *Plant Physiol.* **86**, 319–321.

Glendening, T. M. and Poulton, J. E. (1990). *Plant Physiol.* **94**, 811–818.

Glover, J. R., Chapple, C. C. S., Rothwell, S., Tober, I. and Ellis, B. E. (1988). *Phytochemistry* **27**, 1345–1348.

Gmelin, R. and Kjaer, A. (1970). *Phytochemistry* **9**, 591–593.

Goldner, M., Nairn, C. A., Patel, P. V., Martin, P. M. V., Tan, E. L., Parsons, N. J. and Smith, H. (1987). *Ann. Inst. Pasteur/Microbiol.* **138**, 325–332.

Grob, K. and Mathile, Ph. (1980). *Z. Pflanzenphysiol.* **98**, 235–243.

GrootWassink, J. W. D., Balsevich, J. J. and Kolenovsky, A. D. (1990). *Plant Sci.* **66**, 11–20.

GrootWassink, J. W. D., Nelson, L. A. K., Kolenovsky, A. D., Jain, J. C. and Underhill, E. W. (1988). Abstract, Eucarpia Congress on Genetic Manipulation in Plant Breeding, 11–16 September, 1988, Helsingør, Denmark.

Guignard, L. (1890). *J. Bot.* **4**, 385–394; 412–430; 435–455.

Halkier, B. A. and Møller, B. L. (1989). *Plant Physiol.* **90**, 1552–1559.

Halkier, B. A. and Møller, B. L. (1990). *J. Biol. Chem.* **265**, 21 114–21 121.

Halkier, B. A. and Møller, B. L. (1991). *Plant Physiol.* **95**, in press.

Halkier, B. A., Olsen, C. E. and Møller, B. L. (1989). *J. Biol. Chem.* **264**, 19 487–19 494.

Halkier, B. A., Scheller, H. V. and Møller, B. L. (1988). *CIBA Found. Symp.* **140**, 49–66.

Halkier, B. A., Lykkesfeldt, J. and Møller, B. L. (1991). *Proc. Natl. Acad. Sci. USA* **88**, 487–491.

Harrington, H. M. and Smith, I. K. (1980). *Plant Physiol.* **65**, 151–155.

Hasapis, X. and MacLeod, A. J. (1982). *Phytochemistry* **21**, 1009–1013.

Henderson H. M. and McEwen, T. J. (1972). *Phytochemistry* **11**, 3127–3133.

Hogge, L. R., Reed, D. W., Underhill, E. W. and Haughn, G. W. (1988). *J. Chromatogr. Sci.* **26**, 551–556.

Höglund, A.-S., Lenman, M., Falk, A. and Rask, L. (1991). *Plant Physiol.* **95**, 213–221.

Holm. A., Carlsen, L. and Ettlinger, M. G. (1975). *In* "Organic Sulfur Chemistry" (C. J. M. Stirling, ed.), p. 347. Butterworths, London.

Hultman, E. (1959). *Nature (Lond.)* **183**, 108–109.

Hwu, J. R. and Tsay, S.-C. (1990). *Tetrahedron* **46**, 7413–7428.

Iversen, T-.H. (1970). *Protoplasma* **71**, 451–466.

Jain, J. C., Michayluk, M. R., GrootWassink, J. W. D. and Underhill, E. W. (1989a). *Plant Sci.* **64**, 25–29.

Jain, J. C., Reed, D. W., GrootWassink, J. W. D. and Underhill, E. W. (1989b). *Anal. Biochem.* **178**, 137–140.

Jain, J. C., GrootWassink, J. W. D., Kolenovsky, A. D. and Underhill, E. W. (1990a). *Phytochemistry* **29**, 1425–1428.

Jain, J. C., GrootWassink, J. W. D., Reed, D. W. and Underhill, E. W. (1990b). *J. Plant Physiol.* **136**, 356–361.

Jaroszewski, J. W. and Ettlinger, M. G. (1981). *Phytochemistry* **20**, 819–821.

Josefsson, E. (1971a). *Physiol. Plant.* **24**, 150–159.

Josefsson, E. (1971b). *Physiol. Plant.* **24**, 161–175.

Josefsson, E. (1973). *Physiol. Plant.* **29**, 28–32.

Kindl, H. (1965). *Monatsh. Chem.* **96**, 527–532.

Kindl, H. and Schiefer, S. (1969). *Monatsh. Chem.* **100**, 1773–1787.

Kindl, H. and Underhill, E. W. (1968). *Phytochemistry* **7**, 745–756.

Kjaer, A. (1954). *Acta Chem. Scand.* **8**, 1110.

Kjaer, A. (1980–81). *Food Chem.* **6**, 223–234.

Kjaer, A. and Christensen, B. (1959). *Acta Chem. Scand.* **13**, 1575–1584.

Kjaer, A. and Wagner, S. (1955). *Acta Chem. Scand.* **9**, 721–726.

Koch, B., Nielsen, V. S., Halkier, B. A. and Møller, B. L. (1991). *Arch. Biochem. Biophys.*, **292**, 141–150.

Kozlowska, H. J., Nowak, H. and Nowak, J. (1983). *J. Sci. Food Agric.* 34, 1171–1178.

Kutacek, M., Prochazka, Z. and Veres, K. (1962). *Nature* 194, 393–394.

Langer, P. (1983). *In* "CRC Handbook of Naturally Occurring Food Toxicants" (M. Rechcígl, ed.), pp. 101–129. CRC Press, Boca Raton, FL.

Larsen, P. O. (1981). *In* "The Biochemistry Plants", Vol. 7 (P. K. Stumpf and E. E. Conn, eds), pp. 501–525. Academic Press, New York.

Lee, C. J. and Serif, G. S. (1971). *Biochem. Biophys. Acta* 230, 462–467.

Lefebvre, D. D. (1990). *Plant Physiol.* **93**, 522–524.

Lein, K.-A. (1972) *Angew. Bot.* **46**, 137–159.

Löffelhardt, W. and Kindl, H. (1975). *Z. Naturforsch* **30c**, 233–239.

Lonnerdal, B. and Janson, J.-C. (1973). *Biochim. Biophys. Acta* **315**, 421–429.

Loomis, W. D., Sandstrom, R. P., Pearce, P. D. and Burbott, A. J. (1981). *Phytochemical Society of North America Newsletter*, July 1981.

Lüthy, B. and Mathile, P. (1984). *Biochem. Physiol. Pflanzen.* **179**, 5–12.

Macfarlane Smith, W. H. and Griffiths, D. W. (1988). *J. Sci. Food Agric.* **43**, 121–134.

MacGibbon, D. B. (1975). *NZ J. Sci.* **18**, 217–219.

MacGibbon, D. B. and Allison, R. M. (1970). *Phytochemistry* **9**, 541–544.

MacGibbon, D. B. and Beuzenberg, E. J. (1978). *NZ J. Sci.* **21**, 389–392.

MacLeod, A. J. and Rossiter, J. T. (1986). *Phytochemistry* **25**, 1047–1051.

Maheshwari, P. N., Stanley, D. W., Beveridge, T. J. and van de Voort, F. R. (1981). *J. Food Biochem.* **5**, 39–61.

Marsh, R. E. and Waser, J. (1970). *Acta Crystallogr. Sect. B* **26**, 1030–1037.

Matsuo, M. (1968a). *Chem. Pharm. Bull.* **16**, 1128–1129.

Matsuo, M. (1968b). *Tetrahedron Lett.* **38**, 4101–4104.

Matsuo, M. and Underhill, E. W. (1969). *Biochem. Biophys. Res. Commun.* **36**, 18–23.

Matsuo, M. and Underhill, E. W. (1971). *Phytochemistry* **10**, 2279–2286.

Matsuo, M. and Yamazaki, M. (1966). *Biochem. Biophys. Res. Commun.* **24**, 786–791.

Matsuo, M., Kirkland, D. F. and Underhill, W. (1972). *Phytochemistry* **11**, 697–701.
McFarlane, I. J., Lees, E. M. and Conn, E. E. (1975). *J. Biol. Chem.* **250**, 4708–4713.
Møller, B. L. (1978). *J. Lab. Comp. Radiopharm.* **14**, 663–671.
Møller, B. L. and Conn, E. E. (1979). *J. Biol. Chem.* **245**, 8575–8583.
Nagashima, Z. and Uchiyama, M. (1959). *J. Agric. Chem. Soc. Japan* **33**, 980–984.
Nugon-Baudon, L., Rabot, S., Wal, J. M. and Szylit, O. (1990). *J. Sci. Food Agric.* **52**, 547–559.
Ohtsuru, M. and Hata, T. (1972). *Agric. Biol. Chem.* **36**, 2495–2503.
Ohtsuru, M. and Hata, T. (1973a). *Agric. Biol. Chem.* **37**, 269–275.
Ohtsuru, M. and Hata, T. (1973b). *Agric. Biol. Chem.* **37**, 1971–1972.
Ohtsuru, M. and Hata, T. (1973c). *Agric. Biol. Chem.* **37**, 2543–2548.
Ohtsuru, M. and Hata, T. (1979). *Biochim. Biophys. Acta* **567**, 384–391.
Ohtsuru, M. and Kawatani, H. (1979). *Agric. Biol. Chem.* **43**, 2249–2255.
Ohtsuru, M., Tsuruo, I. and Hata, T. (1969a). *Agric. Biol. Chem.* **33**, 1309–1314.
Ohtsuru, M., Tsuruo, I. and Hata, T. (1969b). *Agric. Biol. Chem.* **33**, 1315–1319.
Ohtsuru, M., Tsuruo, I. and Hata, T. (1969c). *Agric. Biol. Chem.* **33**, 1320–1325.
Palmer, M. V., Yeung, S. P. and Sang, J. P. (1987). *J. Agric. Food Chem.* **35**, 262–265.
Palmieri, S., Iori, R. and Leoni, O. (1986). *J. Agric. Food Chem.* **34**, 138–140.
Palmieri, S., Iori, R. and Leoni, O. (1987). *J. Agric. Food Chem.* **35**, 617–621.
Palmieri, S., Iori, R. and Leoni, O. (1988). *J. Agric. Food Chem.* **36**, 872.
Pessina, A., Thomas, R. M., Palmieri, S. and Luisi, P. L. (1990). *Arch. Biochem. Biophys.* **280**, 383–389.
Petroski, R. J. (1986). *Plant Sci.* **44**, 85–88.
Petroski, R. J. and Tookey, H. L. (1982). *Phytochemistry* **21**, 1903–1905.
Phelan, J. R. and Vaughan, J. G. (1980). *J. Exp. Bot.* **31**, 1425–1433.
Phelan, J. R., Allen, A. and Vaughan, J. G. (1984). *J. Exp. Bot.* **35**, 1558–1564.
Pihakaski, K. and Iversen, T.-H. (1976). *J. Exp. Bot.* **27**, 242–258.
Pihakaski, K. and Pihakaski, S. (1978a). *J. Exp. Bot.* **29**, 335–345.
Pihakaski, S. and Pihakaski, K. (1978b). *J. Exp. Bot.* **29**, 1363–1369.
Rausch, T., Butcher, D. and Hilgenberg, W. (1983). *Physiol. Plant.* **58**, 93–100.
Rodman, J. E. (1981). *In* "Phytochemistry and Angiosperm Phylogeny" (D. A. Young and D. S. Seigler, eds), pp. 43–79. Praeger Publ., New York.
Rossiter, J. T., James, D. C. and Atkins, N. (1990). *Phytochemistry* **29**, 2509–2512.
Sang, J. P., Minchinton, I. R., Johnstone, P. K. and Truscott, R. J. W. (1984). *Can. J. Plant. Sci.* **64**, 77–93.
Schwimmer, S. (1961). *Acta Chem. Scand.* **15**, 535–544.
Scofield, A. M., Rossiter, J. T., Witham, P., Kite, G. C., Nash, R. J. and Fellows, L. E. (1990). *Phytochemistry* **29**, 107–109.
Seigler, D. (1975). *Phytochemistry* **14**, 9–29.
Springett, M. B. and Adams, J. B. (1989). *Food Chem.* **33**, 173–186.
Stewart, J. M., Nigam, S. N. and McConnell, W. B. (1974). *Can. J. Biochem.* **52**, 144–145.
Sumner, J. B. (1925). *J. Biol. Chem.* **65**, 393–395.
Tang, C.-S. (1973). *Phytochemistry* **12**, 769–773.
Tani, N., Ohtsuru, M. and Hata, T. (1974). *Agric. Biol. Chem.* **38**, 1623–1630.
Tapper, B. A. and Butler, G. W. (1967). *Arch. Biochem. Biophys.* **120**, 719–721.
Tapper, B. A. and Butler, G. W. (1972). *Phytochemstry* **11**, 1041–1046.
Thangstad, O. P., Iversen, T.-H., Slupphaug, G. and Bones, A. (1990). *Planta* **180**, 245–248.
Tookey, H. L. (1973). *Can. J. Biochem.* **51**, 1305–1310.
Tookey, H. L., VanEtten, C. H. and Daxenbichler, M. E. (1980). *In* "Toxic Constituents of Plant Foodstuffs", 2nd edn (I. E. Liener, ed.), pp. 103–142. Academic Press, New York.
Tsuruo, I. and Hata, T. (1968a). *Agric. Biol. Chem.* **32**, 479–483.
Tsuruo, I. and Hata, T. (1968b). *Agric. Biol. Chem.* **32**, 1425–1431.
Tsuruo, I., Yoshida, M. and Hata, T. (1967). *Agric. Biol. Chem.* **31**, 18–26.
Uda, Y., Kurata, T. and Arakawa, N. (1986). *Agric. Biol. Chem.* **50**, 2735–2740.
Underhill, E. W. (1967). *Eur. J. Biochem.* **2**, 61–63.

Underhill, E. W. (1980). *In* "Encyclopedia of Plant Physiology, New Series", Vol. 8 (E. A. Bell and B. V. Charlwood, eds), pp. 493–511. Springer-Verlag, Berlin, Heidelberg and New York.
Underhill, E. W. and Chisholm, M. D. (1964). *Biochem. Biophys. Res. Commun.* **14**, 425–430.
Underhill, E. W. and Kirkland, D. F. (1972). *Phytochemistry* **11**, 1973–1979.
Underhill, E. W. and Wetter, L. R. (1969). *Plant Physiol.* **44**, 584–590.
Underhill, E. W., Chisholm, M. D. and Wetter, L. R. (1962). *Can J. Biochem. Physiol.* **40**, 1505–1514.
Underhill, E. W., Wetter, L. R. and Chisholm, M. D. (1973). *Biochem. Soc. Symp.* **38**, 303–326.
Uppstrom, B. (1983). *Sveriges Utsadesforenings Tidskift* **93**, 331–336.
Van Etten, C. H. and Tookey, H. L. (1979). *In* "Herbivores, their Interaction with Secondary Plant Metabolites" (G. A. Rosenthal and D. H. Janzen, eds), pp. 471–500. Academic Press, New York.
Van Etten, C. H. and Tookey, H. L. (1983). *In* "CRC Handbook of Naturally Occurring Food Toxicants" (M. Rechcígl, ed.), pp. 15–30. CRC Press, Boca Raton, FL.
Vaughan, J. G., Gordon, E. and Robinson, D. (1968). *Phytochemistry* **7**, 1345–1348.
Vose, J. R. (1972). *Phytochemistry* **11**, 1649–1653.
Walter, W. and Schaumann, E. (1971). *Synthesis*, 111–130.
Wetter, L. R. (1964). *Phytochemistry* **3**, 57–64.
Wetter, L. R. and Chisholm, M. D. (1968). *Can. J. Biochem.* **46**, 931–935.
Wilkinson, A. P., Rhodes, M. J. C. and Fenwick, G. R. (1984a). *Anal. Biochem.* **139**, 284–291.
Wilkinson, A. P., Rhodes, M. J. C. and Fenwick, G. R. (1984b). *J. Sci. Food Agric.* **35**, 543–552.
Wilkinson, A. P., Rhodes, M. J. C. and Fenwick, G. R. (1988). *J. Agric. Food Chem.* **36**, 871.

9 Prenyltransferases and Cyclases

WILLIAM R. ALONSO and RODNEY CROTEAU

Institute of Biological Chemistry, Washington State University, Pullman, WA 99164–6340, USA

I. INTRODUCTION

Prenyltransferases catalyse the condensation between C-1 of an allylic (prenyl) pyrophosphate and C-4 of isopentenyl pyrophosphate to yield the next higher

METHODS IN PLANT BIOCHEMISTRY Vol. 9
ISBN 0–12–461019–6

C_5-homologue of the allylic substrate, whereas the terpene cyclases (synthases) catalyse, as the generic name suggests, the cyclisation of the various allylic substrates (Fig. 9.1). The reaction mechanisms of both enzyme types are similar, and are considered to involve initial ionisation of the allylic pyrophosphate with electrophilic attack of the resulting allylic carbocation on a double bond (intermolecular in the case of prenyl transfer and intramolecular in the case of cyclisation). In the case of prenyl transfer, deprotonation of the resulting cation yields the next C_5-homologue of the allylic cosubstrate. In the case of cyclisation, rearrangement and/or hydride shift may precede quenching of the cyclic carbocation by deprotonation or nucleophile capture (Fig. 9.2).

There is considerable interest in the prenyltransferases and cyclases responsible for the construction of monoterpenes (C_{10}), sesquiterpenes (C_{15}) and diterpenes (C_{20}) because these enzymes function at key branch points in isoprenoid metabolism (Fig. 9.1) and thus may have regulatory functions (West *et al.*, 1978). The monoterpene, sesquiterpene and diterpene cyclases are also of great mechanistic interest since a very limited number of natural acyclic substrates (C_{10} = geranyl pyrophosphate; C_{15} = farnesyl pyrophosphate; C_{20} = geranylgeranyl pyrophosphate) serve as the universal precursors of several hundred cyclised terpenoid products generated by

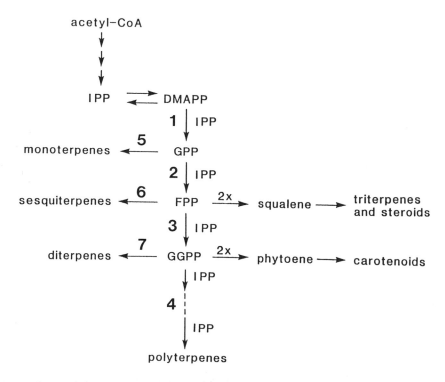

FIG. 9.1. Outline of the major branches of terpenoid biosynthesis. Abbreviations: IPP, DMAPP, GPP, FPP and GGPP are the pyrophosphate esters of isopentenol, dimethylallyl alcohol, geraniol, farnesol and geranylgeraniol, respectively. Important enzymes of the pathway include: (1) geranyl pyrophosphate synthase; (2) farnesyl pyrophosphate synthase; (3) geranylgeranyl pyrophosphate synthase; (4) polyprenyl transferase; (5) monoterpene synthase; (6) sesquiterpene synthase; and (7) diterpene synthase.

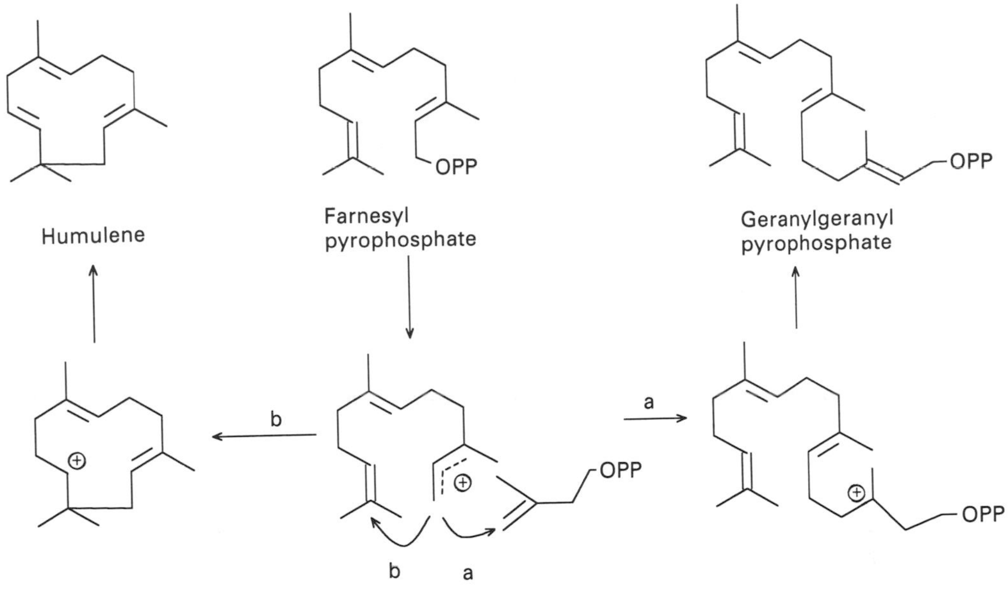

FIG. 9.2. Ionisation of farnesyl pyrophosphate leading to either condensation with isopentenyl pyrophosphate (a) to yield geranylgeranyl pyrophosphate (prenyltransferase) or cyclisation (b) to yield humulene.

what must be several hundred distinct, and catalytically diverse, cyclisation enzymes. Details of monoterpene (Croteau, 1987; Gershenzon and Croteau, 1990), sesquiterpene (Cane, 1990) and diterpene (West, 1981) biosynthesis, and in-depth coverage of the prenyltransferase reaction (Poulter and Rilling, 1981), are provided elsewhere and will not be discussed here. Rather, we shall concern ourselves with practical aspects of the isolation and measurement of prenyl transfer and cyclisation enzymes. Examination of these enzymes of isoprenoid biosynthesis is difficult because of various interfering materials obtained in plant extracts, and because of other limitations of the plant sources, or of the enzymes themselves. We shall describe these problems and point out generally applicable techniques and procedures by which they may be overcome. It need be emphasised, however, that the difficulties encountered in the isolation, assay and purification of enzymes of the different terpenoid classes, and from different plant sources, may vary considerably in detail. Specific protocols are provided, but these should be taken only as starting points for subsequent optimisation.

II. GENERAL CONSIDERATIONS

Prenyltransferases and cyclases are most often operationally soluble proteins possessing molecular weights in the range 50–100 kDa (Croteau and Cane, 1985). These enzymes may be associated with the endoplasmic reticulum (Belingheri *et al.*, 1988) or plastids (Gleizes *et al.*, 1983; Kreuz and Kleinig, 1984) *in situ*, but are easily solubilised (Moesta and West, 1985) and seemingly stable in this state, in spite of their significant hydrophobicity and relatively low pI values (pH 4-5) (Croteau and Cane, 1985). Michaelis constants for the pyrophosphate ester substrates are in the range 1–20 μM, and the only cofactor required is a divalent metal ion, with Mg^{2+} or Mn^{2+} usually preferred (K_m often in the mM range). The pH optimum is generally within one unit of neutrality. In spite of the similarity in gross properties, these enzymes are very sensitive to minor changes in pH, ionic strength, osmolarity and the concentration of reactants, and all conditions must be optimised empirically to provide maximum catalytic rates.

The terpenoid cyclases and prenyltransferases often occur in plant tissues containing high levels of low molecular weight, interfering substances (e.g. oils, pigments, phenolics, resins and waxes) and other enzymes (proteases, phenol oxidases, and competing phosphohydrolases) that can hinder isolation, fractionation and the preliminary demonstration of activity. The use of complexing agents and adsorbents in properly buffered solutions can reduce the destructive influence of many interfering substances, and the addition of specific inhibitors and protectants can circumvent the effects of many competing, or otherwise deleterious, activities.

Another limitation in studying enzymes of terpenoid metabolism is that, like most enzymes involved in the biosynthesis of natural products, they do not occur in very high tissue concentrations. Additionally, the reactions catalysed are rather slow, with turnover numbers in the range $0.01–1.0 s^{-1}$ (Croteau and Cane, 1985; Cane and Pargellis, 1987; Hohn and Plattner, 1989; Munck and Croteau, 1990; Vogeli et al., 1990; Alonso and Croteau, 1991). It is imperative, therefore, to select the type of tissue at the appropriate developmental stage which affords the greatest abundance of the relevant enzyme(s). Even with this, highly sensitive radiochemical assays are often required to monitor these activities.

A singular advantage of studies on isoprenoid biosynthesis is that the end-products which accumulate in plant tissue are often readily measurable by gas-liquid chromatography (GLC) or high performance liquid chromatography (HPLC) techniques (Jennings and Shibamoto, 1980; Croteau and Ronald, 1983). Thus, even if catabolism (Croteau, 1988) cannot be discounted, such measurements can provide an estimate of minimum in vivo flux through a pathway. A measured enzyme activity that is significantly lower than in vivo flux is unlikely to be an accurate reflection of maximum catalytic capability, and can reveal whether there are problems with the extraction or assay of the enzyme. The occurrence of such activity losses, due to inhibition or degradation of the enzyme, can generally be detected by 'mixing experiments' in which an extract free of such problems is employed as a control (Lewinsohn et al., 1991a). A low recovery of activity in the mixture, relative to the arithmetic sum expected from sample plus control, is indicative of a problem, and extraction and/or assay conditions must be altered, and the mixing experiment repeated. In the absence of a suitable control, there is no alternative but to optimise extraction and assay procedures by trial and error.

In herbaceous species, it is the young expanding leaves that accumulate terpenoid products most rapidly, and this tissue is the most enriched in prenyltransferases (Croteau and Purkett, 1989), as well as in monoterpene and sesquiterpene cyclases. Studies of the synthesis of monoterpenes as a function of leaf development have revealed that cyclisation processes (though not necessarily secondary transformations) essentially cease at full leaf expansion because of the diminution of cyclase levels (Croteau *et al.*, 1981). Furthermore, resins, phenolics, and other interfering substances also accumulate during leaf development and further complicate the isolation of cyclases from mature tissue (see below).

Considerable advantage in enzyme isolation accrues if the relevant biosynthetic enzymes are inducible. Following stem wounding, the levels of monoterpene cyclase activity in conifers such as *Abies grandis* increase as much as 10-fold relative to non-wounded control trees (Lewinsohn *et al.*, 1991b). The magnitude of the increase in cyclase activity was dependent on the severity of the wound, and increased cyclase activity was detected in both bark (phloem) and wood (xylem) tissues. In certain plant systems, the intracellular levels of several different enzymes of terpenoid metabolism are known to increase markedly after treatment with fungal spores, or fungal cell wall fragments. For example, in the diterpene series, casbene synthase activity in castor bean seedlings increased more than 10-fold after exposure to fungal spores (Moesta and West, 1985). Similarly, 5-epi-aristolochene synthase activity was not detectable in tobacco cell suspension cultures, yet within 12 h of treatment with a fungal wall hydrolysate, or cellulase, maximum activity of this sesquiterpene cyclase was observed (Vogeli *et al.*, 1990). Although plant cell suspension cultures can be induced to synthesise and excrete diterpenoids and sesquiterpenoids (especially phytoalexins), the undifferentiated cells are not suitable for the study of natural products such as monoterpenes, that are most often synthesised in specialised secretory structures *in vivo* (e.g. oil glands, resin ducts) (Falk *et al.*, 1990).

The presence of oils, resins and phenolic substances (along with phenol oxidases) which inactivate cyclases and other plant enzymes through both covalent and non-covalent interactions, is a particularly severe problem in essential oil and resin bearing plants, and has been discussed in detail by Loomis (Loomis and Battaile, 1966; Loomis *et al.*, 1979) and others (Anderson, 1968; Rhodes, 1977). Many techniques have been devised to minimise such destructive interactions (Loomis and Battaile, 1966; Anderson, 1968; Rhodes, 1977; Loomis *et al* 1979), and the most effective strategies for maximising extractable cyclase activity employ a combination of low temperature, reducing agents, chelators, and various adsorbent substances.

Buffers for the extraction of prenyltransferases and cyclases from herbaceous and woody plants generally contain adsorbents such as polyvinyl(poly)pyrrolidone, which is particularly effective in binding phenolics and thus minimising enzyme inactivation. Generally, high levels of multiple reducing agents (metabisulphite and ascorbate) are included in the medium to keep phenolics in the reduced state, and a sufficiently low pH is maintained to promote their binding to polyvinyl(poly)pyrrolidone. Likewise, XAD (a polystyrene resin) is routinely included in the extraction buffer to adsorb hydrophobic materials, such as oils, resins, waxes and other lipid material released on cell disruption, that can bind to and inactivate the enzymes, either directly or by aggregation. Although proteases are rarely a problem in the isolation of prenyltransferases and cyclases from higher plants, standard inhibitors can be included in the extraction buffer, such as TCPK and PMSF for serine proteases, or antipain for thiol proteases (i.e. papain type), and exogenous proteins such as bovine serum albumin (or casein) can be employed to spare the enzymes, but these materials must be removed in subsequent purification.

III. ISOLATION

A. Objectives and Problems

The primary objective of initial investigations is to maximise the level of extractable activity in order to adequately define reaction parameters, establish a reproducible standardised assay and permit subsequent purification. It is vital to select a tissue at the developmental stage that affords the greatest enrichment of the desired enzymes, and this selection can often be accomplished by simple analytical determination of product accumulation. Such measurements, by GLC or HPLC, also establish minimum pathway flux, a critical benchmark for evaluating extractable activity as outlined above.

For optimal and reproducible enzyme extractions, the isolation buffer constituents must be carefully chosen to minimise the deleterious effects of the various interfering substances that are found in plant tissues. Protein extraction in moderate ionic strength buffers, containing 150 mM sucrose, high levels of multiple reducing agents (metabisulphite and ascorbate), thiol protecting reagents, and both polyvinyl(poly)pyrrolidone and polystyrene as adsorbents has been found to be particularly effective in maximising extractable cyclase and transferase activity from herbaceous and woody plants.

During the past 15 years the techniques for extracting enzymes of isoprenoid metabolism from the leaf epidermis of higher plants have improved markedly. Originally, whole leaf extractions were employed to isolate these enzymes, but such extracts contained high levels of bulk proteins and interfering proteins (e.g. phosphohydrolases), as well as resins and phenolic substances (described previously) and were not particularly useful sources of prenyltransferases or cyclases. A more suitable procedure for members of the Lamiaceae and Asteraceae is the selective extraction of prenyltransferase and cyclase enzymes from glandular trichomes, which are specialised, multicellular, epidermal secretory organs (Gershenzon et al., 1992). The total amount of monoterpene cyclase activity per g of leaf and the specific activity per mg protein in several different types of cell-free extracts derived from peppermint leaves are presented in Table 9.1, as well as data on the total amount of geranyl

TABLE 9.1. Comparative activities of (−)-limonene cyclase and geranyl pyrophosphate phosphohydrolase in different types of cell-free extracts prepared from peppermint leaves.

Enzyme	Activity[a] (Specific activity)[b]		
	Whole leaf homogenate	Leaf surface extract	Extract of isolated secretory cells
(−)-Limonene cyclase	0.82 (0.11)	8.6 (4.3)	301 (727)
Geranyl pyrophosphate phosphohydrolase	20.4 (2.7)	1.3 (0.62)	0.12 (0.29)

[a] Total activity: nmol product h^{-1} (g fresh weight)$^{-1}$.
[b] Specific activity: nmol product h^{-1} (mg protein)$^{-1}$.
Data from Gershenzon et al. (1992).

pyrophosphate phosphohydrolase activity per g of leaf. The cell-free preparations from, purified gland cells clearly yield the highest levels of the pertinent extractable enzymes, yet they contain less total protein and considerably lower levels of interfering materials than do the two alternative extraction methods.

B. Protocol for Extraction of Whole Leaves

The young leaves are harvested, and washed with distilled H_2O and patted dry. The tissue is then frozen in liquid N_2 and ground to a fine powder with a mortar and pestle, and the powder stirred into a cold slurry consisting of one-third tissue weight of polyvinyl(poly)pyrrolidone and an equal tissue weight of hydrated XAD-4 polystyrene resin (Rohm and Haas Corp) in $5\,ml\,g^{-1}$ tissue of 50 mM MES buffer, pH 6.5, containing 5.0 mM dithiothreitol, 10 mM $Na_2S_2O_5$, 10 mM sodium ascorbate, 30 mM $MgCl_2$, and 20% (v/v) glycerol. After stirring for 60 min at 4°C, the mixture is filtered through cheesecloth, and the filtrate centrifuged at 27 000 × g for 20 min (pellet discarded) and then at 105 000 × g for 1 h to provide a supernatant used as the source of these operationally soluble enzymes. These preparations contain only modest amounts of cyclase activity (Table 9.1), and they are contaminated with some phenolics and phosphohydrolases (Table 9.1).

Alternatively, small preparations of 5–10 g of young leaves can be homogenised in an ice-filled glass homogeniser (Ten-Broeck) with a slurry ($4\,ml\,g^{-1}$ tissue) consisting of an equal weight of polyvinyl(poly)pyrrolidone in cold 0.1 M NaH_2PO_4 buffer, pH 6.5, containing 0.25 M sucrose, 50 mM $Na_2S_2O_5$, 10 mM sodium ascorbate, and 1.0 mM dithiothreitol. The crude homogenate is then slurried with an equal tissue weight of hydrated XAD-4 polystyrene resin (Rohm and Haas Corp) for 10 min, filtered through cheesecloth, and the filtrate centrifuged at 27 000 × g for 20 min (pellet discarded) and then at 105 000 × g for 1 h to provide a cell-free extract used as the source of prenyltransferases and cyclases. This procedure provides enzyme solutions that contain relatively lower levels of phenolic materials than do the previous preparations, but the method has limited utility because it is very labour intensive and time consuming, and it cannot be efficiently scaled up to process large amounts of leaves. Protocols similar to those for whole leaves, but employing both polyvinyl-pyrrolidone and polyvinyl(poly)pyrrolidone, have been developed for use with woody stems that are frozen and pulverised prior to extraction (Lewinsohn et al., 1991a).

C. Protocol for Extraction of Leaf Epidermis

Since epidermal oil glands (types of modified trichomes) constitute a major site of terpene biosynthesis in herbaceous plants (Croteau et al., 1981; Croteau and Johnson, 1984), methods were developed to selectively remove the contents of these structures, or of the epidermis itself, to provide extracts enriched in prenyltransferases and cyclases yet largely devoid of substances derived from the mesophyll.

Initially, the leaf epidermis was manually removed from the mesophyll by gentle brushing with a soft bristle toothbrush while submerging the leaves in chilled buffer. Homogenisation of the epidermal fragments, followed by centrifugation, affords a supernatant fraction containing the bulk of the cyclase activity of the leaf tissue, and relatively low levels of competing activities and interfering substances. This tedious and time-consuming procedure was of little preparative value, and the youngest leaves, that contain the highest levels of cyclase per g leaf tissue, were the most difficult to process.

Next, Gershenzon *et al.* (1987) developed a mechanised technique suitable for the selective extraction of the contents of epidermal oil glands from leaves of varying size, toughness and surface topography. This technique facilitates processing 20 g of tissue at one time, and it can be scaled up to accommodate 70 g of leaf tissue by fitting the drive unit (bead-beater base) with a 1 l polycarbonate grinding chamber (both from Bio-spec Products, Bartlesville, OK). The mechanised epidermal abrasion technique significantly reduces the level of extracted competing activities, such as geranyl pyrophosphate phosphohydrolase (Table 9.1), and this method permits large amounts of tissue to be expediently processed. This mechanised extraction process provides substantial enrichment of the relevant enzymes, but results in quite large volumes of buffer solution (15 ml buffer g^{-1} leaf tissue). Moreover, the extract generally contains co-extracted pigments, phenolics and resinous materials that are deleterious to enzyme activity, and that can adversely influence subsequent purification steps. For these reasons, a suitable technique is required to concentrate the dilute protein solution and to remove the interfering substances (see below).

D. Protocol for Extraction of Purified Gland Cells

Recently, the epidermal abrasion technique was modified to obtain intact glandular trichome cells from which prenyltransferases and cyclases are easily extracted (Gershenzon *et al.*, 1992). The method has been applied to leaves of peppermint and several other essential oil producing species. Young peppermint leaves are collected and soaked in ice-cold distilled water for 1 h immediately prior to extraction to maximise turgidity. Each preparation with 20 g leaf tissue requires 130 g glass beads (0.5 mm diameter), 2.5 g polyvinylpyrrolidone (PVP-40), 20 g Amberlite XAD-4 polystyrene resin (Rohm and Haas Corp) and 250 ml extraction buffer (25 mM HEPES, pH 7.3, containing 200 mM sorbitol, 10 mM sucrose, 5 mM dithiothreitol, 0.5 mM KH_2PO_4, 5 mM $MgCl_2$, 10 mM KCl and 0.6% (w/v) methyl cellulose). Surface abrasion is carried out at 4°C in the 300 ml polycarbonate chamber of the beadbeater (Bio-spec Products, Bartlesville, OK) for three 1-min pulses with the rotor speed controlled by a rheostat set at 90 V. The resulting suspension is permitted to cool for 1 min between pulses. The isolated glandular trichome secretory cell clusters are then separated from other constituents of the crude extract by sieving through a series of nylon meshes (Small Parts, Miami, FL). The clusters (approx. 60 μm in diameter) readily pass through meshes of 350 and 105 μm, and are collected on a mesh of 20 μm.

The cell clusters are initially washed on the 20 μm mesh with extraction buffer without methyl cellulose, and subsequently washed with the buffer to be used for sonication [25 mM KH_2PO_4 (pH 6.0), 250 mM sucrose, 10 mM sodium metabisulphite, 1 mM dithiothreitol, 1 mM EDTA, and 10 mM sodium ascorbate] and washed off the mesh into two 15-ml centrifuge tubes. The tubes are placed on ice for 30 min to permit the secretory cell clusters to settle out, and the upper few ml of buffer, containing leaf hairs and other low density contaminants, are removed prior to cell disruption.

Twenty mg of polyvinyl(poly)pyrrolidone and 40 mg Amberlite XAD-4 polystyrene resin (Rohm and Haas Corp.) are added to each tube in order to adsorb the oils and phenolic compounds released on sonication, which is carried out using a microprobe (Braun-Sonic 2000) at maximum power for three 1-min bursts. After sonication, the cell extract is filtered through 20 μm mesh and the filtrate centrifuged at 195 000 $\times$ g for 90 min to provide the supernatant used as the enzyme source.

These cell-free preparations contain the highest levels of prenyltransferase and cyclase activity (per g tissue) that have ever been obtained from mint leaves [$(-)$-limonene cyclase activity in gland sonicates is 30 times that in extracts prepared by the mechanised epidermal abrasion technique (Table 9.1)]. Furthermore, these preparations have much less total protein than do extracts prepared by other methods, and they contain much lower levels of phosphohydrolases (Table 9.1) and other deleterious materials derived from mesophyll cells. This method of enzyme preparation presents a number of other advantages: enzymes are removed from the leaves within intact glandular trichomes and are therefore not directly exposed to the initial isolation medium; the enzymes are obtained in a small volume of buffer; and the method is fully amenable to processing large amounts of leaf tissue without the need for concentration steps.

E. Special Considerations

Any study of higher plant prenyltransferases and cyclases requires a regular, relatively large, and consistent supply of tissue. Many studies have been carried out with common herbs (sage, thyme, mint, etc.) for which seed is commercially available. The plants must be propagated under controlled conditions (i.e. greenhouse or growth chamber) to ensure reproducibility, because the growth environment can have significant influence on enzyme activity levels. An optimal growth environment for each species must be determined empirically; however, our experience with members of the mint family suggests that long warm days (30°C, 16 h) and cool nights (15°C) with high light intensity (15 000 lux) promote terpene biosynthesis.

Many prenyltransferases and cyclases that have been studied in detail have been isolated from microbial systems. The principal physical and chemical properties of these enzymes are comparable to those of herbaceous plants, and the methods used for their isolation and purification are similar, although endogenous proteases may cause a greater problem (see for example Hohn and VanMiddlesworth, 1986; Brinkhaus and Rilling, 1988; Hohn and Plattner, 1989; and a general discussion by Croteau and Cane, 1985).

IV. ASSAY

A. Objectives and Problems

The primary goal in establishing a standard assay is that activity measurements be accurate and reliable. The pH, substrate concentration and concentration of cofactor should be roughly optimised in initial experiments. It is essential to carry out all measurements in the linear range regarding protein concentration and time. The reasons for deviation from linearity must be ascertained, and most often can be attributed to degradative losses of enzyme or inhibition, or losses of substrate to phosphohydrolases or to other competing reactions. It is important, at least in the early phases of the investigation, to determine the complete fate of the substrate in conversion to organic solvent-soluble (see below) and water-soluble products. Since the substrates are pyrophosphate esters, it is easiest to examine residual substrate and derived water-soluble metabolites (e.g. the catalytically inactive monophosphate esters) by ion-exchange chromatography (HPLC) or silica chromatography (TLC or HPLC). The most commonly encountered problem is competing phosphohydrolases and these enzymes may be inhibited by NaF (salts of molybdate, vanadate and arsenate are also occasionally employed) (Banthorpe *et al.*, 1975; Croteau and Karp, 1979); however, caution need be exercised to avoid inhibition of prenyltransferases and cyclases as well. Once the interferences are determined, it may be necessary to reoptimise conditions because the preliminary optimisation may be a result of reducing the interference(s), rather than promoting cyclase or prenyltransferase activity *per se*.

Since there are no rapid (e.g. spectrophotometric) assay methods available to measure prenyltransferase and cyclase activities, these assays are dependent upon the determination of product formation by methods that exploit the chemical or physical properties of target isoprenoids. Hydrophobic reaction products derived from ^{3}H- or ^{14}C-labelled pyrophosphorylated precursors can be extracted with organic solvent (e.g. pentane, diethyl ether) and measured by radioassay after suitable chromatographic purification (Croteau and Cane, 1985). Pyrophosphorylated cyclic products or products of the prenyltransferase reaction are generally hydrolysed to the corresponding alcohols to permit extraction and analysis as above. The allylic pyrophosphate substrates of cyclase and prenyltransferase reactions are generally radiolabeled at C-1 (to simplify the synthesis) to specific activities of 50–200 Ci mol^{-1} with ^{3}H (Hohn and VanMiddlesworth, 1986; Vogeli *et al.*, 1990) or 0.1–0.5 Ci mol^{-1} with ^{14}C (Dogbo and Camara, 1987; Brinkhaus and Rilling, 1988). General procedures for the synthesis and purification of these substrates can be found in Croteau and Karp (1976), Cane *et al.* (1982, 1984), and Davisson *et al.* (1986a, b). A number are now available from Amersham and American Radiolabeled Chemicals.

The mixture and types of cyclases, prenyltransferases and competing activities encountered, as well as inhibitors, are unique to each plant. Thus, the details of incubation conditions and chromatographic separations must be empirically determined for each specific application. In most assays, cyclic terpenoids are generated in nmol amounts and they are often volatile and labile (e.g. to oxidative decomposition or acid-catalysed rearrangement), thus unlabelled carrier compounds should be added as soon as possible to minimise degradation of the labeled product and losses on handling.

B. Incubation Procedures

Prenyltransferases and cyclases are generally assayed in buffers of moderate ionic strength (10–50 mM) in the pH range 6–8. This pH range may reflect the ionic state of the pyrophosphate ester–metal ion complex (Croteau, 1987), rather than an inherent property of the enzyme itself. Often, Good's buffers, such as MES, Tricine, or Mopso, or sodium or potassium phosphate buffers, or combinations of phosphate and Good's buffers are employed; however, phosphate and pyrophosphate salts (which may promote enzyme stability) are often inhibitory, with apparent K_i values of 100 mM and 0.1 mM, respectively. These enzymes require a divalent metal ion for catalytic activity, with Mg^{2+} or Mn^{2+} usually preferred, and they are stabilised by polyhydric alcohols (5–20% sucrose, sorbitol, or glycerol) and thiol protecting reagents such as dithiothreitol (0.5–1 mM) or mercaptoethanol (5–10 mM).

Typically, the enzyme preparation is dialysed or desalted to assay conditions, or an aliquot is diluted into assay buffer. It is important that the incubation time be short relative to the enzyme half-life and that the reaction kinetics be linear. Cyclase reactions are initiated by addition of substrate (to saturating substrate concentrations, generally 10–50 μM) and incubation is often carried out from 15 min to several hours at 27–37°C in Teflon-lined screw-capped vials of 7 ml capacity. The prenyltransferase assay requires two substrates (the prenyl donor and the isopentenyl pyrophosphate acceptor) and the reaction is initiated by addition of both substrates to saturating concentrations (10–500 μM). Incubation is carried out under conditions (time and temperature) similar to those described for cyclases. Since allylic pyrophosphates are prone to solvolytic decomposition, especially at acidic pH, boiled controls should be subjected to identical incubation conditions and subsequent analysis.

C. Isolation of Products from Cyclase Assays

If a complex mixture of cyclic products is anticipated, it is generally advisable to carry out a group separation of hydrocarbons from oxygenated compounds as the initial step of product analysis. This separation is accomplished by the selective adsorption of oxygenated products on silicic acid. After incubation, the reaction mixture is chilled to reduce losses of volatile products, and extracted with an equal volume of pentane (or hexane) which is passed over a short (0.5 × 3 cm) column of silica overlaid with $MgSO_4$. The extraction is repeated with another portion of pentane and the column is washed with this extract and with a further 0.5 ml of pentane. The combined pentane eluate contains the terpene hydrocarbons which do not bind to silicic acid under these conditions. The original reaction mixture is subsequently re-extracted with an equal portion of diethyl ether which is passed over the same silicic acid column to recover the oxygenated products. The column is washed with an additional 1.5 ml of ether to permit complete recovery of compounds at least as polar as geraniol. This procedure allows essentially complete separation of even relatively non-polar oxygen-containing products, such as 1,8-cineole, from monoterpene dienes and trienes (limonene, myrcene, etc.).

To assay the formation of cyclic phosphorylated products and to assess the residual level of phosphorylated substrate in the assay, traces of ether are removed from the extracted reaction mixture under a stream of nitrogen, followed by the addition of 3 units of wheat germ acid phosphatase and 2 units of potato apyrase in 1.0 ml of 0.1 M sodium acetate, pH 5 (or 3 units of calf intestine alkaline phosphatase and 2 units of apyrase in 1.0 ml of 0.1 M Tris-Cl, pH 8.5). The vial is resealed and incubated for several hours at 30°C. The chilled reaction mixture is then extracted twice with 1.5 ml portions of ether to recover the liberated alcohols (the residual aqueous phase should be examined to insure complete hydrolysis), and the combined extract dried by passage through a 0.5 × 3 cm column of anhydrous Na_2SO_4. In preparation for subsequent radiochromatographic analysis, appropriate unlabelled terpene standards are added to the various extracts containing hydrocarbons, oxygenated products, and alcohols liberated by phosphatase treatment, and the samples, after radioassay of an aliquot, are concentrated to a convenient volume under a stream of N_2 at $0°C$.

In assaying the formation of saturated products (1,8-cineole, borneol, patchoulol, etc.), it is often convenient to treat the extracts with OsO_4 to convert the double bonds of olefinic products to the corresponding diols, thereby simplifying subsequent chromatographic analysis. If only one or a few cyclic products is anticipated in an assay, or if preliminary work has revealed it to be unnecessary to examine the complete spectrum of products, it is often sufficient to carry out only the initial solvent partitioning step of the assay. Small amounts of enzyme can be quickly monitored by several microscale assays that have been developed (Hohn and VanMiddlesworth, 1986; Vogeli *et al.*, 1990; Lewinsohn *et al.*, 1991a).

D. Isolation of Products from Prenyltransferase Assays

For the 'acid lability' assay (Holloway and Popjak, 1967; Rilling 1985), which depends upon the solvolysis of enzymatically generated, labelled allylic pyrophosphates in the presence of radioactive isopentenyl pyrophosphate (which is stable in the presence of acid), 1 ml of pentane is added to the ice-chilled incubation mixture, followed by addition of 1 M HCl (to a final concentration of about 0.25 M HCl), and the mixture is shaken for 30 min at $30°C$. Following chilling in ice, addition of NaCl to saturation, vigorous mixing, and centrifugation to separate phases, an aliquot of the pentane layer (containing solvolysis products of allylic pyrophosphates) is taken for determination of radioactivity. Calibration of the assay with 1-^3H labelled C_5- to C_{15}-allylic pyrophosphates (5 nmol) indicates that $>95\%$ of the label originally present as allylic pyrophosphate ester is recovered as pentane-soluble solvolysis products. Following removal of the remaining pentane layer and re-extraction of the reaction mixture with an additional 1 ml of pentane, the combined extract is diluted with about 1 mg each of the appropriate carrier standards, dried over anhydrous Na_2SO_4, and concentrated under vacuum in preparation for subsequent analysis of these solvolysis products. Isopentenyl pyrophosphate isomerase activity (Satterwhite, 1985) is determined by incubation of isopentenyl pyrophosphate alone, in the absence of allylic co-substrate. These controls should always be included if the presence of isomerase is suspected.

For the 'enzymatic hydrolysis' assay (Croteau and Karp, 1979; Poulter and Rilling, 1981; Koyama *et al.*, 1985), which involves the phosphatase-catalysed hydrolysis of the labeled pyrophosphate esters and release of the corresponding prenols, 1 ml of a 0.2 M Tris solution, pH 8.0, containing 2 units each of potato apyrase and calf intestine alkaline phosphatase, is added to the ice-chilled incubation mixture and 1 ml of pentane is added as an overlay, and the sample is incubated overnight at $30°C$. Following enzymatic hydrolysis, the liberated prenols are extracted into diethyl ether (2×1 ml, with vigorous shaking and centrifugation to separate phases as before) and the ether extract treated with a few mg of $NaBH_4$ to convert any aldehydes present to the corresponding alcohols (the latter precaution is necessary because commercially available alkaline phosphatase may contain low but variable amounts of an apparent alcohol oxidase activity). The reaction mixture is passed through a short column of Na_2SO_4, and 1 mg each of appropriate prenol carriers is added prior to aliquot counting and concentration of the sample in preparation for further analysis.

Calibration of the assay with [1-^{3}H]geranyl pyrophosphate (5 nmol) gives about 85% yield of [1-^{3}H]geraniol; the lack of quantitative recovery probably results, at least in part, from loss of label on oxidation to the aldehyde.

E. Chromatographic Separation of Products

A number of chromatographic procedures can be employed for the separation of individual components in mixtures of terpene hydrocarbons and oxygenated terpenoids (as well as phosphatase-derived products); TLC, GLC and HPLC are well-suited for most routine assays. Thin layer chromatography on silica gel G (developed with hexane containing up to 5% ether) resolves some olefins, but argentation chromatography on 12% AgNO$_3$-silica gel G [(developed with hexane–benzene–ether (50/50/1;v/v/v)] is far superior for separating complex olefin mixtures (Gambliel and Croteau, 1982). Thin layer chromatography on silica gel G using hexane containing ethyl acetate (5–30%, depending on the polarity of the products) as the developing solvent is suitable for the separation of many oxygen-containing terpenoids (Battaile *et al.*, 1961). The products can be readily located by spraying with a 0.2% ethanolic solution of 2,7-dichlorofluorescein and visualisation under UV light. Products of interest may be radioassayed by transferring the appropriate section of gel directly into a scintillation vial containing fluor, or the powder may be transferred to a glass wool-plugged Pasteur pipette and the product eluted with dry ether or other solvent for subsequent radioassay or further analysis. Thin layer chromatography radioscanners are unsuitable for volatile terpenoids since evaporative losses during manipulation are variable and unacceptably high.

Assays should be calibrated with labelled standard or, if a standard is not available, by running the initially purified sample back through the entire procedure to define the limits of maximum loss. The recovery of monoterpenes is in the range of 70–95% depending on the nature of the product and the type of assay (Croteau and Karp, 1977a; Gambliel and Croteau, 1982). Without added carrier, the bulk of the monoterpene olefins generated in an assay may be lost during isolation and separation; however, the recovery of sesquiterpenes is somewhat higher.

A particularly powerful technique for the separation and analysis of complex mixtures of isoprenoids is radio-GLC, whereby individual components separated on the chromatographic column are determined directly in a heated flow-through proportional counter, or are similarly monitored in an ambient flow-through counter after conversion to ^{14}CO$_2$ and/or ^{3}H$_2$ in a combustion–reduction train (Croteau and Cane, 1985; Satterwhite and Croteau, 1988). Even crude mixtures containing both hydrocarbons and oxygenated products can be separated and analysed at a sensitivity of a few hundred dpm per component of injected sample, when appropriate chromatography columns and temperature programming are employed.

For the separation of phosphorylated products, TLC is often employed, either in the adsorption (silica gel H) or partition (cellulose) mode (Sofer and Rilling, 1969; Croteau and Ronald, 1983). Also, reversed-phase and ion-exchange chromatography find occasional use, the latter particularly at the semi-preparative scale (Croteau and Karp, 1977b). All of these methods are readily adapted to HPLC.

F. Product Identification and Determination of Labelling Pattern

Since chromatographic coincidence is insufficient to establish the identity (radio-chemical purity) of a labelled product, additional criteria must be applied for confirmation of structure, at least initially. Seldom is enough biosynthetic material available to permit spectroscopic analysis, and thus co-crystallisation to constant specific radioactivity of the suspected product (or a suitable derivative, see below) with an authentic standard is most often employed for this purpose. Olefins are generally converted to crystalline halides or nitrosohalides, or oxidised to alcohols or carbonyl compounds and recrystallised as the respective esters, urethanes, or conventional carbonyl derivatives. Classical methods for the preparation of numerous derivatives of terpenoid olefins, alcohols, aldehydes and ketones have been described (Sterrett, 1975).

Many plant species are known to produce enantiomeric terpenoids, and it therefore may be necessary to determine the stereochemical composition of the cyclic products generated by cell-free extracts. Sage, for example, produces (+)- and (−)-limonene, (+)- and (−)-α-pinene and (+)- and (−)-camphene, as well as (+)- and (−)-bornyl pyrophosphate, the enantiomers being synthesised by separate cyclases (Croteau, 1984; Gambliel and Croteau, 1984). Conversion of enantiomers to diastereomeric derivatives, followed by fractional crystallisation or chromatographic separation, may be employed (Gambliel and Croteau, 1982). Direct separation of enantiomers by GLC or HPLC on chiral phase columns (Konig et al., 1982; Pirkle and Pochapsky, 1987; Satterwhite and Croteau, 1987; Satterwhite and Croteau, 1988; Davin et al., 1991) offers a possible alternative. Capillary radio-gas chromatographic methods (Gross et al., 1980; Croteau and Satterwhite, 1990) must be used for the former if insufficient biosynthetic product is available for mass detection. Radiochemical co-crystallisation offers an unequivocal, but tedious, approach to the resolution of labelled enantiomers when optically pure carriers are available (Gambliel and Croteau, 1984).

Finally, it is necessary to establish the labeling patterns of products derived from specifically labelled precursors in order to confirm basic cyclisation and rearrangement schemes. Failure to follow such fundamental precautions has too often led to claims which have later been withdrawn in the light of more careful work. A detailed description of the procedures which have been used for the degradation of terpenoid metabolites is well beyond the scope of this chapter. However, routes for determining the location of label in several monoterpenes derived from 1-^{3}H-labelled acyclic precursors have been described (Croteau, 1981, 1984). Similarly, routes for the degradation of a number of sesquiterpenes derived from specifically labelled [^{3}H]- and [^{14}C]farnesyl pyrophosphate and other precursors have been reported (Cane, 1981; Cane et al., 1981a, b; Cane and Tillman, 1983).

Relatively few elongation products are anticipated in prenyltransferase assays, and the corresponding prenols, obtained by enzymatic hydrolysis and separation by chain length, are conveniently determined by conversion to the crystalline phenylurethanes (Bates et al., 1963). Alternatively, oxidation of the allylic alcohol to the corresponding aldehyde (e.g. with MnO_2; Attenburrow et al., 1952; Corey, et al., 1968) permits the preparation of carbonyl derivatives such as the semicarbazone (Sterrett, 1975), although one-half the radioactivity of the 1-^{3}H-labelled products will be lost.

V. PURIFICATION

A. Objectives and Problems

The paramount objective of prenyltransferase and cyclase purification is to maximise the yield of homogeneous enzyme by the development of an efficient and reproducible fractionation scheme. This class of enzymes shares many common properties, such as molecular weight, pI, degree of hydrophobicity, etc., that can be exploited in purification. Likewise, there are common pitfalls in the initial extraction step (described previously), as well as similar complications associated with handling and purifying these enzymes. We present here an overview of preliminary considerations pertinent to the fractionation of these enzymes, and subsequently discuss the advantages and limitations of relevant purification techniques.

Prior to the initial testing of purification procedures, it is important to optimise the level of extractable activity and to select a tissue with a high abundance of the desired enzyme (described previously). Another fundamental consideration is the establishment of a standard assay that is streamlined to facilitate the rapid evaluation of numerous samples. For example, performing a group separation of hydrocarbons from oxygenated compounds is often sufficient to monitor the course of purification of an enzyme that produces a single product. An additional preliminary concern is the establishment of suitable procedures for storing the enzyme at various stages of purity. A plethora of methods for storing prenyltransferases and cyclases have been described (Barnard and Popjak, 1981; Cane *et al.*, 1981a; Sasiak and Rilling, 1988; Light and Dennis, 1989; Vogeli *et al.*, 1990). As a rule, precipitation methods are suitable only for protein solutions with concentrations greater than $1 \, mg \, ml^{-1}$. Other relevant techniques include storage at $-20°C$ (or $-80°C$) in pH 6.5–7.0 buffer containing 15–25% polyhydric alcohol, $5 \, mM \, KH_2PO_4$, and $1 \, mM$ dithiothreitol, or lyophilisation (Alonso and Croteau, 1991). There is no method currently available that is suitable for all enzymes of terpenoid metabolism, because storage stability varies enormously with the particular enzyme, as well as with the degree of purification, and the optimal storage conditions must therefore be determined empirically.

B. Prenyltransferase and Cyclase Purification

A diverse assortment of purification techniques have been devised to purify these enzymes because each plant source contains a unique mixture of cyclases, prenyltransferases, competing activities and contaminating materials. Generally, one or two preliminary purification steps are employed to remove interfering substances (and to concentrate the crude homogenate if required), followed by several high-resolution steps to achieve homogeneity with the highest overall recovery of activity. The selectivity and capacity of each technique ultimately determines if it may be used early in the purification scheme, or is relegated to the terminal fractionation steps. High capacity processes, such as precipitation, and ion-exchange, dye–ligand, and hydrophobic interaction chromatography, are often employed early in the isolation scheme, whereas chromatofocusing and isoelectric focusing, hydroxyapatite, affinity and gel permeation chromatography, and gel electrophoresis, which have lower capacities, are typically used in the final stages of purification.

It is generally desirable to sequence purification steps so that the enzyme is simply collected from one step and immediately subjected to the next and does not require intermediate treatments such as dialysis, ultrafiltration, etc. In fact, avoiding intermediate manipulation is often of critical importance in prenyltransferase and cyclase purification because of the lability of these enzymes. For example, 5-epi-aristolochene synthase undergoes substantial losses of activity during ultrafiltration, presumably as a consequence of the high degree of hydrophobicity of this enzyme (Vogeli *et al.*, 1990). A gentle method for small-scale volume reduction is immersion of a dialysis bag containing the partially purified enzyme in powdered sucrose or polyethylene glycol (mol. mass 20 + kDa). For changing buffer conditions, dialysis or desalting on a low molecular weight cut-off gel permeation column is preferred. Enzyme solutions should always be handled at 0–4°C in the presence of thiol protecting reagents and 5–20% polyol, if these conditions do not compromise chromatographic separations.

C. Concentration of the Initial Extract

An efficient technique is often needed for concentration of dilute enzyme extracts. Precipitation steps (with ammonium sulphate, polyethylene glycol, etc.) are often employed and, although these methods lack selectivity, the enzyme precipitate is often remarkably stable. Another useful strategy is to employ a negative precipitation step (e.g. with protamine sulphate or streptomycin sulphate: Frost and West, 1977; Moesta and West, 1985); however, the enzyme of interest remains in a dilute solution that requires subsequent concentration.

Batch-wise anion-exchange chromatography is another technique frequently employed to concentrate dilute enzyme extracts; however, both the proteins of interest and phenolic substances tend to bind tightly to this matrix and elute under similar conditions. Dye–ligand chromatography is more suitable than anion-exchange because the recoveries are better and most interfering materials, including phenolics, can be easily removed (Lanznaster and Croteau, 1991). In the purification of γ-terpinene synthase, Matrex Gel Red A provides about five-fold purification with ten-fold concentration, and removes phenolic materials from the extract that would otherwise reduce resolution and recovery in the subsequent isoelectric focusing step (Alonso and Croteau, 1991). If not removed, phenolic substances can also rapidly damage HPLC columns used for protein purification.

D. Purification to Homogeneity

1. Anion-exchange chromatography

Anion-exchange chromatography is almost always employed as a step in the purification of prenyltransferases and cyclases. This technique affords high capacity and resolution, and is well suited for the purification of prenyltransferases and cyclases because these enzymes generally possess lower pI values than most contaminating proteins. Thus, much of the bulk protein can be removed when anion-exchange chromatography is carried out under the lowest pH conditions at which the prenyltransferases and cyclases bind and still retain activity. Several investigators have employed two anion-exchange steps, once as an early concentration step and again on a different type of matrix, or using a different type of elution, to obtain high resolution (Moesta and West, 1985; Hohn and VanMiddlesworth, 1986; Munck and Croteau, 1990). The selectivity of anion-exchange chromatography is illustrated by the resolution of several cyclase activities (of similar mass) that were not resolvable by other modes of chromatography (Duncan and West, 1981; Dehal and Croteau, 1988; Gijzen *et al.*, 1991).

2. Gel permeation chromatography

Gel permeation chromatography (gel filtration, molecular sieving) is well suited for the purification of labile, hydrophobic enzymes because proteins remain in solution at all times and do not bind to the chromatographic matrix. However, the absence of such binding limits capacity and resolution, and so this method is often employed in the final steps of purification. Gel permeation chromatography is a convenient method for resolving cyclases and prenyltransferases at the extremes of the molecular weight range encountered (50–100 kDa), and this range should be borne in mind when selecting a gel filtration matrix.

3. Hydroxyapatite

Rapid batch-wise adsorption/desorption with hydroxyapatite is useful for concentrating dilute protein solutions early in the purification scheme; however hydroxyapatite columns (crystalline calcium phosphate) have limited capacity and are therefore generally employed later in the purification scheme (Dudley *et al.*, 1986; Cane and Pargellis, 1987; Sasiak and Rilling, 1988; Vogeli *et al.*, 1990). Hydroxyapatite chromatography has proved to be essential in some cases for removing contaminating proteins that are not eliminated by other methods (Munck and Croteau, 1990).

4. Hydrophobic interaction chromatography

Hydrophobic interaction chromatography is used sparingly for the purification of prenyltransferases and cyclases because the method does not always provide sharp resolution and, most importantly, because these hydrophobic enzymes sometimes bind so tenaciously to the matrix that they can only be eluted with considerable loss of activity. Usually, the recoveries are acceptable with the less hydrophobic matrices (C-1 or C-2), although selectivity is reduced (Munck and Croteau, 1990; Alonso and Croteau, 1991). Phenyl-substituted matrices have been used on occasion (Vogeli *et al.*, 1990; Gijzen *et al.*, 1991). A unique feature of hydrophobic interaction chromatography is that this step can be placed anywhere in the purification scheme, because volume reduction or buffer change is not required prior to sample application (i.e. only sufficient salt is added to the solution to ensure binding).

5. Dye–Ligand Chromatography

Dye–ligand interaction chromatography was discussed earlier as an efficient means for the concentration and preliminary purification of prenyltransferases and cyclases

(Lanznaster and Croteau, 1991); however, this powerful technique can also provide exceptionally high-fold purifications very late in the fractionation sequence. The complex interactions between the dyes and proteins are thought to include hydrogen bonding, as well as hydrophobic and electrostatic effects (Scopes, 1987). Consequently, there are many variables, including temperature, pH, ionic strength, and metal ions, that can be exploited to optimise the method for a particular application. Multiple dye–ligand chromatography steps have proved crucial to the purification of two disparate isoprenoid biosynthetic enzymes; casbene synthase (Moesta and West, 1985) and geranylgeranyl pyrophosphate synthase (Brinkhaus and Rilling, 1988).

6. Isoelectric focusing and chromatofocusing

Isoelectric focusing and chromatofocusing are high resolution techniques that separate proteins in solution based on isoelectric point (pI). These methods are sometimes key steps in the purification of prenyltransferases (Sasiak and Rilling, 1988) and cyclases (Munck and Croteau, 1990; Vogeli et al., 1990; Alonso and Croteau, 1991), because these types of enzymes generally focus at pH values that are lower than those of most bulk proteins. A major disadvantage of these methods is that very modest recoveries of activity are often obtained. This shortcoming probably results from extended exposure of the enzyme to pH values near the pI, at which the enzyme is presumably least stable.

7. Affinity chromatography

This technique is potentially one of the most powerful purification methods because it should be possible to design highly selective affinity ligands for prenyltransferase and cyclase binding. The major limitation of this mode of chromatography is that each putative affinity adsorbent must be synthesised in good yield and tested with the protein of interest under a variety of conditions.

An important paper by Bartlett et al. (1985) describes the use of an affinity column of geranylmethylphosphonate-agarose to purify to homogeneity from crude samples the farnesyl pyrophosphate synthase of two diverse organisms. An aminophenethyl pyrophosphate affinity column was employed to purify both isopentenyl pyrophosphate isomerase and geranylgeranyl pyrophosphate synthase from Capsicum chromoplasts (Dogbo and Camara, 1987). Conversely, the cis-polyisoprene (rubber) prenyltransferase did not bind to the geranylmethylphosphonate-agarose column (Bartlett et al., 1985) which was so effective in the isolation of farnesyl pyrophosphate synthases, or to the cis-analogue of this column (Light and Dennis, 1989). Matrices of this type, with the phosphate/pyrophosphate function facing inward, do not work very well for the purification of monoterpene and sesquiterpene cyclases. Because affinity chromatography is a somewhat higher risk venture that requires synthesis and concomitant testing, it is often more prudent to first attempt the desired purification by modes of chromatography that employ commercially available matrices.

8. Non-denaturing PAGE

Nor-denaturing PAGE is a very selective purification technique because fractionation is based on net charge as well as size and shape of the proteins. Recently, a wound-inducible monoterpene synthase, (−)-pinene cyclase from *Abies grandis*, was purified to a very high degree by discontinuous non-denaturing PAGE at neutral pH (Lewinsohn *et al.*, 1992). Other discontinuous non-denaturing PAGE systems that employ higher pH are not useful for this purpose because the recoveries of activity are unacceptably low (Munck and Croteau, 1990; Alonso and Croteau, 1991). The system of Lewinsohn *et al.* (1992) has considerable potential for the purification and analysis of cyclases and prenyltransferases because it affords excellent recovery of activity (50–100%) in addition to achieving outstanding resolution.

E. Special Considerations

1. Criteria for purity

Rarely do prenyltransferase or cyclase purifications provide sufficient protein for analytical ultracentrifugation, or for other analytical methods that require samples in the mg range. Although chromatographic evidence may provide a good indication of purity, analytical SDS-PAGE is generally considered the method of choice in establishing an 'apparently' homogeneous protein. A two-dimensional gel system with non-denaturing PAGE (Lewinsohn *et al.*, 1992) in the first dimension (with coincidence of activity and protein) and SDS-PAGE (Laemmli, 1970) in the second dimension (followed by Coomassie Blue staining, or silver staining for greater sensitivity) can provide the best evidence of purity when limited amounts of protein are available. *N*-Terminal sequence analysis (if unblocked) can usually distinguish different proteins of identical mobility, and may be sufficiently sensitive to reveal contaminating proteins at the 5% level.

2. Characterisation

Early studies with the prenyltransferases and cyclases of plants were carried out with partially purified preparations that were adequate for establishing the mechanistic and stereochemical nature of the reactions catalysed. Furthermore, from this work it became very apparent that this class of enzymes shared many properties (kinetic constants, molecular weight range, pH optimum, pI value, hydrophobicity, cofactor requirement, etc.).

Extraordinary progress in the purification of prenyltransferases and cyclases has been made over the last few years; a number of enzymes from each class have been purified to homogeneity, and prenyltransferase (Anderson *et al.*, 1989; Carattoli *et al.*, 1991) and sesquiterpene cyclase (Hohn and Beremand, 1989) genes have been cloned. The recently developed procedure for obtaining prenyltransferases and cyclases from isolated trichome cells has circumvented many of the problems that were previously encountered with enzyme preparations from essential oil- and resin-producing species, and advances in chromatographic and electrophoretic methods will permit a wide range of these enzymes of isoprenoid metabolism to be purified and studied at the molecular level. Thus, many key questions regarding active site structure and function, and the localisation and regulation of these catalysts, are likely to be addressed in the coming years.

ACKNOWLEDGEMENTS

The research described from the authors' laboratory was made possible by grants from the US Department of Agriculture (91-37302-6311), the National Institutes of Health (GM-31354) and the US Department of Energy (DE-FG06-91ER13869), and by Project 0268 from the Agricultural Research Center, Washington State University.

REFERENCES

Alonso, W. R. and Croteau, R. (1991). *Arch. Biochem. Biophys.* **286**, 511–517.
Anderson, J. W. (1968). *Phytochemistry* **7**, 1973–1988.
Anderson, M. S., Yarger, J. G., Burck, C. L. and Poulter, C. D. (1989). *J. Biol. Chem.* **264**, 19176–19184.
Attenburrow, J., Cameron, A. F. B., Chapman, J. H., Evans, R. M., Hems, B. A., Jansen, A. B. A. and Walker, T. (1952). *J. Chem. Soc.*, 1094–1111.
Banthorpe, D. V., Chaudhry, A. R. and Doonan, S. (1975). *Z. Pflanzenphysiol.* **76**, 143–154.
Barnard, G. F. and Popjak, G. (1981). *Biochim. Biophys. Acta* **661**, 87–99.
Bartlett, D. L., King, C.-H. R. and Poulter, C. D. (1985). *In* "Methods in Enzymology", Vol. 110 (J. H. Law and H. C. Rilling eds), pp. 171–183. Academic Press, New York.
Bates, R. B., Gale, D. M. and Gruner, B. J. (1963). *J. Org. Chem.* **28**, 1086–1089.
Battaile, J., Dunning, R. L. and Loomis, W. D. (1961). *Biochim. Biophys. Acta* **51**, 538–544.
Belingheri, L., Pauly, G., Gleizes, M. and Marpeau, A. (1988). *J. Plant Physiol.* **132**, 80–85.
Brinkhaus, F. L. and Rilling, H. C. (1988). *Arch. Biochem. Biophys.* **266**, 607–612.
Cane, D. E. (1981). In "Biosynthesis of Isoprenoid Compounds", Vol. 1 (J. W. Porter and S. L. Spurgeon, eds), pp. 283–374. Wiley, New York.
Cane, D. E. (1990). *Chem. Rev.* **90**, 1089–1103.
Cane, D. E. and Pargellis, C. (1987). *Arch. Biochem. Biophys.* **254**, 421–429.
Cane, D. E. and Tillman, A. M. (1983). *J. Am. Chem. Soc.* **105**, 122–124.
Cane, D. E., Iyengar, R. and Shiao, M.-S. (1981a). *J. Am. Chem. Soc.* **103**, 914–931.
Cane, D. E., Swanson, S. and Murthy, P. P. N. (1981b). *J. Am. Chem. Soc.* **103**, 2136–2138.
Cane, D. E., Saito, A., Croteau, R., Shaskus, J. and Felton, M. (1982). *J. Am. Chem. Soc.* **104**, 5831–5833.
Cane, D. E., Abel, C. and Tillman, A. M. (1984). *Bioorg. Chem.* **12**, 312–328.
Carattoli, A., Romano, N., Ballario, R., Morelli, G. and Macino, G. (1991). *J. Biol. Chem.* **266**, 5854–5859.
Corey, E. J., Gilman, N. W. and Ganem, B. E. (1968). *J. Am. Chem. Soc.* **90**, 5616–5617.
Croteau, R. (1981). In "Biosynthesis of Isoprenoid Compounds", Vol. 1 (J. W. Porter and S. L. Spurgeon, eds), pp. 225–282. Wiley, New York.
Croteau, R. (1984). *In* "Isopentenoids in Plants" (W. D. Nes, G. Fuller and L. S. Tsai, eds), pp. 31–64. Dekker, New York.
Croteau, R. (1987). *Chem. Rev.* **87**, 929–954.
Croteau, R. (1988). *In* "Flavors and Fragrances: A World Perspective. Proceedings of the 10th International Congress of Essential Oils, Fragrances and Flavors, Washington, DC, U.S.A." (B. M. Lawrence, B. D. Mookherjee and B. J. Willis, eds), pp. 65–84. Elsevier, Amsterdam.
Croteau, R. and Cane, D. E. (1985). *In* "Methods in Enzymology", Vol. 110 (J. H. Law and H. C. Rilling, eds), pp. 383–405. Academic Press, New York.

Croteau, R. and Johnson, M. A., (1984). *In* "Biology and Chemistry of Plant Trichomes" (P. Healy, E. Rodriguez and I. Mehta, eds), pp. 133–186. Plenum, New York.

Croteau, R. and Karp, F. (1976). *Arch. Biochem. Biophys.* **176**, 734–746.

Croteau, R. and Karp, F. (1977a). *Arch. Biochem. Biophys.* **179**, 257–265.

Croteau, R. and Karp, F. (1977b). *Arch. Biochem. Biophys.* **184**, 77–86.

Croteau, R. and Karp, F. (1979). *Arch. Biochem. Biophys.* **198**, 523–532.

Croteau, R. and Purkett, P. T. (1989). *Arch. Biochem. Biophys.* **271**, 524–535.

Croteau, R. and Ronald, R. C. (1983). *In* "Chromatography: Fundamentals and Applications of Chromatographic and Electrophoretic Methods, Part B: Applications" (E. Heftman, ed.), pp. 147–189. Elsevier, Amsterdam.

Croteau, R. and Satterwhite, D. M. (1990). *J. Chromatogr.* **500**, 349–354.

Croteau, R., Felton, M. and Ronald, R.C. (1980). *Arch. Biochem. Biophys.* **200**, 524–533.

Croteau, R., Felton, M., Karp, F. and Kjonass, R. (1981). *Plant Physiol.* **67**, 820–824.

Davin, L. B., Lewis, N. G. and Umezawa, T. (1991). *In* "Recent Advances in Phytochemistry", Vol. 25 (N. H. Fischer, M. B. Isman and H. A. Stafford, eds), pp. 75–112. Plenum, New York.

Davisson, V. J., Zabriskie, T. M. and Poulter, C. D. (1986a). *Bioorg. Chem.* **14**, 46–54.

Davisson, V. J., Woodside, A. B., Neal, T. R., Stremler, K. E., Muehlbacher, M. and Poulter, C. D. (1986b). *J. Org. Chem.* **51**, 4768–4779.

Dehal, S. S. and Croteau, R. (1988). *Arch. Biochem. Biophys.* **261**, 346–356.

Dogbo, O. and Camara, B. (1987). *Biochim. Biophys. Acta* **920**, 140–148.

Dudley, M. W., Green, T. R. and West, C. A. (1986). *Plant Physiol.* **81**, 343–348.

Duncan, J. D. and West, C. A. (1981). *Plant Physiol.* **68**, 1128–1134.

Falk, K. L., Gershenzon, J. and Croteau, R. (1990). *Plant Physiol.* **93**, 1559–1567.

Fall, R. R. and West, C. A. (1971). *J. Biol. Chem.* **246**, 6913–6928.

Frost, R. G. and West, C. A. (1977). *Plant Physiol.* **59**, 22–29.

Gambliel, H. and Croteau, R. (1982). *J. Biol. Chem.* **257**, 2335–2342.

Gambliel, H. and Croteau, R. (1984). *J. Biol. Chem.* **259**, 740–748.

Gershenzon, J., Duffy, M., Karp, F. and Croteau, R. (1987). *Anal. Biochem.* **163**, 159–164.

Gershenzon, J. and Croteau, R. (1990). *In* "Recent Advances in Phytochemistry", Vol. 24 (G. H. N. Towers and H. Stafford, eds), pp. 99–160. Plenum, New York.

Gershenzon, J., McCaskill, D., Rajaonarivony, J. I. M., Mihaliak, C., Karp, F. and Croteau, R. (1992). *Anal. Biochem.* **200**, 130–138.

Gijzen, M., Lewinsohn, E. and Croteau, R. (1991). *Arch. Biochem. Biophys.* **289**, 267–273.

Gleizes, M., Pauly, G., Carde, J.-P., Marpeau, A. and Bernard-Dagan, C. (1983). *Planta* **159**, 373–381.

Gross, D., Gutekunst, H., Blaser, A. and Hambock, H. (1980). *J. Chromatogr.* **198**, 389–396.

Hohn, T. M. and Beremand, P. D. (1989). *Gene*, **79**, 131–138.

Hohn, T. M. and Plattner, R. D. (1989). *Arch. Biochem. Biophys.* **272**, 137–143.

Hohn, T. M. and VanMiddlesworth, F. (1986). *Arch. Biochem. Biophys.* **251**, 756–761.

Holloway, P. W. and Popjak, G. (1967). *Biochem J.* **104**, 57–70.

Jennings, W. and Shibamoto, T. (1980). "Quantitative Analysis of Flavor and Fragrance Volatiles by Glass Capillary Gas Chromatography". Academic Press, New York.

Konig, W. A., Francke, W. and Benecke, I. (1982). *J. Chromatogr.* **239**, 227–231.

Koyama, T., Fujii, H. and Ogura, K. (1985). *In* "Methods in Enzymology", Vol. 110 (J. H. Law and H. C. Rilling, eds), pp. 153–155. Academic Press, New York.

Kreuz, K. and Kleinig, H. (1984). *Eur. J. Biochem.* **141**, 531–535.

Laemmli, U. K. (1970). *Nature* **227**, 680–685.

Lanznaster, N. and Croteau, R. (1991). *Protein Express. Purif.* **2**, 69–74.

Lewinsohn, E., Gijzen, M., Savage, T. J. and Croteau, R. (1991a). *Plant Physiol.* **96**, 38–43.

Lewinsohn, E., Gijzen, M. and Croteau, R. (1991b). *Plant Physiol.* **96**, 44–49.

Lewinsohn, E., Gijzen, M. and Croteau, R. (1992). *Arch. Biochem. Biophys.* **293**, 167–173.

Light, D. R. and Dennis, M. S. (1989). *J. Biol. Chem.* **264**, 18589–18597.

Loomis, W. D. and Battaile, J. (1966). *Phytochemistry* **5**, 423–438.

Loomis, W. D. (1974). *In* "Methods in Enzymology", Vol. 31 (S. Fleischer and L. Packer, eds), pp. 528–544. Academic Press, New York.

Loomis, W. D., Lile, J. D., Sandstorm, R. P. and Burbott, A. J. (1979). *Phytochemistry* **18**, 1049–1054.

Moesta, P. and West, C. A. (1985). *Arch. Biochem. Biophys.* **238**, 325–333.

Munck, S. L. and Croteau, R. (1990). *Arch. Biochem. Biophys.* **282**, 58–64.

Pirkle, W. H. and Pochapsky, T. C. (1987). *Adv. Chromatogr.* **27**, 73–127.

Poulter, C. D. and Rilling, H. C. (1981). *In* "Biosynthesis of Isoprenoid Compounds", Vol. 1 (J. W. Porter and S. L. Spurgeon, eds), pp. 161–224. John Wiley, New York.

Rhodes, M. J. C. (1977). *In* "Regulation of Enzyme Synthesis and Activity in Higher Plants" (H. Smith, ed.), pp. 245–269. Academic Press, London.

Rilling, H. C. (1985). *In* "Methods in Enzymology", Vol. 110 (J. H. Law and H. C. Rilling, eds), pp. 142–152. Academic Press, New York.

Sasiak, K. and Rilling, H. C. (1988). *Arch. Biochem. Biophys.* **260**, 622–627.

Satterwhite, D. M. (1985). *In* "Methods in Enzymology", Vol. 110 (J. H. Law and H. C. Rilling, eds), pp. 92–99. Academic Press, New York.

Satterwhite, D. M. and Croteau, R. (1987). *J. Chromatogr.* **407**, 243–252.

Satterwhite, D. M. and Croteau, R. (1988). *J. Chromatogr.* **452**, 61–73.

Scopes, R. K. (1987). "Protein Purification: Principles and Practice". Springer-Verlag, New York.

Sofer, S. S. and Rilling, H. C. (1969). *J. Lipid Res.* **10**, 183–187.

Sterrett, F. S. (1975). *In* "The Essential Oils", Vol. 2 (E. Guenther ed.), pp. 818–828. Kreiger, Huntington, NY.

Vogeli, U., Freeman, J. W. and Chapell, J. (1990). *Plant Physiol.* **93**, 182–187.

West, C. A. (1981). *In* "Biosynthesis of Isoprenoid Compounds", Vol. 1 (J. W. Porter and S. L. Spurgeon, eds), pp. 375–411. John Wiley, New York.

West, C. A., Dudley, M. W. and Dueber, M. T. (1978). In "Recent Advances in Phytochemistry", Vol. 13 (T. Swain and G. R. Waller, eds), pp. 163–198. Plenum, New York.

10 Cytochrome P-450 Terpene Hydroxylases

CHARLES A. MIHALIAK, FRANK KARP and
RODNEY CROTEAU

*Institute of Biological Chemistry, Washington State University,
Pullman, WA 99164–6340, USA*

I. INTRODUCTION

Terpenes constitute one of the largest groups of plant natural products. They are composed of repeating five-carbon (isoprene) units and occur as a variety of cyclic and

METHODS IN PLANT BIOCHEMISTRY Vol. 9
ISBN 0-12-461019-6

acyclic olefins, and oxygenated derivatives. Monoterpenes, the C_{10} representatives of the isoprenoid family, are widely distributed in the plant kingdom and along with the sesquiterpenes (C_{15}) are predominant components of the essential oils. Diterpenes (C_{20}) include phytol, resin acids, and gibberellins. Triterpenes, such as sterols and complex triterpenoids (i.e. bitter principles such as limonoids and curcurbitacins), are C_{30} compounds or their derivatives, and the tetraterpenes (C_{40}) are represented by the carotenoid pigments. Higher molecular weight terpenes ($> C_{40}$) include the poly-prenols, rubber, gutta and chicle (Goodwin and Mercer, 1983). All terpenes are derived from acyclic, allylic pyrophosphorylated precursors [e.g., geranyl pyrophosphate (C_{10}), farnesyl pyrophosphate (C_{15}), geranylgeranyl pyrophosphate (C_{20})] that give rise to a variety of corresponding cyclic and acyclic terpenoids, many of which are olefinic.

In terpene metabolism by higher plants, cytochrome P-450 hydroxylases are involved in many pathways, including the elaboration of oxygenated mono- and sesquiterpenes of the essential oils, some plant growth hormones, indole and iridoid alkaloids, steroidal compounds and certain phytoalexins. Monoterpene hydroxylation in plants was first reported using a microsomal preparation of *Vinca rosea* seedlings (Meehan and Coscia, 1973). This hydroxylation of geraniol (and nerol) at the C-10 methyl group occurs in the biosynthesis of several indole and iridoid alkaloids and valepotriates (Madyastha *et al.*, 1976; Violon and Vercruysse, 1985). Several specific cytochrome P-450 hydroxylases are involved in oxygenation of the cyclic monoterpene olefins sabinene (Karp *et al.*, 1987), limonene (Karp *et al.*, 1990), and α- and β-pinene (Karp and Croteau, 1988). Hydroxylation of linalool and its derivatives, and of several *p*-menthene alcohols and ketones, has been reported in cell cultures of *Nicotinia tabacum*, but the specific involvement of cytochrome P-450 in these reactions remains to be demonstrated (Suga and Hirata, 1990). Cytochrome P-450-dependent oxygena-tions of sesquiterpenes include hydroxylation of the furanosesquiterpene, ipomeama-rone (Fujita *et al.*, 1982) and of abscisic acid to form phaseic acid (Gillard and Walton, 1976). Diterpene systems are represented by the hydroxylation of ent-kaur-16-ene (Murphy and West, 1969) en route to gibberellins. Plant steroids and related triter-penoid compounds have been observed to interact with cytochrome P-450 as well. The demethylation of obtusifoliol (Rahier and Taton, 1986), and the hydroxylation of digitoxin and similar cardiac glycosides by cell cultures of *Digitalis lanata* (Petersen and Seitz, 1985; Petersen *et al.*, 1988; Petersen and Seitz, 1988) are typical.

Cytochrome P-450-dependent, mixed function oxygenases participate in numerous reactions in plants, vertebrates, insects and microbes including hydroxylation, epoxida-tion, *N*-, *S*- and *O*-demethylation and dealkylation, and, under special circumstances, they carry out peroxidase and isomerase functions. Plant P-450 systems share many similarities, and have many properties in common with the microsomal P-450 systems of animals. Insects and vertebrates utilise these mixed function oxygenases in hormone production, and to detoxify and eliminate an array of xenobiotic compounds. In plants, P-450-dependent detoxification is involved in the metabolism of pesticides (O'Keefe *et al.*, 1987). Microbes employ P-450 to catabolise monoterpenes as a sole carbon and energy source. The most extensive mechanistic studies of a P-450 cytochrome have been carried out with bacterial P-450$_{cam}$ that catalyses hydroxyl-ation of the monoterpene camphor (Gunsalus *et al.*, 1974, 1975). The crystal structure of this protein has been solved (Poulos, 1986).

Several recent reviews on plant cytochrome P-450 hydroxylases are available (West, 1980; Higashi, 1985; Madyastha, 1985; Karp and Croteau, 1988) as are general discussions of cytochrome P-450 (Fleischer and Packer, 1978; Ortiz de Montellano, 1986). The emphasis of this chapter is terpene hydroxylation, but reference will be made to related plant systems when particularly relevant, and to the general P-450 literature.

II. PROPERTIES OF CYTOCHROME P-450 HYDROXYLASES

B. The Catalytic Cycle

Plant cytochrome P-450 terpene hydroxylase systems are comprised of a heme containing protein (cytochrome) and a flavoprotein, (NAD(P)H-cytochrome P-450 reductase) associated with the endoplasmic reticulum or provacuole. Reducing equivalents from NAD(P)H, molecular oxygen and the substrate interact with the hydroxylase to form an oxygenated terpene product, water and oxidised pyridine nucleotide. During the catalytic cycle, the oxidised form of the cytochrome first binds the substrate, followed by the transfer of one electron from NAD(P)H (via cytochrome P-450 reductase) to reduce the heme iron. The substrate-bound, reduced cytochrome P-450 binds a molecule of O_2 to form a reactive complex. A second electron enters the complex via the reductase (or possibly cytochrome b_5) and the dioxygen is cleaved, resulting in the release of one molecule each of water and the hydroxylated product (Ortiz de Montellano, 1986).

B. Characterisation of Cytochrome P-450 Hydroxylases

Demonstration of the involvement of cytochrome P-450 in plant terpene oxygenation requires the localisation of cytochrome P-450 and NAD(P)H-cytochrome P-450 reductase catalytic activity in the light membrane (microsomal) fraction of the cell, the spectral demonstration that cytochrome P-450 is present in the microsomal preparation, the absolute requirement for NAD(P)H and molecular oxygen, the typical monooxygenase stoichiometry (1 mol each of $NADPH + H^+$, O_2, and substrate utilised for each mol of product and water formed), the incorporation of one oxygen atom from $^{18}O_2$ into each mol of product, the expected response to specific inhibitors (in particular, the inhibition by carbon monoxide in the presence of oxygen, and the reversal of CO inhibition by light with a maximum in the photoaction spectrum at 450 nm) and the spectral demonstration of substrate and inhibitor binding. It is unlikely that all of these criteria will be met in any one system, but a combination of the most specific evidence (e.g. photoreversible CO inhibition, substrate binding spectra) is usually sufficient to establish the probable involvement of cytochrome P-450 in terpene hydroxylation. Additional characterisation includes the demonstration of regio- and stereospecific catalysis, the inducibility of the P-450 hydroxylase and reductase, the production of antibodies with the ability to inhibit specific P-450-dependent reactions (which may also be employed for assessment of the immunological relatedness to other P-450 hydroxylases), and measurement of the spin state of the native cytochrome. Other useful parameters for characterisation include temperature and pH optima, reaction kinetics, stability, turnover number, and precise subcellular localisation.

III. EXTRACTION OF PLANT CYTOCHROME P-450

The microsomal fraction is loosely defined as a collection of membranous vesicles that sediment between 15 000 and 100 000 × g when tissue homogenates are centrifuged (Chrispeels, 1980). Preparation of microsomes involves three steps: selection and harvesting of the tissue containing the enzyme(s), homogenisation of the tissue, and isolation of the microsomes.

B. Tissue Selection

Plant materials present unique problems during enzyme isolation due to the presence of pigments. Diagnostic UV-visible spectra are essential in studying cytochrome P-450; however, in many plant tissues chlorophyll, carotenoids and possibly flavonoids are present in much greater molar quantities than cytochrome P-450 and severely interfere with spectral detection of the enzyme. For this reason, almost all studies of cytochrome P-450 from higher plants have been confined to tissues low in interfering pigments. The presence and metabolic importance of P-450 hydroxylases are undoubtedly far more widespread. Tissue sources utilised thus far include etiolated seedlings, non-photosynthetic storage tissues, glandular trichomes isolated from leaf surfaces, and unpigmented cell cultures.

B. Microsome Preparation

1. Buffers

The buffers used for isolation of plant microsomal membranes typically contain the following: 0.05–0.2 M sodium phosphate, Tris-HCl, Mops-NaOH or HEPES, pH 7.0–7.5; 15–30% glycerol, 0.4 M sucrose or 0.3 M sorbitol; 1.0 mM EDTA; 1.0–5.0 mM dithiothreitol (DTT) or 10–15 mM mercaptoethanol; and Amberlite XAD-4 (1 g per g tissue) or another polystyrene resin to adsorb terpenes and phenolics. Several other buffer additions have been found to increase the activity of P-450 enzymes during isolation, including 0.1% bovine serum albumin (BSA), 0.5–1.0% polyvinyl-pyrrolidone (PVP, mol. mass 10 kDa or 40 kDa), 25 mM $Na_2S_2O_5$, 5.0 μM FAD and FMN, 0.05% cysteine, and various proteinase inhibitors (Higashi *et al.*, 1985; Kochs and Grisebach, 1989; O'Keefe and Leto, 1989; Stewart and Schuler, 1989; Karp *et al.*, 1990).

2. Extraction and isolation of microsomes

Many monoterpene P-450 hydroxylases occur in secretory cells of the glandular trichomes on the leaf surface of herbaceous plants. A mechanised technique has been developed (Gershenzon *et al.*, 1987, 1991) whereby secretory cell clusters are isolated by abrading the leaf surface with glass beads (0.5 mm diameter) in a polycarbonate chamber containing leaves, extraction buffer, glass beads and XAD-4 polystyrene resin. The secretory cells can be separated from the leaves, glass beads, and polystyrene resin by filtration through a series of nylon meshes and then captured on a 20 μm nylon mesh. To rupture the cells, the cell clusters are either frozen in liquid N_2 and ground with a mortar and pestle, or are resuspended in buffer and sonicated. Sonication time is varied depending on the volume of cell clusters and buffer. The enzyme is then slurried with XAD-4 resin for 15 min to remove terpenes and phenolics released during sonication and then filtered again through a 20 μm nylon mesh to remove the XAD-4 and cellular debris. The filtrate is used for isolation of the microsomes by centrifugation. Extraction of other plant cytochrome P-450 types from etiolated seedlings, nonpigmented storage tissues, and cell cultures have employed methods such as grinding with a mortar and pestle (Madyastha *et al.*, 1976; Kochs and Grisebach, 1989; Stewart and Schuler, 1989) or homogenisation in a blender (Higashi *et al.*, 1985; O'Keefe and Leto, 1989).

The most common methods used for isolation of the microsomal fraction from crude homogenates are differential centrifugation and precipitation with Mg^{2+}. For differential centrifugation, the homogenate is centrifuged at 10 000–20 000 × g for 20–30 min to remove denser organelles (e.g. nuclei, mitochondria, chloroplasts). The supernatant is then centrifuged at 100 000–195 000 × g for 60–90 min to obtain the microsomal pellet. The microsomal fraction can also be precipitated from a 3000–10 000 × g supernatant of a tissue homogenate by addition of 10–20 mM $MgCl_2$ and centrifugation at 20 000–40 000 × g for 20 min. It is important to check for the presence of cytochrome P-450 in the pellets from low speed centrifugations since the light membranes often bind to heavier organelles and other cellular debris. Passing the crude extract through a small gauge syringe needle or homogenisation with a Ten-Brock homogeniser to obtain a fine dispersion may reduce non-specific binding and minimise deposition in the low g fraction (Karp *et al.*, 1987, 1990).

Most plant cytochrome P-450 systems are located on the endoplasmic reticulum (ER) (Chrispeels, 1980). Sucrose density centrifugation is a useful technique for separation of ER from other light membranes, and the method may serve as a useful first step in protein purification or for subcellular localisation studies. Detailed methods for isolation and identification of plant ER are available (Chrispeels, 1980; Lord, 1983).

3. *Storage*

Microsomal pellets can usually be stored under an inert atmosphere (nitrogen or argon) in the centrifuge tubes in which they are obtained (and without resuspension) at $-70°C$ for several weeks. Alternatively, the pellets may be resuspended and frozen ($-70°C$) in buffer (e.g. 0.05–0.1 M sodium phosphate or Tris-HCl, pH 7.0–7.5, with 20% glycerol, 1 mM EDTA, 2 mM DTT, 5 μM FAD and FMN).

IV. ASSAYS

A. Catalytic Assays

1. *General considerations*

Factors to consider in choosing an appropriate catalytic assay include: (a) unequivocal identification of the natural substrate and its availability; (b) the ability of the preparation to metabolise alternative substrates which yield easily measurable products; and (c) the presence in the preparation of potential interfering materials (i.e. endogenous substrates or inhibitors) and their ease of removal.

2. *Buffers, substrates and cofactors*

A suitable buffer for catalytic assays of plant cytochrome P-450 hydroxylase activity could include: 10–30% glycerol or 0.4 M sucrose; 50–200 mM sodium phosphate or Tris-HCl, pH 7.0–8.0; 0.5–2.0 mM DTT or 10–15 mM mercaptoethanol; and 1.0 mM EDTA. Buffer supplementation with 5.0 μM each FAD and FMN, cytosolic factors, albumin and detergents has been shown to be stimulatory in several cases (Hasson and West, 1976; Karp *et al.*, 1987). The pH optima for catalysis must be determined empirically, but is usually slightly on the basic side of neutrality. At a pH of one unit above or below this optimum, only 50% of maximum activity is typically seen and may be accompanied by partial conversion of cytochrome P-450 to the catalytically inactive cytochrome P-420 species (Rich and Lamb, 1977; Karp *et al.*, 1990). Most terpenes are essentially insoluble in water, therefore the substrate must be prepared in a suitable solvent (e.g., pentane, ethanol, DMSO or acetone) at concentrations such that the solvent constitutes less than 0.5% of the total assay volume. A concentration of 0.5–2.0 mM NADPH is generally employed, or an NADPH regenerating system (1.0 mM NADPH, 5.0 mM glucose-6-phosphate, 1.0 unit glucose-6-phosphate dehydrogenase per ml) is added to the assay.

3. *Assay of a natural substrate*

Microsomal or solublised microsomal protein suspensions should be prepared to a final protein concentration of 25–1000 μg ml^{-1}. The reaction is initiated by the addition of NADPH (and/or the NADPH regenerating system) to the microsome–substrate mixture, and the suspension incubated 5–120 min under aerobic conditions at 20–32°C. The reaction is terminated by protein denaturation using organic solvent (i.e. diethyl ether). Products are then thoroughly extracted with the solvent, internal standard or unlabelled carrier (radiochemical assay) added, and the extract analysed. Drying the organic extract (over anhydrous MgSO$_4$) and the removal of highly polar contaminating materials (with silica) may be required. If the presence of endogenous substrates, products or inhibitors is significant, the levels of these interfering materials may be reduced by treating the microsomal suspension with polyvinylpyrrolidone, anion-exchange resin or polystyrene resin prior to performing the assay.

4. Product determination

Capillary gas–liquid chromatography (GLC), with confirmation of structure by coupled mass spectrometry, has been invaluable in the determination of products obtained from cytochrome P-450 hydroxylations of monoterpenes, particularly when the identity of the product(s) is uncertain. The sensitivity of capillary GLC extends below one nmol of product, and the regio- and stereo-specificity of the P-450 reaction can often be established by this method (Karp *et al.*, 1987, 1990; Karp and Croteau, 1988; Croteau *et al.*, 1991).

Radiolabelled substrates can extend assay sensitivity into the pmol range once the structure of the product has been established. For monoterpene hydroxylases, assay mixtures are extracted with ether and authentic carriers for the anticipated products are added to the extract prior to separation with a gas chromatograph coupled to a gas proportional counter (Satterwhite and Croteau, 1988). Coincidence of the mass and radiochromatographic peaks on two columns of widely differing polarity is strong presumptive evidence for product identity (confirmation is usually provided by preparation of a crystalline derivative). Less time-consuming routine assays may subsequently be developed using thin layer chromatography (on normal or reversed-phase silica gel G plates). After addition of unlabelled carrier, the plates are developed in the appropiate solvent, the corresponding bands are scraped from the plate, and yield determined by liquid scintillation spectrometry (for non-volatile products, a TLC scanner provides a useful alternative). The method suffers from the disadvantages of having to acquire appropriate radiolabelled substrates, unlabelled standard(s), and a preliminary knowledge of product structure and chromatographic properties. Unsuspected products are unlikely to be detected by TLC but the speed of assay, especially during purification of the enzyme, lends the technique great utility (Gillard and Walton, 1976; Fujita and Asahi, 1985; Petersen and Seitz, 1985; Karp *et al.*, 1987).

HPLC analysis has been successfully applied to separate the products of digitoxin hydroxylation. A Hypersil ODS column was used to detect the cardiac glycosides and to compare retention times with authentic reference standards (Peterson and Seitz, 1985).

5. Alternative substrates

Although plant cytochrome P-450 terpene hydroxylases often exhibit rather strict substrate requirements, measurement of activities with alternative subtrates may be possible. Easily assayed alternative substrates can be particularly useful for rapid monitoring of cytochrome P-450 reactions during enzyme purification; however, caution must be exercised to ensure that the activity measured with the alternative substrate is associated with the enzyme of interest, especially if more than one cytochrome P-450 species is present in the preparation. N-Demethylation reactions are often measured colorimetrically with *p*-chloro-*N*-methylaniline (Young and Beevers, 1976; Dohn and Krieger, 1984; Cottrell *et al.*, 1990) or aminopyrine (Benveniste *et al.*, 1982; Fonne-Pfister *et al.*, 1988) as substrates.

6. Hydroperoxide-supported assays

Measurement of P-450 hydroxylase activity in the absence of cytochrome P-450 reductase and NADPH often can be accomplished using organic hydroperoxides to support P-450-dependent oxygenations (Coon *et al.*, 1990). This method has gained little use with plant systems but has utility in certain situations, especially during purification of the enzyme (O'Keefe and Leto, 1989). The procedure is similar to that outlined for the NADPH-supported assay, except that NADPH and thiol reagents (DTT and mercaptoethanol) are omitted. A typical reaction mixture for the assay for cytochrome P-450 limonene hydroxylase from *Mentha spicata* (spearmint) includes the enzyme in 100 mM Tris-HCl (pH 7.4) with 10% glycerol, 100 mM KCl, 0.5 mM EDTA, 300 μM (−)-limonene, and 100 μM of freshly prepared cumene hydroperoxide. The reaction is initiated by addition of the hydroperoxide and the mixture is incubated at 32°C. Following ether extraction, the product (limonene 1,2-epoxide instead of the normal hydroxylation product carveol) is analysed by capillary GLC. Hydroperoxide-supported systems have also been utilised to study *N*-demethylation reactions (Young and Beevers, 1976; O'Keefe and Leto, 1989).

Cumene hydroperoxide should be purified by extraction with dilute alkali prior to use (Hock and Lang, 1944). The linear range for reaction time is shorter with this assay and the nature of the reaction products may be different than with the NADPH-supported assay. Since non-P-450-heme-liganded enzymes may also yield oxygenated products, caution is advised and supportive evidence should be accumulated to demonstrate that the activity observed is attributable to the enzyme under study.

7. Oxygen requirement

The requirement for O_2 can be determined in an inert atmosphere using an O_2-scavenging system. The system consists of 10 units of glucose oxidase, 6 μmol β-D-glucose and 1300 units of catalase per ml (Karp *et al.*, 1987). The assay is conducted under either a N_2 or Ar atmosphere which can be generated in a test tube by flushing through a septum-topped screw cap and then saturating the atmosphere inside the tube with N_2 or Ar. The hydroxylase: substrate: O_2-scavenging system mixture is pre-incubated for 10 min in the closed container after which NADPH is added anaerobically by syringe through the septum cap. Incubation under these conditions should result in little or no detectable hydroxylase activity.

8. Measurement of NADPH oxidation

When enzymatic products are not known or not easily detected by GC or HPLC, the rate of NADPH oxidation can be monitored as an indication of P-450 metabolism. Assays containing enzyme, NADPH (0.1 mM) and 50 μM of substrate are incubated for 3 min at 30°C. The rate of NADPH oxidation is recorded as absorbance change at 340 nm against a reference sample without substrate (Yu, 1987) using an extinction coefficient of 6.2 cm^{-1} mM^{-1} (Strobel and Dignam, 1978).

9. Reaction stoichiometry

The overall stoichiometry for NADPH:O_2:substrate consumption in cytochrome P-450 dependent hydroxylations should be 1:1:1 (Rich and Lamb, 1977). Oxygen uptake in the preparation is measured using an oxygen electrode. An endogenous rate of O_2 consumed per mg protein per min is first established. NADPH is then added and an increase in O_2 consumption is measured. The addition of substrate (after NADPH) results in additional O_2 consumption due to oxygen utilisation for hydroxylation. The substrate-stimulated increase is observed until the substrate is depleted (returning to the basal NADPH-supported rate). A 1:1 stoichiometry of substrate to oxygen consumption should be observed. NADPH oxidation is measured by absorbance change at 340 nm (see Section IV.A.8). Addition of substrate results in an increase in NADPH oxidation. By performing separate experiments for O_2 uptake and for NADPH oxidation under identical conditions, both the stimulation of NADPH oxidation and of O_2 consumption with substrate consumption should show an overall stoichiometry of about 1:1:1 for NADPH:O_2:substrate consumption. Isotopically labelled dioxygen ($^{18}O_2$) may be used to demonstrate the incorporation of one oxygen atom from molecular oxygen into hydroxylated product and the other into water (Murphy and West, 1969).

10. NADPH-cytochrome P-450 reductase

The activity of NADPH-cytochrome P-450 reductase is measured using cytochrome c as an artificial electron acceptor. A typical reaction mixture contains 50–100 μg protein ml^{-1} with 0.05 mM horseheart cytochrome c and 0.5 mM KCN. The reduction of cytochrome c is measured by monitoring the change in absorbance at 550 nm after addition of NADPH. Enzyme activity is calculated using an extinction coefficient of 21 mM^{-1}cm^{-1} (Madyastha et al., 1976; Madyastha and Coscia, 1979; Benveniste et al., 1986; Peterson and Seitz, 1988). A K_m value for NADPH in the low micromolar range is typically determined with this assay (Rich and Lamb, 1977) and the rate of reduction increases with increasing ionic strength (Strobel and Dignam, 1978). Cytochrome P-450 reductase contains equimolar amounts of FMN and FAD, and supplemental addition of these labile prosthetic groups to the reaction buffer often increases enzymatic activity.

B. Spectral assays

1. Carbon monoxide difference spectra

Cytochrome P-450 under reduced conditions forms a unique spectral complex with carbon monoxide. An absorbance maximum occurs at 450 nm when CO is bound to the reduced enzyme, thus allowing the presence of the cytochrome to be confirmed and the molar concentration to be estimated spectrally (Omura and Sato, 1964). A CO difference spectrum (i.e., with and without CO) can be obtained with chlorophyll- and carotenoid-free microsomes, or with solubilised P-450 suspended in a buffer similar to that used for catalytic assay but free of FAD and FMN (the flavins are bleached upon reduction, resulting in negative absorbances relative to the blank).

Using a split beam spectrophotometer, two protein aliquots (sample and blank) are needed to establish a baseline absorbance between 380 and 600 nm. The suspension in the sample cuvette is gently bubbled with CO for 0.5–2 min after which time a few mg of sodium dithionite are added. Cytochromes b_5 and P-450 will be reduced under these conditions, and the CO complex with the reduced P-450 will appear. Addition of sodium dithionite to the reference cuvette allows recording of the CO-reduced minus reduced P-450 difference spectrum (Estabrook and Werringloer, 1978). When using a diode array spectrophotometer, a background absorbance is recorded for the buffer blank. The enzyme suspension is then reduced with sodium dithionite and a scan is recorded. The reduced sample is next gently bubbled with CO and another spectrum is recorded. The absorbance of the reduced spectrum is then subtracted from the CO-reduced spectrum to give the CO difference spectrum.

An extinction coefficient of $91 \, \text{cm}^{-1} \text{mM}^{-1}$ for A_{450} minus A_{490} is used to calculate the concentration of cytochrome P-450 (Omura and Sato, 1964). An additional peak may be noted at 420 nm (representing denatured P-450). The extinction coefficient for cytochrome P-420 (A_{420} minus A_{490}) is $111 \, \text{cm}^{-1} \text{mM}^{-1}$ (Omura and Sato, 1964). Modifications of this technique have been used to measure the P-450 content in radicles and hypocotyls (Hendry *et al.*, 1981) and in intact glandular trichomes (pers. obs. by the authors). Under strictly anaerobic conditions, cytochrome P-450 reduction can be achieved enzymatically by the addition of NADPH (up to 70% of the level achieved with sodium dithionite) and the rate of reduction can be markedly increased by the presence of substrate (Rich and Bendall, 1975; Rich and Lamb, 1977).

In plant microsomes, an absorbance peak observed at 420 nm may not be due solely to cytochrome P-420. If sodium dithionite is added to both the sample and reference cuvettes and only the sample cuvette is treated with CO, the time-dependent development of peaks at 420 and 485 nm may be observed along with a stable peak at 450 nm. This phenomenon has been attributed to the CO inhibition of dithionite-dependent enzymatic degradation of a carotenoid. In the sample cuvette, the degradation is inhibited, while depletion continues with time in the reference (Madyastha *et al.*, 1976).

Note: Sodium dithionite comes in several grades and will rapidly decompose on the shelf. It should be stored desiccated, in the dark, under an inert atmosphere, and preferably frozen. The reducing capability of this reagent in aqueous solution may be tested with a benzyl viologen solution ($5 \, \text{mg ml}^{-1}$) which will turn dark purple when reduced and will not re-clear under anaerobic conditions.

2. Binding spectra

Binding spectra can be used to measure the kinetics of substrate and inhibitor binding to the oxidised enzyme, to monitor the presence of cytochrome P-450, and to obtain information regarding how potential substrates and inhibitors interact with the enzyme. The cytochrome P-450 concentration should be nearly an order of magnitude higher than that used to generate a useful CO difference spectrum, and the monitored absorbance range altered to 350–500 nm. Stock solutions of the compound to be tested should be prepared in DMSO or acetone. The volume of solvent should not exceed 2% of the final suspension of enzyme.

Two characteristic types of spectra are normally observed (Jefcoate, 1978). Type I spectra, which exhibit an absorbance peak near 385 nm and a trough near 420 nm, are typical of substrates for cytochrome P-450. The magnitude of the trough at saturation is about 20% of the 450 nm peak height of a carbon monoxide difference spectrum (Rich and Lamb, 1977). Type II spectra, exhibiting a peak near 430 nm and a broad trough near 400 nm, are more typical of inhibitors. Various intermediate forms between these extremes may be noted (Jefcoate, 1978).

With a split beam spectrophotometer, a baseline is established with aliquots of the microsomal suspension in the sample and reference cuvettes. The ligand (in solvent) is added to the sample and an equal volume of solvent is added to the reference. The difference spectrum develops between 350 and 500 nm and is recorded. More ligand (sample) and solvent (reference) are added in a stepwise fashion, and measurements are repeatedly recorded until no further spectral changes are noted. With a diode array spectrophotometer, reference measurements are made and stored electronically with stepwise increments of solvent. A fresh sample is prepared and the ligand is added (in solvent) in the same volume increments. Solvent-alone measurements are then subtracted from the ligand-containing measurements to give the series of difference spectra. The difference in absorbance between the peak and the trough are plotted against the substrate concentration (by double reciprocal methods) to yield a dissociation constant. The K_s value determined by this means and the K_m should be in reasonable agreement for a given P-450 species (Rich and Lamb, 1977).

The presence of endogenous substrates or inhibitors often hampers efforts to obtain binding spectra with plant P-450 cytochromes. These compounds, while ordinarily sequestered in the cell, may be released on cell breakage and may bind to the enzyme, thereby interfering with subsequent binding studies. Unless measures are taken to adsorb these substances during homogenisation (e.g. with polyvinlypyrrolidone and/or polystyrene), efforts to obtain spectral evidence for substrate or inhibitor complexes with P-450 may be ineffective. Failure to attain binding spectra with a preparation of demonstrated catalytic capability with the tested substrate suggests that enzyme concentration is too low to observe binding, or that only a fraction of the total P-450 in the preparation is involved in the specific hydroxylation under study (Rich and Lamb, 1977). Type II spectra have been observed using ancymidol (Coolbaugh *et al.*, 1978), aniline and pyridine (Rich and Bendall, 1975) and clotrimazole (pers. obs. by the authors). In the later two cases, type I binding with a known substrate could not be obtained.

C. Inhibitors

Carbon monoxide, along with a number of other chemical agents (Testa and Jenner, 1981), can be used to inhibit cytochrome P-450 hydroxylation. Catalytic inhibition by CO, and reversal of inhibition with blue light (450 nm), is the definitive experiment to demonstrate the involvement of cytochrome P-450 in hydroxylation. Chemical inhibitors may act as general inhibitors of a vast array of P-450 enzymes, or with varying degrees of specificity towards particular hydroxylases. Inhibitors of cytochrome P-450 reductase (e.g. cytochrome *c*) can be used to block electron flow to the hydroxylase, thereby disrupting catalysis.

1. CO inhibition and reversal with blue light

Gas mixtures are prepared by the displacement of degassed, distilled water from inverted filtration flasks with appropriate amounts of pure O_2 and CO to yield CO/O_2 ratios from 1:1 to 19:1. To establish the appropriate atmosphere in and above the assay solution, the gas mixture is bubbled through the enzyme preparation. This can be conveniently accomplished by water displacement of the gas mixture through a syringe needle into the sample contained in a Reacti-Vial (Pierce Chemical Co.). An O_2-scavenging system (see Section III.B.7), substrate and cofactor are then added, and the reaction mixture incubated (Karp *et al.*, 1987). Increasing levels of hydroxylase inhibition should be observed in assays between the 1:1 and 19:1 levels of $CO:O_2$. The presence of substrate has been shown to reduce the sensitivity of certain P-450 systems to CO (Canick and Ryan, 1971; Zachariah and Juchau, 1975).

To demonstrate the influence of light at 450 nm in the presence of CO, pairs of vessels (one masked and one transparent) containing the $CO:O_2$ gassed reaction mixtures are used. The paired reactions are initiated by substrate addition, and the mixtures are irradiated using a light source with maximum output near 450 nm. A glass tank containing 10% aqueous $CuSO_4$ placed between the assay vessels and a white light source can serve this purpose. Relief of inhibition should be noted in the transparent vessel, although full reversal is rarely achieved.

2. Chemical inhibitors

Among the more common general inhibitors used with plant P-450 systems are thiol-directed reagents such as *p*-chloro-mercuribenzoate and *N*-ethylmaleimide (Dennis and West, 1967; Soliday and Kolattukudy, 1978), metal ion chelators such as *o*-phenanthroline and 8-hydroxyquinoline (Soliday and Kolattukudy, 1978), and phospholipases (Madyastha *et al.*, 1976).

More specific inhibitors toward cytochrome P-450 include, β-diethyl aminoethyl-diphenylvalerate dihydrochloride (SKF 525a), metyrapone and the substituted imidazoles miconazole and clotrimazole (Meehan and Coscia, 1973; Karp *et al.*, 1987, 1990). Propiconazole and triadimefon are effective inhibitors of P-450-catalysed reactions in plant sterol biosynthesis (Rahier and Taton, 1986), and 1-*n*-decyl- and 1-geranyl imidazoles are effective inhibitors of kaurene hydroxylation (Wada, 1978). Perillyl imidazole is an effective inhibitor of limonene hydroxylation (R. Croteau, unpubl. res.). Aminobenzotriazole (Benveniste *et al.*, 1982; Reichhart *et al.*, 1982) is a highly selective inhibitor of cinnamate-4-hydroxylase, and α-cyclopropyl-α-[*p*-methoxyphenyl]-5-pyrimidine methyl alcohol (Ancymidol) selectively inhibits $(-)$-kaurene oxidation (Coolbaugh *et al.*, 1978).

3. NAD(P)H-cytochrome P-450 reductase inhibitors

Cytochrome *c* serves as an alternative electron acceptor for reducing equivalents from cytochrome P-450 reductase, and oxidized pyridine nucleotides compete with NADPH for the reductase active site, thereby inhibiting the cycle of oxygenation (Salaün *et al.*,

1978; Madyastha and Coscia, 1979; Peterson *et al.*, 1978; Karp *et al.*, 1990). 2'-AMP, 3-aminonicotinamide adenine dinucleotide phosphate (AADP+), menadione, 2,6-dichlorophenolindophenol (DCPIP), 1,4-napthoquinone, and benzoquinone are also known to inhibit normal electron flow (Young and Beavers, 1976; Benveniste *et al.*, 1989; Zimmerlin and Durst, 1990).

V. SOLUBILISATION AND PURIFICATION

A. Solubilization

1. Procedure

Prior to performing any chromatographic fractionation of P-450 hydroxylases, the enzyme must be solubilised from the microsomal membranes. Several detergents have been successfully employed for this purpose, including sodium cholate, Emulgen 911, CHAPS, reduced Triton X-100 (RTX-100), and octyl glucoside. To date, there is no convenient method for choosing the appropriate detergent, or detergent:protein ratio, for solubilisation other than trial and error. Prior to detergent solubilisation, hydroxylase assays in the presence of a series of ionic and non-ionic detergents should be performed to determine the effect of the detergent on enzyme activity. A systematic survey using ionic and non-ionic detergents at detergent:microsomal protein ratios between 0.1:1 and 10:1 should identify suitable conditions for successful solubilisation (often up to 75%, of the activity) of most P-450 hydroxylases (Hjelmeland and Chrambach, 1984; Yanagita and Kagawa, 1986). Typically, the microsomes are resuspended in buffer (similar to assay buffer) containing the appropriate type and concentration of detergent. Alternatively, the detergent is added dropwise to previously resuspended microsomes until the desired detergent concentration is attained. The mixture is stirred for 30–60 min on ice and then centrifuged at 100 000 × g for 60 min. Sonication of the mixture for 30–60 s prior to centrifugation often increases the efficiency of solubilisation. The efficiency of solubilisation is measured by determining the P-450 concentration by CO difference spectrometry and catalytic activity in aliquots taken prior to centrifugation, and of the supernatant and resuspended pellet after centrifugation. The solubilised enzyme preparation is stable for several weeks when stored in buffer under an inert atmosphere at −70°C.

2. Detergent removal

Prior to performing catalytic assays with solubilised P-450, it may be necessary to remove some or all of the detergent used in the solubilisation step, since high detergent levels often inhibit enzymatic activity of terpene hydroxylases (Madyastha *et al.*, 1976; Karp *et al.*, 1987; pers. obs. by the authors). Detergent removal is also a key step in reconstitution of these membrane proteins (see Section V.E) (Furth, 1980). Several detergent removal methods are available, including adsorption with polystyrene resins, phase partitioning, hydrophobic adsorption chromatography, ion-exchange chromatography, and gel permeation chromatography (Furth, 1980; Furth *et al.*, 1984).

Extractive gel supports, which remove a variety of detergents (e.g. CHAPS, SDS, octyl glucoside, Triton, Tween, NP-40), are commercially available (e.g. Extracti-Gel D, Pierce).

B. Purification

A variety of chromatographic techniques have been used successfully for the purification of solubilised plant cytochrome P-450 enzymes, including anion-exchange (DEAE, Mono Q), hydrophobic interaction (hexyl-agarose), hydroxyapatite adsorption, and gel filtration. These fractionation steps are always carried out in the presence of detergent and at a neutral or slightly basic pH (pH 7.0–8.0). Polyethylene glycol (PEG) precipitation (using a final concentration of 9–16% PEG) has also been utilised as a purification step, and a three-phase partition system (polyethylene glycol–Ficoll–dextran) has been described in the purification of a yeast P-450 (Kärenlampi *et al.*, 1986).

Ion-exchange HPLC has been effective for the separation of P-450 isoforms (O'Keefe *et al.*, 1987). A recently developed HPLC matrix, designed as a solid-phase membrane mimic and denoted as immobilised artificial membrane (IAM) chromatography, has been used for the purification of mammalian P-450 types, but as yet has not been successfully employed for the purification of plant cytochrome P-450 systems (Pidgeon *et al.*, 1991).

If a hydroperoxide supported catalytic assay is available (i.e. an assay without reductase), then non-denaturing PAGE may be used to identify the specific peptide with terpene hydroxylase activity. We have recently isolated the (−)-limonene hydroxylase from spearmint by modified Jovin non-denaturing PAGE (Moos *et al.*, 1988) with approximately 30% recovery of enzyme activity. Individual protein bands can be identified by staining the gel using a non-fixing stain (i.e. Quick stain, Zoion Research) and gel slices cut out, equilibrated with assay buffer, and then assayed for oxygenase activity using the hydroperoxide supported reaction. This technique allows conclusive identification of the hydroxylase on a polyacrylamide gel, and may be used as part of a purification scheme or for obtaining the protein for antibody production.

C. Detection of P-450 During Purification

Although plant P-450 systems seem amenable to a number of purification techniques, difficulties often arise in detection of the enzyme in column fractions once the P-450 has been separated from the P-450 reductase. Ideally, a peroxide-supported catalytic assay can be developed, but several indirect methods of detection are also available.

Monitoring for the presence of a heme protein can be accomplished using a sensitive photometric detector capable of measuring absorbances between 405 and 420 nm. Simultaneous monitoring of a column eluate at 280 nm (for total protein) and at 415 nm (for heme protein) is most useful.

Methods have also been developed for detection of cytochrome P-450 proteins on SDS-PAGE gels. Detection is based upon the inherent heme-associated peroxidase activity of cytochrome P-450 (or P-420). The procedure involves staining for peroxidase activity using a benzidine derivative (e.g., 3,3′,5,5′-tetramethylbenzidine) and

H_2O_2 (Thomas *et al.*, 1976). The gel can be destained with 70 mM sodium sulphite then restained by Coomassie or silver staining techniques. Other benzidine derivatives have recently been developed for detection of nanogram levels of heme containing proteins after SDS-PAGE (Francis and Becker, 1984; Kuo and Fridovich, 1988). These stains attain higher levels of sensitivity but are unstable and may not be amenable to destaining.

The presence of P-450 can also be monitored in column fractions by measuring the CO difference spectrum or substrate/inhibitor binding spectrum. This method is preferred to monitoring total heme protein since only P-450 cytochromes will produce these characteristic spectra. Determining binding spectra requires rather high concentrations of protein since the sensitivity of this assay is much lower than that for CO binding, but this approach has the advantage of identifying unique forms of the enzyme that interact only with the substrate of interest.

Immunological detection of P-450 may be an extremely powerful tool, especially for detection of small quantities of P-450 or for identifying P-450 which has been denatured during purification. Antibodies useful for this purpose can be generated against partially purified protein if the protein can be identified by SDS-PAGE. As little as 200 μg of protein is needed to elicit an antigenic response in rabbits with the highly effective but seldom used 'whiffle-ball' technique (Hillam *et al.*, 1974; Ried *et al.*, 1992). Alternatively, antibodies raised against another plant P-450 may be used to detect P-450 in immunoblots of column fractions; however, a Western blot must also be performed to confirm that the antigenic cross-reactivity is specific for the target P-450 being purified.

D. Protein Assays

Accurate estimation of protein from plant sources is complicated by the presence of interfering substances from the tissues themselves, and can be further compromised by buffer constituents used to maintain enzyme activity (Mattoo *et al.*, 1987; Friedenauer and Berlet, 1989). Procedures successfully used include a modified Lowry method (Bensadoun and Weinstein, 1976), the Biuret method (Ellman, 1962) a modified Bradford technique (Fanger, 1987), and the bichinchinoic acid procedure (Smith *et al.*, 1985; Brown *et al.*, 1989). Difficulties encountered in determining a P-450: protein ratio indicative of homogeneity (17–21 nmol P-450 per mg protein) can often be attributed to loss of the heme moiety or to inaccurate protein estimation (Guengerich, 1987).

E. Reconstitution

Since the reductase component of the P-450 system is separated from the hydroxylase protein during purification, the normal NADPH-supported catalytic cycle cannot function. For this reason, it may be desirable to first purify the P-450 reductase, since the presence of this electron transfer protein can be monitored in the absence of cytochrome P-450. The purified reductase may then be used to reconstitute a functional P-450 complex with crude lipid extracts or a pure phospholipid. The separation and subsequent reconstitution of plant cytochrome P-450 and reductase, with a lipid extract, has been achieved with the nerol/geraniol hydroxylase (Madyastha *et al.*, 1976), the cinnamate-4-hydroxylase (Benveniste *et al.*, 1986), and the digitoxin 12-β-hydroxylase (Petersen and Seitz, 1988) systems. The reductase component of the coupled system is sometimes interchangable between species. Thus, the P-450 reductase may be obtained from another source to be combined with the purified hydroxylase. For example, a P-450 reductase from rabbits has been substituted for the plant form (Licht *et al.*, 1980). It is often necessary to remove most of the detergent from the enzyme suspension prior to reconstitution experiments (Furth, 1980).

VI. Induction

Some plant tissues contain low titers of P-450 hydroxylase unless deliberately induced. The phenomenon has been noted in both intact plants and plant cell cultures, and it provides a potential means of enriching for a specific P-450 form. The contrast between constitutive and induced systems can not be overemphasised, since the levels of P-450 found in uninduced plant microsomes is usually <5% of that found in rat liver microsomes (Hendry and Jones, 1984).

Cut injury, fungal infection and exposure of etiolated tissues to light have been found to induce terpene hydroxylase and cinammate-4-hydroxylase (CA4H) activities (Russell, 1971; Benveniste *et al.*, 1978; Fujita *et al.*, 1982; Fujita and Asahi, 1985), and the production of oxygenated terpenoid constituents of certain essential oils has also been noted after irradiation (Yamaura *et al.*, 1989). Wounding (Benveniste *et al.*, 1977), and treatment with various chemical agents (Reichhart *et al.*, 1979; Adele *et al.*, 1981; Zaprometov and Ermakova, 1989) or elicitors (Bolwell and Dixon, 1986; Tanahashi and Zenk, 1990) is effective in increasing levels of CA4H and protopine 6-hydroxylase, and a variety of elicitor preparations have been shown to increase the levels of P-450 cytochromes thought to be involved in later steps of flavonoid and isoflavonoid phytoalexin biosynthesis (Heller and Kühnl, 1985; Kochs and Grisebach, 1989; Kessmann *et al.*, 1990). The most common inducing technique involves 'cut ageing' or wounding accompanied by the uptake of metal ions, alcohols, DMSO, barbiturates or herbicides (Rich and Lamb, 1977; Reichhart *et al.*, 1980; Salaün *et al.*, 1981; Fonne-Pfister *et al.*, 1988). Similar compounds may be effective when imbibed by seedlings (Zaprometov and Ermakova, 1989; Zimmerlin and Durst, 1990). Clofibrate (a glyoxysome inducer) (Salaün *et al.*, 1986), UV light (Hagmann *et al.*, 1983) and osmotic stress (Kochs *et al.*, 1987) have also been reported as plant P-450 inducers. P-450 substrates, including the monoterpene alcohol geraniol, have been shown to increase the cytochrome P-450 content several-fold when fed to mung bean seedlings (Hendry and Jones, 1984). Similar approaches may be useful in inducing terpene hydroxylases.

VII. SUMMARY

The multicomponent cytochrome P-450 systems of plants are present in low abundance, and the presence of co-extracted, interfering substances can further complicate both the isolation and assay of these microsomal proteins. For these reasons, the choice of starting material is important and has played a major role in the success of all studies with plant cytochrome P-450 systems thus far. Work with these available systems will serve as a guide to future experiments with less tractable tissues and with a wide variety of P-450-dependent hydroxylases involved in both primary and secondary plant metabolism.

ACKNOWLEDGEMENTS

Research described from the authors' laboratory was made possible by a grant from the National Science Foundation (DCB-9104983) and by Project 0268 from the Agricultural Research Center, Washington State University.

REFERENCES

Adele, P., Reichhart, D., Salaün, J. P., Benveniste, I. and Durst, F. (1981). *Plant Sci. Lett.* **22**, 39–46.

Bensadoun, A. and Weinstein, D. (1976). *Anal. Biochem.* **70**, 241–250.

Benveniste, I., Salaün, J. P. and Durst, F. (1977). *Phytochemistry* **16**, 69–73.

Benveniste, I., Salaün, J. P. and Durst, F. (1978). *Phytochemistry* **17**, 359–363.

Benveniste, I., Gabriac, B., Fonne, R., Reichhart, D., Salaün, J. P., Simon, A. and Durst, F. (1982). *In* "Cytochrome P-450, Biochemistry, Biophysics and Environmental Implications" (E. Hietanen, M. Laitinen and O. Hänninen, eds.), pp. 201–208. Elsevier Biomedical Press, New York.

Benveniste, I., Gabriac, B. and Durst, F. (1986). *Biochem. J.* **235**, 365–373.

Benveniste, I., Lesot, A., Hasenfratz, M. P. and Durst, F. (1989). *Biochem J.* **259**, 847–853.

Bolwell, G. P. and Dixon, R. A. (1986). *Eur. J. Biochem.* **159**, 163–169.

Brown, R. E., Jarvis, K. L. and Hyland, K. J. (1989). *Anal. Biochem.* **180**, 136–139.

Canick, J. A. and Ryan, K. J. (1971). *Mol. Cell. Endocron.* **6**, 105–115.

Chrispeels, M. J. (1980). *In* "The Biochemistry of Plants: A Comprehensive Treatise", Vol. 1 (N. E. Tolbert, ed.), pp. 389–412. Academic Press, New York.

Coolbaugh, R. C., Hirano, S. S. and West, C. W. (1978). *Plant Physiol.* **62**, 571–576.

Coon, M. J., Blake, R. C., White, R. E. and Nordblom, G. D. (1990). *Methods Enzymol.* **186**, 273–278.

Cottrell, S., Hartman, G. C., Lewis, D. F. V. and Parke, D. V. (1990). *Xenobiotica* **20**, 711–726.

Croteau, R., Wagsahal, K. C., Karp, F., Satterwhite, D. M., Hyatt, D. C. and Skotland, C. B. (1991). *Plant Physiol.*, **96**, 744–753.

Dennis, D. R. and West, C. A. (1967). *J. Biol. Chem.* **242**, 3293–3300.

Dohn, D. R. and Krieger, R. I. (1984). *Arch. Biochem. Biophys.* **231**, 416–423.

Ellman, G. L. (1962). *Anal. Biochem.* **3**, 40–48.

Estabrook, R. W. and Werringloer, J. (1978). *Methods Enzymol.* **52**, 212–220.

Fanger, B. O. (1987). *Anal. Biochem.* **162**, 11–17.

Fleischer, S. and Packer, L. (eds) (1978). "Methods in Enzymology", Vol. 52. Academic Press, New York.

Fonne-Pfister, R., Simon, A., Salaün, J. P. and Durst, F. (1988). *Plant Sci.* **55**, 9–20.

Francis, R. J., Jr. and Becker, R. R. (1984). *Anal. Biochem.* **136**, 509–514.

Friedenaur, S. and Berlet, H. H. (1989). *Anal. Biochem.* **178**, 263–268.

Fujita, M. and Ashai, T. (1985). *Plant Cell Physiol.* **26**, 389–395.

Fujita, M., Oba, K. and Uritami, P. (1982). *Plant Physiol.* **70**, 573–578.

Furth, A. J. (1980). *Anal. Biochem.* **109**, 207–215.

Furth, A. J., Bolton, H., Potter, J. and Piddle, J. D. (1984). *Methods Enzymol.* **104**, 318–328.

Gershenzon, J., Duffy, M. A., Karp, F. and Croteau, R. (1987). *Anal. Biochem.* **163**, 159–164.

Gershenzon, J., McCaskill, D., Rajaonarivony, J., Mihaliak, C., Karp, F. and Croteau, R.

(1991). *In* "Modern Phytochemical Methods" (N. H. Fisher and H. Stafford, eds), pp. 347–370. Plenum Press, New York.

Gillard, D. F. and Walton, D. C. (1976). *Plant Physiol.* **58**, 790–795.

Goodwin, T. W. and Mercer, K. I. (1983). "Introduction to Plant Biochemistry", p. 451. Pergamon Press, New York.

Guengerich, F. P. (1987). *In* "Mammialian Cytochrome P-450", Vol. 1 (F. P. Guengerich, ed.), pp. 1–54. CRC Press, Boca Raton, FL.

Gunsalus, I. C., Meeks, J. R., Lipscomb, J. D., DeBrunner, P. and Münck, E. (1974). *In* "Molecular Mechanisms of Oxygen Activation" (O. Hayaishi, ed.), pp. 559–613. Academic Press, New York.

Gunsalus, I. C., Pederson, T. C. and Sligar, S. C. (1975). *Ann. Rev. Biochem.* **44**, 377–407.

Hagmann, M., Heller, W. and Grisebach, H. (1983). *Eur. J. Biochem.* **134**, 547–554.

Hasson, E. P. and West, C. A. (1976). *Plant Physiol.* **58**, 473–478.

Heller, W. and Kühnl, T. (1985). *Arch. Biochem. Biophys.* **241**, 453–460.

Hendry, G. A. F. and Jones, O. T. G. (1984). *New Phytol.* **96**, 153–159.

Hendry, G. A. F., Houghton, J. D. and Jones, O. T. G. (1981). *Biochem. J.* **194**, 743–751.

Higashi, K. (1985). *In* "P-450 and Chemical Carcinogenesis" (Y. Tagashira and T. Omura, eds), pp. 49–66. Plenum Press, New York.

Higashi, K., Ikeuchi, K., Obara, M., Karasaki, Y., Hirano, H., Gotoh, S. and Koga, Y. (1985). *Agric. Biol. Chem.* **8**, 2399–2405.

Hillam, R. P., Tengerdy, R. J. and Brown, G. L. (1974). *Infect. Immun.* **10**, 458–463.

Hjelmeland, L. M. and Chrambach, A. (1984). *Methods Enzymol.* **104**, 305–347.

Hock, H. and Lang, S. (1944). *Chem. Berichte* **77**, 257–264.

Jefcoate, C. R. (1978). *Methods Enzymol.* **52**, 258–279.

Kärenlampi, S. O., Nikkilä, H. and Hynninen, P. H. (1986). *Biotech. Appl. Biochem.* **8**, 60–68.

Karp, F. and Croteau, R. (1988). *In* "Bioflavour '87" (P. Schreir, ed.), pp. 173–198. Walter de Gruyter and Co., Berlin.

Karp, F., Harris, J. E. and Croteau, R. (1987). *Arch. Biochem. Biophys.* **256**, 179–183.

Karp, F., Mihaliak, C. A., Harris, J. L. and Croteau, R. (1990). *Arch. Biochem. Biophys.* **276**, 219–226.

Kessmann, H., Choudhary, A. D. and Dixon, R. A. (1990). *Plant Cell Rep.* **9**, 38–41.

Kochs, G. and Grisebach, H. (1989). *Arch. Biochem. Biophys.* **273**, 543–553.

Kochs, G., Welle, R. and Grisebach, H. (1987). *Planta* **171**, 519–524.

Kuo, C. F. and Fridovich, I. (1988). *Anal. Biochem.* **170**, 183–185.

Licht, H. J., Madyastha, K. M., Coscia, C. J., and Krueger, R. J. (1980). *In* "Microsomes, Drug Oxidations and Chemical Carcinogenesis", Vol. 1 (M. J. Coon, A. H. Conney, R. W. Estabrook, H. V. Gelboin, J. R. Gillette and P. J. O'Brien, eds), pp. 211–215. Academic Press, New York.

Lord, J. M. (1983). *In* "Isolation of Membranes and Organelles from Plant Cells" (J. L. Hall and A. L. Moore, eds), pp. 119–134. Academic Press, London.

Madyastha, K. M. (1985). *Biol. Mem.* **11**, 21–33.

Madyashta, K. M. and Coscia C. J. (1979). *J. Biol. Chem.* **254**, 2419–2427.

Madyastha, K. M., Meehan, T. D. and Coscia, C. J. (1976). *Biochemistry* **15**, 1097–1102.

Mattoo, R. L., Ishag, M. and Saleemuddin, M. (1987). *Anal. Biochem.* **163**, 376–384.

Meehan, T. D. and Coscia, C. J. (1973). *Biochem. Biophys. Res. Commun.* **53**, 1043–1048.

Moos, M. Jr., Nguyen, N. Y. and Liu, T. Y. (1988). *J. Biol. Chem.* **13**, 6005–6008.

Murphy, P. J. and West, C. A. (1969). *Arch. Biochem. Biophys.* **133**, 395–407.

O'Keefe, D. P., Romesser, J. A. and Leto, K. J. (1987). *Rec. Adv. Phytochem.* **21**, 151–173.

O'Keefe, D. P. and Leto, K. J. (1989). *Plant Physiol.* **89**, 1141–1149.

Omura, T. and Sato, R. (1964). *J. Biol. Chem.* **235**, 2379–2385.

Ortiz de Montellano, P. R. (ed.) (1986). "Cytochrome P-450; Structure, Mechanism and Biochemistry". Plenum Press, New York.

Petersen, M. and Seitz, H. U. (1985). *FEBS Lett.* **188**, 11–14.

Petersen, M. and Seitz, H. U. (1988). *Biochem. J.* **252**, 537–543.

Petersen, M., Seitz, H. U. and Reinhard, E. (1988). *Z. Naturforsch.* **43c**, 199–206.

Pidgeon, C., Stevens, J., Otto, S., Jefcoate, C. and Marcus, C. (1991). *Anal. Biochem.* **194**, 163–173.

Poulos, T. L. (1986) *In* "Cytochrome P-450: Structure, Mechanism and Biochemistry", pp. 505–523. Plenum Press, New York.

Rahier, A. and Taton, M. (1986). *Biochem. Biophys. Res. Commun.* **140**, 1064–1072.

Reichhart, D., Salaün, J. P., Benveniste, I. and Durst, F. (1979). *Arch. Biochem. Biophys.* **196**, 301–303.

Relchhart, D., Salaün, J. P., Benveniste, I. and Durst, F. (1980). *Plant Physiol.* **66**, 600–604.

Reichhart, D., Simon, A., Durst, F., Mathews, J. M. and Ortiz de Montellano, P. R. (1982). *Arch. Biochem. Biophys.* **216**, 522–529.

Rich, P. R. and Bendall, D. S. (1975). *Eur. J. Biochem.* **55**, 333–341.

Rich, P. R. and Lamb, C. J. (1977). *Eur. J. Biochem.* **72**, 353–360.

Ried, J. L., Everard, J. D., Diani, J. and Walker-Simmons, M. K. (1992). *Biotechniques* **12**, 660–668.

Russell, D. W. (1971). *J. Biol. Chem.* **246**, 3870–3878.

Salaün, J. P., Benveniste, I., Reichhart, D. and Durst, F. (1978). *Eur. J. Biochem.* **90**, 155–159.

Salaün, J. P., Benveniste, I., Reichhart, D. and Durst, F. (1981). *Eur. J. Biochem.* **119**, 651–655.

Salaün, J. P., Simon, A. and Durst, F. (1986). *Lipids* **21**, 776–779.

Satterwhite, D. M. and Croteau, R. B. (1988). *J. Chromatogr.* **452**, 61–73.

Smith, P. K., Krohn, R. I., Hermanson, G. T., Mallia, A. K., Gartner, F. H., Provenzano, M. D., Fujimoto, E. K., Goeke, N. M., Olsen, B. J. and Klenk, D. C. (1985). *Anal. Biochem.* **150**, 76–85.

Soliday, C. L. and Kolattukudy, P. E. (1978). *Arch. Biochem. Biophys.* **188**, 338–347.

Stewart, C. B. and Schuler, M. A. (1989). *Plant Physiol.* **90**, 534–541.

Strobel, H. W. and Dignam, J. D. (1978). *Methods Enzymol.* **52**, 89–97.

Suga, T. and Hirata, T. (1990). *Phytochemistry* **29**, 2393–2406.

Tanahashi, T. and Zenk, M. H. (1990). *Phytochemistry* **29**, 1113–1122.

Testa, B. and Jenner, P. (1981). *Drug Metab. Rev.* **12**, 1–117.

Thomas, P. E., Ryan, D. and Levin, W. (1976). *Anal. Biochem.* **75**, 168–176.

Violon, C. J. I. and Vercruysse, A. A. (1985). *Phytochemistry.* **24**, 2205–2209.

Wada, K. (1978). *Agric. Biol. Chem.* **42**, 2411–2413.

West, C. A. (1980). *In* "The Biochemistry of Plants: A Comprehensive Treatise', Vol. 2 (P. K. Stumpf and E. E. Conn, eds), pp. 317–364. Academic Press, New York.

Yamaura, T., Tanaka, S. and Tabata, M. (1989). *Phytochemistry* **28**, 741–744.

Yanagita, Y. and Kagawa, Y. (1986). *In* "Techniques for the Analysis of Membrane Proteins" (C. I. Ragan and R. J. Cherry, eds), pp. 61–76. Chapman and Hall, London.

Young, O. and Beevers, H. (1976). *Phytochemistry* **15**, 379–385.

Yu, S. J. (1987). *J. Chem. Ecol.* **13**, 423–436.

Zachariah, P. K. and Juchau, M. R. (1975). *Life Sci.* **16**, 1689–1693.

Zaprometov, M. N. and Ermakova, S. A. (1989). *Biokhimiya* **54**, 842–847.

Zimmerlin, A. and Durst, F. (1990). *Phytochemistry* **29**, 1729–1732.

11 Carotenoid Biosynthesis

PETER M. BRAMLEY

Department of Biochemistry, Royal Holloway and Bedford New College (University of London), Egham, Surrey, TW20 OEX

I. INTRODUCTION

The carotenoids are one of the most abundant groups of naturally occurring pigments. They are present in all green tissues, where they are constituents of the chloroplast, as well as being responsible for most of the yellow to red colours of flowers and fruits.

The development of improved separation methods for carotenoids, together with the use of physicochemical techniques, has resulted in the structural characterisation of over 500 carotenoids (Straub, 1987; Britton, 1991), which have both trivial and semi-systematic names. The latter convey structural information, as described by Britton

METHODS IN PLANT BIOCHEMISTRY Vol. 9
ISBN 0–12–461019–6

(1991). This widening array of structures presents exciting challenges regarding the elucidation of biosynthetic pathways, the properties of the enzymes which catalyse these conversions, and the mechanisms of desaturation, cyclisation and insertion of oxygen functions which take place during the modification of the first C_{40} carotenoid, phytoene (7,8,11,12,7′,8′,11′,12′-octahydro-ψ,ψ-carotene).

The majority of investigations prior to the late 1960s relied upon quantitative analyses of carotenoids following mutation, treatment with inhibitors or *in vivo* labelling studies. Although the general features of carotenoid biosynthesis were established in this manner (reviewed by Davies, 1980; Goodwin, 1980; Spurgeon and Porter, 1983; Britton, 1990), a detailed understanding of the enzymology of carotenogenesis and its regulation could not be achieved with these experimental approaches. Consequently, a number of cell-free systems have been developed, with a view to the purification and characterisation of carotenogenic enzymes and elucidation of the regulatory mechanisms which control carotenoid biosynthesis in higher plants and microorganisms.

II. PATHWAYS OF CAROTENOID FORMATION

The carotenoids of higher plants and algae are C_{40} tetraterpenoids, biosynthesised by the isoprenoid pathway. The early stages of the sequence, i.e. those by which the C_5

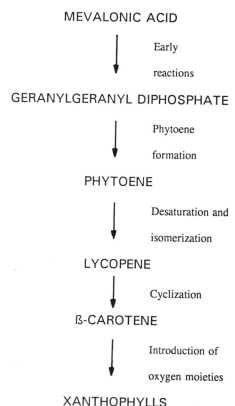

FIG. 11.1. Outline of the stages of carotenoid biosynthesis.

isoprenoid units are formed and used to construct the required chain length of prenyl diphosphate intermediates, are common to the biosynthesis of all isoprenoid classes. Only the later stages, after geranylgeranyl diphosphate (GGDP), are unique to carotenoid biosynthesis (summarised in Fig. 11.1), and will be the subject of this chapter. Details of the earlier stages of isoprenoid biosynthesis can be found in several reviews (Goodwin, 1980; Poulter and Rilling, 1983; Qureshi and Porter, 1983; Poulter, 1990; Alonso and Croteau, 1992).

A. Phytoene Synthesis

The first specific reaction in carotenoid formation is the condensation of two molecules of GGDP to form phytoene. This reaction proceeds via a C_{40} intermediate, prephytoene diphosphate (PPDP). Phytoene is formed directly from PPDP via the stereospecific loss of a proton (Fig. 11.2). In higher plants the phytoene produced is the 15-Z-isomer (Jungalwala and Porter, 1967).

B. Desaturation of Phytoene

15-Z-Phytoene undergoes a series of sequential desaturations (Fig. 11.3) to form phytofluene (7,8,11,12,7′,8′-tetrahydro-ψ,ψ-carotene), ξ-carotene (7,8,7′,8′-tetrahydro-ψ,ψ-carotene), neurosporene (7,8-dihydro-ψ,ψ-carotene) and finally, lycopene

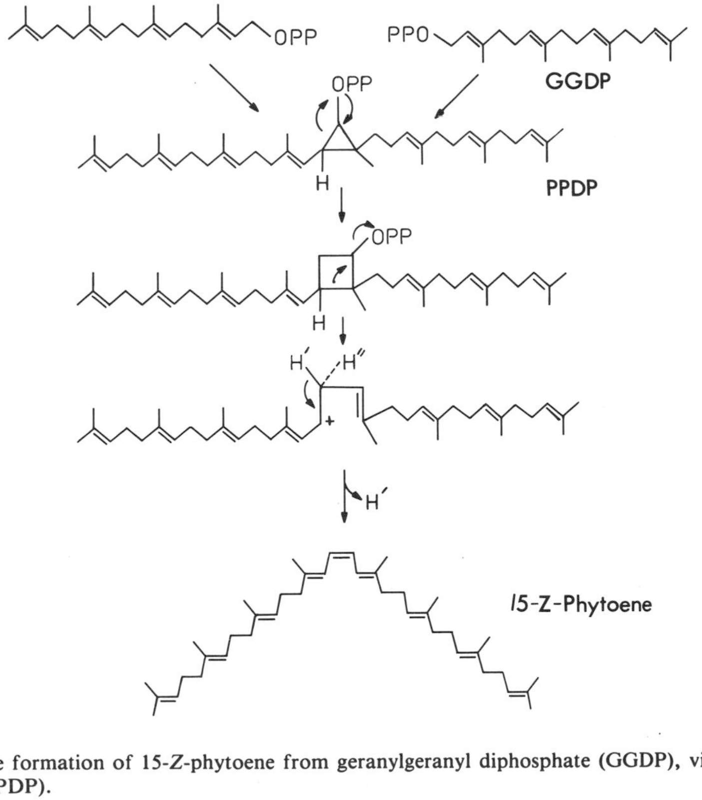

FIG. 11.2. The formation of 15-Z-phytoene from geranylgeranyl diphosphate (GGDP), via prephytoene diphosphate (PPDP).

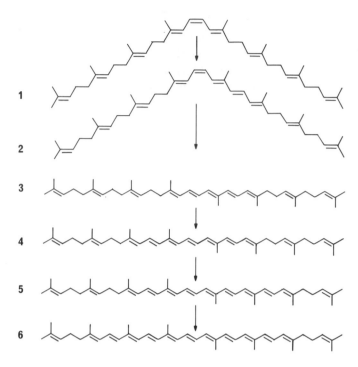

FIG. 11.3. The desaturation of 15-Z-phytoene (**1**) to 15-Z-phytofluene (**2**), all-E-phytofluene (**3**), ζ-carotene (**4**), neurosporene (**5**) and lycopene (**6**).

(ψ,ψ-carotene). At each stage, two hydrogen atoms are lost by *trans*-elimination from adjacent positions (McDermott *et al.*, 1973) to produce a new double bond and extend the conjugated polyene chromophore by two double bonds.

In higher plants and algae the desaturation sequence must include an isomerisation of the 15,15′ double bond, since the more unsaturated carotenes are all-E-isomers. The stage at which this isomerisation occurs may not be the same in all organisms. In both tomato and *Capsicum* chromoplasts isomerisation of phytofluene has been proposed (Kushwaha *et al.*, 1970; Camara *et al.*, 1980), but in a mutant strain of *Scenedesmus obliquus* isomerisation at ζ-carotene or lycopene may be important (Powls and Britton, 1977; Britton *et al.*, 1977a; Humbeck, 1990). In contrast, the halotolerant alga *Dunaliella bardawil* produces 9-Z-phytoene, which can be isomerised to all-E-phytoene prior to desaturation (Ben-Amotz *et al.*, 1988). Finally, the tangerine mutant of tomato accumulates prolycopene, a poly-Z-isomer, instead of the all-E form, as well as a series of Z-isomers of the desaturation intermediates (Clough and Pattenden, 1979, 1983; Frecknall and Pattenden, 1984).

C. Cyclisation

Once a carotenoid end-group has reached the correct level of desaturation, e.g. as in lycopene, cyclisation can take place to give 6-membered β- or ε-rings, at one or both ends of the carotene molecule (Fig. 11.4). It is generally accepted that the rings are

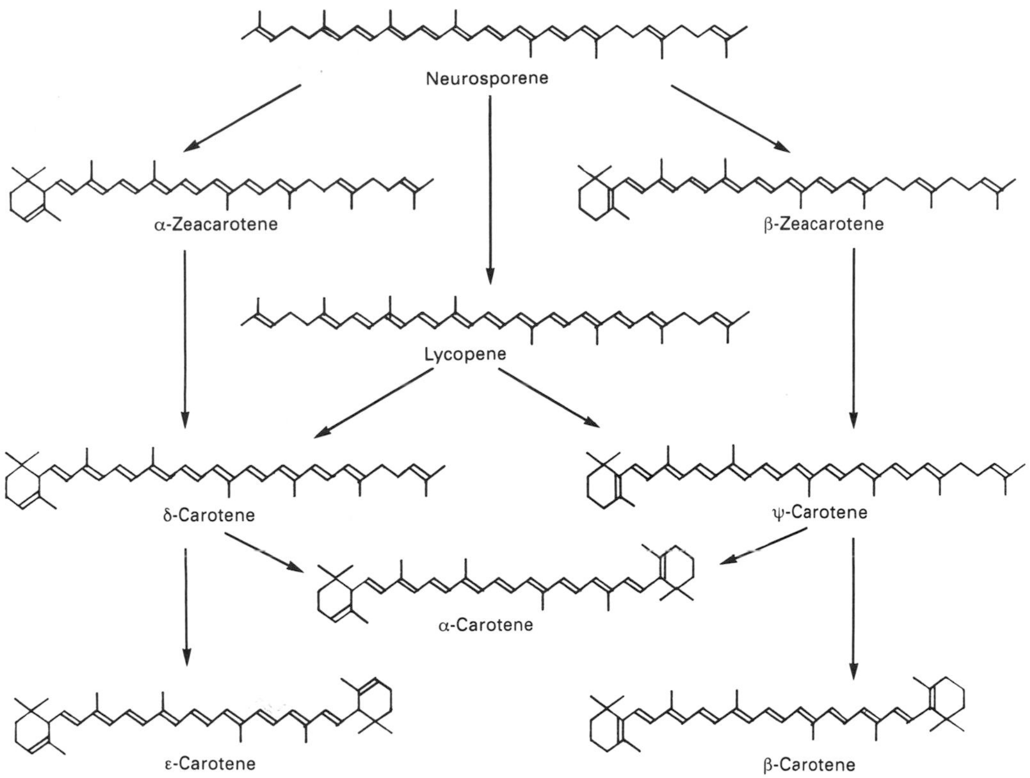

FIG. 11.4. Cyclisation pathways of carotenoids.

formed by proton loss from alternative positions (C-4 or C-6) in the same transient carbonium ion intermediate. The two rings are not interconverted and the stereochemistry of their formation has been elucidated in *Scenedesmus obliquus* (Britton *et al.*, 1977b, c; Britton, 1985a).

D. Xanthophyll Formation

Oxygen functions are thought to be introduced into carotenoids at the end of biosynthetic sequences (Fig. 11.1). The stereochemistry of hydroxylation at C-3 and C-3′, e.g. in zeaxanthin (β,β-carotene-3,3′-diol) or lutein (β, ε-carotene-3,3′-diol) has been investigated (Britton, 1976). The cyclopentane (K) ring of capsanthin (3,3′-dihydroxy-β,K-carotene-6′-one) is thought to be formed by the rearrangement of a 5,6-epoxy-β-ring end-group (Camara, 1980; Camara and Moneger, 1981).

The so-called 'violaxanthin cycle' in which zeaxanthin, antheraxanthin (5,6-epoxy-5,6-dihydro-β,β-carotene-3,3′-diol) and violaxanthin (5,6,5′,6′-diepoxy-5,6,5′,6′-tetrahydro-β,β-carotene-3,3′-diol) are interconverted has been known for some time (Yamamoto, 1979). The cycle involves the light-driven de-epoxidation of violaxanthin to zeaxanthin, via antheraxanthin. Reversal of the sequence is stimulated by the dark. The structures of typical plant xanthophylls are shown in Fig. 11.5.

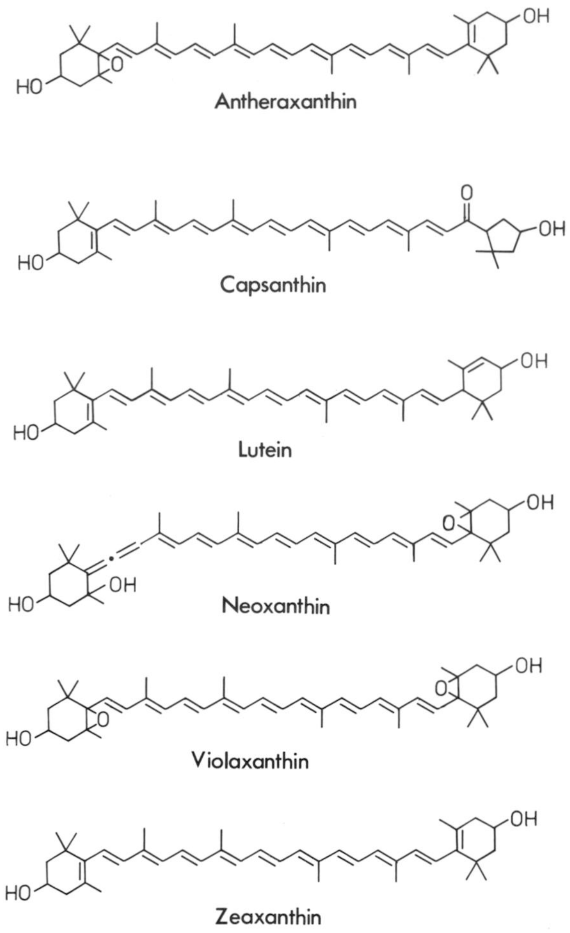

FIG. 11.5. Examples of xanthophylls found in higher plants.

III. ENZYME ASSAYS AND CELL-FREE SYSTEMS

No assays have yet been developed which measure carotenogenic enzyme activities by spectrophotometric or manometric methods. Therefore, it is necessary to use the time-consuming approach of withdrawing samples from incubations and estimating the amount of product, usually by the incorporation of stable or radioactive isotopes.

A. Preparation of Substrates

The most popular substrates used to assay *in vitro* carotenogenesis are those that are commercially available, e.g. [2-^{14}C]mevalonic acid (MVA) and [1-^{14}C]isopentenyl diphosphate (IDP). Stereospecifically labelled MVA can also be purchased, and deuterium oxide (D$_2$O) is useful for studies on the mechanism of cyclisation (Britton *et al.*, 1977b,c). Unfortunately these compounds are precursors of all isoprenoids and so they are not incorporated specifically into the carotenoids.

The direct precursors of carotenoids, or the carotenoids themselves, cannot be purchased labelled with stable or radioactive isotopes, and have to be prepared in the laboratory. Radioactively labelled GGDP can be formed from MVA with enzyme preparations from several organisms and tissues (see Bramley, 1985). Alternatively, it can be chemically synthesised (Upper and West, 1967; Gafni and Schechter, 1978; Soll and Schultz, 1981). PPDP can be prepared from GGDP with a cell extract of *Mycobacterium* (Altman *et al.*, 1972). It has been synthesised from diazogeranyl-geranial and geranylgeraniol, followed by phosphorylation of the resulting alcohol (Dogbo *et al.*, 1987).

Substrate quantities of radioactive carotenoids, with high specific activities, have been prepared from a variety of cell extracts incubated with several precursors. In all cases, it is essential that the substrates are vigorously purified prior to use in enzyme assays. It is particularly important to verify that the correct (usually all-*E*) isomer is used as the substrate, as *Z*-isomers are frequently produced during synthesis and purification (see Section III.B). 15-*Z*-Phytoene can be formed from MVA or IDP with cell extracts of tomato fruits (Kushwaha *et al.*, 1970), *carB* strains of *P. blakesleeanus* (Sandmann *et al.*, 1980; Fraser *et al.*, 1991) or chromoplast stroma of daffodil (Kreuz *et al.*, 1982) or *Capsicum* (Camara *et al.*, 1982).

Tomato plastids, incubated with [1-^{14}C]IDP, produce ^{14}C-labelled 15-*Z*- and all-*E*-phytofluene, as well as all-*E*-ζ-carotene, neurosporene, lycopene and γ-carotene (Papastephanou *et al.*, 1973; Qureshi *et al.*, 1974a, b). [^{14}C]Neurosporene can be isolated from *Rhodopseudomonas spheroides* grown with [2,3-^{14}C]succinate and 7 mM nicotine (Bramley *et al.*, 1977), while [^{14}C]lycopene and γ-carotene are formed from [2-^{14}C]MVA by cell extracts of a *carR* mutant of *Phycomyces* (Bramley *et al.*, 1977). The chemical synthesis of deuterated and tritiated lycopene have been reported (Schwieter *et al.*, 1969; Mayer and Isler, 1971; Eugster, 1979).

^{14}C-Labelled β-carotene can be isolated from cultures of *P. blakesleeanus* (Goodman and Olson, 1969) or *Blakeslea trispora* (Purcell and Walter, 1971), grown on [^{14}C]acetate. Alternatively, cell extracts of *Phycomyces* (Bramley and Davies, 1976) or *Amphidinium carterae* (Swift *et al.*, 1982) can be used with [2-^{14}C]MVA.

The preparation of labelled xanthophylls is not as well documented. One method is the incorporation of ^{14}CO$_2$ into lutein, violaxanthin and neoxanthin of greening leaves (Britton and Goodwin, 1971). Cell extracts of *A. carterae* produce a range of ^{14}C-labelled xanthophylls from [2-^{14}C]MVA (Swift *et al.*, 1982). [^{14}C]Zeaxanthin can also be obtained by incubating *Flavobacterium* cells with [2-^{14}C]MVA (Britton *et al.*, 1980). The synthesis of [^{3}H]antheraxanthin, zeaxanthin and violaxanthin has been reported (Camara and Moneger, 1980, 1981). ^{14}C-Labelled β-cryptoxanthin (3-hydroxy-β-carotene) can be prepared from [^{14}C]phytoene with a membrane preparation of *Aphanocapsa* (Sandmann, 1988).

B. Preparation of Cell-free Systems

A range of techniques has been used to obtain carotenogenic cell extracts from higher plants, algae and cyanobacteria (Bramley, 1990; Table 11.1). The resultant extracts are either used as crude cell-free systems or fractionated to yield organelle preparations

TABLE 11.1. Examples of carotenogenic cell-free systems from higher plants, algae and cyanobacteria.

Tissue/Organism	Comments	Reference
Capsicum annuum, 　*Narcissus*, tomato	Chromoplast preparations	Camara *et al.* (1985) Camara (1985), Beyer *et al.* 　(1985), Porter and 　Spurgeon, (1979)
C. annuum	Xanthophyll interconversions	Camara (1980)
Zea mays	Chloroplasts, producing 　β-carotene	Mayfield *et al.* (1986)
Spinacia oleracea	Chloroplast envelope, 　violaxanthin synthesis	Costes *et al.* (1979)
Lactuca sativa	Violaxanthin cycle	Siefermann and Yamamoto 　(1975)
Amphidinium carterae	Xanthophyll synthesis	Swift *et al.* (1982)
Aphanocapsa	Solubilised fraction producing 　xanthophylls	Sandmann (1988)
Anacystis R2	Xanthophyll formation	Sandmann and Kowalcyk 　(1989)

or membrane fractions. The extraction buffers often include a sulphydryl reagent such as dithiothreitol and may also include a cocktail of protease inhibitors (Gegenheimer, 1990) and polyvinyl pyrrolidone (Loomis and Battaile, 1966) if endogenous proteases or phenolics are present in the extracts, respectively.

In general, plant tissues should be freshly harvested and young; cell breakage of older tissues is more difficult, and often results in low carotenogenic activities. Similarly, ripening, rather than fully ripe, fruit yields higher enzymic activities. Storage of cell extracts is feasible, but the addition of cryoprotectants such as ethylene glycol or glycerol is recommended prior to freezing at − 70°C.

C. Incubation Conditions and Analytical Techniques

Cell extracts are typically incubated with labelled substrates and a mixture of cofactors (see Table 11.2). Alternations in conditions, such at $\pm O_2$, ±light or with different subcellular organelles can often provide useful information regarding the properties of the enzyme under investigation. Water-soluble substrates, e.g. MVA, IDP and GGDP, present few practical difficulties since they readily dissolve in aqueous incubation mixtures. In order to inhibit the hydrolysis of prenyl diphosphates by endogenous phosphatases, potassium fluoride is often added to incubations (e.g. Taylor and Davies, 1982).

Carotenoid substrates are more difficult to use, since they require emulsification in the incubation mixture before addition of the cell extract. This can be achieved by adding the carotenoid, dissolved in light petroleum, acetone or ethanol, to a detergent solution (e.g. Tween 80) which is then shaken vigorously. After removal of the organic solvent, the emulsified carotenoid is added to the cell extract (e.g. Bramley and Davies,

TABLE 11.2. Methods for the preparation of carotenogenic cell extracts from higher plants, algae and cyanobacteria.

Technique	Tissue/Organism	References
Grinding with abrasives	Carrot, Pea fruit	Anderson and Porter (1962), Graebe (1968)
Homogenisation with blender	Pea fruit	Pollard *et al.* (1966)
	Spinach	Subbarayan *et al.* (1970)
	Tomato fruit	Varma and Chichester (1962)
Pressure cell	*Anacystis* R2	Sandmann and Kowalczyk (1989)
Osmotic shock of spheroplasts	*Aphanocapsa*	Clarke *et al.* (1982)
Isolation of chloroplasts	Spinach	Bickel and Schultz (1976), Grumbach (1980), Kreuz and Kleinig (1981)
	Phaseolus vulgaris	Charlton *et al.* (1967), Buggy *et al.* (1974)
Plastid preparations, including acetone powders, membrane fractions and soluble extracts	Tomato fruit	Jungalwala and Porter (1967), Hill *et al.* (1971), Qureshi *et al.* (1974a,b)
	Spinach	Kushwaha *et al.* (1969), Costes *et al.* (1979), Douce and Joyard (1979), Lütke-Brinkhaus *et al.* (1983)
Chromoplast isolation	*Capsicum* fruit	Camara (1980), Camara and Monéger (1981), Camara *et al.* (1982)
	Narcissus flowers	Liedvogel *et al.* (1976), Kreuz *et al.* (1982)

1976; Kushwaha *et al.*, 1976). An alternative procedure is to add the carotenoid, dissolved in ethanol, directly to the incubation (Swift *et al.*, 1982). A convenient and reproducible procedure to avoid the necessity of emulsification is to use a coupled assay, in which the [^{14}C]carotenoid is formed *in situ*, from [2-^{14}C]MVA, and the extract to be assayed is subsequently added directly to the preformed substrate. This approach has been particularly successful in assays of phytoene desaturase, using a cell extract of a *carB* (phytoene-accumulating) strain of *P. blakesleeanus* to produce [^{14}C]phytoene (Sandmann and Bramley, 1985a; El-Jack *et al.*, 1988; Fraser *et al.*, 1991). Coupled assays which utilise fungal mutants blocked at other steps in the carotenoid pathway have also been employed successfully (Sandmann and Bramley, 1985b; El-Jack *et al.*, 1988; Sandmann and Kowalczyk, 1989).

Carotenoids are sensitive to heat, light, oxygen, acids and in some cases, alkali. Consequently, there are considerable practical difficulties associated with their purification and analysis. These problems are heightened with the use of radioactively-labelled substrates since particularly vigorous purification procedures must be employed in order to avoid contamination with other radioactive compounds. Typically, the carotenoids are extracted from incubation mixtures with organic solvents such as light petroleum, chloroform or diethyl ether. The mixture of products is separated by thin layer chromatography and/or HPLC. General protocols for the purification of carotenoids can be found in several reviews (Britton, 1985b, 1991; Goodwin and Britton, 1988) whilst specific protocols applied to a particular study are detailed in the appropriate publications.

Accurate quantitation of radioactivity in the pure carotenoid is obtained by liquid scintillation counting (e.g. Bramley *et al.*, 1974), or by integration of peak areas following detection in HPLC eluates with radioactivity detectors. Deuterated carotenoids are detected by mass spectrometry (e.g. Britton *et al.*, 1977b, c).

IV. PURIFICATION AND PROPERTIES OF CAROTENOGENIC ENZYMES

A. Phytoene Synthase

Numerous crude extracts are able to form phytoene from precursors such as MVA, IDP and GGPP and possible cofactor requirements have been reported (Table 11.3). There is little consensus in the studies, as several pyridine and flavin nucleotides have been shown to stimulate or inhibit phytoene formation. ATP is stimulatory in some extracts when either IDP (Maudinas *et al.*, 1975, 1977; Clarke *et al.*, 1981) or GGDP (Clarke *et al.*, 1982) is used as the substrate. The requirement for a divalent cation has often been reported, with Mn^{2+} being required by the plant enzyme, whilst Mg^{2+} is necessary for the fungal phytoene synthase.

More definitive experiments over the past few years have clarified some of these anomalies. In particular, phytoene synthase has been purified from the chromoplast

TABLE 11.3. Possible cofactor requirements for carotenoid biosynthesis.

Organism	Cofactor requirement Essential	Stimulatory	Reference
		1. Phytoene formation	
Aphanocapsa	—	ATP FAD NAD(P)	Clarke *et al.* (1982)
C. annuum	Mn^{2+}		Dogbo *et al.* (1988)
Flavobacterium	—	NAD(P) FAD NAD(P)H	Brown *et al.* (1975)
N. crassa	Mg^{2+}	—	Spurgeon *et al.* (1979)
P. blakesleeanus	—	ATP	Clarke *et al.* (1981)
Spinach	Light	Boiled cell extract	Subbarayan *et al.* (1970)
Tomato	Mn^{2+}/Mg^{2+}	—SH reagent	Jungalwala and Porter (1967)
	Mn^{2+}	ATP NADP	Mandinas *et al.* (1975, 1977)
		2. Phytoene to lycopene	
Anacystis	$NAD(P)O_2$	—	Sandmann and Kowakzyk (1989)
P. blakesleeanus	—	NADP	Clarke *et al.* (1981)
Spinach	$NADP\ FAD^a$		Subbarayan *et al.* (1970)
Tomato	$FAD^b Mn^{2+}$	Mg^{2+}	Kushwaha *et al.* (1970)
	$NADP^c FAD$	Mg^{2+}/Mn^{2+}	Qureshi *et al.* (1974b)

[a]Phytoene into phytofluene.
[b]Phytofluene into lycopene.
[c]Phytofluene into unsaturated carotenes.

TABLE 11.4. Purification of phytoene synthase from *Capsicum annuum*.

Fraction	Total protein (mg)	Specific activity[a]	Purification factor
Chromoplast stroma	1250	8	1
Polyethylene glycol	275	25	3.12
DEAE-Sephacel	60	80	10
Amino phenethyl diphosphate Sepharose	1.1	3500	437.5
Affi-Gel 501	0.9	4000	500

From Dogbo *et al.*, 1988; reproduced with permission.
[a]Expressed as nmol GGDP incorporated into phytoene per mg protein in 1 h.

stroma of *C. annuum*, to yield a single protein of molecular weight 47.5 kDa, with an absolute requirement for Mn^{2+}, but no other cofactor (Dogbo *et al.*, 1988). This enzyme catalyses a two-step, kinetically coupled reaction (Fig. 11.2), pH optimum 7.6, with K_m values of 0.3 and 0.27 μM for GGDP and PPDP, respectively. The reaction is inhibited by inorganic phosphate. The key step in the purification protocol was affinity chromatography, using aminophenethyl diphosphate as the ligand (Table 11.4).

Phytoene synthase is also located in the stroma of chloroplasts and amyloplasts of *Capsicum*, with the highest activity in the non-photosynthetic plastids (Dogbo *et al.*, 1987). It has also been found in the stroma of pea fruits (Graebe, 1968) and *Triticum* leaves (Kleinig, 1989). In contrast, it is reported to be located on the chloroplast envelope of spinach (Lütke-Brinkhaus *et al.*, 1982) and a peripheral membrane protein of the inner membrane of *Narcissus* chromoplasts (Kreuz *et al.*, 1982). Whether there is a genuine difference in the intracellular location of phytoene synthase between plant species remains to be confirmed, but it does appear that in *Triticum* and *Capsicum*, the protein is encoded by a nuclear gene and is synthesised on 80S ribosomes prior to post-translational processing and entry into the plastid (Camara, 1984).

B. Phytoene Desaturase

To date there has been no report of the purification of a higher plant or algal enzyme capable of converting phytoene into unsaturated carotenes, and so the only information available on the properties of phytoene desaturase of plants has been deduced from studies of crude preparations, often from chromoplasts (Table 11.1). It has been solubilised, with retention of enzymic activity, from *Narcissus* chromoplasts with CHAPS (Beyer *et al.*, 1985) and with Tween 40 from *Aphanocapsa* membranes (Bramley and Sandmann, 1987). Acetone powers of tomato fruit are also active (Porter and Spurgeon, 1979). Detergent-solubilised preparations of fungi have been reported (Bramley and Taylor, 1985; Rau and Mitzka-Schnabel, 1985). In all these cases, phytoene desaturase is membrane-bound; in plants it is thought to be located in the chloroplast envelope, and not the thylakoids (Lütke-Brinkhaus *et al.*, 1982) and on the inner chromoplast membrane of *Narcissus* (Kreuz *et al.*, 1982) and C. *annuum* (Camara *et al.*, 1982).

Although it appears from *in vivo* studies that the conversion of 15-Z-phytoene to all-*E*-lycopene involves four desaturations and one isomerisation (Fig. 11.2), the numbers of enzymes required to catalyse these conversions is still open to question. The use of 15-*Z*- and all-*E*-phytoenes and phytofluenes as substrates in cell extracts of tomato (Kushwaha *et al.*, 1970) and *Capsicum* chromoplasts (Camara, 1981) has shown that isomerisation occurs at the stage of phytofluene in these tissues, but no isomerase has been isolated. It may be that a strained carotene molecule present in the enzyme complex is the true intermediate (Goodwin, 1983). It is important to remember that experiments with geometric isomers of carotenes *in vitro* must be carried out carefully (see Section III.A) in order to avoid the anomalies found between *in vitro* and *in vivo* studies with *Tangella* tomatoes (Qureshi *et al.*, 1974b; Clough and Pattenden, 1979, 1983). The CHAPS-solubilised preparation of *Narcissus* catalyses a series of *Z*-to-*E* isomerisations *in vitro*, although there are no reports of such reactions occurring in the intact flower (Beyer *et al.*, 1989).

There are several candidates for the cofactors required for phytoene desaturase, such as NADP, FAD and/or cytochromes (Table 11.3). Molecular oxygen appears to be required, and metal ions are stimulatory in crude preparations. Early studies with tomato fruit and spinach showed that different cofactors were necessary for the phytoene to phytofluene, and phytofluene to lycopene steps (Kushwaha *et al.*, 1970; Subbarayan *et al.*, 1970). This may be due to separate enzymes being required (see below), or that different cofactors are needed for the dehydrogenation of *E*- and *Z*-isomers. Recent studies with phytoene desaturase of the *Phycomyces* have shown a similar diversity in this respect (P. Fraser and P. M. Bramley, unpub. obs.).

There is some experimental evidence to show that in higher plants, green algae and cyanobacteria, two enzymes are required, the so-called phytoene desaturase, which converts 15-*Z*-phytoene into all-*E*-ζ-carotene, and then ζ-carotene desaturase, which forms all-*E*-lycopene from ζ-carotene, via neurosporene (Fig. 11.2).

The first line of evidence comes from studies on carotene mutants, which are blocked at phytoene or ζ-carotene (reviewed by Bramley and Mackenzie, 1988). Secondly, *in vitro* carotenogenic activities of *Narcissus* and *Aphanocapsa* point to two enzymes being involved. The conversions of phytoene to ζ-carotene and ζ-carotene to lycopene in *Narcissus* have different cofactor requirements and involve the *trans* removal of hydrogens to form ζ-carotene, but *cis* elimination in the ζ-carotene to lycopene sequence. The two stages are also affected differently by the bleaching herbicide, norflurazon (Mayer *et al.*, 1989). Inhibition studies will cell membranes of *Aphano-capsa* also indicate that two enzymes are present in this cyanobacterium (see Sandmann and Boger, 1989). In contrast, genetic studies with *Phycomyces* indicate that a gene codes for a single protein which may aggregate *in vivo* to form an active carotenogenic complex (De la Guardia *et al.*, 1971).

Whatever number of enzymes are responsible for carotene desaturation in photo-synthetic tissues, there is general agreement that the gene(s) is nuclear and the proteins synthesised on 80S ribosomers prior to transport into the plastid (Kirk and Tilney-Bassett, 1978).

C. Other Reactions

No cyclase enzymes have been purified, but lycopene cyclase is membrane bound, probably in the plastid envelope, in both tomato and *Capsicum* fruits (Camara and

Moneger, 1982). Several different cyclisation reactions have been demonstrated with tomato plastids, and the formation of α-, β-, γ- and δ-carotenes (Fig. 11.4) requires FAD (Maudinas *et al.*, 1975). In contrast, no nucleotide cofactor is required for β-carotene synthesis in *Capsicum*, and the intermediate γ-carotene was not observed under optimal rates of conversion, which included the presence of Tween 80 (Camara *et al.*, 1985).

Very few enzymic studies on xanthophyll biosynthesis have been reported. The conversion of zeaxanthin into antheraxanthin and violaxanthin by lettuce chloroplasts requires O_2 and NADPH (Siefermann and Yamamoto, 1975). It was concluded from this study that the epoxidase was an 'external monoxygenase' located in a chloroplast compartment that remains 'neutral' during illumination, i.e. the violaxanthin cycle is a transmembrane system, whereby violaxanthin de-epoxidation occurs on the lumen side of the thylakoid membrane, whilst epoxidation takes place on the stroma side. The hydroxylation of β-carotene to β-cryptoxanthin also requires NADPH and is probably catalysed by a mixed function oxygenase, as the hydroxy group originates from O_2 (Sandmann and Bramley, 1985b).

V. FUTURE STUDIES

Despite the availability of several carotenogenic cell extracts from higher plants, our understanding of the enzymes responsible for carotenoid formation is still rudimentary. The main reason for this is the difficulty in purifying low abundance, membrane-bound proteins. It is important, therefore, to utilise to the full the biochemical and molecular biological approaches which are now available.

In particular, protein separation procedures based on HPLC should be adopted, as they offer significantly better resolution than traditional chromatographic techniques. It should be possible to use affinity chromatography, perhaps with inhibitors of carotenogenesis as ligands, provided that they do not 'leak' from the column. The purification of phytoene synthase included an affinity chromatography step (Table 11.4). Our own efforts to purify phytoene desaturase have successfully used isolelectric focusing as a later step in purification (Fraser, 1991). A comparison with the protocols used for the purification of other membrane-bound isoprenoid enzymes, e.g. HMG-CoA reductase (Kennelly *et al.*, 1983), show that the inherent difficulties can be overcome.

A complementary approach is to utilise the recently cloned and sequenced carotenoid genes as molecular probes. For example, carotenoid genes from *Rhodobacter* (Armstrong *et al.*, 1989, 1990a), *Erwinia* (Misawa *et al.*, 1990), *Aphanocapsa* (Schmidt and Sandmann, 1990), tomato (Ray *et al.*, 1987; Bird *et al.*, 1991) and *Neurospora* (Schmidhauser *et al.*, 1990) have been sequenced. The predicted amino acid sequences of the proteins show some homology between each other (Armstrong *et al.*, 1990b), with the phytoene desaturase primary sequence containing an ADP-binding, $\beta\alpha\beta$-fold (Bartley *et al.*, 1990). The preparation of fusion proteins, followed by antibody production, immunoprecipitation and Western blotting, should assist in the isolation of the appropriate enzymes. Some success with immunolocalisation (Serrano *et al.*, 1990) and molecular weight determinations (Schmidt *et al.*, 1989) has been reported. It must be remembered, however, that the proteins synthesised in the cytoplasm of higher plant cells will be processed during transport and targeting in the plastid. This will alter their molecular weight and possibly their antigenic properties. Thus antibodies to fusion proteins may not interact with the mature, processed enzyme. The definitive experimental approach remains that of isolating a pure, enzymically active protein.

It is most likely that a significant number of carotenoid genes will be cloned and sequenced in the near future, including those from higher plants and algae. The purification of the corresponding enzymes will be far more difficult, but must be achieved if we are to understand the molecular mechanisms underlying carotenogenesis. It should then be possible to turn our attention to the complex regulatory mechanisms associated with plastid development and the concomitant synthesis, transport, targeting and turnover of these enzymes.

REFERENCES

Alonso, D. W. and Croteau, R. (1992). Chapter 9, this volume.

Altman, L. J., Ash, L., Kowerski, R. C., Epstein, W. W., Larsen, B. R., Rilling, H. R., Muscio, F. and Gregonis, D. E. (1972). *J. Am. Chem. Soc.* **94**, 3257–3259.

Anderson, D. G. and Porter, J. W. (1962). *Arch Biochem. Biophys.* **97**, 509–519.

Armstrong, G. A. Alberti, M., Leach, F. and Hearst, J. E. (1989). *Mol. Gen. Genet.* **216**, 254–268.

Armstrong, G. A., Schmidt, A., Sandmann, G. and Hearst, J. E. (1990a). *J. Biol. Chem.* **265**, 8329–8338.

Armstrong, G. A. Alberti, M. and Hearst, J. E. (1990b). *Proc. Natl. Acad. Sci. USA* **87**, 9975–9979.

Bartley, G. E., Schmidhauser, T. J., Yanofsky, C. and Scolnik, P. A. (1990). *J. Biol. Chem.* **265**, 16020–16024.

Ben-Amotz, A., Lers, A. and Avron, M. (1988). *Plant Physiol.* **86**, 1286–1291.

Beyer, P., Weiss, G. and Kleinig, H. (1985). *Eur. J. Biochem.* **153**, 341–346.

Beyer, P., Mayer, M. and Kleinig, H. (1989). *Eur. J. Biochem.* **184**, 141–149.

Bickel, H. and Schultz, G. (1976). *Phytochemistry* **15**, 1253–1255.

Bird, C. R., Ray, J. A., Fletcher, J. D., Boniwell, J. M., Bird, A. S. Teulieres, C., Blain, I., Bramley, P. M. and Schuch, W. (1991). *BioTechnology*, **9**, 635–639.

Bramley, P. M. (1985). *Adv. Lip. Res.* **21**, 243–279.

Bramley, P. M. (1990). *In* "Carotenoids: Chemistry and Biology" (N. I. Krinsky, M. M. Mathews-Roth and R. F. Taylor, eds), pp. 185–194. Plenum, New York.

Bramley, P. M. and Davies, B. H. (1976). *Phytochemistry* **15**, 1913–1916.

Bramley, P. M. and Mackenzie, A. (1988). *Curr. Top. Cell, Regln.* **29**, 291–343.

Bramley, P. M. and Sandmann, G. (1987). *Phytochemistry* **26**, 1935–1939.

Bramley, P. M. and Taylor, R. F. (1985). *Biochim. Biophys. Acta* **839**, 155–160.

Bramley, P. M., Davies, B. H. and Rees, A. F. (1974). *In* "Liquid Scintillation Counting", Vol. 3 (M. A. Crook and P. Johnson, eds), pp. 76–85. Heyden, London.

Bramley, P. M., Than, A. and Davies, B. H. (1977). *Phytochemistry* **16**, 235–238.

Britton, G. (1985a). *Pure Appl. Chem.* **57**, 701–708.

Britton, G. (1985b). *Methods Enzymol.* **111**, 113–149.

Britton, G. (1990). *In* "Carotenoids: Chemistry and Biology" (N. I. Krinsky, M. M. Mathews-Roth and R. F. Taylor, eds), pp. 167–184. Plenum, New York.

Britton, G. (1991). *In* "Methods in Plant Biochemistry", Vol. 7 (B. V. Charlwood and D. V. Banthorpe, eds), pp. 473–518. Academic Press, London.

Britton, G. and Goodwin, T. W. (1971). *Methods Enzymol.* **18C**, 654–701.

Britton, G., Powls, R. and Schulze, R. M. (1977a). *Arch. Microbiol.* **113**, 281–284.

Britton, G., Lockley, W. J. S., Patel, N. J., Goodwin, T. W. and Englert, G. (1977b). *J. Chem. Soc., Chem. Commun.*, 655–656.

Britton, G., Lockley, W. J. S., Patel, N. J. and Goodwin, T. W. (1977c). *FEBS Lett.* **79**, 281–283.

Britton, G., Goodwin, T. W., Brown, D. J. and Patel, N. J. (1980). *Methods Enzymol.* **67F**, 264–270.

Brown, D. J., Britton, G. and Goodwin, T. W. (1975). *Biochem. Soc. Trans.* **3**, 741–742.

Buggy, M. J., Britton, G. and Goodwin, T. W. (1974). *Phytochemistry* **13**, 127–129.

Camara, B. (1980). *Biochem. Biophys. Res. Commum.* **93**, 113–117.

Camara, B. (1981). PhD Thesis, Université Pierre et Marie Curie, Paris.

Camara, B. (1984). *Plant Physiol.* **74**, 112–116.

Camara, B. (1985). *Pure Appl. Chem.* **57**, 675–677.

Camara, B. and Monéger, R. (1980). *In* "Biogenesis and Function of Plant Lipids" (P. Mazliak, P. Benveniste, C. Costes and R. Douce, eds), pp. 363–367. Elsevier, Amsterdam.

Camara, B. and Monéger, R. (1981). *Biochem. Biophys. Res. Commun.* **99**, 1117–1122.

Camara, B. and Monéger, R. (1982). *Physiol. Vég.* **20**, 757–773.

Camara, B., Payan, C., Escoffier, A. and Monéger, R. (1980). *C. R. Hebd, Séances Acad. Sci., Ser. D* **291**, 303–306.

Camara, B., Bardat, F. and Monéger, R. (1982). *Eur. J. Biochem.* **127**, 255–258.

Camara, B., Dogbo, O., d'Harlingue, A., Kleinig, H. and Monéger, R. (1985). *Biochem. Biophys. Acta* **836**, 262–266.

Charlton, J. M., Treharne, K. J. and Goodwin, T. W. (1967). *Biochem. J.* **105**, 205–212.

Clarke, I. E., Sandmann, G. and Bramley, P. M. (1981). Proceedings of the 6th International Symposium on Carotenoids, Liverpool.

Clarke, I. E., Sandmann, G., Bramley, P. M. and Böger, P. (1982). *FEBS Lett.* **103**, 17–21.

Clough, J. M. and Pattenden, G. (1979). *J. Chem. Soc., Chem. Commun.*, 616–619.

Clough, J. M. and Pattenden, G. (1983). *J. Chem. Soc., Perkin Trans. 1*, 3011–3018.

Costes, C., Burghoffer, C., Joyard, J., Block, M. and Douce, R. (1979). *FEBS Lett.* **103**, 17–21.

Davies, B. H. (1980). *In* "Pigments in Plants" (F. C. Czygan, ed.), pp. 31–56. Gustav Fischer, Stuttgart.

De la Guardia, M. D., Aragon, C. M. G., Murillo, F. J. and Cerdá-Olmedo, E. (1971). *Proc. Natl. Acad. Sci. USA* **68**, 2012–2015.

Dogbo, O., Bardat, F., Laferrière, A., Quennemet, J., Brangeon, J. and Camara, B. (1987). *Plant Sci.* **49**, 89–101.

Dogbo, O., Laferrière, A., d'Harlingue, A. and Camara, B. (1988). *Proc. Natl. Acad. Sci. USA* **85**, 7054–7058.

Douce, R. and Joyard, J. (1979). *In* "Plant Organelles", Vol. 9 (G. Reid, ed.), pp. 47–59. Ellis Horwood, Chichester.

El-Jack, M., Mackenzie, A., and Bramley, P. M. (1988). *Planta* **174**, 59–66.

Eugster, C. H. (1979). *Pure Appl. Chem.* **51**, 463–506.

Fraser, P. (1991). PhD Thesis. University of London.

Fraser, P., Mackenzie, A. and Bramley, P. M. (1991). *Phytochemistry*, **30**, 3971–3976.

Frecknall, E. A. and Pattenden, G. (1984). *Phytochemistry* **23**, 1707–1710.

Gafni, Y. and Shechter, I. (1978). *Anal. Biochem.* **92**, 248–252.

Gegenheimer, P. (1990). *Methods Enzymol.* **182**, 174–193.

Goodman, De, W. S. and Olson, J. A. (1969). *Methods Enzymol.* **115**, 462–475.

Goodwin, T. W. (1980). "The Biochemistry of Carotenoids", Vol. 1. Chapman and Hall, London.

Goodwin, T. W. (1983). *Biochem. Soc. Trans.* **11**, 473–483.

Goodwin, T. W. and Britton, G. (1988). *In* "Plant Pigments" (T. W. Goodwin, ed.), pp. 61–131. Academic Press, New York.

Graebe, J. E. (1968). *Phytochemistry* **7**, 2003–2020.

Grumbach, K. H. (1980). *In* "Biogenesis and Function of Plant Lipids" (P. Mazliak, R. Benveniste, C. Costes and R. Douce, eds), pp. 421–426. Elsevier, Amsterdam.

Hill, H. M., Calderwood, S. K. and Rogers, L. J. (1971). *Phytochemistry* **10**, 2051–2058.

Humbeck, K. (1990). *Planta* **182**, 204–210.

Jungalwala, F. B. and Porter, J. W. (1967). *Arch. Biochem. Biophys.* **119**, 209–219.

Kennelly, P. J., Brandt, K. G. and Rodwell, V. W. (1983). *Biochemistry* **22**, 2784–2788.

Kirk, J. T. O. and Tilney-Bassett, R. (1978). "The Plastids". Freeman and Co., San Francisco.

Kleinig, H. (1989). *Ann. Rev. Plant Physiol. Plant Mol. Biol.* **40**, 39–59.

Kreuz, K. and Kleinig, H. (1981). *Planta* **153**, 578–581.

Kreuz, K., Beyer, P. and Kleinig, H. (1982). *Planta* **154**, 66–69.

Kushwaha, S. C., Subbarayan, C., Beeler, D. A. and Porter, J. W. (1969). *J. Biol. Chem.* **244**, 3635–3642.

Kushwaha, S. C., Suzue, G., Subbarayan, C. and Porter, J. W. (1970). *J. Biol. Chem.* **245**, 4708–4717.

Kushwaha, S. C., Kates, M. and Porter, J. W. (1976). *Can. J. Biochem.* **54**, 816–823.

Liedvogel, B., Sitte, P. and Falk, H. (1976). *Cytobiology* **12**, 155–174.

Loomis, W. D. and Battaile, J. (1966). *Phytochemistry* **5**, 423–438.

Lütke-Brinkhaus, F., Liedvogel, B., Kreuz, B. and Kleinig, H. (1982). *Planta* **156**, 176–180.

McDermott, J. C. B., Britton, G. and Goodwin, T. W. (1973). *Biochem. J.* **134**, 1115–1117.

Maudinas, B., Bucholtz, M. L., Papastephanou, C., Katiyar, S. S., Briedis, A. V. and Porter, J. W. (1975). *Biochem. Biophys. Res. Commun.* **66**, 430–436.

Maudinas, B., Bucholtz, M. L., Papastephanou, C., Katiyar, S. S., Briedis, A. V. and Porter, J. W. (1977). *Arch. Biochem. Biophys.* **180**, 354–362.

Mayer, H. and Isler, O. (1971). *In* "Carotenoids" (O. Isler, ed.), pp. 325–575. Birkhauser-Verlag, Basel.

Mayer, M. P., Bartlett, D. L., Beyer, P. and Kleinig, H. (1989). *Pestic. Biochem. Physiol.* **34**, 111–117.

Mayfield, S. P., Nelson, T., Taylor, W. C. and Malkin, R. (1986). *Planta* **169**, 23–32.

Misawa, N., Nakagawa, M., Kobayashi, K., Yamano, S., Izawa, Y., Nakamura, K. and Harashiva, K. (1990). *J. Bacteriol.* **172**, 6704–6712.

Papastephanou, C., Barnes, F. J., Briedis, A. V. and Porter, J. W. (1973). *Arch. Biochem. Biophys.* **157**, 415–425.

Pollard, C. J., Bonner, J., Haagen-Smit, A. J. and Nimmo, C. C. (1966). *Plant Physiol.* **41**, 66–70.

Porter, J. W. and Spurgeon, S. L. (1979). *Pure Appl. Chem.* **51**, 609–622.

Poulter, C. D. (1990). *In* "Biochemistry of Cell Walls and Membranes in Fungi" (P. J. Kuhn, A. D. J. Trinci, M. J. Jung, M. W. Gorosey and L. G. Copping, eds), pp. 169–188. Springer-Verlag, New York.

Poulter, C. D. and Rilling, H. C. (1983). *In* "Biosynthesis of Isoprenoid Compounds", Vol. 1 (S. L. Spurgeon and J. W. Porter, eds), pp. 161–224. Wiley, New York.

Powls, R. and Britton, G. (1977). *Arch. Microbiol.* **115**, 175–179.

Purcell, A. and Walter, W. M. Jr. (1971). *Methods Enzymol.* **18**, 701–706.

Qureshi, N. and Porter, J. W. (1983). *In* "Biosynthesis of Isoprenoid Compounds", Vol. 1 (S. L. Spurgeon and J. W. Porter, eds), pp. 47–94. Wiley, New York.

Qureshi, A. A., Andrews, A. G., Qureshi, N. and Porter, J. W. (1974a). *Arch. Biochem. Biophys.* **162**, 93–107.

Qureshi, A. A., Qureshi, N., Kim, M. and Porter, J. W. (1974b). *Arch. Biochem. Biophys.* **162**, 117–125.

Rau, W. and Mitzka-Schnabel, U. (1985). *Methods Enzymol.* **110**, 253–267.

Ray, J., Bird, C., Maunders, M., Grierson, D. and Schuch, W. (1987). *Nucl. Acid Res.* **15**, 10587.

Sandmann, G. (1988). *Methods Enzymol.* **167**, 329–335.

Sandmann, G. and Böger, P. (1989). *In* "Target Sites of Herbicide Action" (P. Böger and G. Sandmann, eds), pp. 25–44. CRC Press, Boca Raton, FL.

Sandmann, G. and Bramley, P. M. (1985a). *Planta* **164**, 259–263.

Sandmann, G. and Bramley, P. M. (1985b). *Biochim. Biophys. Acta* **843**, 73–77.

Sandmann, G. and Kowalczyk, S. (1989). *Biochem. Biophys. Res. Commun.* **163**, 916–921.

Sandmann, G., Hilgenberg, W. and Böger, P. (1980). *Z. Naturforsch.* **35c**, 927–930.

Schmidhauser, T. J., Lauter, F. R., Russo, V. E. A. and Yanofsky, C. (1990). *Mol. Cell. Biol.* **10**, 5064–5070.

Schmidt, A. and Sandmann, G. (1990). *Gene* **91**, 113–117.

Schmidt, A., Sandmann, G., Armstrong, G. A., Hearst, J. E. and Böger, P. (1989). *Eur. J. Biochem.* **184**, 375–378.

Schwieter, U., Englert, G., Rigassi, N. and Vetter, W. (1969). *Pure Appl. Chem.* **20**, 365–420.

Serrano, A., Gimenez, P., Schmidt, A. and Sandmann, G. (1990). *J. Gen. Microbiol.* **136**, 2465–2469.

Sierfermann, D. and Yamamoto, H. Y. (1975). *Arch. Biochem. Biophys.* **171**, 70–77.

Soll, J. and Schultz, G. (1981). *Biochem. Biophys. Res. Commun.* **99**, 907–912.

Spurgeon, S. L. and Porter, J. W. (1983). *In* "Biosynthesis of Isoprenoid Compounds", Vol. 2 (J. W. Porter and S. L. Spurgeon, eds), pp. 1–122. Wiley, New York.

Spurgeon, S. L., Turner, R. V. and Harding, R. W. (1979). *Arch. Biochem. Biophys.* **195**, 23–29.

Straub, O. (1987). "Key to Carotenoids", 2nd edn. Birkhauser-Verlag, Basel.

Subbarayan, C., Kushwaha, S. C., Suzue, G. and Porter, J. W. (1970). *Arch. Biochem. Biophys.* **137**, 547–557.

Swift, I. E., Milborrow, B. V. and Jeffrey, S. W. (1982). *Phytochemistry* **21**, 2859–2864.

Taylor, R. F. and Davies, B. H. (1982). *Can. J. Biochem.* **60**, 675–683.

Upper, C. D. and West, C. A. (1967). *J. Biol. Chem.* **242**, 3285–3292.

Varma, T. N. R. and Chichester, C. O. (1962). *Arch. Biochem. Biophys.* **96**, 265–269.

Yamamoto, H. Y. (1979). *Pure Appl. Chem.* **51**, 639–648.

12 Enzymes of Chlorophyll and Heme Biosynthesis

ALISON G. SMITH[1] and W. TREVOR GRIFFITHS[2]

[1]Department of Plant Sciences, University of Cambridge, Cambridge, CB2 3EA, UK

[2]Department of Biochemistry, University of Bristol, Bristol BS8 1TD, UK

METHODS IN PLANT BIOCHEMISTRY Vol. 9
ISBN 0-12-461019-6

I. INTRODUCTION

Chlorophyll is one of a group of tetrapyrrole molecules which are essential to plant metabolism. Other members of the group include heme, siroheme (the prosthetic group of nitrite and sulphite reductases) and the phytochrome chromophore (Fig. 12.1), and as such these molecules are found in most compartments within the cell: chloroplasts, mitochondria, associated with the endoplasmic reticulum, and in the cytosol. All the compounds have the common tetrapyrrole structure, differing from one another by the nature of the ring substituents, the metal ion which is coordinated into the circular

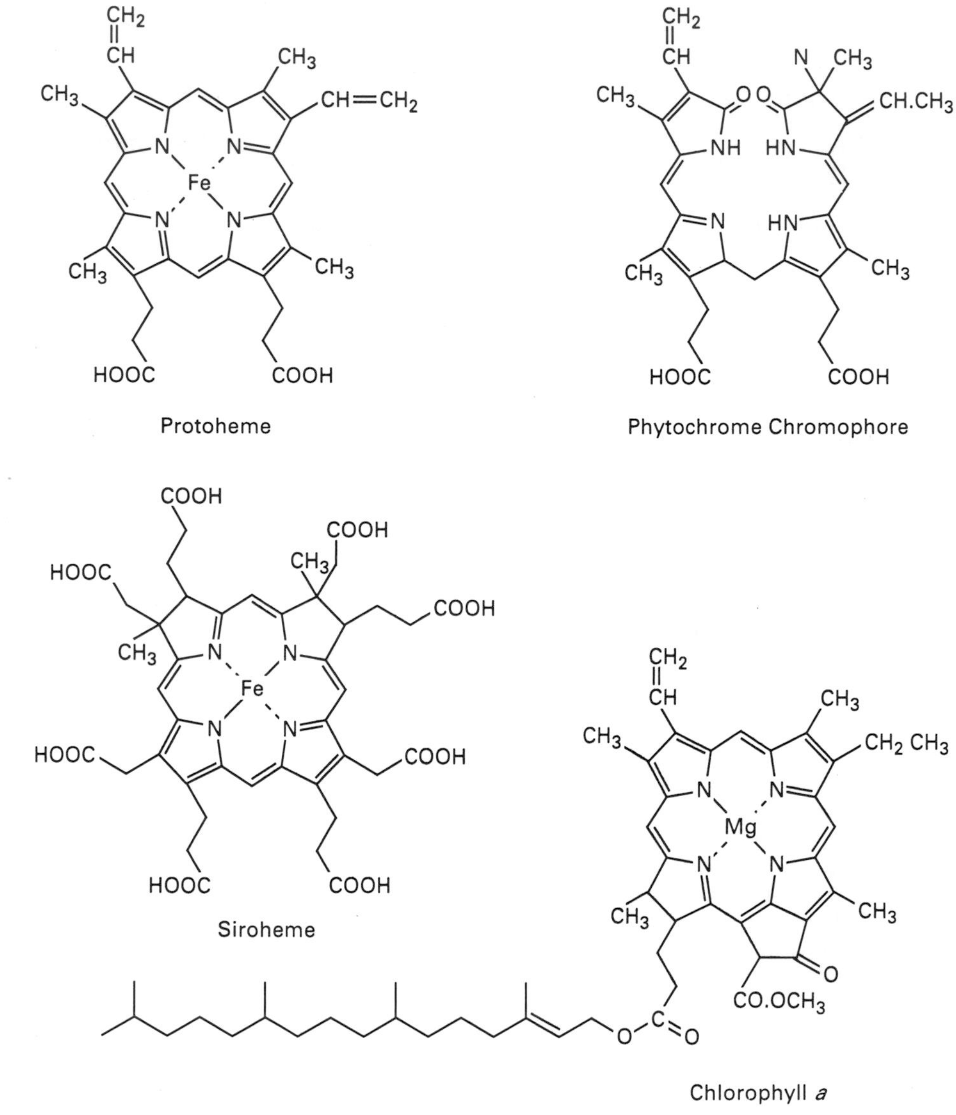

FIG. 12.1. Structure of major tetrapyrrole compounds of higher plants.

molecules, and the presence of adducts such as isoprenoid chains. It is likely that all of these molecules are associated with proteins *in vivo*, and this alters their behaviour, so that the same tetrapyrrole molecule can have quite different properties depending on the protein with which it is associated: for instance the redox potentials of proto-heme in the various cytochrome *b*s, or the absorption maxima of chlorophyll *a* in different chlorophyll proteins.

II. BIOSYNTHESIS OF CHLOROPHYLL AND HEME

The synthesis of all cellular tetrapyrrole molecules proceeds along a common route, with a number of branch-points leading to the different end-products (Fig. 12.2). The tetrapyrrole biosynthesis pathway has been the subject of a number of recent reviews (e.g. Beale and Weinstein, 1990; Kannangara, 1990), and so will be dealt with only briefly here. Furthermore, since most of the studies have concentrated on the synthesis of chlorophyll, and to a lesser extent heme, this chapter will be confined to the enzymes concerned with this part of the pathway.

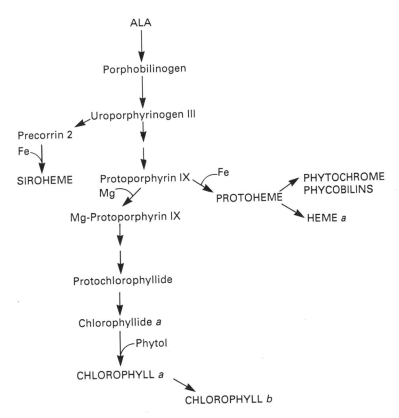

FIG. 12.2. Overview of the pathway of tetrapyrrole synthesis in plants and algae showing the major end-products.

A. Formation of Protoporphyrin IX and Heme

The first committed precursor of the tetrapyrrole pathway, 5-aminolevulinic acid (ALA), is made from the intact carbon skeleton of glutamate in three enzymic steps (Fig. 12.3a), where glutamate is first activated by ligation to a chloroplast-encoded tRNA^{Glu} molecule, then reduced to glutamate-1-semialdehyde (or a chemically equivalent molecule), followed by a transamination step to yield ALA. This is the only known example of the involvement of an aminoacyl-tRNA molecule in a reaction which does not result in the formation of a peptide bond. This route of ALA synthesis is referred to as the C$_5$ pathway, to distinguish it from the synthesis of ALA in animals and yeast, which is from succinyl CoA and glycine, and catalysed by a single enzyme, ALA synthase (Fig. 12.3b). The only photosynthetic eukaryote in which this last enzyme has been unequivocally demonstrated is the phytoflagellate *Euglena gracilis* (Beale *et al.*, 1981; Weinstein and Beale, 1983), and it is now generally considered that ALA for all cellular tetrapyrroles in plants and most algae is manufactured by the C$_5$ pathway. The evidence for this comes from the preferential incorporation of glutamate rather than glycine into tetrapyrroles (e.g. Castelfranco and Jones, 1975; Oh-hama *et al.*, 1982; Chibbar and van Huystee, 1983; Schneegurt and Beale, 1986) and the sensitivity of cellular tetrapyrrole synthesis to gabaculine, a specific inhibitor of the C$_5$ pathway (Section III.A) (Gardner and Gorton, 1985; May *et al.*, 1987; Werck-Reichhardt *et al.*, 1988). It now appears that the C$_5$ pathway is

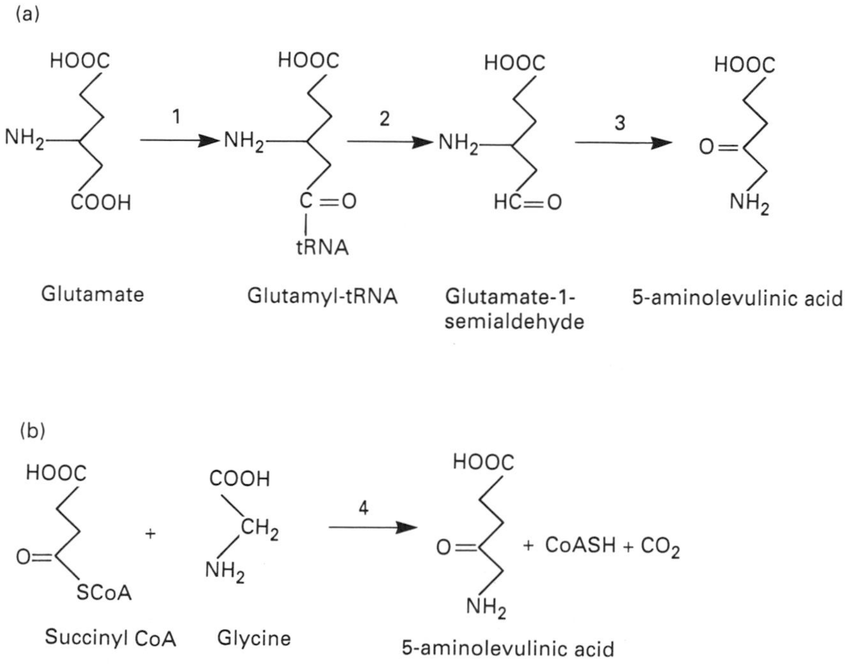

(a)

(b)

FIG. 12.3. Synthesis of the tetrapyrole precursor 5-aminolevulic acid (ALA). (a) The C$_5$ pathway, involving the three enzymes glutamyl-tRNA synthetase (1), glutamyl-tRNA reductase (2) and glutamate-1-semialdehyde aminotransferase (3). (b) Shemin's pathway catalysed by ALA synthase (4).

much more universal, and therefore likely to be more ancient, than the ALA synthase route, since not only does it operate in cyanobacteria and prochlorophytes, but also in most of the phototrophic bacterial groups (Avissar *et al.*, 1989), and in *Escherichia coli, Bacillus subtilis, Clostridium thermoaceticum* and *Methanobacterium thermoautrophicum* (Avissar and Beale, 1989a; Li *et al.*, 1989; O'Neill *et al.*, 1989), none of which make chlorophyll. The exceptions which utilise ALA synthase include species such as *Rhodospirillum rubrum, Rhodobacter spheroides* (Avissar *et al.*, 1989) and *Rhizobium meliloti* (Leong *et al.*, 1982), which are all members of the α-subgroup of purple bacteria. This is interesting, given the proposal that this subgroup is the progenitor of mitochondria (Yang *et al.*, 1985), the site of ALA synthase in animals and yeast.

The next part of the pathway to form protoporphyrin IX, the last common intermediate between heme and chlorophyll, is probably mechanistically identical in all organisms (Fig. 12.4). Two ALA molecules are condensed to form the pyrrole porphobilinogen (PBG), then four PBGs are joined to produce the linear tetrapyrrole molecule hydroxymethylbilane (HMB). This is an unstable molecule and will spontaneously cyclise to form the symmetrical isomer uroporphyrinogen I (urogen I), with a $t_{1/2}$ of 4 min at pH 8. However, *in vivo*, HMB is cyclised by urogen III synthase, with the reversal of ring D, to form the biological isomer urogen III. The pyrrole substituents are next modified by oxidative decarboxylation, and the methylene bridges oxidised to form the fully conjugated molecule protoporphyrin IX, which is now able to chelate a metal ion. Insertion of Fe^{2+} produces protoheme, which is then subsequently modified to form other hemes. Alternatively, the chelation of a Mg^{2+} ion leads to the synthesis of the chlorophylls (Section II.B). The linear chromophore of phytochrome is thought to be derived from protoheme by a pathway similar to that of the phycobilin pigments of cyanobacteria, red algae and cryptophytes (Brown *et al.*, 1990), while siroheme is produced by the methylation of urogen III to form precorrin-2, followed by oxidation and chelation of iron. Precorrin-2 is also the precursor of the cobalt-containing corrinoid, vitamin B_{12} (Warren and Scott, 1990).

B. Formation of Chlorophyll

Intermediates involved in the chlorophyll biosynthetic pathway between protoporphyrin IX and chlorophyll *a* are not so well established as those involved in the earlier stages of porphyrin formation for the simple reason that they have not been so widely studied. The first reaction specific to chlorophyll formation is the chelation of protoporphyrin IX by Mg^{2+} followed by methylation at the 13-propionate side-chain to form Mg protoporphyrin monomethyl ester (Fig. 12.5). A complex series of reactions follows involving oxidative cyclisation of the 13-propionate side-chain culminating in the formation of the additional isocyclic ring E in the product — Mg-2,4-divinyl phaeoporphyrin a_5 monomethyl ester. This compound, alternatively named divinyl protochlorophyllide (DVPChlide), under normal diurnal growth conditions in higher plants is reduced at C-17/C-18 in the light by the photoenzyme protochlorophyllide reductase, resulting in the chlorin divinyl chlorophyllide (DVChlide). Finally, this is successively reduced at C-8 to form monovinyl chlorophyllide (MVChlide) followed by esterification of the C-17 propionate group with the isoprenoid alcohol, phytol, to give chlorophyll *a*.

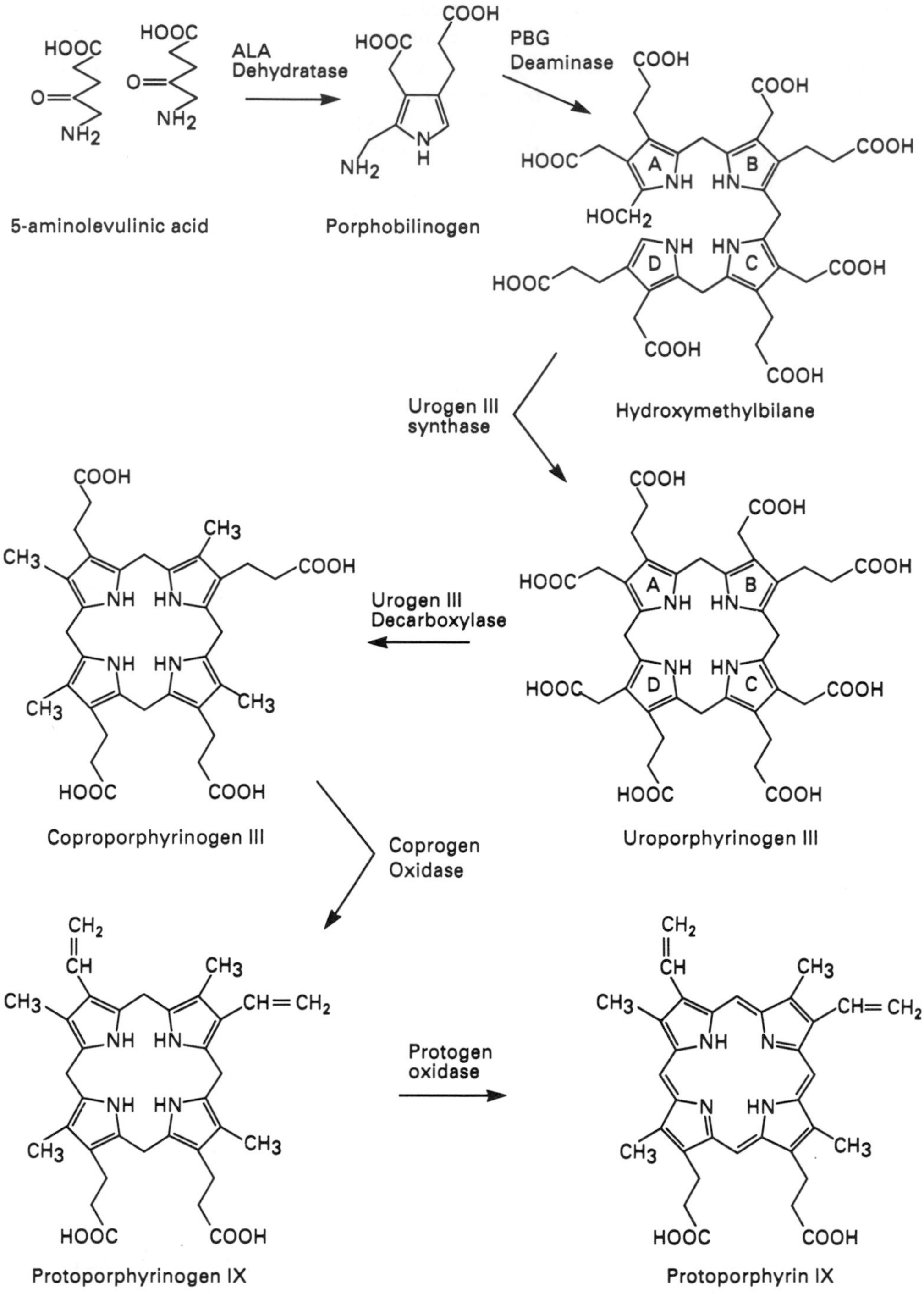

FIG. 12.4. The pathway of synthesis of the heme and chlorophyll precursor, protoporphyrin IX, from ALA.

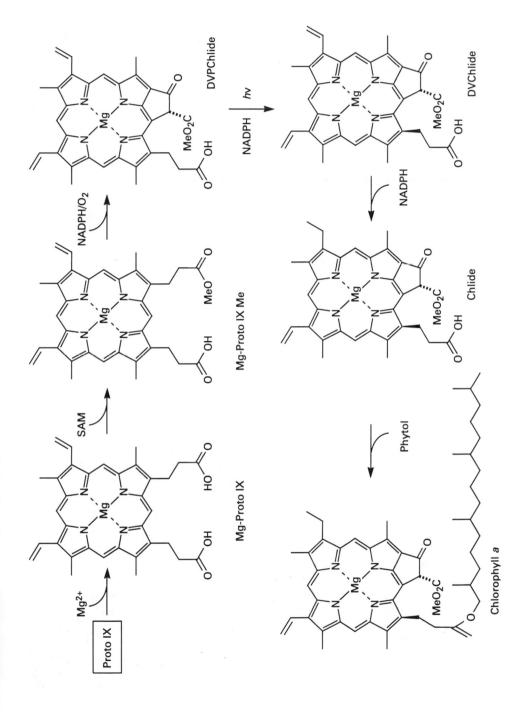

FIG. 12.5. The Mg-branch of the tetrapyrrole biosynthetic pathway.

Chlorophyll a is invariably accompanied within photosynthetic tissues, by a range of chemically modified forms of the pigment such as chlorophyll a' (the 13^2S epimer) and phaeophytin a (minus Mg^{2+}). A variety of other minor forms involving modifications of the various substituents on the pyrrole macrocycle have also been reported as components of photosynthetic tissues. Of these perhaps the most significant, at least quantitatively, is the 8-vinyl analogue of chlorophyll a found in a mutant (ON 8147) of *Zea mays* (Bazzaz, 1981) and in some marine phytoplankton (Chisholm *et al.*, 1988). Whether the remaining components are of any metabolic significance or are merely artefacts generated during analysis of the extremely labile native compounds remains to be established.

Chlorophyll b is also always associated with chlorophyll a in chloroplasts of higher plants. In chlorophyll b the 7-ethyl group has been oxidised to a formyl ($-CHO$) function. Chlorophyll b fulfills an accessory, light-harvesting function in the photosynthetic process, a role taken over largely by the various 'c' type chlorophylls in some algae, especially the chromophytes (see Hiller *et al.*, 1991). In the c type chlorophylls the 13-propionate side-chain is replaced by an acrylic residue with the c_1 and c_2 forms related structurally to the mono- and divinyl analogues of chlorophyll a, respectively. As will become apparent later in this section, the enzymes or indeed the intermediates involved in the formation of the b or c type chlorophylls or the various chlorophyll a analogues referred to above remain completely uncharacterised.

III. ENZYMES COMMON TO HEME AND CHLOROPHYLL SYNTHESIS

Isolated chloroplasts from plants and algae will incorporate exogenous [^{14}C]glutamate or [^{14}C]ALA into chlorophyll and heme, and all the enzymes of this pathway have been detected in chloroplasts. The subcellular site of the synthesis of tetrapyrroles found elsewhere in the cell, such as mitochondrial heme, or the phytochrome chromophore, remains less certain, although, as mentioned above (Section II.A), the available evidence is that ALA for all cellular tetrapyrroles is made via the C_5 pathway. Because this involves a chloroplast-encoded tRNA molecule, it can be presumed to be confined to the plastid. Smith (1988) has shown that the majority, if not all, of the next two enzymes of the pathway, ALA dehydratase and PBG deaminase (PBGD; Fig. 12.4), are confined to the plastids, even in *Arum* spadices, a tissue where most tetrapyrrole synthesis would be expected to be for mitochondrial heme (ap Rees *et al.*, 1983). Furthermore, although plant mitochondria appear to have ferrochelatase (Little and Jones, 1976) and protoporphyrinogen IX (protogen) oxidase activity (Jacobs and Jacobs, 1987; Matringe *et al.*, 1989; A. G. Smith, unpubl. obs.) coproporphyrinogen III (coprogen) oxidase cannot be detected in mitochondria from etiolated pea leaves or *Arum* spadices (A. G. Smith and G. Elder, in prep.). Taken together, these results imply that the majority of the enzymes of the tetrapyrrole biosynthesis pathway are exclusively plastidic, even in non-photosynthetic tissue. Thus although this chapter deals mainly with the purification and characterisation of these enzymes from leaf tissue, or single-celled algae, most of the enzymes involved in heme synthesis can be assayed using the same protocols in roots (Smith, 1986) and other non-photosynthetic tissue.

A. The C_5 Pathway of ALA Synthesis

The C_5 pathway of ALA synthesis is sufficiently complex that to cover it completely would warrant an entire chapter. This section will therefore necessarily be brief, and will not consider ALA synthase; for information on the assay and characterisation of this enzyme from *Euglena gracilis*, the reader is referred to Foley *et al.* (1982). However, the assay of the complete pathway from glutamate to ALA will be discussed, in addition to the each of the individual enzymic steps, since this is generally much more straightforward and, in crude extracts, this is often the only means to be certain that the activity measured is specific for ALA synthesis.

The C_5 pathway has been the subject of extensive research since its discovery in the early 1970s, and although there are still some areas of uncertainty, the sequence of events shown in Fig. 12.3a is generally accepted to be correct. Soluble extracts from spinach and barley chloroplasts were first shown to be capable of ALA synthesis from glutamate, in the presence of ATP, Mg^{2+} and NADPH (Gough and Kannangara, 1976, 1977), and this has since been demonstrated in many different species of plants and algae. The soluble extracts from barley (Wang *et al.*, 1981) and *Chlamydomonas* (Wang *et al.*, 1984) could be fractionated into at least three different macromolecular components by affinity chromatography, each of which was necessary to reconstitute ALA synthesis from glutamate. Then in 1984, the startling discovery was made that one of the components was an RNA molecule, and that ALA synthesis in barley chloroplasts was sensitive to ribonuclease A and snake venom phosphodiesterase (Kannangara *et al.*, 1984). The RNA component from barley was purified and sequenced, and found to be a $tRNA^{Glu}$ with a UUC anticodon (Schön *et al.*, 1986). The first enzyme of the pathway is thus glutamyl-tRNA synthetase, and the formation of glutamyl-tRNA, in an ATP-dependent reaction identical to that for the synthesis of aminoacylated tRNAs for protein synthesis, activates the glutamate for the subsequent energetically unfavourable reactions. The $tRNA^{Glu}$ is encoded by the chloroplast genome (Berry-Lowe, 1987), and since, in the vast majority of plants and algae, there is only one *trn*E in chloroplast DNA (Sugiura, 1989), it must also be involved in chloroplast protein synthesis. In barley chloroplasts, although a number of glutamate-accepting tRNAs have been detected, only one will participate in ALA synthesis, the other two being $tRNA^{Gln}$ species, which are subsequently trans-amidated to glutaminyl-$tRNA^{Gln}$, after first being misacylated with glutamate (Schön *et al.*, 1988). In the cyanobacterium *Synechocystis* 6803, two forms of $tRNA^{Glu}$, differing by a base modification at the first anticodon position (O'Neill and Söll, 1990), can be separated by high performance liquid chromatography (HPLC). However, both charged species support ALA and protein synthesis *in vitro* (Schneegurt and Beale, 1988; O'Neill and Söll, 1990). There must therefore be some means of controlling the entry of glutamyl-$tRNA^{Glu}$ into the two pathways.

The second step of the C_5 pathway entails the NADPH-dependent reduction of the α carbon of glutamate. The glutamyl-tRNA reductase is the only such enzyme known which requires a tRNA as a cofactor. It is extremely sensitive to inhibition by proto-heme, and has been proposed to be the site of feedback inhibition of the pathway of ALA synthesis which occurs in the dark (Huang and Wang, 1986; Wang *et al.*, 1987). The product of the reductase is a compound whose precise chemical nature is uncertain. It was proposed initially that it was glutamate-1-semialdehyde (GSA) (Kannangara and Gough, 1978). Subsequent workers have shown that chemically synthesised GSA is identical to the product of the reductase step (Houen *et al.*, 1984; Mau *et al.*, 1987; Gough *et al.*, 1989), and is transaminated by the last enzyme to form ALA. Furthermore, if the transaminase is inhibited by gabaculine (see below) in greening barley leaves (Kannangara and Schouboe, 1985), or an ALA-synthesising extract from *Chlamydomonas* (Mau *et al.*, 1987), then the compound that accumulates has very similar properties to synthetic GSA. However, Jordan (1990) has challenged this conclusion on the grounds of the extreme lability of α-aminoaldehydes such as GSA. He proposes instead that the intermediate is a cyclic form of GSA, 2-hydroxy, 3-aminotetrahydropyran-1-one, which is chemically equivalent to GSA, but much more stable, and may well have been the actual product of the previous syntheses of GSA. However, whatever the exact identity of the intermediate, since most of the literature refers to it as GSA, the same name will be used in this chapter.

The final step of the C_5 pathway, catalysed by the enzyme GSA aminotransferase, also has unusual characteristics. Unlike most other aminotransferases, it requires no exogenous amino donor, and will convert GSA to ALA in the absence of any added cofactor. It remains unclear, however, whether the amino group in ALA is the same as that in GSA, or whether the reaction is intermolecular, involving two GSA molecules. This question was studied in *Chlamydomonas*, using glutamate specifically labelled with ^{13}C or ^{15}N. Although a large proportion of the ALA produced contained two heavy atoms (Mau and Wang, 1988), providing support for intermolecular transamination, the possibility of isotopic redistribution after the production of ALA was not ruled out. Another unresolved question is the requirement for a vitamin B_6 cofactor. GSA aminotransferase is extremely sensitive to gabaculine and aminooxyacetate, which are considered to be pyridoxal antagonists. However, although the *Chlorella* enzyme was shown to require pyridoxal phosphate (Avissar and Beale, 1989b) and the enzyme from *Synechococcus* PCC 6301 was stimulated by pyridoxamine phosphate (Bull *et al.*, 1990), the purified barley enzyme showed no requirement for either cofactor, nor could a tightly bound form be detected spectrophotometrically (Grimm *et al.*, 1989). It is conceivable that the higher plant GSA aminotransferase is different from either the algal or cyanobacterial enzymes. However, given that the primary amino acid sequences of the barley and *Synechococcus* enzymes are 72% identical (Kannangara, 1990), this seems unlikely. Indeed, evidence has recently been presented for the presence of a vitamin B_6-derived cofactor in the enzyme from pea leaves (Nair *et al.*, 1991).

1. Assay of ALA synthesis from glutamate

The formation of ALA from glutamate by cell-free extracts, either form whole cells or isolated plastids, has been described in spinach (Gough and Kannangara, 1976), barley (Gough and Kannangara, 1977), cucumber (Weinstein and Castelfranco, 1978), *Chlamydomonas* (Wang *et al.*, 1984), *Chlorella* (Weinstein and Beale, 1985) and *Euglena* (Mayer *et al.*, 1987). Extracts are generally prepared in a buffer containing 100 mM Tricine, pH 7.5–8.3, 150 mM glycerol, 1–15 mM $MgCl_2$ and 1 mM dithiothreitol (DTT). Some workers also include pyridoxal phosphate at 20 μM. Approximately 3–5 mg of protein from these extracts would then be assayed. A typical reaction (e.g. from *Chlorella*: Weinstein and Beale, 1985) would contain 1 mM glutamate, 5 mM ATP, 1 mM NADPH, 5 mM levulinic acid (to inhibit the next enzyme in the pathway, ALA dehydratase; Section III.B.1) in a total volume of 0.5 ml. After incubation at 30°C for 20 min, the reaction is stopped by the addition of sodium dodecyl sulphate (SDS) to 5% and acidification with citric acid. The ALA formed is purified by chromatography on Dowex columns, followed by reaction with ethylaceto-acetate to form ALA-pyrrole (Mauzerall and Granick, 1956). This can then be quantified after the addition of an equal volume of modified Ehrlich's reagent by measuring the absorption at 553 nm (Urata and Granick, 1963). Alternatively, a radioactive assay can be used, where approximately 2.5 μCi of [^{14}C]glutamate per ml of reaction is substituted for the unlabelled substrate. In this case, carrier ALA must be added before the Dowex step. Then, after ALA-pyrrole formation, the amount of ALA formed is measured in a scintillation counter. The authenticity of the pyrrole can be checked by thin layer chromatography (TLC) (Wang *et al.*, 1984), paper chromatography (PC) (Rieble and Beale, 1988) or HPLC (Chang *et al.*, 1990).

2. Fractionation of the components of the C_5 pathway

Details of the purification of each of the individual components are given in the following sections. However, a brief discussion will be made here of a simple fractionation of the macromolecular components, since this is frequently desired, for instance to determine whether the C_5 pathway is operating, or to separate the components from one another for reconstitution assays during more complete purification. Extracts, prepared as described above, are first subjected to centrifugation at 264 000 × *g* for 90 min and treated with ammonium sulphate to precipitate proteins, and leave RNA in the supernatant. The protein fraction after desalting, is then subjected to serial affinity chromatography on a combination of reactive Blue Sepharose, heme-Sepharose and/or ADP-Sepharose columns (Wang *et al.*, 1981, 1984; Weinstein *et al.*, 1987). The aminotransferase does not bind to any of the matrices, while Blue Sepharose retains both the glutamyl-tRNA synthetase and reductase activities. However, only the reductase binds to heme-Sepharose, presumably reflecting the inhibition of the reductase activity by protoheme (Wang *et al.*, 1987). Similarly, the reductase binds to ADP-Sepharose, a support frequently used to retain proteins with NADP binding sites. After affinity chromatography, all of the different fractions are required to reconstitute ALA-synthesising activity from glutamate, but this is generally comparable to the activity of the unfractionated extract. For the *Chlorella* enzyme, in order to stabilise the enzyme fractions and obtain maximum yield, it was found to be important to include a high concentration of glycerol (1 M) during this procedure (Weinstein *et al.*, 1987). Further examples of separation are given in Chang *et al.* (1990) and Chen *et al.* (1990a, b).

3. tRNA

Because of the lability of RNA and its sensitivity to ribonucleases, which are ubiquitous, it is essential to prepare all glassware, plasticware and solutions RNase-free (see Sambrook *et al.*, 1989). In addition, gloves should be worn throughout the procedure, and changed frequently. The basic procedures for isolation of RNA are also described and explained in Sambrook *et al.* (1989) or any other molecular biology manual.

Low molecular weight RNA can be prepared by the method of Farmerie *et al.* (1982). Soluble extracts, either of whole tissues or purified plastids, are prepared in an appropriate buffer (e.g. 10 mM Tris-HCl, pH 7.5, 10 mM Mg acetate, 100 mM NaCl, 10 mM 2-mercaptoethanol) and if required, proteins can be removed by ammonium sulphate precipitation. The solution is made 1% (w/v) with SDS and extracted two or three times with phenol (equilibrated in buffer above), and then chloroform-isoamyl alchohol (24:1; v/v). Nucleic acids are precipitated from the aqueous phase with ethanol, and the tRNA fraction isolated by DEAE-cellulose chromatography. Further purification of the tRNA^{Glu} has been achieved by reversed-phase HPLC (Kannangara *et al.*, 1984; Schön *et al.*, 1986; Peterson *et al.*, 1988) or by affinity chromatography on a column of immobilised tRNA^{Phe}, whose anticodon (GAA) is the complement of the anticodon of tRNA^{Glu} (Schneegurt and Beale, 1988). The presence of tRNA^{Glu} is assayed by its ability to support ALA synthesis from gluta-mate in a reconstituted system with purified or partially purified enzymes, or in an extract previously depleted of RNA, either by ammonium sulphate fractionation (Schneegurt and Beale, 1988) or by pre-treatment with ribonuclease (Weinstein and Beale, 1985). It may also be assayed by its glutamate-acceptor activity (Schön *et al.*, 1986), but care must be taken to ensure that this is not due to the presence of tRNA^{Gln}, which can be misacylated with glutamate (Section III.A).

4. *Glutamyl-tRNA synthetase*

Glutamyl-tRNA synthetase (also called ligase; EC 6.1.1.17) activity is usually assayed by determining the amount of acid-precipitable tRNA^{Glu} charged with radioactive ([^{14}C] or [^{3}H]) glutamate. A typical reaction mixture would contain: 50 mM Tricine pH 8.3, 10 mM MgCl$_2$, 2 mM DTT, 1 mM ATP, 1–5 µg tRNA and 2.5 µCi [^{3}H]gluta-mate in a total volume of 15 µl (Chang *et al.*, 1990). After incubation at 29°C for 30 min, the reaction mixture is spotted onto a glass-fibre filter and unincorporated glutamate is removed by washing in 5% trichloroacetic acid (TCA). The amount of label in TCA-precipitable material is determined by scintillation counting.

Glutamyl-tRNA synthetases have been purified and characterised from a number of different plant and algal sources, including wheat (Ratinaud *et al.*, 1983), barley (Bruyant and Kannangara, 1987) and *Chlamydomonas* (Chen *et al.* 1990a; Chang *et al.*, 1990). In wheat three different enzyme activities were observed, representing the cytosolic, mitochondrial and plastid isozymes, only the latter of which is involved in both protein and ALA synthesis. Its K_m values were 0.2 µM for the tRNA and 10 µM for glutamate. After purification by ammonium sulphate fractionation and chromato-graphy on hydroxyapatite and phosphocellulose, its native molecular weight was 110 kDa, but on SDS-PAGE, a band of 55 kDa was detected, indicating that the functional enzyme was a dimer. A similar subunit size (of 59 kDa) was observed for the synthetase from barley chloroplasts (Bruyant and Kannangara, 1987) and *Chlamydomonas* (62 kDa; Chen *et al.*, 1990a). The native molecular weight of the latter enzyme, purified by DEAE-cellulose, phosphocellulose, and FPLC on Mono Q and Mono S, was also 62 kDa, implying that the enzyme is a monomer. In contrast, Chang *et al.* (1990) reported that the *Chlamydomonas* glutamyl-tRNA synthetase, purified by ammonium sulphate fractionation, chromatography on DEAE and Blue B affinity columns and native gel electrophoresis, was a dimer of two subunits of 32.5 kDa. Although Chen *et al.* (1990a) showed that their enzyme could misacylate tRNA^{Gln} with glutamate (a characteristic of plastid ligases) and Chang *et al.* (1990) demonstrated that their preparation would support ALA synthesis in a reconstituted system, both purifications started with total soluble proteins. It remains possible therefore that, as for wheat, one or other of the enzymes purified was not the plastid one involved in ALA synthesis. This discrepancy requires further investigation. In all the studies, only one plastid glutamyl-tRNA synthetase activity has been found, implying that a single enzyme is responsible for the synthesis of glutamyl-tRNA for ALA and protein synthesis. In this context it is interesting that Chang *et al.* (1990) observed that their enzyme from *Chlamydomonas* was almost completely inhibited by 5 μM heme (Huang and Wang, 1986).

5. Glutamyl-tRNA reductase

The reaction catalysed by glutamyl-tRNA reductase is the conversion of glutamyl-tRNA to GSA (or a chemical equivalent; Section IV.A). In order to assay it, it is necessary to prepare the substrate with radiolabelled [^{14}C]- or [^{3}H]glutamate and pure tRNA^{Glu}, using purified or partially purified glutamyl-tRNA synthetase (Huang and Wang, 1986; Jahn *et al.*, 1991). Then the product, GSA, must be separated from free glutamate, using Dowex ion-exchange chromatography (Kannangara and Gough, 1978; Jahn *et al.*, 1991) and/or reversed-phase HPLC (Mau *et al.*, 1987; Chen *et al.*, 1990b). Generally, however, most workers employ coupled assays (using purified or partially purified glutamyl-tRNA synthetase or GSA aminotransferase, from which reductase activity has been removed by affinity chromatography) either to generate glutamyl-tRNA, or to convert the GSA to ALA, or both (e.g. Avissar and Beale, 1989a; Chen *et al.*, 1990b).

Of the three enzymes in the C$_5$ pathway, glutamyl-tRNA reductase has proved the most difficult to purify, principally as a result of its lability (Kannangara *et al.*, 1988) and the difficulty in eluting it in an active form from the affinity columns to which it binds strongly. To date, the only report of a successful purification is from *Chlamydomonas* (Chen *et al.*, 1990b), in a procedure involving chromatography on DEAE-cellulose, phosphocellulose, phenyl-Sepharose, affi-gel blue and FPLC on Mono S and Superose 12. The enzyme was purified 2000-fold, but with a yield of only 0.3%. It had a native and subunit molecular weight of 130 kDa, implying that it is a monomer.

6. GSA aminotransferase

The last enzyme in the C$_5$ pathway, GSA aminotransferase (EC 5.4.3.8), was the first to be identified and has been studied extensively in a number of different plant and algal species. The assay involves determining the amount of ALA (measured as described above in Section IV.A.1) produced from GSA. The substrate is made chemically by ozonolysis of vinyl γ-aminobutyric acid followed by Dowex ion-exchange chromatography (Gough *et al.*, 1989), and is stable frozen in aqueous solution for several months. A typical reaction mixture in 1 ml would contain: 0.1 M

Tricine-NaOH, pH 7.9, 0.3 M glycerol, 1 mM DTT, 25 mM $MgCl_2$ and 25 μM GSA. If the enzyme preparation is a crude extract, 5 mM levulinic acid is also included. After a 20 min incubation at 28°C, the reaction is stopped by the addition of TCA, and the amount of ALA determined spectroscopically after formation of the pyrrole (Section III.A.1). Alternatively, 0.1 μCi of [^{14}C]GSA can be substituted for the non-radioactive substrate, and the amount of ALA can be determined by scintillation counting (Grimm et al., 1989).

Purification of the enzyme from both barley and the cyanobacterium *Synechococcus* has been achieved by taking advantage of the fact that GSA aminotransferase is not retained by a variety of different affinity matrices, while many other proteins, including the other C_5 pathway enzymes, bind to at least one. Passage of soluble proteins from greening barley chloroplasts or *Synechococcus* cells through Blue-Sepharose, Red-agarose and chlorophyllin-Sepharose, left the majority of aminotransferase activity in the run-through fraction. After electrophoresis on a non-denaturing gel a single band was observed, which retained GSA aminotransferase activity after elution from the gel. On SDS-PAGE a number of bands were visible, but there was a major one of 46 kDa, the intensity of which corresponded to distribution of GSA aminotransferase activity in different fractions (Grimm et al., 1989).

B. ALA Dehydratase

1. Reaction

The reaction catalysed by ALA dehydratase (EC 4.2.1.24) is the condensation of two molecules of ALA to form the monopyrrole, porphobilinogen (PBG). The K_m for ALA is generally 0.3–1.5 mM, and the enzyme from all sources is strongly inhibited in a competitive manner by the substrate analogue levulinic acid, and by succinyl acetone (Brumm and Friedmann, 1981). The mechanism has been studied extensively in the animal and bacterial enzymes. In single turnover experiments, it was shown that the first ALA molecule to bind to the enzyme contributed the propionate group of PBG (Jordan and Seehra, 1980a, b; Jordan and Gibbs, 1985), while the second substrate molecule formed the acetic acid side of PBG. The first substrate molecule forms a Schiff base with the enzyme (Nandi and Shemin, 1968), and an essential lysine has been identified at the active site as the residue involved (Nandi, 1978; Gibbs and Jordan, 1986). In addition, ALA dehydratase contains a number of essential sulphydryl groups, which must be kept reduced for activity. This explains the remarkable sensitivity of all ALA dehydratases to oxygen or to thiol reagents, such as N-ethylmaleimide or lead ions. In animals, ALA dehydratase is a homo-octamer, with eight Zn^{2+} bound to it, one per subunit (Tsukamoto et al., 1979; Gibbs and Jordan, 1981). These do not appear to function in catalysis, but appear to stabilise the enzyme (Hasnain et al., 1985), and if they are removed, then added Zn^{2+} is necessary for activity. In many bacteria and plants, there is no requirement for exogenous Zn^{2+}, and indeed Zn^{2+} inhibits at mM concentrations (Schneider, 1970; Shibata and Ochiai, 1977). Instead, maximum activity requires Mg^{2+} ions, although there is residual activity of ~30% in their absence (Nandi and Waygood, 1967; Shibata and Ochiai, 1977; Liedgens et al., 1980).

2. Assay

As mentioned above, the plant enzyme requires Mg^{2+} for maximal activity, and reducing agents are also essential. Its pH optimum is about 8, but since a number of workers have reported that Tris buffers inhibit activity, it is advisable to use a buffer such as HEPES in initial experiments. The reaction is set up on ice in a microfuge tube and typically would contain: 50 mM HEPES, pH 8.2, 6 mM $MgCl_2$, 1 mM DTT (or 5 mM 2-ME), 2.5 mM ALA, and extract, in a total volume of 0.25 ml. After 30 min incubation at 37°C, the reaction is stopped by the addition of 0.25 ml of 10% TCA, 2.7% $HgCl_2$, the latter being necessary to stabilise the colour yield. After centrifugation to remove precipitated protein, an equal volume of modified Ehrlich's reagent (Mauzerall and Granick, 1956) is added to the supernatant. The absorbance of the solution at 555 nm is then determined after 10–15 min, and the amount of PBG formed is calculated using $6.2 \times 10^4 M^{-1} cm^{-1}$ as the absorption coefficient for PBG pyrrole. It is also possible to assay the activity of ALA dehydratase in a non-denaturing polyacrylamide gel. For this, electrophoresis is carried out at 4°C, to minimise denaturation of the enzyme. After electrophoresis, the gel is incubated for 2 h at 37°C in assay buffer (50 mM HEPES, pH 8.2, 6 mM $MgCl_2$, 1 mM DTT, 2.5 mM ALA), and then for 10 min in 10% TCA, 2.7% $HgCl_2$. Colour is developed by soaking the gel in modified Ehrlich's reagent, and the position of ALA dehydratase in the gel is revealed by the bright pink colour of the PBG pyrrole.

3. Purification

In general in plant tissues, ALA dehydratase (ALAD) activity is quite labile, and deteriorates on storage at 4°C or −20°C, especially at low protein concentrations. This makes purification of the enzyme difficult and it is essential to include protease inhibitors throughout the procedure. Additionally, because of the active sulphydryl groups, reducing agents such as 2-mercaptoethanol or DTT must be included, since oxidation of the thiols is often irreversible.

Purification of the enzyme has been attempted from a number of plant tissues, including wheat leaves (Nandi and Waygood, 1967), tobacco leaves (Shetty and Miller, 1969), soybean callus (Tigier et al., 1970), spinach leaves (Schneider, 1970) and radish seedlings (Shibata and Ochiai, 1977; Huault and Bruyant, 1984), as well as from Chlorella (Tamai et al., 1979) and Euglena (Stella and Batlle, 1978). However, only the spinach enzyme has been purified to homogeneity (Schneider, 1970; Liedgens et al., 1980). The method employed acetone precipitation, DEAE ion-exchange chromatography, gel filtration on Sephacryl S-300 and finally preparative non-denaturing gel electrophoresis. After electro-elution of the active enzyme from the gel, SDS-PAGE revealed three bands of approximately 50 kDa. Since the native molecular weight from gel filtration was 324 kDa, this implies that spinach ALAD is an hexamer, in contrast to the octameric animal enzyme. The subunit is also larger than the 35 kDa reported for the enzymes from animal (Anderson and Desnick, 1979; Gibbs et al., 1985a), yeast (Myers et al., 1987) and E. coli (Echelard, et al., 1988). Because of the considerable lability of the plant enzyme, and the difficulties this posed for purification of large amounts, Schneider and co-workers sought ways to improve their original procedure.

They developed an affinity chromatography technique using monoclonal antibodies to the spinach enzyme (purified by conventional means) bound to a Sepharose bead matrix (Liedgens *et al.*, 1980). Columns of this immunoabsorbens could be used to purify ALA dehydratase, with a yield of 70–80%, from acetone precipitates of crude spinach leaf extracts. The protein eluted from the column was homogeneous and gave a single band of 50 kDa on SDS-PAGE. However, although the spinach monoclonal antibodies do cross-react with the enzyme from other sources (Schneider and Liedgens, 1981), the cross-reactivity is not sufficient to use them to purify these other ALA dehydratases (K. N. Singh, J. M. Jenkins and A. G. Smith, unpubl. obs.). An alternative rapid procedure has been developed for the purification of ALAD from *Euglena*. After ammonium sulphate fractionation of a crude extract, and then DEAE ion-exchange chromatography, the active fractions are passed first through a dye–ligand column (Red C1, Cambio, Cambridge, UK), to which ALAD does not bind, and then onto a metal chelate column, to which it does bind. Finally, the enzyme is subjected to FPLC on Mono Q and Superose 12. The yield is 12% with a 360-fold purification (J. M. Jenkins, K. N. Singh and A. G. Smith, in prep.). The *Euglena* enzyme has a similar native and subunit molecular weight to that from spinach, and also requires Mg^{2+} for maximal activity.

C. PBG Deaminase

1. Reaction

PBG deaminase (PBGD; EC 4.3.1.8), also known as hydroxymethylbilane synthase, catalyses the additon of four molecules of PBG, with the loss of each free amino group, to form the linear tetrapyrrole hydroxymethylbilane. This compound is unstable and will spontaneously cyclise under aerobic conditions to the non-biological isomer uroporphyrinogen I, and so *in vivo* the next enzyme, urogen III synthase, must be present in sufficient activity to convert all the HMB produced to urogen III. The enzyme mechanism has been the subject of extensive research over the years, but has only relatively recently been elucidated fully (for an historical account, see Jordan, 1990). The monomeric enzyme has a dipyrromethane cofactor (Hart *et al.*, 1987; Jordan and Warren, 1987) covalently attached to a cysteine residue on the enzyme (Hart *et al.*, 1988).

The first PBG (which forms ring A, Fig. 12.4) is added to this dipyrrole, and then the next three substrate molecules (rings B to D) are joined sequentially, until a hexapyrrole is formed, at which point hydrolysis of the link between the cofactor and substrate occurs, releasing HMB. Although most of these studies used the enzyme from *E. coli*, it seems likely that this mechanism is universal, since the dipyrromethane cofactor has been identified in PBGD from other sources, including barley and spinach (Warren and Jordan, 1988). However the K_m for PBG does appear to vary: 1–20 μM for the bacterial and mammalian enzymes (e.g. Jordan and Shemin, 1973; Anderson and Desnick, 1980; Hart *et al.*, 1986) but much higher for algal deaminases; 90 μM for the enzyme from *Chlorella regularis* (Shioi *et al.*, 1980) and 195 μM for that of *Euglena* (Battersby *et al.*, 1979a). The K_m of PBGD from higher plants has not been investigated.

2. Assay

As mentioned above, HMB (Section II.A), the product of PBGD, is unstable and spontaneously cyclises to urogen I. The enzyme is therefore assayed by oxidising this fully to uroporphyrin, which absorbs strongly at 406 nm. Enzymatically produced PBG is available commercially, but is generally very expensive. Alternatively, it can be made in a relatively straightforward chemical synthesis procedure (Battersby *et al.*, 1977). A typical reaction (Williams *et al.*, 1981) would contain 0.2 M sodium phosphate, pH 8.0, 0.6 mM EDTA, 0.44 mM PBG (11 mM stock dissolved in reaction buffer) and enzyme extract, in a total volume of 0.5 ml. After incubation for 30 min at 37°C in the dark, 270 μl 1% KI, 0.5% I_2 is added and the solution left for 2 min at room temperature to oxidise all the porphyrinogen. Then 70 μl 1% sodium metabisulphite is added to remove any unreacted I_2, followed by 70 μl 60% TCA and centrifugation to precipitate proteins. The amount of uroporphyrin is determined from the A_{406} of the supernatant, using $5.28 \times 10^5 \, M^{-1} \, cm^{-1}$ as the absorption coefficient (Rimington, 1960). The requirement for an oxidation step in the assay precludes the use of too high a concentration of reducing agent in the reaction mix, and so if a large volume of tissue extract must be added to obtain measurable activity, it is better to use DTT in the tissue extraction, rather than 2-ME, since it is effective at lower concentrations.

3. Purification

PBGD has been purified from several sources including spinach (Higuchi and Bogorad, 1975), human erythrocytes (Anderson and Desnick, 1980), *Euglena* (Battersby *et al.*, 1983), *E. coli* (Hart *et al.*, 1986; Jordan *et al.*, 1988a), and most recently pea (Spano and Timko, 1991). In all cases, both the native and subunit molecular weight have been shown to be 35–40 kDa, implying that PBGD is a monomer. Generally, the purified enzyme is relatively stable for a number of weeks at −20°C, but for optimum recovery and yield, it is advisable to include protease inhibitors throughout the purification procedures. All the purification methods take advantage of the fact that PBGD is relatively heat stable, while the next enzyme in the pathway, urogen III synthase, is completely inactivated after 10–15 min at 55–60°C. Heat treatment also serves to remove a number of other proteins, and can be a very effective purification step. The subsequent steps then vary depending on the organism and tissue used, but generally involve a combination of conventional and FPLC chromatography. For the pea enzyme the starting material was chloroplasts purified on Percoll gradients (Spano and Timko, 1991), rather than total soluble proteins, which greatly facilitated the rest of the procedure. In addition, the authors found an excellent affinity step in Reactive Red 120 Sepharose, to which PBGD from both pea and *E. coli* bound strongly, and from which the enzyme could be eluted with either a salt gradient or 100–200 μM PBG. Their final preparation was purified 1000-fold, and had a specific activity of 4.88 μmol h^{-1} mg^{-1} protein, which is comparable, although somewhat less than the specific activity of 28 μmol h^{-1} mg^{-1} for the enzyme from *Euglena* (Battersby *et al.*, 1983) and 43 μmol h^{-1} mg^{-1} for the *E. coli* PBGD (Jordan *et al.*, 1988a). This latter enzyme was purified from an over-producing recombinant strain, transformed with an expression plasmid containing the

E. coli PBGD gene (*hemC*). Seventy mg of enzyme could be isolated from a 50 l culture. These large amounts have allowed detailed mechanistic studies to be performed, and crystallisation of the enzyme has also been achieved.

D. Uroporphyrinogen III Synthase

1. Reaction

The reaction catalysed by urogen III synthase is extremely complex, involving as it does the cyclisation of the linear tetrapyrrole HMB, with the concomitant rearrangement of ring D, so that the acetate and propionate side-chains are reversed in comparison with those on rings A–C (Fig. 12.4). The enzyme mechanism has been studied extensively by a number of physicochemical techniques, much of the work being carried out on the enzyme from *Euglena*. The true substrate is the linear tetrapyrrole; the enzyme will not use urogen I (Battersby *et al.*, 1979b, 1983). The mechanism has been proposed to proceed through a spiro intermediate (Mathewson and Corbin, 1963), and a chemically synthesised analogue of the intermediate has been shown to act as a competitive inhibitor (Stark *et al.*, 1986). The enzyme from all sources appears to be a monomer, and there is no requirement for a cofactor; although the partially purified rat liver enzyme was initially reported to have a bound folate (Kohashi *et al.*, 1984), purified urogen III synthase from *Euglena* (Hart and Battersby, 1985) showed no evidence for such a cofactor. Furthermore, after denaturation by SDS-polyacrylamide gel electrophoresis, which would remove a noncovalently attached cofactor such as folate, both the *Euglena* enzyme, and that from rat liver (Smythe and Williams, 1988), could be renatured, with total recovery of enzyme activity.

Because of the lability of HMB, a number of workers have proposed that urogen III synthase and PBGD act *in vivo* as a complex, so that no free HMB is released. This is supported by the observation that the presence of urogen III synthase alters the sedimentation of wheat germ PBGD (Higuchi and Bogorad, 1975), and in *Euglena* affects the K_m of PBGD (Battersby *et al.*, 1979a). While it is relatively straightforward to assay the two enzymes independently, and both have been purified separately from one another from some sources, difficulties are sometimes encountered in the separation of urogen III synthase from PBGD. The reverse can always be achieved by heat treatment of the extract, which totally inactivates urogen III synthase (Section III.C.3).

2. Assay

There are two methods for the assay of urogen III synthase. The first uses the true substrate HMB, which is synthesised chemically (Battersby *et al.*, 1982). The final step, the hydrolysis of the HMB ester, should be carried out just prior to use. The reaction mix, incubated at 25°C in a total volume of 0.5 ml, contains 0.2 M Tris-HCl, pH 8.25, 0.6 mM EDTA, and tissue extract, in a total volume of 0.5 ml. It is initiated by the addition of HMB, to a final concentration of 83 μM, and at 30-s intervals 50 μM aliquots are removed into KI/I$_2$ solution. The amount of uroporphyrin in each aliquot is determined from the A_{406} after oxidation of the excess I$_2$ and deproteinisation (Hart and Battersby, 1985), and the initial rate of uroporphyrin production is estimated from plots of A_{406} against time. It is necessary to determine the non-enzymic rate of ring-closure of the HMB in the absence of added tissue extract, and then to subtract this from that obtained with enzyme.

This assay, although quantitative, requires a supply of HMB, which is difficult to prepare, and is not commercially available. An alternative, semi-quantitative method, which is rapid and more convenient, and thus more suitable for use during enzyme purification studies, is the 'lag' assay (Hart and Battersby, 1985). This is a coupled assay with PBGD, and relies on the fact that, in the absence of urogen III synthase, the HMB produced by PBGD cyclises to urogen relatively slowly, with a $t_{1/2}$ of about 4 min at pH 8.0. However, if urogen III synthase is there as well, then the ring closure is rapid and there is a linear production of urogen with time—i.e. there is no lag. The amount of urogen III synthase activity in an extract is estimated by the difference in amount of uroporphyrin produced in 3 min by a partially purified deaminase preparation, with and without addition of the extract. The amount of PBGD in the extract itself may also be determined, by assaying the activity in a heated sample (to inactivate urogen III synthase, Section III.C.3), but in general, this is much less than the activity of the partially purified deaminase and contributes very little to the production of urogen in 3 min.

The partially purified deaminase can be prepared from the same organism and tissue as the urogen III synthase of interest, but this is not essential, and it may be more convenient to use a tissue with high PBGD activity. Light-grown *Euglena* cells are an excellent source, having a specific activity of 65 nmol h^{-1} mg^{-1} protein. Alternatively, the availability of large quantities of recombinant enzyme from *E. coli* (Jordan *et al.*, 1988b; Miller *et al.*, 1989) may make this the enzyme of choice in the future. Generally, it is sufficient to prepare a soluble protein extract, which is first heat-treated (Section III.C.3) to remove any urogen III synthase activity, and then subjected to ammonium sulphate fractionation. The *Euglena* enzyme precipitates between 40 and 70%. If required, the PBGD can be purified further on an ion-exchange column. It is then stable for months stored in aliquots at $-20°C$. The urogen III synthase assay is carried out by setting up two identical deaminase assays (Section III.C.2), each containing sufficient partially purified deaminase to give an A_{406} of 0.1–0.2 after 3 min incubation. Tissue extract is then added to one tube, and both are incubated at 37°C for precisely 3 min. The reaction is terminated and the amount of uroporphyrin produced is estimated exactly as described for the deaminase assay. This method is linear up to a stimulation of 0.25 absorbance units, but cannot really be used to determine maximum catalytic activities or specific activity. Quantitation can be improved, however, if the products are analysed for the individual uroporphyrin isomers, since with deaminase alone, the product will be uroporphyrin I, while with urogen III synthase as well, uroporphyrin III will be produced. In this case, a single reaction with partially purified deaminase and tissue extract is set up, the reaction is allowed to proceed for 30 min and it is terminated and the products oxidised as in Section III.C.2. After removal of the protein, the supernatant is applied to a reverse-phase HPLC column, and eluted with an acetonitrile gradient (Rossi and Curnow, 1986). Uroporphyrin III is retained longer on the column, because its two adjacent propionic acid groups makes it more hydrophobic than uroporphyrin I.

3. Purification

Urogen III synthase has been purified to homogeneity from *Euglena* (Hart and Battersby, 1985; Gumpel and Smith, 1990), human erythrocytes (Tsai *et al.*, 1987) and *E. coli* (Alwan *et al.*, 1988) and in each case was found to have a native and subunit molecular weight of 28–31 kDa, implying that it is a monomer. The bacterial enzyme was isolated from an overproducing strain transformed with a plasmid containing the *E. coli hemD* gene, so that 4 mg of protein was obtained from 1100 ml of culture. In most organisms, though, the amount of urogen III synthase is very low, so that the yield of enzyme is considerably less than this. Furthermore, purification is often hampered by problems in separating PBGD activity from that of urogen III synthase, which makes it difficult to use the lag assay to monitor the purification process. In *Euglena*, the two enzymes can be separated using ammonium sulphate fractionation, urogen III synthase precipitating at a lower concentration (30–50%) than PBGD. Purification of 1700-fold with a yield of 8% was then achieved using DEAE-Sepharose ion-exchange chromatography, chromatofocussing on Mono P, and finally gel filtration on Superose 12 (Gumpel and Smith, 1990), resulting in a preparation with a specific activity of 1.33 μmol h^{-1} mg^{-1}. In order to obtain maximal yield of enzyme, it was essential to include protease inhibitors in all buffers. This procedure resulted in 260 μmg enzyme from 30 l of *Euglena* culture, a considerable improvement over the previous published method where 550 l was needed to obtain the same amount of enzyme (Hart and Battersby, 1985)! The enzyme is stable stored in liquid N$_2$ for several months.

E. Uroporphyrinogen III Decarboxylase

1. Reaction

Urogen decarboxylase (EC 4.1.1.37) catalyses the removal of the carboxyl group from each of the four acetate side-chains in urogen to yield coprogen. The reaction proceeds in a step wise fashion with the hepta-, hexa- and pentaporphyrinogens as intermediates, and evidence from proton nuclear magnetic resonance (NMR) suggests that the order of decarboxylation is ring D, A, B then C (Jackson *et al.*, 1980). Urogen decarboxylase from every source examined will utilise urogen I in addition to urogen III, with similar K_m values (in the 10–100 nM range), but generally the type III isomer is a better substrate: for instance in human erythrocytes the rate of reaction of urogen decarboxylase is twice as fast with urogen III than with urogen I (de Verneuil *et al.*, 1983). Furthermore, each of the carboxyl intermediates (of either isomer I or III) will also function as substrates, although some also act as inhibitors, depending on the concentration. From studies on the purified enzymes from human and yeast, which are both monomeric, it is clear than a single protein catalyses the decarboxylation of all four acetate groups (de Verneuil *et al.*, 1983; Felix and Brouillet, 1990), but it is uncertain whether the reactions occur at the same active centre. Both the human and yeast enzymes are sensitive to sulphydryl reagents and to inhibition by metal ions such as Cu^{2+}, which are thought to act by binding to an essential thiol group. The tobacco enzyme is also inhibited by a number of heavy metal ions (Chen and Miller, 1974). However, no evidence for the presence of a cofactor has been obtained for urogen decarboxylase from any source.

2. Assay

Urogen decarboxylase activity can be measured either by determining the amount of coprophyrinogen formed, or by the rate of decarboxylation of urogen, in other words the sum of all the porphyrinogens containing four to seven carboxyl groups. In each case the assay conditions are the same, but the analysis of the products differs: the first method uses fluorimetry, the second HPLC. A typical assay mixture would contain: 50 mM sodium phosphate, pH 7.2, 10 mM DTT, 0.2 μM urogen I or III (prepared by chemical reduction of uroporphyrin I or III, see below) and enzyme extract, in a total volume of 1 ml. It is essential to include a blank with no enzyme. The incubation is carried out in a tightly closed tube (which may be under an argon or nitrogen atmosphere, but this is not essential) at 37°C in the dark for up to 30 min. The reaction is stopped by the addition of TCA to 5% and then the porphyrinogens are oxidised by the addition of 60 μl of 30% hydrogen peroxide and exposure to light for 20 min. The amount of coproporphyrin can be estimated by fluorescence spectroscopy (λ_{exc} = 406 nm, λ_{em} = 610 nm) and comparison with a standard curve (Juknat et al., 1989). Alternatively, the porphyrins are converted to their methyl esters and then analysed by reversed-phase HPLC, which allows separation of each of the decarboxylation products (Rossi and Curnow, 1986). Generally, the fluorometric assay is 10 times more sensitive than the HPLC method, which is also much slower, and so the former is generally used for protein purification studies. However, if detailed investigation of the enzyme reaction is required, the ability to measure each of the reaction intermediates makes the HPLC assay the method of choice.

N.B. Preparation of porphyrinogens for enzyme assays.   The true intermediates of the pathway are the reduced porphyrinogens, which are rather unstable and oxidise easily in the light under aerobic conditions to the corresponding porphyrins. It is the fully-conjugated porphyrins which are commercially available (e.g. from Porphyrin Products, Logan, UT, USA), so the substrates for the last three enzymes of protoporphyrin IX synthesis are prepared by chemical reduction of the corresponding porphyrin to the porphyrinogen with freshly prepared sodium amalgam (Labbe et al., 1985). The procedure must be carried out under a stream of nitrogen, and under very dim light, in order to minimise reoxidation. Generally, the yield of reduced substrate is 80–90%, and can be quantified from the amount of fluorescence of the solution; the porphyrins are highly fluorescent, while the porphyrinogens are not. The colourless porphyrinogens can be stored under paraffin oil in the dark at −20°C for a number of months, but should not be used if they have a brownish-red colour, which indicates an increase in the amount of porphyrin present.

3. Purification

Urogen decarboxylase has been investigated in only two photosynthetic organisms: tobacco leaves (Chen and Miller, 1974) and *Euglena gracilis* (Juknat et al., 1989). In the latter case, the enzyme was purified 400-fold after ammonium sulphate fractionation and gel filtration, to yield an enzyme with a specific activity of 0.77 nmol h^{-1} mg^{-1}. The native molecular weight on gel filtration was 54 kDa, slightly larger than the proteins from human erythrocytes or yeast, both of which were estimated to have a native molecular weight of 46 kDa. The human enzyme was purified to homogeneity by a series of conventional chromatographic steps (de Verneuil *et al.*, 1983). It has a specific activity of 9.9 μmol h^{-1} mg^{-1}, and gave a single band on SDS-PAGE, the same size as the native protein, indicating that it is a monomer. Urogen decarboxylase from yeast was purified to homogeneity (with a specific activity of 1.7 μmol h^{-1} mg^{-1}) using a method which included an acid-precipitation step and affinity chromatography on uroporphyrin I linked to Affi-Gel 102 (Felix and Brouillet, 1990). Acid precipitation removes many proteins from the crude lysate, while leaving over 90% of the urogen decarboxylase in the supernatant. The affinity chromatography step was successful with or without prior reduction of the affinity ligand to urogen by sodium amalgam, and provided a 400-fold purification. However, the protein concentration of the eluted enzyme was extremely low (0.1–0.2 μg ml^{-1}) and in order to obtain enzyme activity, it was necessary to add bovine serum albumin to 0.05%.

F. Coproporphyrinogen Oxidase

1. Reaction

Coprogen oxidase (EC 1.3.3.3) catalyses the oxidative decarboxylation of the two propionate groups on rings A and B to vinyl groups. The eukaryotic enzyme requires molecular oxygen and two molecules of CO_2 are released. Studies on the animal enzyme have demonstrated that the reaction proceeds in a stepwise fashion, with the formation of a vinyl group at position 2 (ring A) first, followed by decarboxylation at position 4. The intermediate, harderoporphyrinogen (2-vinyl, 4-propionate porphyrinogen), does not dissociate from the enzyme, and in the presence of coprogen, exogenously supplied harderoporphyrinogen is not decarboxylated because it does not equilibrate with the enzyme-bound molecule (Elder and Evans, 1978a; Jackson *et al.*, 1980). Two mechanisms for the propionate to vinyl conversion have been proposed (for a review see Dailey, 1990). One invokes a β-hydroxypropionate intermediate (Sano, 1966), while the other proposes that the reaction proceeds via hydride ion removal with simultaneous decarboxylation (Seehra *et al.*, 1983). The only mechanistic study of the enzyme from photosynthetic organisms was in *Euglena*, where as for the animal enzyme, harderoporphyrinogen rather than isoharderoporphyrinogen (4-vinyl, 2-propionate porphyrinogen) was decarboxylated (Cavaliero *et al.*, 1974). The enzyme from tobacco leaves was reported to be sensitive to metal ions such as Fe^{2+} and Co^{2+}, which activated it at low concentrations (0.1 mM), but totally abolished activity above 0.5 mM (Hsu and Miller, 1970). This was also observed for crude preparations of the enzyme from mammalian sources (Batlle *et al.*, 1965), but the purified bovine liver enzyme was insensitive to metal ions and to chelating agents (Yoshinaga and Sano, 1980). No evidence was found for bound cofactors, but two or more tyrosine residues appear to be required for activity. Although the enzyme from yeast contains one Fe atom per enzyme subunit (Camadro *et al.*, 1986), this did not appear to be involved in catalysis. In yeast, coprogen oxidase is cytosolic (Camadro *et al.*, 1986), while in mammals it is present in the intermembrane space of mitochondria (Elder and Evans, 1978a; Grandchamp et al., 1978), but in both cases there is some evidence for loose association of the enzyme with a membrane. Interestingly, the activity of the purified, but not crude, enzyme from yeast, bovine liver and mouse liver were stimulated by phospholipids and certain non-ionic detergents (Yoshinaga and Sano, 1980; Camadro et al., 1986; Bogard et al., 1989). In etiolated pea leaves and spadices of *Arum*, coprogen oxidase activity could not be demonstrated in purified mitochondria which had retained their outer membrane (A. G. Smith and G. Elder, in prep.), while it was readily measurable in purified plastids. Although Hsu and Miller (1970) reported that tobacco leaf coprogen oxidase was mainly present in mitochondria, their mitochondria were unlikely to be pure, since they were prepared simply by differential centrifugation, and indeed are described as also containing grana and broken chloroplasts. The K_m for coprogen of the tobacco enzyme was 36 μM, comparable to the bovine liver K_m of 48 μM (Yoshinaga and Sano, 1980). Both of these were determined with a fluorimetric assay for the enzyme. In contrast, using the radiochemical assay (see Section III.F.2) K_m values of 0.1–0.3 μM were found for the enzymes from several sources, including mouse liver (Bogard et al., 1989) and yeast (Camadro et al., 1986).

2. Assay

There are a number of published methods for the assay of coprogen oxidase. Two, which have been used successfully to detect the enzyme in plant tissue, will be described here. The fluorimetric method, developed by Labbe et al. (1985) is a coupled assay. The product of coprogen oxidase is oxidised further by protogen oxidase to protoporphyrin IX, which has a strong fluorescence emission at 632 nm. The protogen oxidase is most conveniently obtained from yeast mitochondrial membranes (YM), since these are free of coprogen oxidase. The membranes are prepared by centrifugation of cell-free yeast extract at 25 000 $\times$ g (Labbe et al., 1985), and can be stored frozen at 30 mg ml^{-1} protein at $-70°$C for several months without loss of activity. The assay conditions must be reducing in order to minimise non-enzymic oxidation of the coprogen and protogen, but there must also be sufficient molecular oxygen for the reaction to proceed. The reaction therefore is carried out in well aerated buffer in a stirred cuvette, but with a high concentration of DTT (6 mM). EDTA is also included to inhibit metal chelatases which might utilise the protoporphyrin IX produced. The reaction mix of 3 ml is set up in a 4 ml fluorimetric cuvette, and contains 0.1 M potassium phosphate, pH 7.6 (air-saturated by vigorous shaking), 1 mM EDTA, 6 mM DTT, 50 μl YM (10–15 nmol protogen oxidised h^{-1}) and tissue extract. The cuvette is placed in a fluorimeter with a thermostatted cell holder maintained at 30°C, and the contents are stirred to keep a constant concentration of O_2. Excitation is at 410 nm, and emission at 632 nm. After measuring a baseline rate of fluorescence, the reaction is started by the addition of 50 μl of coprogen solution (prepared by chemical reduction of coproporphyin; see Section III.E.2). The increase in protoporphyrin IX fluorescence is measured for 2–10 min, and the enzyme activity is calculated by reference to calibration curves for the fluorescence of a standard solution of protoporphyrin IX. It is important, particularly for crude tissue extracts, to determine the rate of non-enzymic oxidation, by including controls of boiled enzyme. If desired, fluorescence spectra may also be taken, to ensure that the fluorescence

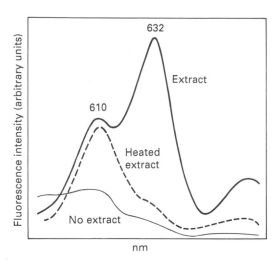

FIG. 12.6.   Spectra of coprogen oxidase assays (Section III.F.2) after 10 min incubation with enzyme extract, heated extract or no extract. The peak at 610 nm is coproporphyrin III, while that at 632 nm is protoporphyrin IX.

measured is due to protoporphyrin IX and not to coproporphyrin. They can be distinguished from one another by the fact that the λ_{em} of coproporphyrin is 610 and not 632 nm (Fig. 12.6).

The other method for measuring the activity of coprogen oxidase is a stopped radioactive assay and relies on the determination of the amount of $^{14}CO_2$ released during the reaction (Elder and Evans, 1978b). [^{14}C]Coproporphyrin III, ^{14}C-labelled in the carboxyl carbons of the 2- and 4-propionate groups, can be prepared by the stepwise modification of the vinyl groups of protoporphyrin IX. This is used as a substrate immediately after reduction with sodium amalgam (Section III.E.2). The reaction is set up in a 7 ml glass vial containing a small glass tube (0.8 ml) supported on a polypropylene cup (made by cutting off the bottom of a microfuge tube). The tube contains 150 μl of 0.1 M Hyamine in methanol, which traps the released CO_2. The reaction mix of 100 μl is 75 mM Tris-HCl, pH 7.2, 10 mg ml^{-1} bovine serium albumin, 1 mM KOH, 10 μl substrate solution (1–2 mM in 125 mM Tris-HCl, 100 mM sodium thioglycollate) and up to 50 μl tissue extract. All of the components except the substrate are placed in the bottom of the glass vial, taking care to avoid the tube containing the Hyamine, and then the vial is sealed with a rubber cap. The reaction is initiated by the addition of the substrate through the rubber cap using a microsyringe, and allowed to proceed for 1 h at 37°C, with gentle shaking. The reaction is terminated by freezing in methanol/dry ice, and then 50 μl of 0.1 M Na_2CO_3, followed by 50 μl of 5 M HCl is injected into the reaction mix, again being careful to avoid the central tube. The vials are shaken at room temperature for 45 min, and then the amount of $^{14}CO_2$ trapped in the Hyamine solution is determined by scintillation counting. Since two molecules of CO_2 are released for each coprogen oxidised, the rate of reaction is half the rate of CO_2 production.

3. Purification

The only report of a purification of coprogen oxidase from a plant source is that of Hsu and Miller (1970). They purified the tobacco leaf enzyme 70-fold using differential centrifugation, ammonium sulphate precipitation and hydroxyapatite chromatography, and their enzyme preparation had a specific activity of $0.144 \, \mu$mol h^{-1} mg^{-1}. Coprogen oxidase has, however, been purified to homogeneity from bovine liver (Yoshinaga and Sano, 1980), yeast (Camadro et al., 1986) and mouse liver (Bogard et al., 1989), with specific activities of 3.5–$7.0 \, \mu$mol h^{-1} mg^{-1}. For maximal yield of enzyme, it was found to be essential to include protease inhibitors, and, for the mouse liver enzyme, to have 1 mM DTT in all buffers. In each case the native enzyme had a molecular weight of approximately 70 kDa, but whereas on SDS-PAGE the yeast and mouse liver enzymes showed single bands of 35 kDa, suggesting that they were dimers, the bovine liver coprogen oxidase was reportedly a monomer. The procedures used to purify the bovine liver and yeast enzymes were very similar: ammonium sulphate fractionation, followed by chromatography on ion-exchange and gel filtration columns. For the mouse liver enzyme, however, Bogard et al. (1989) found a very useful purification step in Blue-Sepharose chromatography (over 100-fold purification). Although this was originally developed as an affinity column for proteins which bound NAD, it had proved very useful for the purification of the later enzymes of heme biosynthesis, including protogen oxidase and ferrochelatase, suggesting that they share common substrate-binding sites.

G. Protoporphyrinogen Oxidase

1. Reaction

The reaction catalysed by protogen oxidase (EC 1.3.3.4) is the six-electron oxidation of protogen IX to the fully conjugated protoporphyrin IX, in a step which requires molecular oxygen. Since the conversion occurs chemically at neutral pH, the existence of an enzyme to catalyse this reaction was initially in some doubt. However, the partial purification of protogen oxidase from yeast conclusively established its role (Poulson and Polglase, 1975). Since then, the dependence of the reaction on protein concentration, and its sensitivity to heat denaturation and changes in pH, have been demonstrated in both mammalian and bacterial systems, and in plant tissues (Jacobs and Jacobs, 1984). Yeast and mammalian protogen oxidase is tightly bound to the inner mitochondrial membrane (Poulson and Polglase, 1975; Deybach et al., 1985) and in plants, protogen oxidase activity is associated with both plastid and mitochondrial membranes (Jacobs and Jacobs, 1984). The mechanism of the protogen oxidase reaction is unknown at present. Three moles of O_2 per mole of protogen appear to be utilised, suggesting that H_2O_2, rather than H_2O, is produced. The enzyme will use as substrates only protogen IX or, at much lower activity, mesoporphyrinogen IX (which has ethyl rather than vinyl groups at positions 2 and 4); implying a highly specific binding site and the K_m for protogen IX is in the micromolar range. However, there are insufficient data to propose a model for catalysis (see Dailey, 1990, for a review).

Recently, there has been increased interest in protogen oxidase, since it is the site of action of a group of herbicides, the diphenyl ethers (DPEs), such as acifluorofen-methyl. These compounds cause peroxidative damage of membranes and other cellular components in treated plants, but only in the light. Since DPEs do not themselves absorb light, their action must be mediated through photodynamic pigments of cellular origin. Initially, these were postulated to be carotenoids, but it was subsequently shown that DPE-treated tissues accumulate protoporphyrin IX (Matringe and Scalla, 1988; Witowski and Halling, 1988), and that DPEs inhibit protogen oxidase from most plant, animal and yeast sources, with an IC_{50} of 4–10 nM (Matringe *et al.*, 1989). In contrast, the herbicides had much less effect on either Mg- or ferrochelatases (IC_{50} > 100 μM). This somewhat surprising result, the accumulation of the *product* of the enzyme which is inhibited, is presumably due to two factors: firstly the substrate of the enzyme, protogen, can be rapidly oxidised to protoporphyrin IX in a non-enzymic reaction, and secondly, the chemically produced protoporphyrin IX is not accessible to the chelatase enzymes. An explanation for these observations comes from the finding that there is no detectable coprogen oxidase activity in plant mitochondria (A. G. smith and G. Elder, in prep.), but protogen oxidase is present (e.g. Jacobs and Jacobs, 1987). This implies that protogen (synthesised principally, or entirely, within the plastid) is delivered to plant mitochondria for the synthesis of mitochondrial heme. The likely association of protogen oxidase with the inner mitochondrial membrane may provide the means of transporting the tetrapyrrole molecule into the organelle, where ferrochelatase is present (Little and Jones, 1976). However, if protogen oxidase is inhibited by DPEs, protogen may be unable to enter the mitochondria, so that it accumulates in the cytoplasm and is oxidised chemically to protoporphyrin IX. This cannot be metabolised further since the Mg- and ferrochelatases are confined to the plastids and mitochondria. In this context, it is interesting that protogen oxidase from soybean root mitochondria is insensitive to DPE inhibition, and soybeans are much more tolerant to DPEs than most plants (Jacobs *et al.*, 1990).

2. *Assay*

Protogen oxidase activity can be measured in a fluorometric assay identical to that for coprogen oxidase, except that the substrate is protogen, and no yeast membranes are included (Labbe *et al.*, 1985). The assay can be continuous, or a stopped assay (Jacobs *et al.*, 1990) and it is essential to include a heated extract control, in order to determine the rate of chemical oxidation of the substrate.

3. *Purification*

Protogen oxidase has been purified to homogeneity from both barley etioplasts and mitochondria (Jacobs and Jacobs, 1987). Triton X-100 extracts of organelles were chromatographed on DEAE-Bio-Gel A and then hydroxyapatite, to yield preparations with specific activities of 0.7–0.9 μmol h^{-1} mg^{-1}. SDS-PAGE revealed a prominent band of 36 kDa and several minor ones. However, these latter could be removed by rechromatography on DEAE-Bio-Gel A. Gel filtration of the preparations showed a single peak which corresponded to 210 kDa, but because of the presence of Triton micelles, this may not be a true measure of the native molecular weight of the protein.

In addition to these physical characteristics, both the mitochondrial and etioplast enzymes had the same pH optimum (5–6) and K_m for protogen (5 μM). Protogen oxidase has also been purified from mammalian sources (e.g. Siepker *et al.*, 1987) and yeast (Labbe-Bois and Labbe, 1990). Although the K_m of these enzymes for protogen was similar to that reported for barley protogen oxidase, the pH optima were higher (pH 7–7.5) and the subunit molecular weights were 65 and 56 kDa, respectively.

IV. ENZYMES SPECIFIC TO THE CHLOROPHYLL BRANCH OF THE PATHWAY

As already mentioned, few of the chlorophyll-specific biosynthetic enzymes have been fully characterised, with the exception of the light-requiring protochlorophyllide reductase and the methyl transferase catalysing the methylation of Mg-protoporphyrin (Mg-proto). Reactions in this section of the pathway have been suggested over the years on the basis of identification of the various intermediates in photosynthetic organisms. Such intermediates are often forced to accumulate in whole cells, e.g. by disruption of the pathway genetically (Granick, 1950; Lascelles, 1966; Gough, 1972) or by inhibitors (Jones, 1966). The conclusions from such studies have recently been largely confirmed by demonstrating, using cell-free preparations, the interconversion of some of the proposed intermediates, largely as expected (see Beale and Weinstein, 1990).

All these intermediates, like chlorophyll *a*, are highly coloured. Consequently, spectroscopic (absorption and fluorescence) techniques play major parts in analysis of activities of the chlorophyll biosynthetic enzymes, whether studied *in vivo* or *in vitro*. The obvious drawback of spectroscopic analysis however is the inevitable presence in photosynthetic tissues or extracts derived therefrom, of mixtures of coloured compounds with overlapping spectral properties. Consequently, separation techniques to produce spectrally pure material are also an essential complement to spectroscopic work carried out on extracts of chlorophyll biosynthetic tissues. Furthermore, for work *in vitro* with isolated enzyme preparations, a supply of the substrate for the particular enzyme under investigation has to be sought. Fortunately, a range of pigment mutants of various photosynthetic bacteria—notably *Rhodobacter spheroides*—have been described blocked at specific steps in the chlorophyll biosynthetic pathway and as such accumulate the specific intermediate preceeding the block, thereby providing excellent sources of such compounds (Saunders, 1973; Marrs, 1981; Hunter and Coomber, 1988). Again, ALA feeding of dark-grown higher plants (cereals or cucumber cotyledons) has also been employed as a method for producing, for example, protochlorophyllide-enriched tissue. This pigment can be purified from such a source and subsequently used in studies on the later stages of chlorophyll *a* biosynthesis *in vitro* (Griffiths, 1978). If substrate requirements are not met from such natural sources resort may be made to the cooperation of synthetic porphyrin chemists to synthesize, to order, intermediates for testing *in vitro*. Amongst the considerable advantages gained from using biologically generated substrates as above is the fact that the material, besides being available in large quantities, is also chirally compatible with the biosynthetic enzymes. Radiolabelled substrates, when available, and cofactors, if known, have also been used occasionally to demonstrate particular chlorophyll biosynthetic reactions *in vitro*.

Reliable cell-free preparations capable of carrying out individual reactions of the chlorophyll-specific branch invariably employ isolated plastids, or preparations derived from them, as enzyme source. Whilst it is perfectly feasible (see below) to demonstrate qualitatively the operation of chlorophyll biosynthetic reactions from work *in vivo* with whole tissue, e.g. intact leaves, incubations *in vitro* are inevitably followed by extraction of total pigments. The resulting pigment mixture is then subjected to a separation process to generate fractions suitable for spectral analysis — the whole operation repeated after varying time intervals to generate a discontinuous estimate of the progress of a particular reaction.

Similar principles have been employed in assaying all the enzymes of this part of the chlorophyll biosynthetic pathway, and as such they can be introduced collectively. They all involve the use of tissue homogenates, isolated plastids or fractions prepared from them as source of enzyme. Choice of starting tissue is largely governed by the particular enzyme to be studied and may be either dark- or light-grown material, the only prerequisite being that the tissue has to be young and therefore metabolically active. Simple plant forms such as algae have been employed only relatively infrequently in studies pertinent to this section.

For etiolated plant tissue, dark-grown cereals, such as 7-day-old barley or wheat, offer considerable advantages over dicotyledonous plants such as peas, beans and spinach. The former can sustain much more tissue growth and development in darkness and therefore provide a ready source of considerable quantities of material. For light-grown or greening tissue, however, this advantage of cereals is lost. Green dicotyledonous tissues also tend to be less fibrous than green monocotyledonous leaves and therefore much easier to homogenise and fractionate. This becomes particularly important for those systems (see below) whose activity appears to have an obligate requirement for intact plastids. Working with etiolated tissue in the absence of any light gives rise to obvious logistic problems. However, sources of illumination, normally green light, which does not influence photomorphogenic development, have been described and are readily available (see Smith, 1975).

Regarding the choice of medium for enzyme or organelle isolation, once more the choice is largely governed by the objective, for example whether intact organelles are required or not. Generally the medium provides: the osmoticum (normally sucrose), a suitable buffer to avoid drastic pH changes during tissue homogenisation, and various additives such as thiol-reducing agents, protease inhibitors, etc., designed to maximise recovery of activity. A wide choice of power homogenisers is available for smashing plant material. In the laboratory of one of the authors (WTG) considerable enthusiasm is shown for the modified Waring blender arrangement incorporating disposable blades (Kannangara *et al.*, 1977) for tissue homogenisation when the aim is to isolate intact plastids. After homogenisation and filtering, the brei is finally fractionated by differential centrifugation yielding the various cell-free preparations ready for enzymatic activity determination.

After incubation *in vitro*, the reactions are quantified by extraction of total pigments. Such an extraction is normally carried out with an organic solvent that is miscible with water, such as acetone, followed by transfer of the pigments to a second, but water-immiscible solvent like ether, which can then be backwashed with water to remove impurities. The extract is then rendered suitable for spectral analysis by removal of coloured contaminants, typically carotenoids and chlorophylls. Fortunately, the biosynthetic intermediates, being free acids (see Fig. 12.5), and therefore polar, can be readily phase-separated from the non-polar chlorophylls and carotenoids by a process of differential solvent extraction. In a typical procedure (e.g. Treffry, 1970) the phase separation might involve partition between a lighter non-polar solvent such as petroleum ether and an immiscible denser polar phase such as aqueous acetone or methanol. Under slightly alkaline conditions (e.g. on addition of 0.01 mM NH_4OH) ionisation of acid groups present in intermediates is facilitated, rendering them even more polar and soluble in the aqueous phase and thereby enchancing their separation from the non-polar interfering material. Recently, solid phase separation techniques, using commercially available matrices, have been introduced into pigment separations in chlorophyll biosynthesis work. Reference to the use of these will be made in the appropriate sections (see below).

Spectroscopic quantitation of pigment extracts for estimating enzyme activity can be achieved by either absorbance or fluorescence measurements, as the chlorophylls and their precursors display intense and characteristic light absorption and emission — particularly in organic solvents. Qualitative identification and quantitation in the case of absorption measurements is made by reference to published spectral characteristics (see Table 12.1) whereas for the fluoresence analysis calibration curves, constructed using pure compounds measured under conditions identical to those of the unknown estimations, must be employed. As often happens, initial extracts recovered during chlorophyll biosynthesis work contain complex mixtures of coloured components. Consequently, these have to be further separated into fractions containing material in a pure form that is suitable for spectroscopic quantitation. Two chromatographic techniques that have been used extensively here are TLC and HPLC.

Thin layer chromatography used in the analytical mode is useful for qualitative description of the composition of an unknown mixture. The preparative TLC procedure, coupled with spectroscopic analysis of separated material recovered from the developed plate, can provide quantitative information on the composition of unknown mixtures. Some data on the behaviour of chlorophyll a and related compounds on the various TLC systems are given in a recent review (Shioi, 1991).

The use of HPLC techniques, which are particularly sensitive when coupled with a fluorimetric detection system, continues to increase in the analysis of chlorophyll and related pigment mixtures. The introduction of diode array detectors and the more recent use of continuous flow fast atom bombardment-mass spectrometry (FAB-MS) (van Breemen et al., 1991), has contributed to the technique becoming particularly powerful. Reversed-phase (C_{18}) solid absorbent, eluted with a range of solvent mixtures, is used most frequently for chlorophyll separations (Mantoura and Lewellyn, 1983) whereas some of the chemically very similar biosynthetic precursors, e.g. mono- and divinylprotochlorophyllides, have been separated in a highly specific manner on columns of polyethylene eluted with acetone–water mixtures (see Shioi, 1991).

Separation techniques similar to those outlined above have also been used for pigment isolation in isotope feeding experiments designed to study aspects of chlorophyll biosynthesis. Thus, incorporation into chlorophyll and related compounds of both stable (^{18}O) and radioactive (^3H, ^{14}C) isotopes have helped to establish mechanisms for some of the biosynthetic reactions, e.g. those concerned with cyclisation (Walker et al., 1989a,b) and protochlorophyllide reductase (Valera et al., 1987; Begley and Young, 1989).

TABLE 12.1. Spectral characteristics of the chlorophylls and related porphyrins of higher plants.

	Absorbance			Fluorescence		
Pigment	Solvent	α-band (nm)	(E mM)	Solvent	Soret excitation max. (nm)	Emission max. (nm)
Protoporphyrin IX	Eth	633	5.2[a]	Eth	406–408	644–646[d]
Mg-Proto IX and methyl ester	Eth	589	18.2[a]	Eth	422	600–602[d]
Protochlorophyllide	Eth	623	35.6[a]	Eth	432	623[c]
Protophaeophytin	Eth	638	2.2[a]			
Chlorophyll a	Acet	663	76.8[b]	Eth	428	666[b]
Phaeophytin a	Eth	666	60.1[c]	Eth	408	672[b]
Chorophyll b	Acet	646	47[b]	Eth	453	646[b]
Phaeophytin b	Eth	654	43.6[c]	Eth	434	658[b]

Eth, anhydrous diethyl ether; Acet, acetone–water (8:2; v/v).
[a] Jones, O. T. G. (1969). *In* "Data for Biochemical Research" (R. M. C. Dawson *et al.*, eds), pp. 318–325. Clarendon Press, Oxford.
[b] Porra, R. J. (1991). *In* "Chlorophylls" (H. Scheer, ed.), pp. 32–57. CRC Press, Boca Raton, FL.
[c] Lichtenthaler, H. K. (1987). *Methods Enzymol.* **148**, 350–382.
[d] Ellsworth, R. K. and Murphy, S. J. (1979). *Photosynthetica* **13**, 392–400.

Regarding purification of the chlorophyll-specific biosynthetic enzymes, the discussion has, of necessity, to be very brief, as the only enzymes to have been purified are protochlorophyllide reductase and Mg-proto methyl transferase. The current status of the purification of the remaining enzymes will be discussed in the appropriate sections dealing with individual enzyme system of the chlorophyll branch.

A. Magnesium Chelatase

1. Reaction

The chelation of Mg^{2+} by the enzyme Mg-proto chelatase represents the first specifically committed reaction in the biosynthesis of chlorophylls in all photosynthetic organisms. Several attempts at characterisation of the system have been made over the past few decades with systems of both bacterial (Gorchein, 1972) and plant (Richter and Rienits, 1982) origin. These met with limited success, however, due to the poor recovery of the activity in the cell-free preparations in either bacterial sphaeroplasts or wheat etioplasts. Recently, however, an *in vitro* plastid system from greening cucumber cotyledons, displaying Mg^{2+} chelation rates approaching the chlorophyll synthesis rate of the original tissue, has been described, and a second system using isolated pea chloroplasts was even more active (Walker and Weinstein, 1991a). In this latter system, activity was still present in broken chloroplasts (Walker and Weinstein, 1991b). Mg-chelatase activity in both the pea and cucumber plastids required ATP, the optimum being determined for the cucumber system as 4 mM. The cucumber plastid system has a requirement for ATP, 4.0 mM being optimal. Furthermore, ATP hydrolysis occurs during the reaction, fulfilling an as yet uncharacterised role. A range of porphyrin substrates can be chelated with protoporphyrin IX, the natural substrate, optimal at 5 μM. Magnesium chelation in plastids is sensitive to various thiol-modifying reagents, both permeants and impermeable forms, implying the possibility of an outer membrane location of the enzyme. Surprisingly, the chelatase is also sensitive to the potent ferrochelatase inhibitor, *N*-methylprotoporphyrin at the μmolar level (50% inhibition at 2 μM), despite the fact that this compound has no effect on chlorophyll synthesis in algal cells (Beale and Chen, 1983). Similarly the reported inhibition of the chelatase by 2,2'-dipyrridyl (50% inhibition at 2 mM) is surprising considering that when intact plants are treated with this reagent it leads to an accumulation of Mg-proto IX (Duggan and Gassman, 1974). An interesting difference exists between the product of Mg^{2+} chelation in higher plant preparations and sphaeroplasts of *Rhodobacter* — Mg-proto is formed in the former, but the methyl ester is the product of the bacterial system. To explain this it is suggested that the chelatase is obligatorily linked with the methylase in the bacteria.

2. Assay

As already mentioned, the most active chelatase preparation to date is the intact plastid system isolated from cotyledons of 7-day-old dark-grown cucumber greened for approximately 24 h. Under optimal conditions this system gives rates of approximately 1.5 nmol porphyrin chelated per mg plastid protein per hour, a rate which is linearly sustained for about 50 min. The assay system used in this work, modified from earlier procedures (Castelfranco *et al.*, 1979), involves incubation of the intact plastids at 1.0 mg protein per ml in a pH 7.8 buffer, 0.5 M sorbitol, 50 mM tricine, 1 mM DTT, 1 mM MgCl$_2$, 1 mM EDTA, 0.1% BSA to which is added the substrates, 1.5 μM protoporphyrin IX (as a DMSO solution) and 4.0 μM MgATP as a regenerating system (20 mM phosphoenol pyruvate and 1.6 units of pyruvate kinase). Total incubation volume is 250 μl and the reaction is allowed to proceed with shaking for 20 min at 30°C in aluminium foil covered tubes. Termination of the reaction is by addition of 0.75 ml of ice-cold acetone followed by centrifugation of the precipitated protein, which, after resuspending in 250 μl of 0.12 N HCl, is re-extracted with a further 0.75 ml of acetone. The two extracts are bulked followed by addition of 4 ml of 75% acetone and removal of esterified pigments by two washes with hexane. Finally, emission of the acetone solution is measured at 595 nm (excitation 419 nm) and Mg-proto concentration (free acid and monomethyl ester) is calculated by reference to a calibration curve preconstructed from authentic Mg-proto IX, which is available (together with several of the unusual compounds mentioned in this section) from Porphyrin Products Ltd, Logan, UT, USA.

3. Purification

Until recently, no success in the purification of the Mg-chelatase system beyond the intact plastid or sphaeroplast stage has been reported. However, Walker and Weinstein (1991b) found that activity was still retained in broken pea chloroplasts, disrupted by hypotonic treatment followed by two freeze-thaw cycles. Furthermore, after centrifugation at 12 000 g for 3 min, to produce a supernatant and pellet fraction, activity could be reconstituted — but both supernatant and pellet fractions were required, and these were shown to be proteinaceous, since boiling either fraction destroyed activity. Despite these encouraging results further purification of Mg-chelatase remains very much one of the challenges facing workers in chlorophyll synthesis.

B. Mg-Proto IX Methyl Transferase

1. Reaction

The transferase, *S*-adenosyl-L-methionine:Mg proto *O*-methyl transferase (EC 2.1.1.11) substitutes a Me group from SAM for the hydrogen on the C$_{13}$ propionate group of the Mg porphyrin (Fig. 12.5). Being relatively easy to assay (see below) it has been studied in a wide range of photosynthetic organisms. The solubility of the enzyme appears to vary depending on the source, being membrane bound in *R. sphaeroides* (Gibson *et al.*, 1963), membrane associated in greening barley (Shieh *et al.*, 1978), soluble in wheat and maize (Hiuchigeri *et al.*, 1981) and with both soluble and insoluble forms in *Euglena* (Ebon and Tait, 1969). Similarly, the reaction mechanism of the enzyme has, rather surprisingly, been found to be different depending on its origin. In *Euglena* binding of the two substrates is random, in *R. sphaeroides* the enzyme has an ordered mechanism with the porphyrin binding first, whereas the wheat enzyme has a ping-pong mechanism, SAM binding first followed by release of *S*-adenyosyl-homocysteine before binding of the porphyrin (see Beale and Weinstein, 1990).

K_m values for the substrates have been determined for the enzymes from various sources and these vary from approximately 20 to 40 μM for the porphyrin and slightly higher for S-adenosyl methionine. Activation of the enzyme by thiol agents, e.g. cysteine and dithiothreitol, and its sensitivity to thiol inhibitors such as PCMB and the heavy metals Cu^{2+}, Ag^{2+}, suggests that transferase is a thiol enzyme. The enzyme activities quoted for the various tissues is much greater than that required to sustain chlorophyll synthesis; furthermore, growth in the light or darkness affected this activity only marginally.

2. Assay

Although several assays of the enzyme have been described they are all based upon monitoring the incorporation of ^{14}C from [Me-^{14}C]SAM into Mg-proto to form the methyl ester. The assay is relatively simple due to the ease of separation of the substrate from the product using solubility-based procedures.

In a typical procedure using 3 h greened 7-day-old dark-grown barley, the finely cut seedlings are homogenised in a medium of 0.5 M sucrose, 0.2 M Tris-HCl, 5 mM $MgCl_2$ and 4 mM 2-mercaptoethanol, pH 7.5. This is filtered through four layers of cheesecloth and the homogenate used directly for assay or fractionated further. The enzyme extract is assayed by adding (μmol) Mg-proto (0.3) 0.4 μCi[Me-^{14}C]SAM (0.1), mercaptoethanol (4), Tris HCl (20) in a total volume of 1 ml, pH 7.5, and incubated with shaking at 37°C for 1 h. At the end of the incubation the porphyrins are extracted (Gibson et al., 1963) twice with 5 ml ethyl acetate–acetic acid (3:1; v/v), the organic phase neutralised with sodium acetate, washed with water and the porphyrins extracted in 3 N HCl. The acid layer is then washed with ethyl acetate, neutralised, and the porphyrins finally extracted into fresh ethyl acetate and the radioactivity of the extract measured. Identification of the product as the methyl ester is carried out by TLC (Rebeiz et al., 1975).

3. Purification

A 20-fold purification of the enzyme has been achieved with ammonium and protamine sulphate fractionation of the greening barley preparation, whereas a 2000-fold purification has been claimed by affinity chromatography of the wheat etioplast enzyme (Richards et al., 1987). No molecular characterisation of the protein has yet been achieved.

C. Mg-Proto IX Monomethyl Ester Oxidative Cyclase

1. Reaction

The conversion of the C-13 methyl propionate group in Mg-proto into the isocyclic ring of protochlorophyllide (Fig. 12.7) is a complex process involving concerted six-electron oxidation, incorporation of an oxygen atom and formation of a new C—C bond. Our present ideas on the cyclisation are due mainly to the collaborative efforts of the groups of P. A. Castelfranco and K. M. Smith at the University of California at Davis. Between them they have synthesized all the intermediates implicated in the process (see Fig. 12.7) and demonstrated their enzymic interconversions with plastid preparations from greening cucumber cotyledons. The first important outcome of

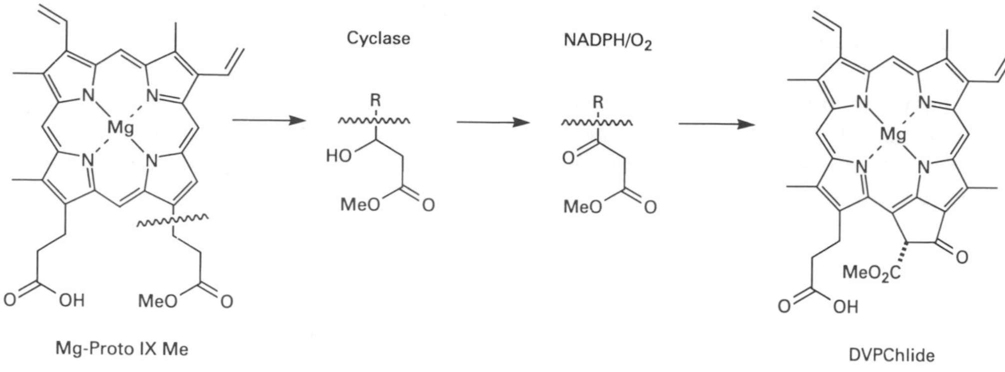

FIG. 12.7. Cyclisation reactions in chlorophyll synthesis.

these studies was the need for a reappraisal of the original scheme for cyclisation proposed by Granick (1950) which involved an initial dehydrogenation of the propionate side-chain to an acrylate residue. Contrary to this, the synthetic (*trans*) acrylate was found not to be converted in this system, whereas the 13-β-hydroxyl-propionate and 13-β-keto propionate analogues of Mg-proto are efficiently converted into protochlorophyllide (Fig. 12.7). Formation of protochlorophyllide, irrespective of substrate, is dependent on the presence of both NADPH and molecular oxygen, suggesting a mixed function oxidase type of mechanism with the $^{18}O_2$ directly incorporated as the carbonyl group of the formed isocyclic ring (see Walker *et al.*, 1988, and references therein). Activity of the system is sensitive to the thiol reagents but unlike the chelatase the cyclase activity measured in intact plastids is insensitive to the non-permeant PCMBS, implying an intrachloroplast localisation.

In earlier studies conversion of Mg-proto to protochlorophyllide could only be demonstrated with whole plastids (Nasrulhaq-Boyce *et al.*, 1987). Recently, however, it has been possible to fractionate the plastids and reconstitute cyclase activity in membrane preparations by addition of supernatant fractions, even after purification of the latter by ammonium sulphate fractionation and phenyl-Sepharose chromatography (Walker *et al.*, 1991).

2. Assay

Several procedures for assaying the cyclase have been described, with different levels of demand for non-routine instrumentation. The procedure to be described here is based on that described in Walker *et al.* (1988) and which should be possible to reproduce in most biochemical laboratories. The assay is carried out with cucumber plastids prepared essentially as already described for the chelatase assay.

Incubations containing approximately 5 mg of plastid protein are carried out in 1 ml of buffer containing 0.5 M sorbitol, 10 mM HEPES, 20 mM TES, 1 mM $MgCl_2$, 1 mM EDTA and 0.5 mg BSA at pH 7.7 together with cofactors: NADPH (5 mM) as a regenerating system, SAM (1 mM) and 10 μM porphyrin substrate, which can be a DMSO solution of either unesterified Mg-proto (preferred) or its dimethyl ester. The reaction mixture is shaken in the dark at 30°C for 1 h, after which PChlide formed is extracted into ether and estimated spectrophotometrically (Chereskin, *et al.*, 1982). Confirmation of the identity of the product should be sought using an HPLC procedure (see earlier). Typical rates for the system are about 0.25 nmol chlorophyllide formed per mg protein per h.

3. Purification

Several protein components are expected to be involved in the cyclisation process, but as yet none have been purified to the extent of unambiguous identification.

D. 8-Vinyl Reductase

1. Reaction

This converts the 8-vinyl group on ring II of the macrocycle to an ethyl group (see Fig. 12.5). Superficially, it might appear that identifying the 8-vinyl reductase enzyme and the point at which it acts in the pathway would be easy assignments. In practice, however, this has not been so, as considerable doubt still surrounds this reaction. It currently appears that reduction of the 8-vinyl group is possible at any number of steps in the pathway between Mg-proto and chlorophyll. By this means one can account for the reported presence of mono- and divinyl analogues of all these latter intermediates in various plant tissues, admittedly mostly in small amounts and under special conditions (Rebeiz and Lascelles, 1982). The problem remains of deciding on the site at which reduction occurs under normal physiological conditions.

It had originally been assumed that 8-vinyl reduction occurs at the level of PChlide. This assumption was based on the original identification of the reduced form of the porphyrin, i.e. monovinyl protochlorophyllide (MVPChlide), as the pigment accumulated in dark-grown plants. Whereas this observation cannot be denied, it is now felt that the accumulation of MVPChlide in such cases is a consequence of the particular experimental conditions used, i.e. a prolonged period in darkness (typically 7–12 days) together with low substrate specificity of the 8-vinyl reductase, resulting effectively in complete reduction of all 8-vinyl groups, irrespective of the parent porphyrin. Thus, any DVPChlide would be expected to be converted into MVPChlide. However, under normal conditions of growth the 12 h or so in darkness would not permit sufficient accumulation of divinyl porphyrin to enable its reduction to the monovinyl form (Griffiths, 1991). It is the opinion of these authors therefore that the physiological substrate of the reductase is DVChlide, but due to its broad specificity, related compounds, e.g. DVPChlide and even Mg-proto, can be reduced given the appropriate conditions. Rather surprising, however, is the recent report claiming the monovinyl form as a better substrate than the divinyl analogue for the cyclase system in vitro (Walker et al., 1989), especially when it is considered that the early work identified DVPChlide as the normal product of the cyclase reaction in vitro (Chereskin et al., 1983). Clarification of the physiological timing of reduction however must await purification of the enzyme and evaluation of its real substrate specificity.

Equally as uncertain is the hydrogen donor or coenzyme of the reductase. Initial studies using crude chloroplasts or homogenates from etiolated wheat and Mg-proto as substrate indicated NADPH as a coenzyme for the 8-vinyl reduction, but shortly afterwards this coenzyme identification was changed to NADH (Ellsworth, 1972; Ellsworth and Hsing, 1973). More recently a soluble plastid fraction from wheat etioplasts has been purified by affinity chromatography to yield a reductase preparations which can utilise both NADH and NADPH as hydrogen donors and furthermore reduced both Mg-proto and DVPChlide (Kwan et al., 1986).

2. Assay

All reported assays of the 8-vinyl reductase are based on measuring incorporation of radiolabel from NAD(P)H (prepared enzymatically) (Kwan *et al.*, 1986) or chemically (Ellsworth and Hsing, 1973) into an 8-vinyl porphyrin. The choice of porphyrin appears limited only by availability and can be any of the Mg-branch intermediates with an 8-vinyl group. The method outlined here is based on that of Kwan *et al.* (1986) using wheat etioplast extracts as source of reductase, which gave rates of 1.6 nmol $h^{-1} mg^{-1}$ protein with Mg-proto monomethyl ester and NADH as substrates. A pH optimum of 7.7 has been reported for the enzyme and the K_m value for the porphyrin estimated as 25 μM.

At the end of the incubation at room temperature, normally 2 h, the reaction is terminated and free base porphyrins extracted by addition of ethyl acetate–acetic acid (3:1; v/v) — 10 ml per 2.5 ml incubation mixture — followed by neutralising with 15 ml of 10% (w/v) sodium acetate. The porphyrins are extracted by separation and centrifugation of the organic layer with an equal volume of 10% (w/v) aqueous HCl and then the porphyrins are transferred into fresh ethyl acetate by neutralising with $NaHCO_3$. Finally, the porphyrins are methylated with diazomethane, purified by chromatography on columns of sucrose and the protoporphyrin dimethyl ester band eluted into ether and aliquots assayed by absorption spectrophotometry and radioactivity estimated by scintillation counting (see Ellsworth and Hsing, 1973).

3. Purification

Some 72-fold purification of the reductase from a soluble fraction of wheat etioplasts has been reported (Richards *et al.*, 1987) by affinity chromatography using Zn-proto monomethyl ester as a ligand bound covalently to the matrix Sepharose 4B.

E. NADPH-Protochlorophyllide Oxidoreductase

1. Reaction

Protochlorophyllide reductase catalyses the light-dependent reduction of the porphyrin pyrrole ring D, C-17/C-18 double bond during chlorophyll synthesis by higher plants. The same reaction in photosynthetic bacteria and simpler plants such as algae or gymnosperms can be achieved without intervention of light, but details of this system have not been established. In contrast, the light-requiring reductase has been well characterised (see Griffiths, 1991). The enzyme, which is widely distributed in plastid-containing tissues, shows the unusual property of being present at very high levels in dark-grown tissues containing etioplasts, but at very much reduced levels in chloroplast-containing green tissues. Consequently, the bulk of the studies on the enzyme have been carried out with etiolated tissue as the source material.

Catalytically, the enzyme, which is membrane associated, binds NADPH and PChlide to form a ternary enzyme–substrates complex which is stable in the dark. Exposure to light, however, results in hydrogen transfer from the coenzyme to the porphyrin with formation of $NADP^+$ and Chlide and this is followed by dissociation of the products releasing the enzyme for further catalysis as shown below

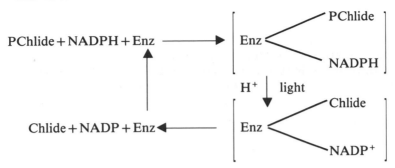

(where Enz represents protochlorophyllide reductase)

The reductase shows an absolute specificity for NADPH as reductant for which a K_m of 35 μM has been estimated in etioplast membranes. The overall reduction is stereospecific with hydride transfer occurring exclusively from the A face of the planar nicotinamide ring and added stereospecifically and exclusively to the C_{18} ring (Begley and Young, 1989). A broad porphyrin specificity range is displayed by the reductase with both DVPChlide and MVPChlide reduced equally and effectively, K_m values 0.96 and 2.6 μM, respectively (Whyte, 1989). Other features of the porphyrin molecule which are essential for activity as substrates for the reductase include a central metal atom which can be either Mg^{2+} or Zn^{2+} and a free carboxyl group at C-13^3 (Griffiths, 1980). The reductase is sensitive to thiol reagents like DTNB and the substituted maleimides (NPM and NEM), high ionic strength (e.g. 80% inhibition at 150 mM KCl), and is particularly sensitive to the flavin inhibitor quinacrine which causes 50% inhibition at 10 μM.

2. Assay

As already mentioned plastids from etiolated plants are particularly enriched sources of the reductase, with cereal seedlings grown for approximately 7 days in darkness providing convenient material. Etioplasts are readily recovered from such tissue by homogenisation in a buffer containing 0.5 M sucrose, 0.2% BSA, 1 mM $MgCl_2$, 20 mM TES, 20 mM HEPES and 1 mM EDTA adjusted with KOH to pH 7.2, with cysteine (5 mM) added immediately prior to homogenisation. After filtering, the homogenate is centrifuged transiently to 9000 × g and the resulting pellet resuspended in fresh buffer without cysteine and recentrifuged, first at 500 × g for 1.5 min, and any pellet discarded, and then at 2500 × g for 10 min to sediment the etioplast fraction. This is resuspended in the initial buffer and aliquots, equivalent to approx. 50 μg protein, taken for reductase assay. The assay is based upon continuous spectrophotometric measurement of Chlide formation (at 672 nm) in response to flash illumination in the presence of excess of the two substrates, PChlide and NADPH. It is carried out in a spectrophotometer cuvette in the presence of approximately 200 μM NADPH, which is added either as the reduced coenzyme or more conveniently as $NADP^+$ in the presence of an enzyme-reducing system, together with about 5 μM PChlide. An optimum, predetermined, amount of Triton X-100 is added, which facilitates the attainment of maximum rates. The PChlide is usually added as a methanolic solution or as an aqueous solution in cholate (Griffiths, 1978). Flash excitation of the reaction mixture is carried out at a flash rate of 1 ms xenon flash per 20 s (Mecablitz photographic lamp) with the excitation filtered through a 1 cm layer of saturated $CuSO_4$ solution with the photomultiplier protected from this by two layers of Kodak Wratten No. 70 gelatine filters transmitting only light of wavelength greater than 650 nm. Progress of the reaction is observed as stepwise increase in absorbance at 672 nm (see Griffiths, 1978) as a result of the Chlide formed during each flash. Rates of Chlide synthesis are calculated from the slope of this increase, assuming an ε mM at 672 nm of 91 l mmol^{-1} cm^{-1} for Chlide in aqueous environment. Rates of 1.5 nmol min^{-1} mg^{-1} protein are typical under these conditions.

3. Purification

The reductase is by far the major protein present in the internal membranes of plastids, where it can comprise more than 80% of the total protein in some purified membrane preparations. Therefore, isolation of such membranes represents a significant purification of the enzyme in its native form. In practice, due to the lability of the enzyme, it is difficult to isolate an enzyme preparation with a specific activity higher than that present in purified membranes. Low concentrations of the non-ionic detergent Triton X-100 can be effectively used to solubilise the enzyme from membrane preparations without too much loss of activity in purification attempts (Apel *et al.*, 1980; Beer and Griffiths, 1981) but as yet no reliable estimate of the molecular weight of the native enzyme is available. In contrast, on SDS-PAGE the denatured enzyme behaves as a single or doublet of related peptides of molecular mass of 35–37 kDa, depending on species.

F. Chlorophyll Synthetase

1. Reaction

This enzyme system catalyses esterification of the C_{17} propionate group of the tetrapyrrole with a fatty alcohol which, in chlorophyll *a*, is the diterpenoid phytol. Whereas the esterification may take place at different steps in the tetrapyrrole pathway, such as the level of PChlide or Chlide, different alcohol residues like geranyl geraniol (C_{20}) or phytol can also be used for the reaction. It is also possible for non-terpenoid alcohols to participate in esterification, resulting in production of 7-alkyl chlorophyll derivatives which have been detected as minor components in some higher plant tissues (Rebeiz *et al.*, 1983). Also, in photosynthetic bacteria, bacteriochlorophylls esterified with phytol, geranyl geraniol or farnesol (C_{15}) are well established.

This section will confine itself to a discussion of esterification in chlorophyll *a* formation by higher plants. It was originally thought that the enzyme chlorophyllase was responsible for the esterification due to its ability to catalyse esterification and hydrolysis of Chlide and various fatty alchohol esters *in vitro*. It has since been shown, however, that in aqueous solutions hydrolysis is the favoured reaction so that now chlorophyllase is viewed as a catabolic enzyme. The enzyme 'chlorophyll synthetase' (Rudiger *et al.*, 1977) is now firmly established as catalysing the esterification reaction during chlorophyll *a* synthesis, with its activity having been detected in a range of plant tissues such as cereals, spinach and beans. In such tissues the enzyme is associated with plastid internal membranes and can catalyse esterification of Chlide with either phytol pyrophosphate (PPP) or its tri-unsaturated analogue, geranyl geranyl pyrophosphate (GGPP). Esterification with GGPP, which occurs primarily in greening etiolated tissues, is followed by a series of stepwise NADPH-dependent hydrogenations proceeding from the distal end of the molecule to generate chlorophyll with a phytol side-chain (Schoch and Schaffer, 1978) as shown below:

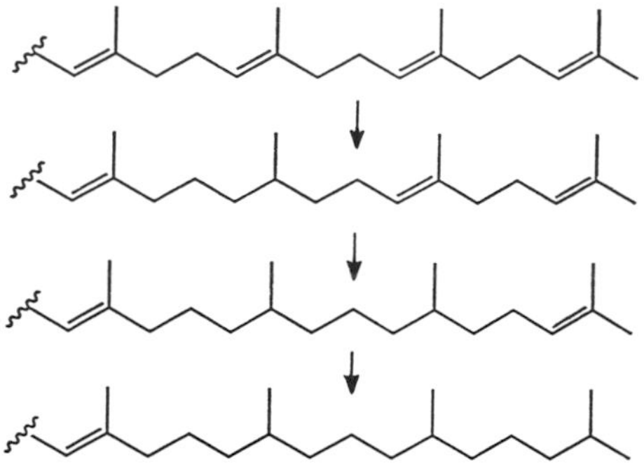

All these forms have been detected *in vivo* as transient intermediates during chlorophyll formation in greening bean leaves (Schoch, 1978). Their accumulation can be enhanced by feeding etiolated wheat seedlings with the herbicide aminotriazole which appears to inhibit the hydrogenations (Rudiger and Benz, 1979). The synthetase can also utilise PPP for esterification to form chlorophyll *a* directly, a reaction which is favoured in green tissues. It has been possible to reconstitute the synthetase activity with plastid membrane preparations from etiolated oat seedlings (Benz and Rudiger, 1981a,b). This system has been used to test the substrate specificity of the synthetase for both alcohol and porphyrin (solubilised in 0.1% cholate). Interestingly, both Chlide *a* and *b* are efficiently esterified (with GGPP) whereas PChlide and bacteriochorophyllide *a* are not.

As already implied, the synthetase is probably a peripheral protein of plastid inner membranes. Such membranes also accommodate the enzymes producing the porphyrin, that is PChlide reductase. However, the terpenoid pathway producing the second esterifying substrate is located in the plastid envelope. How the latter arrives at the site of esterification has not yet been established.

2. Assay

Detection of the chlorophyll synthetase activity is best achieved by measuring the incorporation of [^{3}H]GGPP into esterified cholate using illuminated etioplast membranes. The procedure followed is based upon that described by Rudiger *et al.* (1980). Synthesis of all *trans*-[1-^{3}H]geranyl geraniol can be achieved by reducing *trans*-geranyl geranial with [G-^{3}H]NaBH$_4$. The procedure gave a product with a specific activity of 34 Ci mol^{-1}.

Etioplasts are isolated from 100 g of 7-day-old dark-grown wheat by a standard procedure (Griffiths, 1978) but using a medium containing 0.45 M sorbitol as osmoticum. Osmotic lysis is achieved by resuspending the pellet in phosphate buffer, pH 7.1, followed by homogenisation in a Potter–Elvehjem homogeniser. An aliquot containing approximately $10 \mu g$ PChlide is added to 3 ml of buffer containing $0.17 \mu M$ $MgCl_2$, $0.08 \mu M$ $MnCl_2$, $0.07 \mu M$ NaF and 3 mM NADPH and illuminated with white fluorescent light to phototransform the endogenous PChlide to Chlide. The suspension is then sonicated for $2 \times 10 s$ in an ice bath to break up any intact plastids and finally incubated at 28°C for 1 h.

The reaction is terminated by addition of acetone (6 ml) followed by demetallation of the pigments by addition of acid, extraction into ether followed by purification of the esterified pigments by chromatography on a powdered sugar column and analysis of the mixture by HPLC and scintillation counting (Rudiger *et al.*, 1980). Whereas absolute esterification rates are difficult to assess from this procedure, an indication of the relative extent of esterification by different substrates can be achieved.

3. Purification

No purification of the synthetase system, beyond the plastid membrane stage, has yet been achieved.

ACKNOWLEDGEMENTS

Financial support is acknowledged from the Science and Engineering Research Council and the Agricultural and Food Research Council of the UK.

REFERENCES

Alwan, A. F. and Jordan, P. M. (1988). *Biochem. Soc. Trans.* **16**, 965–966.
Anderson, P. M. and Desnick, R. J. (1979). *J. Biol. Chem.* **254**, 6924–6930.
Anderson, P. M. and Desnick, R. J. (1980). *J. Biol. Chem.* **255**, 1993–1999.
ap Rees, T., Bryce, J. H., Wilson, P. M. and Green, J. H. (1983). *Arch. Biochem. Biophys.* **227**, 511–521.
Apel, K., Santel, H. K., Redlinger, T. E. and Falk, H. (1980). *Eur. J. Biochem.* **111**, 251–258.
Avissar, Y. J. and Beale, S. I. (1989a). *J. Bacteriol.* **171**, 2919–2924.
Avissar, Y. J. and Beale, S. I. (1989b). *Plant Physiol.* **89**, 852–859.
Avissar, Y. J. and Beale, S. I. (1990). *J. Bacteriol.* **172**, 1656–1659.
Avissar, Y. J., Ormerod, J. G. and Beale, S. I. (1989). *Arch Microbiol.* **151**, 513–519.
Batlle, A. M. delC., Benson, A. and Rimington, C. (1965). *Biochem. J.* **97**, 731–740.
Battersby, A. R., McDonald, E., Williams, D. C. and Wurziger, H. K. W. (1977). *J. Chem. Soc., Chem. Commun.*, 113–115.
Battersby, A. R., Fookes, C. J. R., Gustafson-Potter, K., Matcham, G. W. J. and McDonald, E. (1979a). *J. Chem. Soc., Chem. Commun.*, 1155–1158.
Battersby, A. R., Fookes, C. J. R., Matcham, G. W. J., McDonald, E. and Gustafson-Potter, K. E. (1979b). *J. Chem. Soc., Chem. Commun.*, 316–319.
Battersby, A. R., Fookes, C. J. R., Gustafson-Potter, K. E., McDonald, E. and Matcham, G. W. J. (1982). *J. Chem. Soc., Perkin Trans.*, 2427–2444.
Battersby, A. R., Fookes, C. J. R., Matcham, G. W. J., McDonald, E. and Hollenstein, R. (1983). *J. Chem. Soc., Perkin Trans. I*, 3031–3040.
Bazzaz, M. B. (1981). *Photobiochem. Photobiophys.* **2**, 199–207.
Beale, S. I. (1990). *Plant Physiol.* **93**, 1273–1279.
Beale, S. I. and Chen, N. C. (1983). *Plant Physiol.* **71**, 263–268.

Beale, S. I. and Weinstein, J. D. (1990). *In* "Biosynthesis of Heme and Chlorophylls" (H. A. Dailey, ed.), pp. 287–391. McGraw-Hill, New York.

Beale, S. I., Foley, T. and Dzelzkalns, V. (1981). *Proc. Natl. Acad. Sci. USA* **78**, 1666–1669.

Beer, N. S. and Griffiths, W. T. (1981). *Biochem. J.* **195**, 83–92.

Begley, T. P. and Young, H. (1989). *J. Am. Chem. Soc.* **111**, 3095–3096.

Benz, J. and Rudiger, W. (1981a). *Z. Pflanzenphysiol.* **102**, 95–100.

Benz, J. and Rudiger, W. (1981b). *Z. Naturforsch.* **36c**, 51–57.

Berry-Lowe, S. (1987). *Carlsberg Res. Commun.* **52**, 197–210.

Bogard, M., Camadro, J.-M., Nordmann, Y. and Labbe, P. (1989). *Eur. J. Biochem.* **181**, 417–421.

Brown, S. B., Houghton, J. D. and Vernon, D. I. (1990). *J. Photochem. Photobiol.* **5**, 3–23.

Brumm, P. J. and Friedman, H. C. (1981). *Biochem. Biophys. Res. Commun.* **102**, 854–859.

Bruyant, P. and Kannangara, C. G. (1987). *Carlsberg Res. Commun.* **52**, 99–109.

Bull, A. D., Breu, V., Kannangara, C. G., Rogers, L. J. and Smith, A. J. (1990). *Arch. Microbiol.* **154**, 56–59.

Camadro, J.-M., Chambon, H., Jolles, J. and Labbe, P. (1986). *Eur. J. Biochem.* **579**, 587–87.

Castelfranco, P. A. and Jones, O. T. G. (1975). *Plant Physiol.* **55**, 485–490.

Castelfranco, P. A., Weinstein, J. D., Schwatz, S., Pardo, A. D. and Wezelman, B. E. (1979). *Arch. Biochem. Biophys.* **192**, 592–598.

Cavaleiro, J. A. S., Kenner, G. W. and Smith, K. M. (1974). *J. Chem. Soc., Perkin Trans. I*, 1188–1194.

Chang, T.-E., Wegmann, B. and Wang, W.-Y. (1990). *Plant Physiol.* **93**, 1641–1649.

Chen, M.-W., Jahn, D., Schön, A., O'Neill, G. P. and Söll, D. (1990a). *J. Biol. Chem.* **265**, 4054–4057.

Chen, M.-W., Jahn, D. and O'Neill, G. P. (1990b). *J. Biol. Chem.* **265**, 4058–4063.

Chen, T. C. and Miller, G. W. (1974). *Plant Cell Physiol.* **15**, 993–1005.

Chereskin, B. M., Castelfranco, P. A., Dallas J. L. and Straub, K. M. (1983). *Arch. Biochem. Biophys.* **226**, 10–18.

Chereskin, B. M., Wong, Y.-S. and Castelfranco, P. A. (1982). *Plant Physiol.* **70**, 987–993.

Chibbar, R. N. and van Huystee, R. B. (1983). *Phytochemistry* **22**, 1721–1723.

Chisholm, S. W., Olson, R. J., Zettler, E. R., Goericke, R., Waterbury, J. B. and Welschmeyer, N. A. (1988). *Nature* **334**, 340–343.

Dailey, H. A. (1990). *In* "Biosynthesis of Heme and Chlorophylls" (H. A. Dailey, ed.), pp. 123–161. McGraw-Hill, New York.

de Verneuil, H., Sassa, S. and Kappas, A. (1983). *J. Biol. Chem.* **258**, 2454–2460.

Deybach, J.-C., da Silva, V., Grandchamp, B. and Nordmann, Y. (1985). *Eur J. Biochem* **149**, 431–435.

Duggan, J. and Gassman, M. (1974). *Plant Physiol.* **53**, 206–215.

Ebon, J. G. and Tait, G. H. (1969). *Biochem. J.* **111**, 573–582.

Echelard, Y., Dymetryszyn, J., Drolet, M. and Sasarman, A. (1988). *Mol. Gen. Genet.* **214**, 503–508.

Elder, G. H. and Evans, J. O. (1978a). *Biochem. J.* **172**, 345–347.

Elder, G. H. and Evans, J. O. (1978b). *Biochem. J.* **169**, 205–214.

Ellsworth, R. K. (1972). *In* "The Chemistry of Plant Pigments", Vol. III (C. O. Chichester, ed.), pp. 85–102. Academic Press, New York.

Ellsworth, R. K. and Hsing, A. S. (1973). *Biochim. Biophys. Acta* **313**, 119–129.

Farmerie, W. G., Delehanty, J. and Barnett, W. E. (1982). *In* "Methods in Chloroplast Molecular Biology" (M. Edelman, R. B. Hallick and N.-H. Chua, eds), pp. 335–346. Elsevier, Amsterdam.

Felix, F. and Brouillet, N. (1990). *Eur. J. Biochem.* **188**, 393–403.

Foley, T., Dzelzkalns, V. and Beale, S. I. (1982). *Plant Physiol.* **70**, 219–226.

Gardner, G. and Gorton, H. L. (1985). *Plant Physiol.* **77**, 540–543.

Gibbs, P. N. B. and Jordan, P. M. (1981). *Biochem. Soc. Trans.* **9**, 232–233.

Gibbs, P. N. B. and Jordan, P. M. (1986). *Biochem. J.* **236**, 447–451.

Gibbs, P. N. B., Chaudhry, A.-G. and Jordan, P. M. (1985). *Biochem. J.* **230**, 25–34.

Gibson, K. D., Neuberger, A. and Tait, G. H. (1963). *Biochem. J.* **88**, 325–334.

Gorchein, A. (1972). *Biochem. J.* **127**, 97–106.

Gough, (1972). *Biochim. Biophys. Acta* **286**, 36–54.

Gough, S. P. and Kannangara, C. G. (1976). *Carlsberg Res. Commun.* **41**, 183–190.

Gough, S. P. and Kannangara, C. G. (1977). *Carlsberg Res. Commun.* **42**, 459–464.

Gough, S. P., Kannangara, C. G. and Bock, K. (1989). *Carlsberg Res. Commun.* **54**, 99–108.

Grandchamp, B., Phung, N. and Nordmann, Y. (1978). *Biochem, J.* **176**, 97–102.

Granick, S. (1950). *Harvey Lectures* **44**, 220–245.

Griffiths, W. T. (1978). *Biochem. J.* **174**, 681–692.

Griffiths, W. T. (1980). *Biochem. J.* **186**, 267–278.

Griffiths, W. T. (1991). *In* "Chlorophylls" (H. Scheer, ed.), pp. 433–450. CRC Press, Boca Raton, FL.

Grimm, B., Bull, A. D., Welinder, K., Gough, S. P. and Kannangara, C. G. (1989). *Carlsberg Res. Commun.* **54**, 67–79.

Gumpel, N. J. and Smith, A. G. (1990). *Biochem. Soc. Trans.* **18**, 500–501.

Hart, G. J. and Battersby, A. R. (1985). *Biochem. J.* **232**, 151–160.

Hart, G. J., Abell, C. and Battersby, A. R. (1986). *Biochem. J.* **240**, 273–276.

Hart, G. J., Miller, A. D., Leeper, F. J. and Battersby, A. R. (1987). *J. Chem. Soc., Chem. Commun.*, 1762–1765.

Hart, G. J., Miller, A. D. and Battersby, A. R. (1988). *Biochem, J.* **252**, 909–912.

Hasnain, S. S., Wardell, E. M., Garner, C. D., Schlosser, M. and Beyersmann, D. (1985). *Biochem. J.* **230**, 625–633.

Higuchi, M. and Bogorad, L. (1975). *Ann. NY Acad. Sci.* **244**, 401–408.

Hiller, R. G., Anderson, J. M. and Larkum, A. W. D. (1991). *In* "Chlorophylls" (H. Scheer, ed.), pp. 529–548. CRC Press, Boca Raton, FL.

Hinchigeri, S. B., Chan, J. C. S. and Richards, W. R. (1981). *Photosynthetica* **15**, 351–359.

Houen, G., Gough, S. P. and Kannangara, C. G. (1984). *Carlsberg Res. Commun.* **48**, 567–572.

Hsu, W. P. and Miller, G. W. (1970). *Biochem. J.* **117**, 215–220.

Huang, D-D. and Wang, W-Y. (1986). *J. Biol. Chem.* **261**, 13451–13455.

Huault, C. and Bruyant, P. (1984). *Physiol. Veg.* **22**, 209–213.

Hunter, C. N. and Coomber, S. A. (1988). *J. Gen. Microbiol.* **134**, 1491–1497.

Jackson, A. H., Sancovich, H. A. and Ferramola, A. M. (1980). *Bioorg. Chem.* **9**, 71–120.

Jacobs, J. M. and Jacobs, N. J. (1984). *Arch. Biochem. Biophys.* **229**, 312–319.

Jacobs, J. M. and Jacobs, N. J. (1987). *Biochem. J.* **244**, 219–224.

Jacobs, J. M., Jacobs, N. J., Borotz, S. E. and Guerinot, M. L. (1990). *Arch. Biochem. Biophys.* **280**, 369–375.

Jahn, D., Michelson, U. and Söll, D. (1991). *J. Biol. Chem.* **266**, 2542–2548.

Jones, O. T. G. (1966). *Biochem. J.* **101**, 153–160.

Jordan, P. M. (1990). *In* "Biosynthesis of Heme and Chlorophylls" (H. A. Dailey, ed.), pp. 55–121. McGraw-Hill, New York.

Jordan, P. M. and Gibbs, P. N. B. (1985). *Biochem. J.* **227**, 1015–1020.

Jordan, P. M. and Seehra, J. S. (1980a). *FEBS Lett.* **114**, 283–286.

Jordan, P. M. and Seehra, J. S. (1980b). *J. Chem. Soc., Chem. Commun.*, 240–242.

Jordan, P. M. and Shemin, D. (1973). *J. Biol. Chem.* **248**, 1019–1024.

Jordan, P. M. and Warren, M. J. (1987). *FEBS Lett.* **225**, 87–92.

Jordan, P. M., Thomas, S. D. and Warren, M. J. (1988a). *Biochem. J.* **254**, 427–435.

Jordan, P. M., Warren, M. J., Williams, H. J., Stolowich, N. J., Roessner, C. A., Grant, S. K. and Scott, A. I. (1988b). *FEBS Lett.* **235**, 189–193.

Juknat, A. A., Seubert, A., Seubert, S. and Ippen, H. (1989). *Eur. J. Biochem.* **179**, 423–428.

Kannangara, C. G. (1990). *In* "Cell Culture and Somatic Cell Genetics of Plants", Vol. 7: The Molecular Biology of Plastids and Mitochondria (L. Bogorad and I. K. Vasil, eds). Academic Press, New York.

Kannangara, C. G. and Gough, S. P. (1978). *Carlsberg Res. Commun.* **43**, 185–194.

Kannangara, C. G. and Schouboe, A. (1985). *Carlsberg Res. Commun.* **50**, 179–191.

Kannangara, C. G. Gough, S. P., Hansen, S., Rasmussen, J. N. and Simpson, D. J. (1977). *Carlsberg Res. Commun.* **42**, 431–439.

Kannangara, C. G., Gough, S. P., Oliver, R. P. and Rasmussen, S. K. (1984). *Carlsberg Res. Commun.* **49**, 417–437.

Kannangara, C. G. Gough, S. P., Bruyant, P., Hoober, J. K., Kahn, A. and von Wettstein, D. (1988). *Trends Biochem. Sci.* **13**, 139–143.

Kohashi, M., Clement, R. P., Tse, J. and Piper, W. N. (1984). *Biochem. J.* **220**, 755–765.

Kwan, L. Y.-M., Darling, D. L. and Richards, W. R. (1986). *In* "Regulation of Chloroplast Differentiation" (G. Akoyunoglou and H. Senger, eds), pp. 57–62. Alan R. Liss, New York.

Labbe, P., Camadro, J.-M. and Chambon, H. (1985). *Anal. Biochem.* **149**, 248–260.

Labbe-Bois, R. and Labbe, P. (1990). *In* "Biosynthesis of Heme and Chlorophylls" (H. A. Dailey, ed.), pp. 235–285. McGraw-Hill, New York.

Lascelles, J. (1966). *Biochem. J.* **100**, 175–182.

Leong, S. A., Ditta, G. S. and Helinski, D. R. (1982). *J. Biol. Chem.* **257**, 8724–8730.

Li, J.-M., Brathwaite, O., Cosloy, S. D. and Russell, C. S. (1989). *J. Bacteriol.* **171**, 2547–2552.

Liedgens, W., Grützmann, R. and Schneider, H. A. W. (1980). *Z. Naturforsch.* **35c**, 958–962.

Little, H. N. and Jones, O. T. G. (1976). *Biochem. J.* **156**, 309–314.

Mantoura, R. F. C. and Llewellyn, C. A. (1983). *Anal. Chim. Acta* **151**, 297–314.

Marrs, B. (1981). *J. Bacteriol.* **146**, 1003–1012.

Mathewson, J. H. and Corbin, A. H. (1963). *J. Am. Chem. Soc.* **83**, 135–137.

Matringe, M. and Scalla, R. (1988). *Plant Physiol.* **86**, 619–622.

Matringe, M., Camadro, J.-M., Labbe, P. and Scalla, R. (1989). *Biochem. J.* **260**, 231–235.

Mau, Y.-H. and Wang, W.-Y. (1988). *Plant Physiol.* **86**, 793–797.

Mau, Y.-H., Wang, W.-Y., Tamura, R. and Chang, T.-E. (1987). *Arch. Biochem. Biophys.* **255**, 75–79.

Mauzerall, D. and Granick, S. (1956). *J. Biol. Chem.* **219**, 435–446.

May, T. B., Guikema, J. A., Henry, R. L., Schuler, M. K. and Wong, P. P. (1987). *Plant Physiol.* **84**, 1309–1313.

Mayer, S. M., Beale, S. I. and Weinstein, J. D. (1987). *J. Biol. Chem.* **262**, 12541–12549.

Miller, A. D., Packman, L. C., Hart, G. J., Alefounder, P. R., Abell, S. and Battersby, A. R. (1989). *Biochem. J.* **262**, 119–124.

Myers, A. M., Crivellone, M. D., Koerner, T. J. and Tzagoloff, A. (1987). *J. Biol. Chem.* **262**, 16822–16829.

Nair, S. P., Harwood, J. L. and John, R. A. (1991) *FEBS Lett.* **283**, 4–6.

Nandi, D. L. (1978). *Z. Naturforsch.* **33c**, 799–800.

Nandi, D. L. and Shemin, D. (1968). *J. Biol. Chem.* **243**, 1236–1242.

Nandi, D. L. and Waygood, E. R. (1967). *Can. J. Biochem.* **45**, 327–336.

Nasrulhaq-Boyce, A., Griffiths, W. T. and Jones, O. T. G. (1987). *Biochem. J.* **243**, 23–29.

O'Neill, G. P. and Söll, D. (1990). *J. Bacteriol.* **172**, 6363–6371.

O'Neill, G. P., Chen, M. W. and Söll, D. (1989). *FEMS Microbiol. Lett.* **60**, 255–260.

Oh-hama, T., Seto, H., Otake, N. and Miyachi, S. (1982). *Biochem. Biophys. Res. Commun.* **105**, 647–652.

Peterson, D., Schön, A. and Söll, D. (1988). *Plant Mol. Biol.* **11**, 293–299.

Poulson, R. and Polglase, W. J. (1975). *J. Biol. Chem.* **250**, 1269–1274.

Ratinaud, M. H., Thomes, J. C. and Julien, R. (1983). *Eur. J. Biochem.* **135**, 471–477.

Rebeiz, C. A. and Lascelles, J. (1982). *In* "Photosynthesis: Energy Conversion by Plants and Bacteria", Vol. I (Govindjee, ed.) pp. 689–780. Academic Press, New York.

Rebeiz, C. A., Matheis, J. R., Smith, B. B., Rebeiz, C. C. and Dayton, D. F. (1975). *Arch. Biochem. Biophys.* **166**, 446–465.

Rebeiz, C. A., Wu, S.M., Kuhadja, M., Dariell, H. and Perkins, E. J. (1983). *Mol. Cell. Biochem.* **57**, 97–125.

Richards, W. R., Fung, M., Wessler, A. N. and Hinchigeri, S. B. (1987). *In* "Progress in Photosynthesis Research", Vol. IV (J. Biggins, ed.), pp. 475–82. Martinus-Nijhoff, Boston.

Richter, M. L. and Rienits, K. G. (1982). *Biochim. Biophys. Acta* **717**, 255–264.

Rieble, S. and Beale, S. I. (1988). *J. Biol. Chem.* **263**, 8864–8871.

Rimington, C. (1960). *Biochem. J.* **75**, 620–623.

Rossi, E. and Curnow, D. H. (1986). *In* HPLC of Small Molecules: A Practical Approach"
(C. K. Lim, ed.). pp. 261–303. IRL Press, Oxford.

Rudiger, W. and Benz. J. (1979). *Z. Naturforsch.* **34c**, 1055–1060.

Rudiger, W., Hedden, P., Kost, H. P. and Chapman, D. J. (1977). *Biochem. Biophys. Res. Commun.* **74**, 1268–1272.

Rudiger, W., Benz. J. and Guthoff, C. (1980). *Eur. J. Biochem.* **109**, 193–200.

Sambrook, J., Fritsch, E. F. and Maniatis, T. (1989). "Molecular Cloning: A Laboratory Manual", 2nd edn. Cold Spring Harbor Laboratory Press, Cold Spring Harbor, MI.

Sano, S. (1966). *J. Biol. Chem.* **241**, 5276–5283.

Saunders, V. A. (1973). Ph.D. Thesis, University of Bristol.

Schneegurt, M. A. and Beale, S. I. (1986). *Plant Physiol.* **81**, 965–971.

Schneegurt, M. A. and Beale, S. I. (1988). *Plant Physiol.* **86**, 497–504.

Schneider, H. A. W. (1970). *Z. Pflanzenphysiol. Bd.* **62**, 328–342.

Schneider, H. A. W. and Liedgens, W. (1981). *Z. Naturforsch.* **36c**, 44–50.

Schoch, S. (1978). *Z. Naturforsch.* **33c**, 712–717.

Schoch, S. and Schaffer, W. (1978). *Z. Naturforsch.* **33c**, 408–412.

Schön, A., Krupp, G., Gough, S., Berry-Lowe, S., Kannangara, C. G. and Söll, D. (1986). *Nature* **322**, 281–284.

Shön, A., Kannangara, C. G., Gough, S. P. and Söll, D. (1988). *Nature*, **331**, 187–190.

Seehra, J. S., Jordan, P. M. and Akhtar, M. (1983). *Biochem. J.* **209**, 709–718.

Shetty, A. S. and Miller, G. W. (1969). *Biochem. J.* **114**, 331–337.

Shibata, H. and Ochiai, H. (1977). *Plant Cell Physiol.* **18**, 421–429.

Shieh, J., Miller, G. W. and Psenak, M. (1978). *Plant Cell Physiol.* **19**, 1051–1059.

Shioi, Y. (1991). *In* "Chlorophylls" (H. Scheer, ed.), pp. 59–88. CRC Press, Boca Raton, FL.

Shioi, Y., Nagamine, M., Kuroki, M. and Sasa, T. (1980). *Biochim. Biophys. Acta* **616**, 300–309.

Siepker, L. J., Ford, M., de Kock, R. and Kramer, S. (1987). *Biochim. Biophys. Acta* **913**, 349–358.

Smith, A. G. (1986). *In* "Regulation of Chloroplast Differentiation" (G. Akoyunoglou and H. Senger, eds), pp. 49–54. Alan R. Liss, New York.

Smith, A. G. (1988). *Biochem. J.* **249**, 423–428.

Smith, H. (1975). *In* "Phytochrome and Photomorphogenesis" (H. Smith, ed.), pp. 214–223. McGraw-Hill, London.

Smythe, E. and Williams, D. C. (1988). *Biochem. J.* **253**, 275–279.

Spano, A. J. and Timko, M. P. (1991). *Biochim. Biophys. Acta* **1076**, 29–36.

Stark, W. M., Hart, G. J. and Battersby, A. R. (1986). *J. Chem. Soc., Chem. Commun.*, 465–467.

Stella, A. M. and Batlle, A. M.delC. (1978). *Plant Sci. Lett.* **11**, 87–92.

Sugiura, M. (1989). *Ann. Rev. Cell Biol.* **5**, 51–70.

Tamai, M., Shioi, Y. and Sasa, T. (1979). *Plant Cell Physiol.* **20**, 435–444.

Tigier, H. A., Batlle, A. M.delC. and Locascio, G. A. (1970). *Enzymologia* **38**, 43–56.

Treffry, T. (1970). *Planta* **91**, 279–284.

Tsai, S.-F., Bishop, D. F. and Desnick, R. J. (1987). *J. Biol. Chem.* **262**, 1268–1273.

Tsukamoto, I., Yoshinaga, T. and Sano, S. (1979). *Biochim. Biophys. Acta* **570**, 167–178.

Urata, G. and Granick. S. (1963). *J. Biol. Chem.* **238**, 811–820.

Valera, V., Fung, M., Wessler, A. and Richards, W. W. (1987). *Biochem. Biophys. Res. Commun.* **148**, 515–520.

van Breemen, R. B., Canjura, F. L. and Schwartz, S. J. (1991). *J. Chromatogr.* **542**, 373–383.

Walker, C. J. and Weinstein, J. D. (1991a). *Plant Physiol.* **95**, 1189–1196.

Walker, C. J. and Weinstein, J. D. (1991b). *Proc. Natl Acad. Sci. USA* **88**, 5789–5793.

Walker, C. J., Mansfield, K. E., Rezzano, I. N., Hanamoto, C. M., Smith, K. M. and Castelfranco, P. A. (1988). *Biochem. J.* **255**, 685–692.

Walker, C. J., Mansfield, K. E., Smith, K. M. and Castelfranco, P. A. (1989a). *Biochem. J.* **255**, 599–602.

Walker, C. J., Mansfield, K. E., Smith, K. M. and Castelfranco, P. A. (1989a). *Biochem. J.* **257**, 599–602.

Walker, C. J., Castelfranco, P. A. and Whyte, B. J. (1991). *Biochem. J.* **276**, 691–697.

Wang, W.-Y., Gough, S. P. and Kannangara, C. G. (1981). *Carlsberg Res. Commun.* **46**, 243–257.

Wang, W.-Y., Huang, D.-D., Stachon, D., Gough, S. P. and Kannangara, C. G. (1984). *Plant Physiol.* **74**, 569–575.

Wang, W.-Y., Huang, D.-D., Chang, T.-E., Stachon, D. and Wegman, B. (1987). *In* "Progress in Photosynthetic Research", Vol. 4 (J. Biggins, ed.), pp. 423–430. Martinus Nijhoff, Dordrecht.

Warren, M. J. and Jordan, P. M. (1988). *Biochem. Soc. Trans.* **16**, 963–965.

Warren, M. J. and Scott, A. I. (1990). *Trends Biochem. Sci.* **15**, 486–491.

Weinstein, J. D. and Beale, S. I. (1983). *J. Biol. Chem.* **258**, 6799–6807.

Weinstein, J. D. and Beale, S. I. (1985). *Arch. Biochem. Biophys.* **239**, 87–93.

Weinstein, J. D. and Castlefranco, P. A. (1978). *Arch. Biochem. Biophys.* **186**, 376–382.

Weinstein, J. D., Mayer, S. M. and Beale, S. I. (1987). *Plant Physiol.* **84**, 244–250.

Werck-Reichhart, D., Jones, O. T. G. and Durst, F. (1988). *Biochem. J.* **249**, 473–480.

Whyte, B. J. (1989). Ph.D. Thesis, University of Bristol.

Williams, D. C., Morgan, G. S., McDonald, E. and Battersby, A. R. (1981). *Biochem. J.* **193**, 301–310.

Witowski, D. A. and Halling, B. P. (1988). *Plant Physiol.* **87**, 632–637.

Yang, D., Oyaizu, Y., Oyaizu, H., Olsen, G. J. and Woese, C. R. (1985). *Proc Natl Acad. Sci. USA* **82**, 4443–4447.

Yoshinaga, T. and Sano, S. (1980). *J. Biol. Chem.* **255**, 4722–4726.

13 Enzymology of Indole Alkaloid Biosynthesis

VINCENZO DE LUCA

Plant Biology Research Institute, Department of Biological Sciences, University of Montréal, 4101 Sherbrooke Street East, Montréal, Québec, Canada, H1X 2B2

METHODS IN PLANT BIOCHEMISTRY Vol. 9
ISBN 0-12-461019-6

I. INTRODUCTION

The monoterpenoid indole alkaloids, which occur in the family Apocynaceae, constitute one of the medicinally important groups of secondary metabolites produced by
flowering plants. The extensive research of the past 30 years has substantially increased
our knowledge of the variety of chemical structures produced and more recently, of
the biosynthesis of indole alkaloids. As a result of this extensive research, the structures
of over 1500 indole alkaloids have been elucidated (Ganzinger and Hesse, 1976).

Interest in the biosynthesis of indole alkaloids was first stimulated by chemical
speculations on the mode of formation of the carbon skeleton of this class of
compounds. Early experimental work demonstrated that tryptophan could serve as a
precursor of the tryptamine moiety of indole alkaloids (Leete *et al.*, 1965), whereas
several proposals (reviewed in Parry, 1972) were made concerning the origin of the
the 'C$_9$–C$_{10}$' unit which is incorporated into three major skeletal forms (Fig. 13.1) and
results in the different classes of known indole alkaloids. Investigations with intact
plants and with plant tissues (reviewed in Battersby, 1971) involving radioactive tracers
of mevalonic acid established that secologanin is the ultimate precursor for the
'C$_9$–C$_{10}$' unit common to the majority of indole alkaloids.

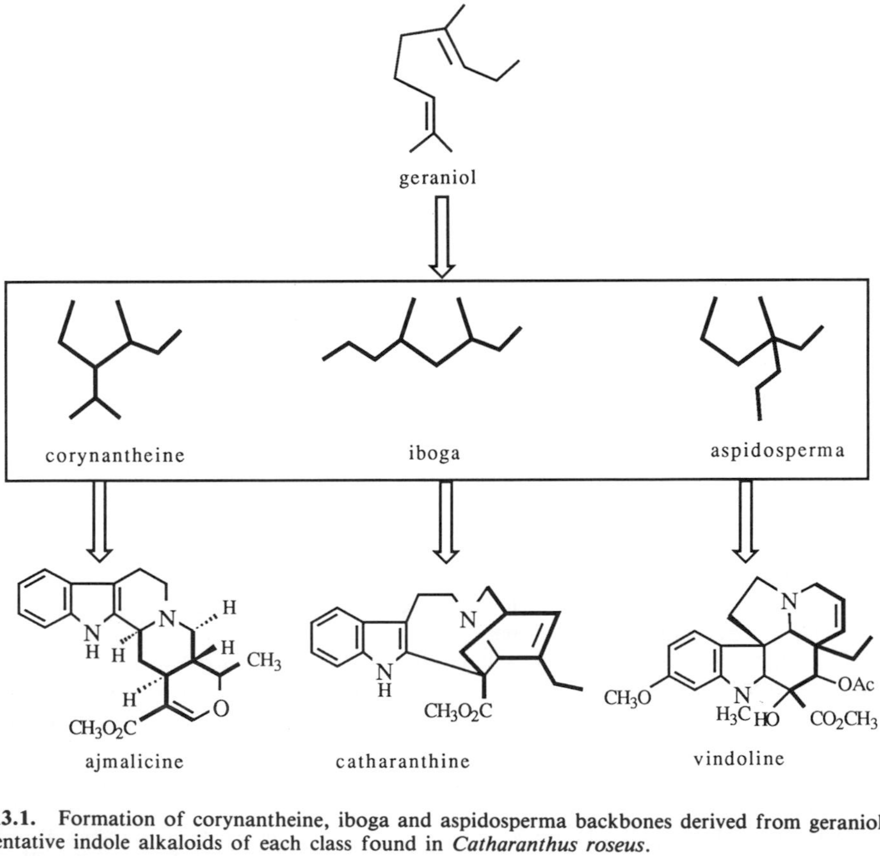

FIG. 13.1. Formation of corynantheine, iboga and aspidosperma backbones derived from geraniol and
representative indole alkaloids of each class found in *Catharanthus roseus*.

Several reviews have appeared on biosynthetic studies with radioactive tracers (Battersby, 1971; Parry, 1972; Cordell, 1974), on the enzymology of the early steps of indole alkaloid biosynthesis (Madyastha and Coscia, 1979a), on the discovery and preliminary studies of the 'strictosidine synthetase' set of enzymes (Stöckigt, 1980; Zenk, 1980; Scott *et al.*, 1981) and on the production of indole alkaloids by plant cell suspension cultures (Zenk, *et al.*, 1977, Kurz and Constabel, 1985; De Luca and Kurz, 1988; Van der Heijden *et al.*, 1989). The main emphasis of this chapter is on the enzymology of indole alkaloid biosynthesis, with special attention focused on vindoline biosynthesis in *Catharanthus roseus*.

A. Enzymology of Indole Alkaloid Biosynthesis

Essentially all the information that is available on the biosynthesis of indole alkaloids is based on experimentation with one species, *Catharanthus roseus*. The studies of indole alkaloid biosynthesis in this species are so numerous for several reasons: (1) the plant is easy to grow and to maintain; (2) it produces a broad spectrum of compounds including the commercially important dimeric indole alkaloids, vinblastine and vincristine; (3) the plant is easy to culture under *in vitro* conditions; (4) cell suspension cultures can be induced to produce several different classes of indole alkaloids as well as the respective pathway-specific enzymes; and (5) the extraordinary complexity of these alkaloids has continued to stimulate the interest of several generations of researchers.

II. ENZYMES FOR THE FORMATION OF THE C_9–C_{10} UNIT, SECOLOGANIN

A. Cytochrome P-450-Dependent Monoterpene Hydroxylase

The first step in the biosynthesis of secologanin from the mevalonate pathway requires the conversion of geraniol via a cytochrome P-450-dependent hydroxylase (Meehan and Coscia, 1973) (Fig. 13.2, reaction 1). Protein extracts from *Catharanthus roseus*, as well as from other indole alkaloid or monoterpene glucoside synthesising plants, contain geraniol hydroxylase activity, unlike plants which produce neither of these secondary metabolites (Madyastha and Coscia, 1979a). This enzyme reaction represents the first specific step for the non-tryptamine portion of indole alkaloid biosynthesis in *C. roseus*.

1. Properties

The characteristic behaviour of geraniol hydroxylase and its separation into a cytochrome P-450 component and a NADPH-specific cytochrome *c* reductase component suggests that this enzyme belongs to the cytochrome P-450 superfamily of monooxygenases. Hydroxylase activity is absolutely dependent on NADPH and O_2. Enzyme activity is stimulated by dithiothreitol (DTT) (Madyastha *et al.*, 1976) whereas carbon monoxide, a typical cytochrome P-450 hydroxylase inhibitor, successfully inhibits this reaction. Inhibition is reversed by treatment with light between 420 and 450 nm (Meehan and Coscia, 1973). The partially purified hydroxylase converts

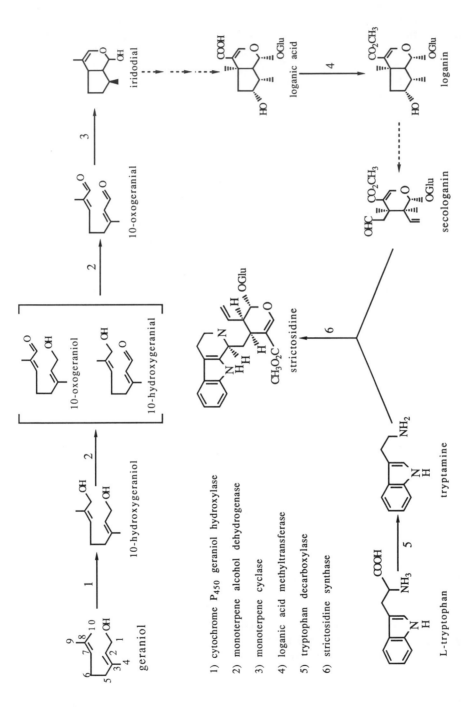

FIG. 13.2. Enzymatic steps in the formation of strictosidine from L-tryptophan and geraniol precursors. Dotted lines represent uncharacterised enzyme reactions.

geraniol (K_m 5.5 μM) and its *cis* isomer, nerol (K_m 11 μM) to their corresponding 10-hydroxy products, whereas the corresponding pyrophosphate derivatives are not accepted as substrates (Madyastha and Coscia, 1979a). The enzyme shows a broad pH optimum with geraniol as substrate where maximum activity is observed at pH 8.

2. Enzyme assay

Protein from *Catharanthus* extracts is added to a NADPH regenerating system (5 mM glucose 6-phosphate, 0.5 mM ADP, 2.5 units of glucose-6-phosphate dehydrogenase), as well as 0.5 μCi of [1-^{3}H]geraniol (0.066 mCi μmol^{-1}, 1 mM DTT, 0.1 M phosphate (pH 7.2) or Tris-HCl (pH 7.8) buffer, in a final volume of 1.5 ml. The reaction is terminated after 30 min incubation at 25°C by addition of 2.5 ml of methanol and both tritiated substrates and products are extracted in 5 ml of CHCl$_3$. The CHCl$_3$ is evaporated with a stream of nitrogen and the hydroxylated product is separated from its substrate by thin layer chromatography (TLC) on silica gel (solvent system: benzene–acetone–ethyl acetate, 2:1:1; v/v/v). The substrate and product can be visualised on film by autoradiography or alternatively, after chromatography with authentic standard (10-hydroxygeraniol), the product is visualised by using I$_2$ vapours, and tritiated product is removed to be quantitated by liquid scintillation spectrometry (Madyastha *et al.*, 1976).

The NADPH-dependent cytochrome *c* reductase and cytochrome P-450 are assayed by previously established procedures for cytochrome P-450 monooxygenases (Ernster *et al.*, 1962 and Omura and Sato, 1964, respectively).

3. Purification

The hydroxylase can be isolated from membranes of 5 day-old etiolated *Catharanthus* seedlings (Meehan and Coscia, 1973), as well as from those of cell suspension cultures (Spitsberg *et al.*, 1981). Membranes are isolated by extraction of seedlings in a buffer (0.1 M Tris-HCl, pH 7.6, 10 mM KCl, 10 mM MgCl$_2$, 10 mM EDTA, 10 mM sodium metabisulphite and 1 mM DTT) containing 0.4 M sucrose. After mixing with polyclar AT to bind phenolic compounds (Loomis, 1974), the extract is filtered and the filtrate is centrifuged at 3000 × g to remove cellular debris. The supernatant is centrifuged at 20 000 × g for 20 min and the resulting pellet contains the majority of the hydroxylase activity. Detailed localisation studies show the enzyme to be associated with provacuolar membranes (Madyastha *et al.*, 1977), rather than in the endoplasmic reticulum where cytochrome P-450 hydroxylases are commonly found (Nebert, 1979).

Geraniol hydroxylase is partially purified by solubilising it from the membranous pellet with sodium cholate, followed by ultracentrifugation (Madyastha *et al.*, 1976). The solubilised enzyme can be separated into two characteristic components, cytochrome P-450 and NADPH cytochrome *c* reductase, by ion-exchange chromatography on DEAE cellulose columns. Hydroxylase activity is reconstituted by combining fractions containing cytochrome P-450 and NADPH cytochrome *c* reductase together with total *Catharanthus* lipids.

Cytochrome P-450 and NADPH cytochrome *c* reductase were recently purified to homogeneity (Meijer *et al.*, 1990) from cell suspension cultures of *C. roseus* grown on alkaloid production medium. The two proteins are separated by DEAE-Sephacel ion-exchange chromatography and the NADPH cytochrome c reductase is purified to homogeneity (subunit molecular weight 74 kDa) by affinity chromatography on 2′,5′-ADP-Sepharose (Madyastha and Coscia, 1979a, b; Meijer *et al.*, 1990). Pure cytochrome P-450 (subunit 66 kDa) is obtained by further chromatography on hydroxylapatite ultrogel, ω-aminooctyl agarose and TSK Phenyl-5PW columns (Meijer *et al.*, 1990).

B.   Biosynthesis of Loganic Acid

The iridoid glucoside, loganic acid, is produced from 10-hydroxygeraniol by a number of rearrangements which have been investigated by *in vivo* tracer experiments (reviewed by Cordell, 1974). The most recent results suggest that loganic acid is either synthesised by a route involving 9,10-dioxogeranial and iridotrial (Balsevich and Kurz, 1983) or by one involving 10-oxogeranial or 10-oxoneral and iridodial (Uesato *et al.*, 1986a,b). It has been proposed that a non-specific oxidoreductase may participate in the early stages of biosynthesis (Madyastha and Coscia, 1979a), but recent studies with a highly specific monoterpene alcohol oxidoreductase (Ikeda *et al.*, 1991) suggest that the latter pathway (Uesato *et al.*, 1986a,b) is involved in the biosynthesis of loganic acid.

C.   Acyclic Monoterpene Primary Alcohol:NADP⁺ Oxidoreductase
      (EC 1.1.1.183)

1.   Properties

Acyclic monoterpene primary alcohol:NADP$^+$ oxidoreductase catalyses the reversible oxidation of 10-hydroxygeraniol to produce 10-oxogeranial through 10-oxogeraniol or 10-hydroxygeranial in the presence of NADP$^+$ (Fig. 13.2, reaction 2). The oxidation of 10-hydroxygeraniol to produce the 10-oxogeraniol and 10-hydroxygeranial intermediates in a ratio of 7:9 (Uesato *et al.*, 1986b) suggests that this dehydrogenase has no specificity for the oxidation of either hydroxyl group of 10-hydroxygeraniol. The enzyme has been isolated and characterised from cell suspension cultures of *Rauwolfia serpentina* (Ikeda *et al.*, 1991). The purified enzyme is a monomer of molecular weight 44 kDa which contains zinc ions and has a pI of 5.4. The enzyme is stabilised with DTT and glycerol and rapidly loses activity when they are omitted from the solution. The enzyme is inactivated by iodoacetamide and N-ethylmaleimide, which suggests that sulphhydryl groups are required for activity. The K_m values for NADP$^+$/ NADPH coenzymes are 25 and 5.5 μM, respectively, with nerol (oxidation direction) or neral (reduction direction) as substrate, whereas NAD$^+$/NADH coenzymes are not acceptable for either reaction. The oxidoreductase shows a much higher affinity for geraniol than for 10-hydroxygeraniol (K_m 100 and 700 μM, respectively). The enzyme will accept as substrates primary allylic alcohols with chain lengths longer than six carbons, whereas secondary and tertiary alcohols as well as ethanol are not acceptable. The pH optima for the oxidation of 10-hydroxygeraniol and its reverse reaction are pH 9 and pH 7–10, respectively (Ikeda *et al.*, 1991).

   Substrate interaction and product inhibition kinetics have established the mechanism of the oxidoreductase reaction as a sequential ordered bi–bi mechanism in which NADP$^+$ is the first substrate to bind and NADPH is the second product to be released from the enzyme. Stereospecificity studies with (4R)- and (4S)-[4-^{2}H]NADPH using gas chromatography–mass spectroscopy (GC-MS) indicate that the enzyme transfers exclusively the *pro-R* hydrogen at C-4 of NADPH to neral to produce nerol (Ikeda *et al.*, 1991) which is characteristic for zinc-containing dehydrogenases.

2. Enzyme assay

Acyclic monoterpene primary alcohol dehydrogenase is assayed spectrophotometrically at 340 nm by measuring the production of NADPH while oxidising 10-hydroxygeraniol, or the production of NADP$^+$ while reducing 10-hydroxygeranial (Ikeda *et al.*, 1991). The reaction mixture contains 0.4 μmol of NADP$^+$, 2 μmol of 10-hydroxygeraniol and enzyme in a total of 1 ml containing 20 μmol of Bis-Tris-Propane-HCl buffer (pH 9) at an assay temperature of 25°C. The reverse reaction contains the same buffer (pH 7), 0.1 μmol of NADPH, 1 μmol of 10 hydroxygeranial and enzyme in a total volume of 1 ml at the same assay temperature.

3. Purification

The enzyme is extracted from frozen cells of *Rauwolfia serpentina* in 20 mM Tris-HCl, pH 8, 1 mM EDTA, 1 mM DTT, 1 mM PMSF, and Polyclar (AT 20 g per 800 ml) under a nitrogen atmosphere and is filtered through Miracloth. After ammonium sulphate fractionation, the 30–70% cut is dissolved in buffer containing 20% ammonium sulphate and is applied to Butyl Toyopearl hydrophobic chromatography. Active fractions eluting at 0% (NH$_4$)$_2$SO$_4$ are stabilised with 20% glycerol, concentrated by Amicon PM10 ultrafiltration and dialysed. The sample is then further purified by DEAE Toyopearl column chromatography followed by AF-Red Toyopearl dye affinity chromatography and Cellulofine GCL-1000sf gel filtration chromatography. The final homogeneous enzyme preparation is produced in a 10% yield and has a specific activity (45 nkat mg^{-1}) 86-fold higher than that of crude extract. An *N*-terminal amino acid sequence of the purified oxidoreductase reveals that it is quite different from those of other dehydrogenases isolated from microorganisms, plants and animals (Ikeda *et al.*, 1991).

D. Monoterpene Cyclase

1. Purification and characterisation

Monoterpene cyclase catalyses the formation of iridodial from 10-oxogeranial (Fig. 13.2, reaction 3). The enzyme is extracted from *Rauwolfia serpentina* cell suspension cultures in 20 mM Tris-HCl, pH 7.2, containing 5 mM 2-mercaptoethanol, and is partially purified by ammonium sulphate fractionation (35–70%) followed by DEAE-Sephacel, Phenyl-Superose and Mono Q anion-exchange chromatography. The purified enzyme has a specific activity (7.9 nkat mg^{-1}) 440-fold higher than crude extracts (Uesato *et al.*, 1987). Enzyme-mediated cyclisation is accelerated more effectively by NADPH than by NADH with a pH optimum of 7. The enzyme appears to be a tetramer with a molecular weight of 118 kDa as judged by gel filtration chromatography and a subunit of 28.7 kDa determined by SDS-PAGE.

2. Enzyme assay

The assay contains 60 nmol [1-^{3}H]10-oxogeranial (5 μCi of unknown specific activity), 0.25 μmol NADPH, 5 mM 2-mercaptoethanol, 20 mM Tris-HCl, pH 7.2, and enzyme (Uesato et al., 1987). After incubation at 25°C for 5–20 min, the products are separated by TLC on silica gel (CH$_2$Cl$_2$–EtOH, 95:5; v/v). Labelled iridodial is removed and quantitated by liquid scintillation spectrophotometry.

E. S-Adenosyl-L-Methionine:Loganic Acid Methyltransferase

1. Purification and properties

The final step in loganin biosynthesis involves the O-methylation of the carboxyl group of loganic acid by S-adenosyl-L-methionine:loganic acid methyltransferase (Fig. 13.2, reaction 4). The enzyme can be partially purified from etiolated 6- to 8-day-old developing seedlings of Catharanthus roseus which display maximal enzyme activity (Madyastha et al., 1973). Seedlings are extracted in 0.2 M phosphate, pH 7.2, 1 mM DTT, 10 mM sodium metabisulphite, and polyclar AT (20% w/w) and the slurry is filtered. The filtrate is centrifuged at 10 000 × g for 20 min and the supernatant is desalted on Bio-Gel P-10 followed by acetone precipitation and fractionation of the enzyme by ammonium sulphate precipitation (30–65%). The enzyme is desalted and further purified by DEAE cellulose ion-exchange chromatography. The partially purified enzyme does not require divalent cations for activity and it catalyses methyl group transfer from S-adenosyl-L-methionine to loganic acid to form loganin.

Simple kinetic analyses show apparent K_ms for S-adenosyl-L-methionine and for loganic acid of 0.06 and 12.5 mM, respectively. Substrate specificity studies reveal that both loganic and secologanic acid are equally well accepted as substrates, whereas removal of the hydroxyl group or epimerisation at position 7 of loganic acid results in complete loss of methyltransferase activity.

2. Assay

The methyltransferase is assayed by measuring the transfer of the ^{14}CH$_3$-group from the methyl donor S-adenosyl-L-methionine to loganic acid. The assay mixture contains 3 μmol of loganic acid, 200 nmol of S-adenosyl-[^{14}CH$_3$]L-methionine (140 000 dpm), 4 μmol dithiothreitol, 30 μmol of phosphate buffer and the reaction is initiated with enzyme in a final reaction volume of 0.5 ml (Madyastha et al., 1973). After a 3 h incubation at 32°C, the reaction is stopped with 1 ml methanol containing 0.5 mg loganin. The mixture is centrifuged and the supernatant is subjected to preparative TLC on silica gel (chloroform–methanol, 7:3; v/v). The loganin band is extracted from silica with methanol, followed by filtration to remove the silica, and the filtrate is evaporated to dryness. The residue is taken up in water, is equilibrated to pH 8–9 and is applied to a column of Dowex-1-formate. The column is eluted with 3 ml of water and the neutral fraction is chromatographed by descending paper chromatography (Whatman No. 1 developed with 1-butanol–acetic acid–water, 15:1:4; v/v/v). The chromatogram is scanned with a radiochromatogram scanner; the radioactive spot is removed and quantitated by liquid scintillation spectrophotometry.

III. THE ENZYMATIC SYNTHESIS OF TRYPTAMINE

A. Tryptophan Decarboxylase (EC 4.1.1.27)

1. Properties

The tryptamine moiety of indole alkaloids is known to be derived from tryptophan (Leete et al., 1965) and it has been speculated that the enzyme tryptophan decarboxylase (Fig. 13.2, reaction 5) catalyses a rate-limiting step in indole alkaloid biosynthesis (Knobloch et al., 1981; Sasse et al., 1983). Tryptophan decarboxylase catalyses the pyridoxal phosphate- (Noé et al., 1984) and pyrroquinoline quinone-mediated (Pennings et al., 1989a) decarboxylation of tryptophan to produce tryptamine. The decarboxylase appears transiently in developing seedlings of Catharanthus roseus (De Luca et al., 1986) and of Cinchona ledgeriana (Aerts et al., 1990), respectively, and its presence in Catharanthus coincides with the accumulation of monoterpenoid indole alkaloids in this plant (De Luca et al., 1986). The enzyme can also be transiently induced in cell suspension cultures by transfer of cells to an alkaloid production medium (Knobloch et al., 1981) or by treatment of cells with fungal elicitors (Eilert et al., 1987), but induction does not always coincide with accumulation of indole alkaloids. This discrepancy between intact plant and tissue culture suggests that the synthesis of indole alkaloids will not necessarily be controlled by the supply of tryptamine precursor.

Tryptophan decarboxylase from C. roseus is a cytosolic soluble enzyme (De Luca and Cutler, 1987), which occurs as a dimeric protein (molecular weight 115 kDa) consisting of two identical subunits of 54 kDa and with a pI of 5.9 as determined by isoelectric focussing (Noé et al., 1984). Substrate specificity studies indicate that unlike aromatic amino acid decarboxylases from animals, only L-tryptophan and 5-hydroxytryptophan will be decarboxylated, but not L-phenylalanine, L-tyrosine or L-dihydroxyphenylalanine. The enzyme has a pH optimum of 8.5, shows a K_m of 75 and 1300 μM for L-tryptophan and for 5-hydroxytryptophan, respectively, and a K_i of 310 μM for the product, tryptamine (Noé et al., 1984).

Polyclonal antibodies raised against tryptophan decarboxylase have been used to investigate the transient appearance of this enzyme during C. roseus seedling development. Results clearly indicate that the appearance of enzyme activity is strictly correlated with the developmentally regulated accumulation of tryptophan decarboxylase antigen, whereas the loss of enzyme activity is correlated with the loss of this antigen and the appearance of processed inactive forms of the enzyme (Fernandez et al., 1989a). In vitro studies with crude protein extracts containing tryptophan decarboxylase activity suggest that the enzyme can exist in vivo in different conformational states, a dimeric stable form in equilibrium with a monomeric form which is susceptible to irreversible inactivation (Fernandez et al., 1989b). When the equilibrium is shifted to monomer production and inactivation, the enzyme is quickly degraded by a proteolytic process which requires ATP. These data suggest that tryptophan decarboxylase activity is highly regulated by transcriptional, translational and post-translational controls whose roles are to regulate the production of tryptamine from tryptophan in *C. roseus*.

2.   Enzyme assay

(a)   Radiometric assays.   The radiochemical assay of tryptophan decarboxylase is based on the original method developed by Gibson *et al.* (1972). The reaction mixture contains 1.7 nmol L-[methylene-^{14}C]tryptophan (59 mCi mmol^{-1}, 1 μmol pyridoxal phosphate, and enzyme extract in a total of 0.1 ml Tris-HCl, pH 8, at 30°C for 1 h. The reaction is stopped by raising to pH 10 with 1 N NaOH and [2-^{14}C]tryptamine is extracted from unreacted substrate with 0.25 ml ethyl acetate. The ethyl acetate layer is separated from the reaction mixture by centrifugation. An aliquot (100 μl) is placed in scintillation cocktail and the product is quantitated by liquid scintillation spectrophotometry (Knobloch *et al.*, 1981; De Luca *et al.*, 1986). Alternatively the ethyl acetate layer is evaporated and product is analysed by TLC (acetone–methanol–diethylamine, 7:2:1; v/v/v) combined with autoradiography. An alternative radiometric assay for this enzyme is based on the measurement of $^{14}CO_2$ evolved as a result of enzymatic decarboxylation of L-[carboxyl-^{14}C]tryptophan (adapted from Leinweber and Walker, 1967).

(b)   HPLC assay.   Tryptophan decarboxylase is also assayed by high performance liquid chromatography (HPLC) (Pennings *et al.*, 1987). This method requires a high performance liquid chromatograph with florescence detection where wavelength settings are 280 nm for excitation and 354 nm for emission. The enzyme assay contains 1 mM L-tryptophan, 0.4 mM pyridoxal-5'-phosphate, 3 mM DTT, 0.1 M Tris-HCl, pH 8, and enzyme in a final volume of 0.1 ml. The assay is incubated at 35°C for 1 h and the reaction is terminated by cooling in ice water and by adding 0.1 ml methanol. A known amount of internal standard, 5-methyltryptamine, is added and peak height ratios between tryptamine and internal standard are compared after HPLC on a LiChrosorb RP-8 Select B column (7 μm particle size) from Merck.

3.   Purification

Cell suspension cultures (1 kg of cells) of *C. roseus* are transferred to 20 l of an alkaloid production medium and are grown in an air lift fermenter for 48 h, resulting in a 10-fold increase of tryptophan decarboxylase enzyme activity (Noé *et al.*, 1984). The cells are harvested and extracted by homogenisation in the presence of 0.1 M Tris-HCl, pH 7.5, in a DYNO-MILL with glass beads. After centrifugation, the supernatant is fractionated with ammonium sulphate and the pellet (42–55% $(NH_4)_2SO_4$) is dissolved, desalted and chromatographed on a DEAE Sephacel anion-exchange column. After elution with a linear gradient of KCl, active fractions are concentrated by ultrafiltration, desalted and purified by hydroxyapatite chromatography followed by gel filtration on ACA-34 and TSK-4000 columns, Mono Q high performance anion-exchange chromatography and a final gel filtration on TSK 3000. This protocol gives a homogeneous protein in a 5% yield.

Tryptophan decarboxylase can also be induced and purified from cell suspension cultures of *C. roseus* which have been treated with fungal elicitors (Fernandez *et al.*, 1989a). Elicited cell cultures contain 15- to 20-fold higher levels of tryptophan decarboxylase than untreated cells (Eilert *et al.*, 1987). In addition, cells from the elicited culture can be freeze-dried without any loss of enzyme activity, thus facilitating their storage. The enzyme from elicited cultures can be purified to homogeneity using similar column chromatography steps to those reported above, except that a six-fold improvement in recovery of pure tryptophan decarboxylase is obtained.

4. *Molecular studies*

Full length tryptophan decarboxylase cDNA clones have been isolated (De Luca *et al.*, 1989) by using anti-tryptophan decarboxylase antibodies to screen a cDNA expression library prepared from poly(A)$^+$ RNA of 7-day-old, light-induced *C. roseus* seedlings. The amino acid sequence deduced from this cDNA clone encodes a protein of molecular weight 56 kDa, which is similar to the size estimated previously by SDS-PAGE (Noé *et al.*; 1984, Noé and Berlin, 1985; Fernandez *et al.*, 1989a). The protein sequence of tryptophan decarboxylase shows that 39% of the amino acid residues are identical to those of L-DOPA decarboxylase from *Drosophila melanogaster*. This cDNA clone has been inserted into an *Agrobacterium tumefasciens* vector, and transgenic plants (*Nicotiana tabacum*) expressing high decarboxylase activities and accumulating high levels of tryptamine have been produced (Songstad *et al.*, 1990).

IV. THE FORMATION OF STRICTOSIDINE

A. Strictosidine Synthase (EC 4.3.3.2)

The chemical condensation of tryptamine and secologanin produces a mixture of two epimers, strictosidine and vincoside. However, only vincoside was believed to be the correct biological product of the enzymatic reaction (reviewed in Cordell, 1974). The identification of the correct stereochemistry of vincoside (Blackstock *et al.*, 1971; De Silva *et al.*, 1971; Mattes *et al.*, 1975), together with cell-free studies with the enzyme strictosidine synthase, have since proven strictosidine, rather than vincoside, to be the universal precursor to indole alkaloids with (*S*)-3-α-stereochemistry (Stöckigt, 1978) as well as those with 3-β-configuration (Nagakura *et al.*, 1979) (reviewed in Stöckigt, 1980).

1. *Properties*

The enzyme strictosidine synthase (Fig. 13.2, reaction 6) catalyses the stereospecific condensation of secologanin and tryptamine to produce H-3-α-(*S*)-strictosidine, the key intermediate in monoterpenoid indole alkaloid biosynthesis. The enzyme is present in cell suspension cultures and seedlings of *Catharanthus roseus* (Scott and Lee, 1975). Partially purified *C. roseus* cell culture preparations of strictosidine synthase (Treimer and Zenk, 1978; Mizukami *et al.*, 1979) show the enzyme to have a molecular weight of 34–38 kDa with a pI of 4.6. The enzyme isolated from *Rauwolfia serpentina* cell cultures (Hampp and Zenk, 1988) has molecular weights of 30 and 35 kDa by high performance gel filtration chromatography and by SDS-PAGE, respectively, with a pI of 4.5. The discrepancy in size appears to be due to a 5.3% carbohydrate content of this protein which would affect the molecular weight determination by SDS-PAGE. The enzymes from both sources show similar pH optima (pH 6.5–6.8) and similar K_m values for tryptamine and secologanin (2–4 mM). Antibodies raised against the *Rauwolfia* strictosidine synthase cross-react with the enzyme from *Catharanthus*, but not with the enzyme of other Rubiaceae species (Hampp and Zenk, 1988). Isoenzymes of strictosidine synthase have been reported in cell cultures of *C. roseus* (Pfitzner and Zenk, 1989).

Strictosidine synthases isolated from *Catharanthus* (Treimer and Zenk, 1979) and *Rauwolfia* (Hampp and Zenk, 1988) cell cultures show high substrate specificity for tryptamine and secologanin. Of several aromatic substrates tested, low levels of enzyme activity are observed with 7-methyl-, 6-hydroxy-, 7-fluoro-, 5-fluoro-, and 5-hydroxytryptamine, whereas no activity is observed with L- or D-tryptophan or several other substituted amines. Only secologanin and closely related derivatives (dihydro-, 2'-*O*-methyl- and 3'-*O*-methyl-secologanin) are acceptable as substrates, whereas secologanic acid is not (Treimer and Zenk, 1979). Strictosidine synthase activity occurs in cell cultures of plant species which produce indole alkaloids but is completely absent from cultures of plants which do not synthesise them (Treimer and Zenk, 1978).

2. Enzyme assays

(a) *[side chain-2-¹⁴C]Tryptamine-based enzyme assay.* Strictosidine synthase activity is determined by measuring the formation of [^{14}C]strictosidine after incubation with [*side chain*-2-^{14}C]tryptamine and secologanin at 30°C for 1 h (Mizukami *et al.*, 1979). The reaction mixture contains 2.5 mM [2-*side chain*-^{14}C]tryptamine (0.4 Ci mol^{-1}), 10 mM secologanin, 10 mM 2-mercaptoethanol, 50 mM Tris-HCl, pH 7.5, in a final volume of 1.0 ml. The presence of endogenous β-glucosidase activity in crude extracts requires the inclusion in the assay of 25 mM δ-D-gluconolactone inhibitor in order to prevent the further metabolism of the strictosidine product. The reaction is stopped by raising the pH to 10 with 10% K_2CO_3. Reaction products are extracted by repeated extraction (three times) with ethyl acetate. After evaporation, the ethyl acetate extract is analysed by silica gel G TLC with development in acetone–methanol–diethylamine (7:2:1; v/v/v). The strictosidine product is removed for quantitation by liquid scintillation spectrophotometry.

(b) *[indole-2-³H]Tryptamine-based enzyme assay.* This assay is based on the elimination of the hydrogen atom at C-2 of the indole ring during the formation of strictosidine from secologanin and tryptamine (Treimer and Zenk, 1978). The reaction mixture contains 0.26 μmol [*indole*-2-^{3}H]tryptamine (2.1 Ci mol^{-1}), 0.8 μmol secologanin, 20 μmol phosphate buffer (pH 7.0) and enzyme in a total volume of 0.2 ml. The reaction mixture is incubated for 45 min at 35°C and HO^{3}H is recovered by sublimation and is counted directly by liquid scintillation spectrophotometry. This method successfully eliminates the requirement for TLC.

A useful variation of this assay has recently been reported (Pfitzner and Zenk, 1989).

In this procedure, the assay is stopped by addition of 0.3 ml of a charcoal suspension (2 g per 50 ml H_2O). After shaking the mixture for 1 min, the charcoal along with 99% of the unreacted tryptamine is eliminated by centrifugation and an aliquot of the supernatant is counted by liquid scintillation spectrophotometry.

(c) Spectrophotometric assay. The treatment of strictosidine with strong acids at 100°C produces a yellow colour with an absorbance maximum at 348 nm and this is the basis for a spectrophotometric assay for strictosidine synthase (Walton *et al.*, 1987). The reaction mixture contains 100 mM phosphate buffer, pH 7, 1.3 mM tryptamine, 4 mM secologanin and 50 mM δ-D-gluconolactone in a final volume of 0.5 ml. The reaction mixture is incubated for 2 h at 30°C and is stopped with 25 μl of 3.6 M NH_4OH and substrates as well as products are extracted with 2 × 1.5 ml ethyl acetate. The ethyl acetate fraction is re-extracted with 5 ml 180 mM NH_4OH in order to remove secologanin substrate. An aliquot of the ethyl acetate is removed, evaporated and 1 ml of 5 N H_2SO_4 is added to the residue. After heating for 45 min at 100°C the absorbance at 348 nm is measured in a spectrophotometer. This method is less sensitive than the two radiometric assays but eliminates the requirement for radioactive substrates.

(d) HPLC assay. Strictosidine synthase activity can also be determined by HPLC with UV detection (Pennings *et al.*, 1989b). The reaction mixture contains 0.1 M phosphate buffer, pH 6.8, 1 mM tryptamine, 5 mM secologanin, 3 mM DTT, enzyme and 100 mM δ-D-gluconolactone in a final volume of 0.1 ml. After incubation at 30°C for 30 min, the reaction is stopped by immersion in ice water and by addition of 0.1 ml 5% trichloroacetic acid. A known amount of internal standard, codeine-HCl, is added and the samples are centrifuged. The supernatant is then submitted to HPLC on a Lichrosorb RP-8 Select B column (4 × 250 mm, 7 μm particle size) from Merck and the eluant consists of 7 mM sodium dodecyl sulphate and 25 mM NaH_2PO_4 in methanol–water (68:32; v/v) at pH 6.2. The major advantage of this method is the ability to measure simultaneously the production of strictosidine and the disappearance of tryptamine.

3. Purification

Strictosidine synthase was first isolated from *C. roseus* tissue cultures, and purified to apparent homogeneity by two different groups (Mizukami *et al.*, 1979; Treimer and Zenk, 1978). This enzyme has recently been repurified to apparent homogeneity (Pfitzner and Zenk, 1989), but is difficult to characterise because it occurs in multiple isomeric forms. Homogeneous strictosidine synthase has also recently been isolated from cell suspension cultures of *R. serpentina* (Hampp and Zenk, 1988). The enzyme from *Rauwolfia* cell cultures is at least 20 times more abundant than in cell cultures of *Catharanthus*. The specific activity of homogeneous enzyme from *Rauwolfia* (184 nkat mg^{-1}) and from the recently purified enzyme from *Catharanthus* (104 nkat mg^{-1}) is 30-fold and 17-fold higher than the enzyme purified from *Catharanthus* (5.9 nkat mg^{-1}) in 1979, reflecting the higher purity of these preparations. In addition, the enzyme from *Rauwolfia* (half-life 36 days at 37°C) shows much higher stability than the enzyme from *Catharanthus* (half-life 5 h at 37°C) (Hampp and Zenk, 1988; Pfitzner and Zenk, 1989).

Strictosidine synthase from *R. serpentina* is extracted by dispersing at room temperature 2.5 kg of frozen tissue in 2.5 l of 25 mM phosphate buffer pH 6.5 and cell debris is removed by filtration through cheesecloth, followed by continuous flow filtration and two other filtration steps coupled to treatment with Amberlite XAD-2 resin to decrease alkaloid content (Hampp and Zenk, 1988). The extract is applied directly to Cellufine DEAE-AH anion-exchanger and enzyme is eluted after extensive washing with extraction buffer containing 1 M NaCl. The enzyme is purified to homogeneity by Phenyl-Sepharose CL-4B hydrophobic interaction chromatography followed by high performance ion-exchange chromatography on a DEAE (SP500) column and and high performance gel filtration on a TSK (G3000SW) column. This protocol produces 1.9 mg of pure enzyme in 35% yield.

The enzyme from *C. roseus* is purified by a procedure which includes ammonium sulphate fractionation, gel filtration on Ultrogel AcA44 (LKB), Matrex gel red A, Chromatofocusing and semi-preparative PAGE. This protocol produces 0.28 mg of apparently homogeneous enzyme in 6% yield (Pfitzner and Zenk, 1989).

4. *Molecular studies*

Strictosidine synthase which has been purified to homogeneity from *R. serpentina* cell cultures (Hampp and Zenk, 1988), has been used to generate amino acid sequences of tryptic peptides and to isolate full length cDNA clones to this enzyme (Kutchan *et al.*, 1988; Kohrer *et al.*, 1989). The enzyme from *Catharanthus* has also been cloned (McKnight *et al.*, 1990) by using labelled probes from the sequence of *R. serpentina* (Kutchan *et al.*, 1988) to screen the same cDNA library used for the cloning of tryptophan decarboxylase (De Luca *et al.*, 1989). The two strictosidine synthases show a high degree of similarity, as predicted by their amino acid sequences which encode proteins of the same length (344 residues; molecular weight 38 kDa) and which possess 80% amino acid sequence similarity (274 residues are identical). The two proteins possess similar glycosylation sites and both contain a signal peptide which targets the enzyme to the plant vacuole. Enzymatically active strictosidine synthase from *R. serpentina* has been expressed in *Escherichia coli* and can be used to synthesise unlimited quantities of virtually pure strictosidine which is secreted into the medium of the bacterial broth (Kutchan, 1989). Enzymatically active, glycosylated and correctly targeted strictosidine synthase from *C. roseus* has been expressed in transgenic tobacco plants (McKnight *et al.*, 1991).

V. BIOSYNTHESIS OF *CATHARANTHUS* ALKALOIDS FROM STRICTOSIDINE

A. Glucoalkaloid-specific β-Glucosidase

B. Cathenamine Reductase

1. *Characterisation and properties*

Strictosidine occupies a central position in the biosynthesis of the major classes of monoterpenoid indole alkaloids of *Catharanthus* as well as in the four plant families

(Apocynaceae, Loganiaceae, Rubiaceae and Nyssaceae) which produce these alkaloids. The first step in the conversion of Corynanthe alkaloids requires the loss of glucose, resulting in the formation of a reactive aglycone which in turn opens to form a series of reactive intermediates (reviewed in Stöckigt, 1980). This step is catalysed by a

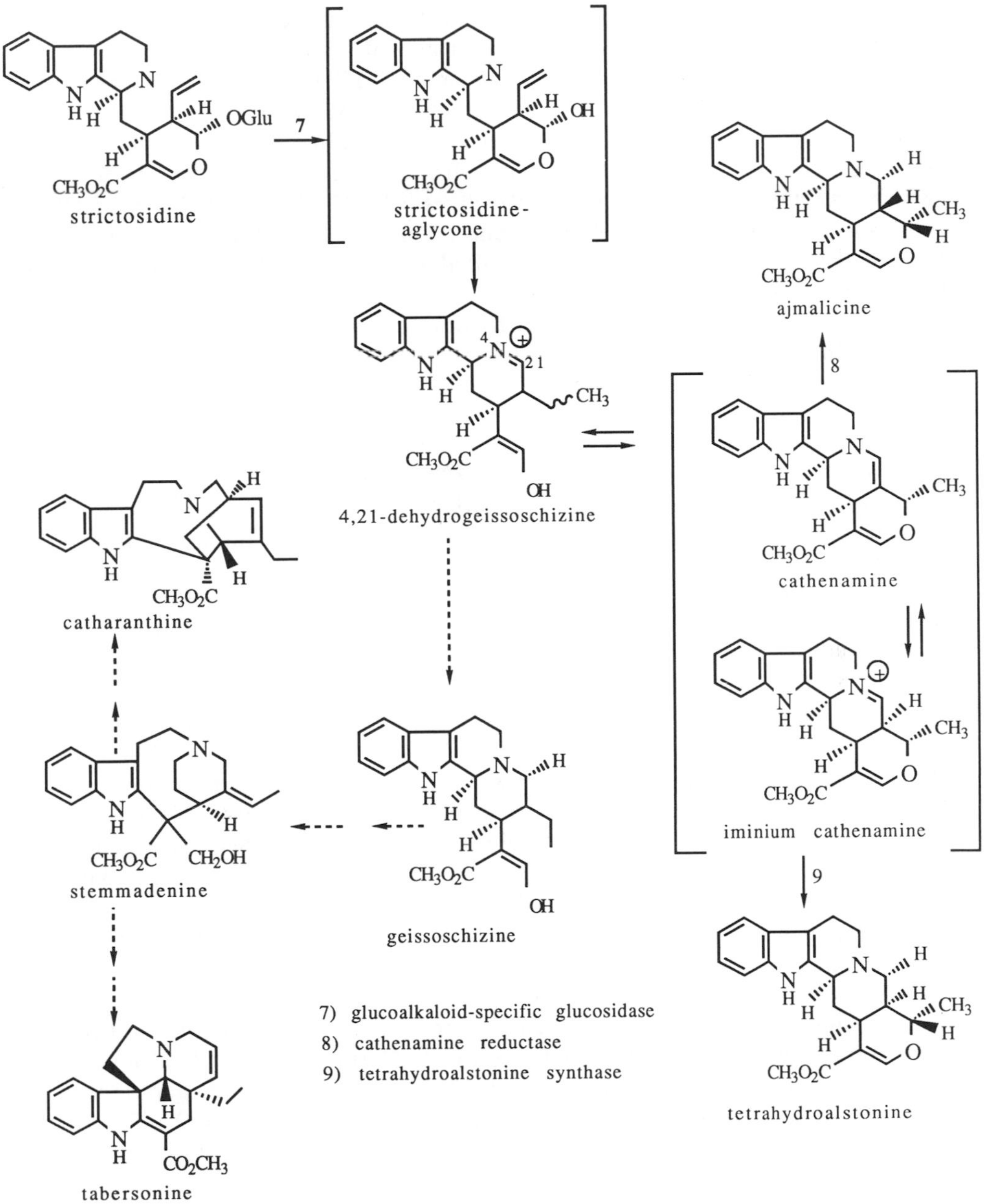

7) glucoalkaloid-specific glucosidase
8) cathenamine reductase
9) tetrahydroalstonine synthase

FIG. 13.3. Enzymatic steps in the formation of ajmalicine, tetrahydroalstonine, tabersonine and catharanthine from strictosidine. Dotted lines represent uncharacterised enzyme reactions.

glucoalkaloid-specific β-glucosidase (Fig. 13.3, reaction 7), which is separated into two charge isoforms by DEAE cellulose ion-exchange chromatography (Hemscheidt and Zenk, 1980). These glucosidases show a high specificity for strictosidine, since the 3-β-(R) epimer (vincoside) is a poor substrate and strictosidine lactam, vincoside lactam, secologanin, other closely related substrates and artificial substrates (p-nitrophenyl-β-D-glucoside) are not hydrolysed by this enzyme. The glucoalkaloid-specific β-glucosidase isoforms have a molecular weight of 55 kDa as determined by gel filtration on a calibrated Sephadex G-150 superfine column and pH optima of 5.0–5.5.

In vivo feeding experiments as well as enzymatic studies with cell-free extracts from *Catharanthus* cell cultures demonstrate (Stöckigt, 1980; Zenk, 1980) that β-glucosidase-mediated production of the aglycone of strictosidine is unstable and results in the non-enzymatic formation of reactive intermediates including 4,21-dehydrogeissoschizine. At this point the pathway diverges where this intermediate undergoes cyclisation to form cathenamine or is reduced enzymatically to geissoschizine, which serves as the precursor for more complex alkaloids such as catharanthine or vindoline (Fig. 13.3).

Remarkably, two enzymes have been characterised, one which reduces cathenamine to ajmalicine (cathenamine reductase) (Fig. 13.3, reaction 8) and one which reduces the iminium form of cathenamine to tetrahydroalstonine (tetrahydroalstonine synthase) (Fig. 13.3, reaction 9) (Hemscheidt and Zenk, 1985). The two enzymes have a similar pH optima (pH 6.6) and identical molecular weights of 81 kDa as determined by chromatography on a calibrated column of Sephadex G-100 superfine. Separation of cathenamine-reducing activities can be achieved by DEAE cellulose ion-exchange chromatography after gel filtration on an Ultrogel AcA-44 column. The tetrahydroalstonine synthesising activity can be partially purified by ammonium sulphate fractionation, gel filtration and Matrex green dye–ligand chromatography. Substrate specificity studies reveal that NADPH serves as a reducing cofactor but not NADH. The K_m for cathenamine is 62 μM.

2. Enzyme assays

(a) Glucoalkaloid-specific β-glucosidase. The reaction mixture contains 5 μmol phosphate, pH 6.3, 300 pmol [3-^{3}H]strictosidine (specific activity 34 mCi mmol^{-1}) enzyme in a total volume of 50 μl (Hemscheidt and Zenk, 1980). The reaction is incubated for 30 min at 30°C, after which it is stopped with 0.5 ml ice-cold phosphate buffer, pH 7. The aglycone product is extracted into benzene–n-hexane (3:1; v/v) from the aqueous phase. The product is counted by liquid scintillation spectrophotometry and corrections are made for a 5% carry-over of strictosidine.

(b) Cathenamine reductase. The reaction mixture contains 10 μl of cathenamine (2 mg cathenamine per ml dimethylformamide), 500 nmol NADPH, 20 μmol phosphate buffer, pH 6.6, and enzyme in a final volume of 210 μl. The reaction is incubated for 20 min at 30°C and is stopped by the addition of 20 μl 1 M Na$_2$CO$_3$. The mixture is shaken for 30 s and 20 μl aliquots are submitted to HPLC on a Merck Li Chro Cart

RP 18 column (10 μm, 250 × 4 mm). Equilibration and elution is carried out in 1% $(NH_4)_2CO_3$–acetonitrile (45:55; v/v), with detection at 280 nm.

C. The Biosynthesis of Catharanthine and Vindoline in *Catharanthus roseus*

Catharanthine and vindoline are members of the Iboga and Aspidosperma alkaloids, respectively. Catharanthine is believed to be derived from geissoschizine via the intermediates stemmadenine and allocatharanthine, whereas vindoline is derived from stemmadenine via tabersonine (reviewed in Cordell, 1974; Stöckigt, 1978). Our knowledge about the reactions involved in the transformation of these intermediates into Iboga and Aspidosperma alkaloids are based completely on *in vivo* feeding experiments and the enzymology remains to be elucidated.

D. The Biosynthesis of Vindoline from Tabersonine

The biosynthesis of vindoline from tabersonine involves three hydroxylations, one *O*-methylation, one *N*-methylation and one *O*-acetylation. A sequence of enzyme reactions which convert tabersonine to vindoline is proposed based on the accumulation of tabersonine as well as four out of five vindoline intermediates in dark-grown *C. roseus* seedlings and the quantitative conversion of tabersonine as well as those intermediates into vindoline when seedlings are transferred to light (Balsevich *et al.*, 1986; De Luca *et al.*, 1986).

E. *S*-Adenosyl-L-methionine:16-hydroxytabersonine-16-*O*-methyltransferase

1. Properties

The enzyme involved in the *O*-methylation of 16-hydroxytabersonine (Fig. 13.4, reaction 10) catalyses the second step in the conversion of tabersonine to vindoline (Balsevich *et al.*, 1986; De Luca *et al.*, 1986), contrary to a previous hypothesis which indicated this to be the second to last step in this pathway (Fahn *et al.*, 1985a,b). Preliminary identification of this *O*-methyltransferase has been carried out with crude desalted (Sephadex G-25) leaf extracts from *C. roseus* plants (Fahn *et al.*, 1985b). The biosynthetic pathway proposed by Fahn *et al.*, is partially based on the ability of this enzyme to catalyse the methylation of 16-*O*-demethyl-17-*O*-deacetylvindoline. The 16-hydroxytabersonine substrate, however, remains to be tested.

2. Enzyme assay

The *O*-methyltransferase assay contains 1.5 nmol [^{3}H]*S*-adenosyl-L-methionine (38 kBq nmol^{-1}), 0.1 M phosphate buffer, pH 8, and enzyme in a final volume of 0.2 ml. After incubation for 2 h at 30°C, the reaction is stopped with base (NaOH), and the mixture is extracted with 0.4 ml ethyl acetate. The organic extract which contains the methylated product is quantified by liquid scintillation spectrophotometry.

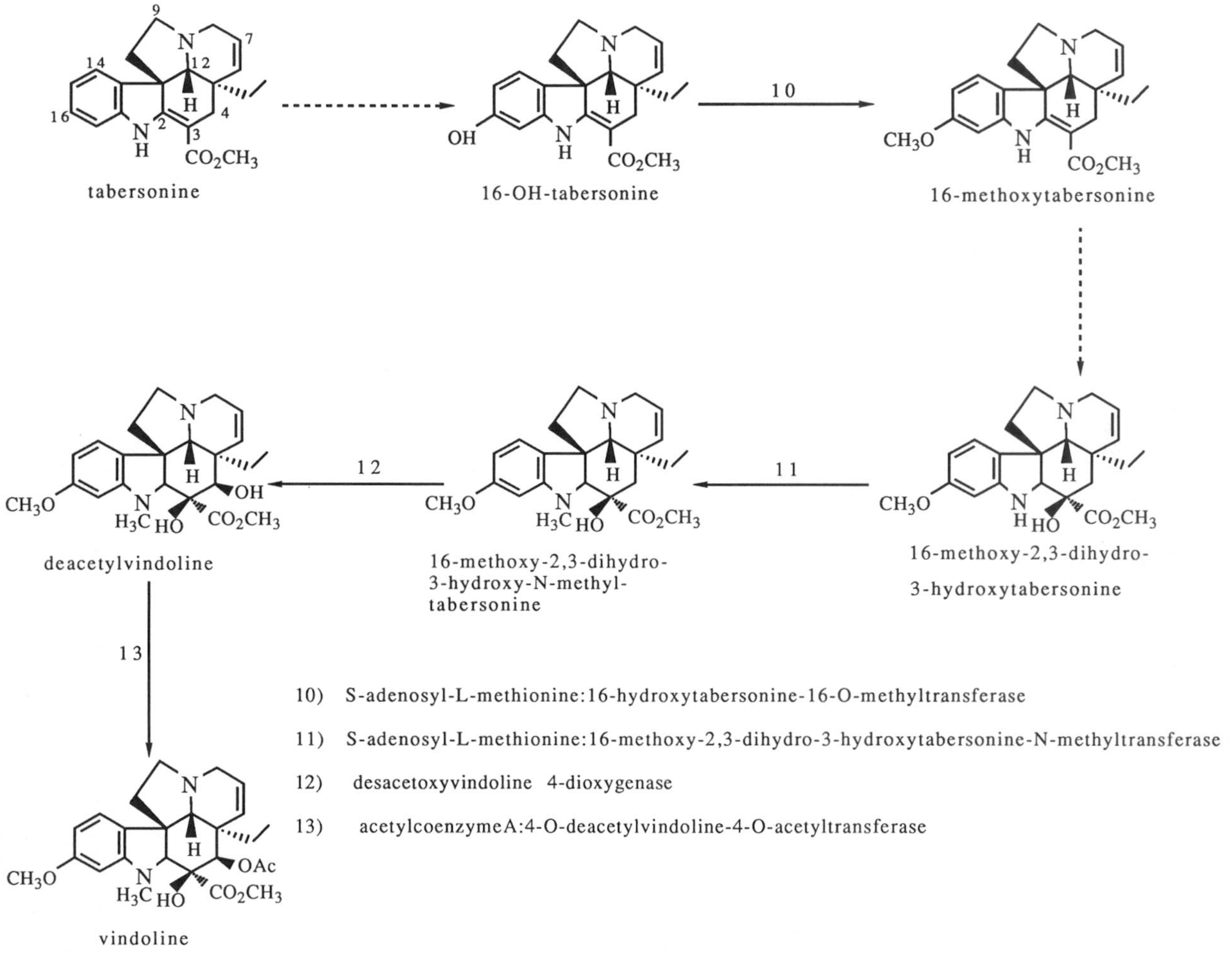
tabersonine
16-OH-tabersonine
16-methoxytabersonine
16-methoxy-2,3-dihydro-3-hydroxytabersonine
16-methoxy-2,3-dihydro-3-hydroxy-N-methyl-tabersonine
deacetylvindoline
vindoline
10
11
12
13
10) S-adenosyl-L-methionine:16-hydroxytabersonine-16-O-methyltransferase
11) S-adenosyl-L-methionine:16-methoxy-2,3-dihydro-3-hydroxytabersonine-N-methyltransferase
12) desacetoxyvindoline 4-dioxygenase
13) acetylcoenzymeA:4-O-deacetylvindoline-4-O-acetyltransferase

FIG. 13.4 Enzymatic steps in the formation of vindoline from tabersonine. Dotted lines represent uncharacterised enzyme reactions.

F. S-Adenosyl-L-methionine:16-methoxy-2,3-dihydro-3-hydroxytabersonine-N-methyltransferase

1. Properties

The third to last step in vindoline biosynthesis involves the S-adenosyl-L-methionine-dependent N-methylation of 16-methoxy-2,3-dihydro-3-hydroxytabersonine to produce desacetoxyvindoline (Fig. 13.4, reaction 11) (De Luca et al., 1987). The enzyme can be extracted and partially purified from C. roseus leaves by filtration of extracts on Sephadex G-100 where N-methyltransferase elutes in the void volume and is associated with chlorophyll-containing fractions (De Luca and Cutler, 1987). Furthermore, studies with isolated lysed Catharanthus protoplasts and their fractionation on sucrose density gradients suggest that the N-methyltransferase is localised in membranes of chloroplast thylakoids (De Luca and Cutler, 1987). This result suggests that an intermediate involved in the late stages of vindoline biosynthesis must be transported to the chloroplast for N-methylation to occur and the N-methylated product must be transported out of the chloroplast to the cytoplasm where the last two steps in vindoline biosynthesis occur (De Luca and Cutler, 1987; De Carolis et al., 1990). The enzyme shows a high degree of substrate specificity for its indole alkaloid substrate, suggesting that the 2,3 double bond of tabersonine must be reduced before it is acceptable for N-methylation and that this alkaloid is the natural substrate for this N-methyltransferase.

2. Enzyme assay

The enzyme can be assayed radiometrically by using S-adenosyl-[^{14}CH$_3$]L-methionine as the methyl donor (De Luca and Cutler, 1987). The enzyme assay contains 5 μM 2,3-dihydro-3-hydroxytabersonine, 8.8 μM S-adenosyl-[^{14}CH$_3$]L-methionine (60 mCi mmol^{-1}) and enzyme in a total of 100 μl of 0.1 M Tris-HCl, pH 8. After incubation 30°C for 30 min the reaction is stopped with 50 μl of 1 N NaOH and the labelled alkaloid product is separated from the unreacted methyl donor by extraction with ethyl acetate. The organic phase is counted for radioactivity by liquid scintillation spectrophotometry.

G. Desacetoxyvindoline Dioxygenase

1. Properties

The enzyme which catalyses the C-4 hydroxylation of desacetoxyvindoline (2,3-dihydro-3-hydroxy-N(l)-methyltabersonine) to the 3,4-dihydroxy derivative is a 2-oxoglutarate-dependent dioxygenase (Fig. 13.4, reaction 12) (De Carolis et al., 1990). The enzyme has typical properties for this class of dioxygenase which shows an absolute requirement for 2-oxoglutarate and whose activity is enhanced by ascorbate. The enzyme exhibits a pH optimum between 7 to 8 and has an apparent molecular weight of 45 kDa as determined by Superose 12 high performance gel filtration chromatography. The 4-hydroxylase exhibits strict specificity for position 4 of desacetoxyvindoline and this activity is eliminated with indole alkaloid substrates missing the

N-methyl group or containing the 2,3 double bond. These substrate specificity studies suggest and confirm previous studies (Balsevich *et al.*, 1986; De Luca *et al.*, 1986) that hydroxylation at position 3 and N-methylation occur prior to hydroxylation at position 4. The appearance of 4-hydroxylase is developmentally regulated and is only induced in developing etiolated seedlings after light treatment (De Carolis *et al.*, 1990).

2. *Enzyme assays*

The 4-hydroxylase can be assayed by measuring the formation of 2,3-dihydro-3,4-dihydroxy-N(1)-[^{14}CH$_3$]tabersonine (De Carolis *et al.*, 1990). In order to perform this assay, the radiolabelled substrate for the hydroxylase is synthesised enzymatically using the N-methyltransferase described in Section V.F. The assay mixture contains 0.56 nmol 2,3-dihydro-3-hydroxy-N(1)-[^{14}CH$_3$]tabersonine (containing 44 600 dpm), 10 mM 2-oxoglutarate, 7.5 mM ascorbate, 0.1 M HEPES-HCl, pH 7.5, and protein in a final volume of 0.2 ml. After incubation at 30°C for 1 h, the reaction is stopped by addition of 0.1 ml 1 N NaOH and the alkaloids are extracted with 0.5 ml ethyl acetate. The organic layer is evaporated to dryness and dissolved in 10 μl of methanol. The unreacted substrate and hydroxylated product are separated by silica gel G TLC (methanol–ethyl acetate, 1:9; v/v). The reaction can be visualised by autoradiography, or both substrate and product can be removed for quantitation by liquid scintillation spectrophotometry.

The 4-hydroxylase is also assayed by measuring the release of ^{14}CO$_2$ when using [1-^{14}C]2-oxoglutarate as a cosubstrate. This assay is only useful after partial purification of the enzyme since contaminating enzymes in crude extracts compete for this substrate. The assay mixture contains 9.2 μM [1-^{14}C]2-oxoglutarate (specific activity 50 mCi mmol^{-1}), 5 μM unlabelled alkaloid substrate, 7.5 mM ascorbic acid, 0.5 mg catalase, 100 mM HEPES-HCl, pH 7.5, and protein in a final volume of 1 ml. After incubation at 30°C for 1 h, the reaction is stopped by addition of 0.1 ml 6 N HCl and dioxygenase activity is measured by measuring the liberation of ^{14}CO$_2$. The gas is trapped in filter paper imbibed with base, the filter paper is dried and the ^{14}CO$_2$ is quantitated by liquid scintillation spectrophotometry (as described by Hashimoto and Yamada, 1986).

H. Acetylcoenzyme A:4-*O*-deacetylvindoline 4-*O*-acetyltransferase

1. *Properties*

The final step in vindoline biosynthesis is carried out by an *O*-acetyltransferase which catalyses the reversible transfer of acetate from acetylcoenzyme A to deacetylvindoline (Fig. 13.4, reaction 13) (De Luca *et al.*, 1985; Fahn *et al.*, 1985a). The appearance of this enzyme is developmentally regulated in *C. roseus* seedlings and enzyme activity appears only after light treatment of etiolated seedlings (De Luca *et al.*, 1985). The enzyme shows a pH optimum of 8–9, an apparent molecular weight of 45 kDa determined by high performance gel filtration, and an apparent pI of 4.6 determined by chromatofocusing (De Luca *et al.*, 1985). The purified enzyme is resolved into five charge isoforms by isoelectric focusing with isoelectric points between 4.3 and 5.4 (Fahn and Stöckigt, 1990; Power *et al.*, 1990) with each isoform having a different specific activity (Fahn and Stöckigt, 1990). The enzyme is a dimer composed of two unequal subunits with molecular weights of 21 and 33 kDa, respectively (Power *et al.*, 1990), as determined by SDS-PAGE. The enzyme purified by Fahn and Stöckigt, (1990) also yields a dimer composed of two unequal subunits with molecular weights of 20 and 26 kDa, respectively, using the same technique. Substrate saturation studies with partially purified enzyme (De Luca *et al.*, 1985) and enzyme purified to homogeneity (Fahn and Stöckigt, 1990; Power *et al.*, 1990) give similar K_m values of 5.4 vs 6.5 and 0.7 vs 1.3 μM for acetylcoenzyme A and deacetylvindoline, respectively. Product inhibition studies show that coenzyme A is a powerful inhibitor (K_i 8 μM) (De Luca *et al.*, 1985) where it acts as a competitive inhibitor against acetylcoenzyme A and as a non-competitive inhibitor against deacetylvindoline (Fahn and Stöckigt, 1990). In contrast, the enzyme is not inhibited significantly by vindoline concentrations up to 2 mM (De Luca *et al.*, 1985; Fahn and Stöckigt, 1990).

The enzyme exhibits a high degree of substrate specificity for its indole alkaloid substrate (De Luca *et al.*, 1985; Fahn *et al.*, 1985a). The enzyme's specificity for highly substituted derivatives supports previous *in vivo* experiments (Balsevich *et al.*, 1986; De Luca *et al.*, 1986), which suggest that this enzyme catalyses the last step in vindoline biosynthesis.

2. Enzyme assays

(b) Radioisotopic assay. The enzyme is assayed radioisotopically (De Luca *et al.*, 1985). The assay mixture contains 4.4 μM [1-^{14}C]acetylcoenzyme A (54 mCi mmol^{-1}), 5 μM deacetylvindoline, 100 mM HEPES, pH 7.5, and enzyme in a final 0.1 ml. After incubation for 5 min at 30°C, the reaction is terminated by addition of 50 μl of 1 N NaOH. The reaction product is extracted from the unreacted [1-^{14}C]acetylcoenzyme A with 0.25 ml ethyl acetate and 0.1 ml of the organic solvent is counted by liquid scintillation spectrophotometry. Alternatively the organic extract is taken to dryness and the radioactive product is submitted to TLC on silica gel G (methanol–ethyl acetate, 1:9; v/v) followed by autoradiography. A similar assay has been developed using tritiated acetyl coenzyme A (Fahn *et al.*, 1985a).

(a) Spectrophotometric assay. The enzyme is also assayed spectrophotometrically (Power, 1989) according to an assay developed for bacterial kanamycin acetyltransferase (Radika and Northrup, 1984). In this assay, free coenzyme A reacts with aldithriol to form a chromophore that absorbs at 324 nm with a molar extinction coefficient of 19 800. The enzyme assay consists of 60 μM acetylcoenzyme A, 60 μM deacetylvindoline, 120 μM aldithriol, 10 mM ascorbate, 100 mM HEPES, pH 7.6, and enzyme in a total volume of 1 ml. The increase in absorbance is measured spectrophotometrically at 324 nm. Since aldithriol reacts with mercapto groups, care should be taken to remove DTT or 2-mercaptoethanol from protein extracts prior to enzyme assay.

3. Purification

Deacetylvindoline-O-acetyltransferase can be purified to homogeneity using young leaves of *C. roseus* as an enzyme source (Power *et al.*, 1990). Leaves (1 kg) are mixed with Polyclar AT (10% w/w) and are homogenised in a Waring blender in 4l of extraction buffer (0.1 M HEPES, pH 7.6, 5 mM EDTA, 10 mM 2-mercaptoethanol and 0.1 mM PMSF). The homogenate is filtered through Miracloth and the filtrate is centrifuged at 10 000 × g for 10 min to remove cellular debris. The resulting supernatant is mixed with Dowex 1-X 4 (10% w/v) and is filtered through Whatman No. 1 filter paper. The resulting filtrate is fractionated with ammonium sulphate (46–70%) and the enzyme is purified by chromatography on Sephacryl S-200 gel filtration, followed by Q-Sepharose anion-exchange chromatography, HA-Ultrogel chromatography and affinity chromatography on an Agarose-hexane-coenzyme A type 5 column. The most crucial step in this purification involves the affinity step, which results in an increase in specific activity from 0.54 to 36 pkat μg^{-1} protein.

The enzyme has also been purified by Fahn and Stöckigt (1990), who developed a similar protocol for purification of this enzyme. The enzyme is purified by dye–ligand chromatography using Procion Red HE-3b followed by coenzyme A affinity chromatography and hydrophobic interaction chromatography. Homogeneous enzyme is finally obtained after preparative isoelectric focusing resulting in five charge isoforms with specific activities varying between 12.6 and 181.7 nkat mg^{-1} protein.

REFERENCES

Aerts, R. J., Van Der Leer, T., Van Der Heijden, R. and Verpoorte, R. (1990). *J. Plant Physiol.* **136**, 86–89.
Balsevich, J. and Kurz, W. G. W. (1983). *Planta Med.* **49**, 79–84.
Balsevich, J. De Luca, V. and Kurz, W. G. W. (1986). *Heterocycles* **24**, 2415–2421.
Battersby, A. R. (1971). *In* "The Alkaloids", Specialist Periodical Reports, Vol. I (J. E. Saxton, ed.), pp. 31–112. The Chemical Society, London.
Blackstock, W. P., Brown, R. T., and Lee, G. K. (1971). *Chem. Commun.*, 908–910.
Cordell, G. A. (1974). *Lloydia* **37**, 219–298.
De Carolis, E., Chan, F., Balsevich, J. and De Luca, V. (1990). *Plant Physiol.* **94**, 1323–1329.
De Luca, V. and Cutler, A. (1987). *Plant Physiol.* **85**, 1099–1102.
De Luca, V. and Kurz W. G. W. (1988). *In* "Monoterpenoid Indole Alkaloids" (F. Constable and I. Vasil, eds.), pp. 385–401. Academic Press, London.
De Luca, V., Balsevich, J. and Kurz, W. G. W. (1985). *J. Plant. Physiol.* **121**, 417–428.
De Luca, V., Balsevich, J., Tyler, R. T., Eilert, U., Panchuk, B. D. and Kurz, W. G. W. (1986). *J. Plant Physiol.* **125**, 147–156.
De Luca, V., Balsevich, J., Tyler, R. T. and Kurz, W. G. W. (1987). *Plant Cell Rep.* **6**, 458–461.
De Luca, V., Marineau, C., Brisson, N. (1989). *Proc. Natl. Acad. Sci. USA* **86**, 2582–2586.
De Silva, K. T. D., Smith, G. N. and Warren, K. E. (1971). *Chem. Commun.*, 905–907.
Eilert, U., De Luca, V., Constabel, F. and Kurz, W. G. W. (1987). *Arch. Biochem. Biophys.* **254**, 491–497.
Ernster, L., Seikevitz, P. and Palade, G. E. (1962). *J. Cell Biol.* **15**, 541–560.
Fahn, W. and Stöckigt, J. (1990). *Plant Cell Rep.* **8**, 613–616.
Fahn, W., Gundlach, H., Deus-Neumann, B. and Stöckigt, J. (1985a). *Plant Cell Rep.* **4**, 333–336.
Fahn, W., Laussermair, E., Deus-Neumann, B. and Stöckigt, J. (1985b). *Plant Cell Rep.* **4**, 337–340.
Fernandez, J. A., Owen, T. G., Kurz, W. G. W. and De Luca, V. (1989a). *Plant Physiol.* **91**, 79–84.
Fernandez, J. A., Kurz, W. G. W. and De Luca, V. (1989b). *Biochem. Cell Biol.* **67**, 730–734.
Ganzinger, D. and Hesse, M. (1976). *Lloydia* **39**, 326–349.
Gibson, R. A., Barret, G. and Wightman, F. (1972). *J. Exp. Bot.* **23**, 775–786.

Hampp, N. and Zenk, M. H. (1988). *Phytochemistry* **27**, 3811–3815.
Hashimoto, T. and Yamada, Y. (1986). *Plant Physiol.* **81**, 619–625.
Hemscheidt, T. and Zenk, M. H. (1980). *FEBS Lett.* **110**, 187–191.
Hemscheidt, T. and Zenk, M. H. (1985). *Plant Cell Rep.* **4**, 216–219.
Ikeda, H., Esaki, N., Nakai, S., Hashimoto, K., Uesato, S., Soda, K. and Fujita, T. (1991). *J. Biochem. (Japan)* **109**, 341–347.
Knobloch, K. H., Hansen, B. and Berlin, J. (1981). *Z. Naturforsch.* **36c**, 40–43.
Kohrer, K., Kutchan, T. and Domdey, H. (1989). *DNA* **8**, 143–147.
Kurz, W. G. W. and Constabel, F. (1985). *CRC Crit. Rev. Biotech.* **2**, 105–118.
Kutchan, T. M. (1989). *FEBS Lett.* **257**, 127–130.
Kutchan, T. M., Hampp, N., Lottspeich, F., Bayreuther, K. and Zenk, M. H. (1988). *FEBS Lett.* **237**, 40–44.
Leete, E., Ahmad, A. and Kempis, I. (1965). *J. Am Chem. Soc.* **87**, 4168–4170.
Leinweber, F. J. and Walker, L. A. (1967). *Anal. Biochem.* **21**, 131–134.
Loomis, W. D. (1974). *Methods Enzymol.* **31A**, 528–544.
Madyastha, K. M. and Coscia, C. J. (1979a). *In* "Enzymology of Indole Alkaloid Biosynthesis" (T. Swain and G. R. Waller, eds), pp. 85–129. Plenum Press New York.
Madyastha, K. M. and Coscia, C. J. (1979b). *J. Biol. Chem.* **254**, 2419–2427.
Madyastha, K. M., Guarnaccia, R., Baxter, C. and Coscia, C. J. (1973). *J. Biol. Chem.* **248**, 2497–2501.
Madyastha, K. M., Meehan, T. D. and Coscia, C. J. (1976). *Biochemistry* **15**, 1097–1102.
Madyastha, K. M., Ridgeway, J. E., Dwyer, J. G. and Coscia, C. J. (1977). *J. Cell Biol.* **72**, 302–313.
Mattes, K. C., Hutchinson, C. R., Springer, J. P. and Clardy, J. (1975). *J. Am. Chem. Soc.* **97**, 6270–6271.
McKnight, T. D., Roessner, C. A., Devagupta, R., Scott, A. I. and Nessler, C. L. (1990). *Nucl. Acids Res.* **18**, 4939.
McKnight, T. D., Bergey, D. R., Burnett, R. J. and Nessler, C. L. (1991). *Planta* (submitted).
Meehan T. D. and Coscia, C. J. (1973). *Biochem. Biophys. Res. Commun.* **53**, 1043–1048.
Meijer, A. H., Pennings, E. J. M., De Waal, A. and Verpoorte, R. (1990). *In* "Progress in Plant Cellular and Molecular Biology" (H. J. J. Nijkamp, L. H. W. Van Der Plas and J. Van Aartrijk, eds), pp. 769–774. Kluwer Academic Publishers, Dordrecht, Boston and New York.
Mizukami, H., Nordlov, H., Lee, S. L. and Scott, A. I. (1979). *Biochemistry* **18**, 3760–3763.
Nakagura, N., Rueffer, M. and Zenk, M. H. (1979). *J. Chem. Soc., Perkin Trans. I*, 2308.
Nebert D. W. (1979). *Mol. Cell. Biochem.* **72**, 27–46.
Noé, W. and Berlin, J. (1985). *Planta* **166**, 500–504.
Noé, W., Mollenschott, C. and Berlin, J. (1984). *Plant Mol. Biol.* **3**, 281–288.
Omura, T. and Sato, R. (1964). *J. Biol. Chem.* **239**, 2370–2378.
Parry R. J. (1973). *In* "The Biosynthesis of *Catharanthus* Alkaloids" (W. I. Taylor and N. R. Farnsworth, eds), pp. 141–191. Marcel Dekker, Inc., New York.
Pennings, E. J. M., Hegger, I., Van der Heijden, R., Duine, J. A. and Verpoorte, R. (1987). *Anal. Biochem.* **165**, 133–136.
Pennings, E. J. M., Groen, B. W., Duine, J. A. and Verpoorte, R. (1989a). *FEBS Lett.* **255**, 97–100.
Pennings, E. J. M., Van den Bosch, R. A., Van der Heijden, R., Stevens, L. H., Duine, J. A. and Verpoorte, R. (1989b). *Anal. Biochem.* **176**, 412–415.
Pfitzner, U. and Zenk, M. H. (1989). *Planta Med.* **55**, 525–530.
Power, R. (1989). M.Sc. Thesis, University of Saskatchewan, Saskatoon, Saskatchewan, Canada, pp. 51–52.
Power, R., Kurz, W. G. W. and De Luca, V. (1990). *Arch. Biochem. Biophys.* **279**, 370–376.
Radika, K. and Northrup, D. B. (1984). *Arch. Biochem. Biophys.* **233**, 272–285.
Sasse, F., Buchholtz, M. and Berlin, J. (1983). *Z. Naturforsch.* **38c**, 916–922.
Scott, A. I. and Lee, S. L. (1975). *J. Am. Chem. Soc.* **97**, 6906–6909.
Scott, A. I., Lee, S., Culver, M. G., Wan, W., Hirata, T., Guéritte, F., Baxter, R. L., Nordlov, H., Dorschel, C. A., Mizukami, H. and Mackenzie, N. E. (1981). *Heterocycles* **15**, 1257–1274.

Songstad, D. D., De Luca, V., Brisson, N., Kurz, W. G. W. and Nessler, C. L. (1990). *Plant Physiol.* **94**, 1410–1413.

Spitsberg, V., Coscia, C. J. and Krueger, R. J. (1981). *Plant Cell Rep.* **1**, 43–47.

Stöckigt, J. (1978). *J. Chem. Soc., Chem. Commun.*, 1097–1099.

Stöckigt, J. (1980). *In* "Indole and Biogenetically Related Alkaloids" (J. D. Phillipson and M. H. Zenk, eds), pp. 113–141. Academic Press, London.

Treimer, J. F. and Zenk, M. H. (1978). *Phytochemistry* **17**, 217–232.

Treimer, J. F. and Zenk, M. H. (1979). *Eur. J. Biochem.* **101**, 225–233.

Uesato, S., Kanomi, S., Iida, A., Inouye, H. and Zenk, M. H. (1986a). *Phytochemistry* **25**, 839–842.

Uesato, S., Ogawa, Y., Inouye, H., Saiki, K. and Zenk, M. H. (1986b). *Tetrahedron Lett.* **27**, 2893–2896.

Uesato, S., Ikeda, H., Fujita, T., Inouye, H. and Zenk, M. H. (1987). *Tetrahedron Lett.* **28**, 4431–4434.

Van del Heijden, R., Verpoorte, R. and Ten Hoopen, H. J. G. (1989). *Plant Cell Tissue Organ Culture* **18**, 231–280.

Walton, N. J., Skinner, S. E., Robins, R. J. and Rhodes, M. J. (1987). *Anal. Biochem.* **163**, 482–488.

Zenk, M. H. (1980). *J. Nat. Prod.* **43**, 438–451.

Zenk, M., El-Shagi, H., Arens, H., Stöckigt, J., Weiler, E. W. and Deus, B. (1977). *In* "Plant Tissue Culture and its Biotechnological Applications" (W. Barz, E. Reinhard and M. H. Zenk, eds), pp. 27–44. Springer-Verlag, Berlin.

14 Nicotine and Tropane Alkaloids

TAKASHI HASHIMOTO and YASUYUKI YAMADA

Department of Agricultural Chemistry, Faculty of Agriculture, Kyoto University, Kyoto 606–01, Japan

I. INTRODUCTION

Nicotine and tropane alkaloids contain in their molecules a five-membered pyrrolidine ring. In nicotine and its direct metabolites (e.g. nornicotine, myosmine and cotinine), a pyridine ring is attached at C-2 of the pyrrolidine ring, whereas in tropane alkaloids the pyrrolidine ring combines with a piperidine ring to form a bicyclic tropane.

METHODS IN PLANT BIOCHEMISTRY Vol. 9
ISBN 0–12–461019–6

As expected from the presence of the common pyrrolidine ring in their structures, both types of alkaloids share a common early biosynthetic pathway originating from ornithine and/or arginine, and diverging at *N*-methylpyrrolinium cation (Fig. 14.1). A pathway involving δ-*N*-methylornithine had been proposed for tropane alkaloid biosynthesis (Leete, 1979) to account for the apparent asymmetric incorporation of [2-^{14}C]ornithine into the tropine moiety of hyoscyamine (Leete, 1964) but now is disfavoured over the pathway shown in Fig. 14.1 that involves a symmetrical diamine putrescine (Hashimoto *et al.*, 1989b,c,d; Leete, 1991). Since polyamines are also derived from putrescine, the pathways specific to nicotine and tropane alkaloids start by the *N*-methylation of putrescine, a reaction unique to alkaloid-producing plants.

Biosynthesis of putrescine can take two routes; the ornithine-derived route catalysed by ornithine decarboxylase (ODC; EC 4.1.1.17) and the arginine-derived route catalysed initially by arginine decarboxylase (ADC; EC 4.1.1.19)(cf. Chapter 18, this volume). These two pathways are not mutually exclusive and each plant species shows different ratio of ODC to ADC activities. Incorporation of labelled ornithine and arginine into alkaloids, the effects of specific inhibitors of the decarboxylases on alkaloid synthesis, and changes of enzyme activities during induction of alkaloid synthesis, indicate that the contribution of the arginine pathway is greater than that of the ornithine pathway for the supply of putrescine to alkaloid biosynthesis (Tiburcio *et al.*, 1985; Tiburcio and Galston, 1986; Hashimoto *et al.*, 1989d; Robins *et al.*, 1991).

Although also occurring in other families, tropane alkaloids are typically found in the Solanaceae (Romeike, 1978). In particular, tropane esters with tropic acid or related acids (e.g. hyoscyamine and scopolamine) are restricted to the Solanaceae. Nicotine and its derivatives are found, usually in trace amounts, in several families, but are main components of the two solanaceous genera, *Nicotiana* and *Duboisia* (Evans, 1980). The latter species are interesting from a biosynthetic point of view in that they produce both nicotine and tropane alkaloids. Distribution of known biosynthetic enzymes among species is well correlated with the distribution of these alkaloids, indicating that the enzymes specifically function in the biosynthesis of alkaloids. Oxidative deamination of *N*-methylputrescine in alkaloid biosynthesis is catalysed by a diamine oxidase, but the oxidases found in alkaloid-producing species differ from ubiquitous diamine oxidase in that the enzymes that function in alkaloid biosynthesis are highly specific to *N*-methylated diamines (Mizusaki *et al.*, 1972; Hashimoto *et al.*, 1990).

Nicotine and tropane alkaloids are mainly produced in the root. This had been initially suggested by reciprocal grafting experiments between alkaloid-producing and non-producing solanaceous plants (reviewed by Waller and Nowacki, 1979). Recent development of plant organ culture technologies and characterisation of several bio-synthetic enzymes have confirmed the root-specific expression of alkaloid biosynthesis. Immunohistochemical localisation of a scopolamine synthesising enzyme at the pericycle even suggests that the developmentally young roots without secondary growth, such as small branch roots and cultured roots, produce alkaloids more actively than the mature roots of intact plants (Hashimoto *et al.*, 1991). Fast growing cultured roots, produced either by addition of appropriate concentration of auxin in the culture medium or by transformation with *Agrobacterium rhizogenes*, are therefore ideal materials as enzyme sources (Hashimoto and Yamada, 1991). One kilogram of fresh roots can be routinely harvested within one month of culture in ordinary laboratories, and may be stored frozen at −20°C without significant loss of enzyme activity for later use in enzyme purification.

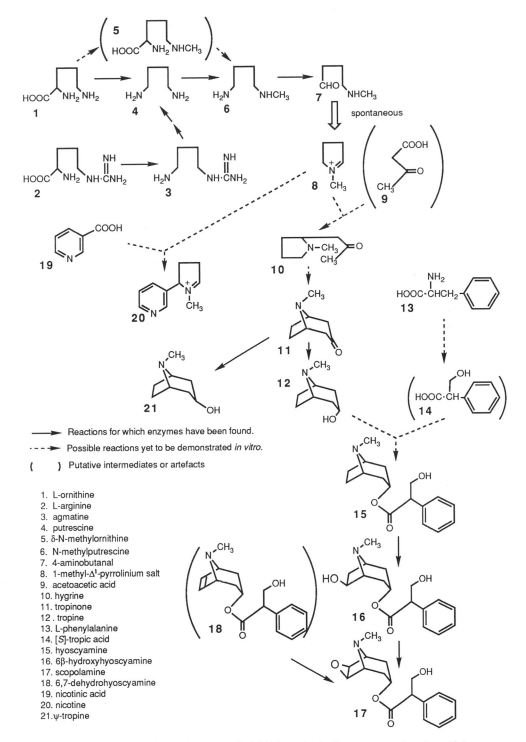

FIG. 14.1. Pathways of nicotine and tropane alkaloid biosynthesis. Key: ⟶, reactions for which enzymes have been found; ---→, possible reactions yet to be demonstrated *in vitro*; ( ), putative intermediates or artefacts; 1, L-ornithine; 2, L-arginine; 3, agmatine; 4, putrescine; 5, δ-N-methylornithine; 6, N-methyl-putrescine; 7, 4-aminobutanal; 8, 1-methyl-Δ'-pyrrolinium salt; 9, acetoacetic acid; 10, hygrine; 11, tropinone; 12, tropine; 13, L-phenylalanine; 14, [S]-tropic acid; 15, hyoscyamine; 16, 6β-hydroxyhyo-scyamine; 17, scopolamine; 18, 6,7-dehydrohyoscyamine; 19, nicotinic acid; 20, nicotine; 21, ψ-tropine.

Not many enzymes in the nicotine/tropane pathways have been isolated and characterised. In particular we lack information on the enzymes catalysing two important condensation reactions: nicotine formation from N-methylpyrrolinium cation and nicotinic acid (or its derivative), and hyoscyamine formation from tropine and [S]-tropic acid (or its derivative). Although several reports claim to have demonstrated these enzyme activities in cell-free systems (Kaczkowski, 1964; Jindra and Staba, 1968; Friesen and Leete, 1990), these results have not been substantiated by others. In this chapter, we have chosen three enzymes that have been relatively well characterised: one from the early biosynthetic pathway common to nicotine and tropane alkaloids, and two from the tropane-specific pathway. Steady advances in the techniques in enzyme purification and molecular cloning are leading to the deeper understanding of properties and regulation of enzymes involved in the biosyntheses of these important alkaloids.

II. PUTRESCINE N-METHYLTRANSFERASE

A. Reaction

Putrescine N-methyltransferase (PMT; EC 2.1.1.53) transfers the methyl group from S-adenosylmethionine (SAM) to an amino group of putrescine, the first reaction specific to nicotine and tropane alkaloid biosynthesis, thereby diverging the flow of putrescine away from entering into polyamine metabolism. The much lower abundance of conjugated forms of N-methylputrescine than of putrescine in cultured roots of *Hyoscyamus albus* (Hashimoto *et al.*, 1989c) suggests that once N-methylated, the diamine becomes a poor substrate for the conjugation reaction and remains in the unconjugated form to be utilised preferentially for alkaloid biosynthesis.

B. Assay

Two assay methods for PMT have been reported but the following method based on dansylation of the reaction product followed by separation by high performance liquid chromatography (HPLC) is most convenient for routine measurements. This method was originally developed by Feth *et al.* (1985) and subsequently modified by us (Hibi *et al.*, 1992).

The reaction mixture of $50\,\mu l$ contains, at a final concentration, 4 mM putrescine dihydrochloride, 4 mM SAM, 50 mM potassium phosphate buffer (pH 8.0) and enzyme. The small scale of reaction volume is convenient because the following derivatisation reaction can be done in the same 1.5 -ml Eppendorff tube. After 30 min of incubation at 30°C, the reaction is stopped by heating at 100°C for 3 min. The reaction product N-methylputrescine is then dansylated. To the incubation mixture, $200\,\mu l$ of 65 mM borate-KOH buffer (pH 10.5) is added, followed by addition of $150\,\mu l$ of dansylchloride solution (5.4 mg ml^{-1} acetonitrile, freshly prepared). The mixture is heated for 15 min at 56°C in the dark, centrifuged briefly and then applied directly to the HPLC column (10–$20\,\mu l$). Because dansylchloride and dansylated amines decompose in the light, their unnecessary exposure to light should be avoided.

Dansylated amines are analysed with HPLC on a 5-μm LiChrospher 100 RP-18 endcapped column (4 mm $\times$ 25 cm; Merck) at a flow rate of 1.0 ml min^{-1} at 40°C. The isocratic mobile phase is a mixture of an aqueous 2% H_3PO_4 solution (adjusted to pH 5.2 with triethylamine) and acetonitrile at a ratio of 35:65 (v/v). Use of a fluorescent detector (excitation at 365 nm and emission at 510 nm) gives much higher sensitivity than detection at 217 nm.

The other assay method measures the radioactivity in the reaction product formed from amine substrate and radioactive SAM labelled at the methyl group to be transferred (Mizusaki *et al.*, 1971). This method is useful when studying substrate specificity of PMT. For the reaction (50-μl scale), the cold SAM component in the above reaction mixture for HPLC analysis is replaced with 1 mM [*methyl*-^{14}C or -^{3}H]SAM (4 mCi mmol^{-1} or 147 kBq mmol^{-1}). After the reaction is stopped by the addition of 20 μl of 10% NaOH solution saturated with NaCl, 1 ml of chloroform is added and the mixture is vortexed for 1 min. The mixture is then briefly centrifuged, after which 0.5 ml of the organic phase is transferred to a counting vial containing 3 ml of a toluene scintillator mixture. When different substrate amines are compared for reactivity, the remaining aqueous phase is re-extracted with 1 ml of chloroform to check whether respective methylated products are efficiently transferred to the organic phase.

C. Purification

PMT is an extremely difficult enzyme to purify, largely because the enzyme behaves abnormally during column chromatography, always being eluted as a broad peak from ion-exchange columns and hydrophobic columns (Hibi *et al.*, 1992). The enzyme also tends to be inactivated in the presence of ammonium sulphate. Tobacco PMT has been purified 30-fold by precipitation with ammonium sulphate and by gel filtration (Mizusaki *et al.*, 1971). We also have succeeded only in partial purification (80-fold) by combinations of ion-exchange, hydrophobic, SAH-affinity and electrofocusing columns (unpubl. res.).

D. Properties and Regulation

PMTs from tobacco roots (Mizusaki *et al.*, 1971) and from cultured roots of *Hyoscyamus albus* (Hibi *et al.*, 1992) show marked specificity toward putrescine (1,4-diaminobutane). Shorter 1,3-diaminopropane and longer 1,5-diaminopentane (cadaverine) are poorer substrates (*H.albus*) or not accepted as substrates (tobacco). No activities are found with monoamines in *H. albus* enzymes.

When various plant organs and cultured tissues of *Atropa belladonna* and *H. niger* were assayed, we found strong PMT activities only in branch roots and cultured roots, and weak activities in main roots (Hibi *et al.*, 1992), thus supporting our previous conclusion that tropane alkaloids are mainly synthesised at the pericycle cells in the developmentally young root tissues (Hashimoto *et al.*, 1991). PMT activity is absent in the tobacco leaf but present in the tobacco root, where the activity is increased ten-fold by decapitation (Mizusaki *et al.*, 1971).

III. TROPINONE REDUCTASE

A. Reaction

There exist two types of tropinone reductase (TR) that reduce the keto group of tropinone stereospecifically to an hydroxy group using NADPH as a cofactor. One type (TR-I) forms tropine (3α-hydroxytropane), while the other type (TR-II) produces ψ-tropine (3β-hydroxytropane). These TRs with distinct stereospecificity are present simultaneously in a given solanaceous species, with a varying activity ratio between the two TRs (Hashimoto et al., 1992). TR-I supplies the tropane moiety of major tropane alkaloids in the Solanaceae. Although ψ-tropine derivatives such as tigloidine and calystergines (Romeike, 1978; Goldmann et al., 1990) are reported in several solanaceous plants that possess relatively high TR-II activities in their cultured roots (Hashimoto et al., 1992), the role of TR-II is not clear in many other species in which ψ-tropine derivatives have been reported to be present only in small quantities.

B. Assay

Because most crude enzyme extracts from tropane alkaloid producing plants contain both TR activities, it is important to distinguish between the two activities in the enzyme assays. This can be best accomplished by gas–liquid chromatography (GLC) separation of tropine and ψ-tropine. The reaction mixture of 1 ml contains, at a final concentration, 4 mM tropinone, 2 mM NADPH, 100 mM potassium phosphate buffer (pH 5.8) and enzyme. After incubation for 1 h at 30°C, the reaction is stopped by the addition of 100 μl of 25% NH₄OH after which 1 ml of the alkali solution is transferred to an Extrelut-1 column (Merck). Alkamines are then eluted from the column by 6 ml of chloroform and the chloroform extract is evaporated to dryness. The dry residue is then dissolved in 50 μl of the dioxane solution containing 0.1% (v/v) nicotine as an internal standard. Alkaloids are separated by GLC equipped with an apolar capillary column (e.g. OV-101) at 120°C. The order of elution is tropinone, tropine, ψ-tropine and nicotine, and some tailing of the alkaloid peaks is observed. With a polar capillary column (e.g. PEG-20M), alkaloids are separated without displaying the peak tailing, but the column deteriorates quickly and irreversibly under present conditions.

After two TRs are separated by column chromatography (see beow), the assay method of choice is to measure the consumption of NADPH at 340 nm (Koelen and Gross, 1982; Dräger et al., 1988). The concentration of NADPH is lowered to 0.2 mM to maintain low background absorbance. The reference reaction without tropinone should be always included to offset non-enzymatic decomposition of NADPH in the relatively acidic buffer at 30°C. The molar absoption coefficient of 6200 is used to calculate the amount of NADPH consumed.

C. Purification

TR-I is much more unstable than TR-II in a simple buffer containing reducing agent (Koelen and Gross, 1982; Dräger et al., 1988). Addition of 10% gycerol to the buffer helps stabilise both reductases to some extent, and addition of 0.1 mM NADPH is essential for stability of TR-I during purification (Hashimoto et al., 1992). After precipitation between 45% and 75% saturation of ammonium sulphate, the two TRs are separated completely and reproducibly by a DEAE-Sepharose Fast Flow column with a KCl gradient from 0 M to 0.4 M, and then are separately purified by a series of column chromatography on hydroxyapatite, Phenyl-Superose HR 10/10 and Mono P (Hashimoto *et al.*, 1992). By this procedure, TR-I and TR-II have been purified from cultured roots of *Hyoscyamus niger* 1800-fold and 2600-fold, respectively, to apparent homogeneity.

D. Properties and Regulation

The pH optima for reduction reaction catalysed by both TRs are between 5.8 and 6.8 (Koelen and Gross, 1982; Dräger *et al.*, 1988; Couladis *et al.*, 1991; Hashimoto *et al.*, 1992). TR-I isolated from *Datura stramonium* is reported to catalyse the oxidation of tropine at somewhat alkaline conditions with pH optimum of 9.5 (Koelen and Gross, 1982) but the reverse reaction has not been demonstrated for TR-II. The K_m values for tropinone are significantly lower in TR-II (35 μM; Dräger *et al.*, 1988) than in TR-I (830 μM for *Datura* enzyme and 340 μM for *Hyoscyamus* enzyme; Koelen and Gross, 1982 and Hashimoto *et al.*, 1992). The higher affinity of TR-II toward tropinone might restrict the *in vivo* tropinone pool from being utilised efficiently for hyoscyamine synthesis. Both reductases reduce not only tropinone but also several other ketones with a six-membered ring system, but the substrate specificities of two TRs are distinct (Dräger *et al.*, 1988; Hashimoto *et al.*, 1992). For example, TR-II is highly active on *N*-alkyl-4-piperidones but TR-I is nearly inactive.

TR-I is most active in branch roots with small or no activity present in other plant organs of *A. belladonna*, *H. niger* and *D. stramonium*, the results consistent with the active alkaloid biosynthesis in the developmentally young roots (Hashimoto *et al.*, 1992). On the other hand, although strong TR-II activity is found in branch roots, other plant organs (especially stems) often show considerable TR-II activities (Hashimoto *et al.*, 1992).

IV. HYOSCYAMINE 6β-HYDROXYLASE

A. Reaction

Hyoscyamine 6β-hydroxylase (H6H; EC 1.14.11.11) was initially discovered as a 2-oxoglutarate-dependent dioxygenase that catalyses the 6β-hydroxylation of hyoscyamine in the biosynthetic pathway leading to scopolamine (Hashimoto and Yamada, 1986). The reaction product 6β-hydroxyhyoscyamine is then epoxidised *in vivo* to scopolamine by the removal of 7β-hydrogen (Leete and Lucast, 1976; Hashimoto and Yamada, 1989). Although H6H also shows a low level (about 2% of the hydroxylation activity) of this epoxidation activity, the low *in vitro* activity seemed insufficient to explain the normally very low accumulation of 6β-hydroxyhyoscyamine in scopolamine-producing plants (Hashimoto *et al.*, 1989a; our unpubl. res.). Our recent experiment using transgenic tobacco plants expressing *H. niger* H6H gene indicates, however, that H6H functions *in vivo* as an efficient bifunctional enzyme that is responsible for the conversion of hyoscyamine to scopolamine by way of 6β-hydroxyhyoscyamine in plants (our unpubl. res.; see below for more details).

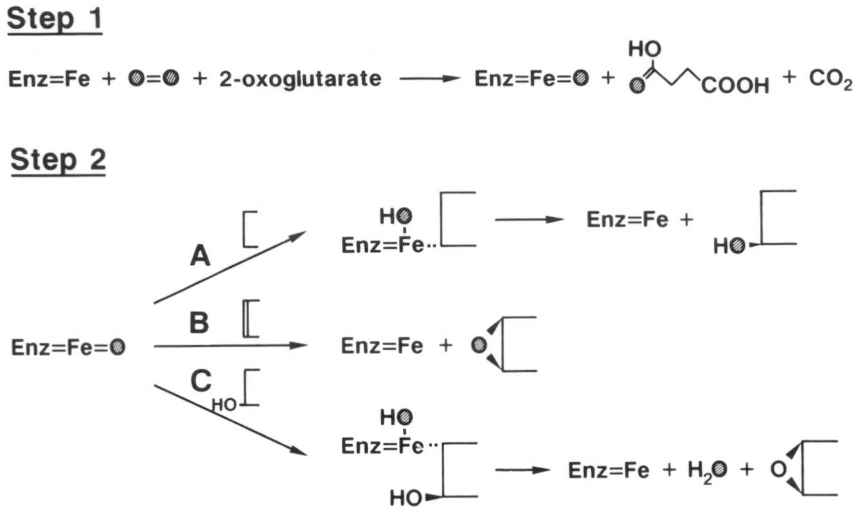

Step 1

Enz=Fe + ⦿=⦿ + 2-oxoglutarate ⟶ Enz=Fe=⦿ + [structure: HO—⦿—COOH] + CO₂

Step 2

FIG. 14.2. Possible reaction mechanism of hydroscyamine 6-β-hydroxylase.

6,7-Dehydrohyoscyamine had been proposed to be a precursor of scopolamine *in vivo*, based on its conversion to scopolamine when fed to plants (Foder *et al.*, 1959). A feeding experiment with [6-*hydroxy*-^{18}O]6β-hydroxyhyoscyamine, however, demonstrated that the unsaturated alkaloid, which has not been isolated from plants, is not a precursor of scopolamine (Hashimoto *et al.*, 1987). Superfluous oxidation reaction catalysed by H6H is responsible for the coversion of exogenously fed 6,7-dehydrohyoscyamine to scopolamine (Hashimoto and Yamada, 1987).

B. Assay

Two assay methods are available; one is specific to H6H, and the other is generally applicable to all 2-oxoglutarate-dependent dioxygenases.

The H6H-specific method measures the formation of 6β-hydroxyhyoscyamine with GLC. The reaction mixture of 1 ml is made up with 50 mM Tris-HCl buffer (pH 7.8 at 30°C), 0.4 mM ferrous sulphate, 4 mM sodium ascorbate, 1 mM 2-oxoglutaric acid (neutralised with NaOH before use), 2 mg ml^{-1} catalase (e.g. C-10 of Sigma) and the enzyme. For crude enzyme preparations, inclusion of 5% (v/v) acetone in the assay mixture increases the enzyme activity by up to 40%. The reaction is stopped after 1 h of incubation at 30°C by the addition of 100 μl of 1 M sodium carbonate buffer (pH 10.5). Then, 1 ml of this alkaline reaction mixture is loaded on an Extrelut-1 column (Merck) to which 6 ml of chloroform is added. The chloroform extract is evaporated to dryness and then dissolved in 100 μl of a mixture of 1,4-dioxane and *N,O*-bis(trimethylsilyl) acetamide (4:1; v/v) that contains tricosane at a concentration of 0.5 mg ml^{-1} as the internal standard. The trimethylsilyl-derivatisation reaction usually completes within 30 min at room temperature. The resulting derivatised alkaloids are analysed with GLC (Hashimoto and Yamada, 1986).

The other method measures the decarboxylation of 2-oxoglutarate. The composition of the assay mixture (1 ml) is the same as described above except that cold 2-oxoglutaric acid is replaced with 2-oxo[1-^{14}C]glutarate (1 mM, 2 kBq ml^{-1}). This mixture is incubated in stoppered test tubes (18 × 105 mm). A wire, to which a 1 × 4-cm rectangle of Whatman 3MM filter paper moistened with NCS reagent (Amersham) has been attached, is inserted through the stopper of the tube. The reaction was stopped by an injection of 200 μl of 25% (w/v) trichloroacetic acid. The tubes then are shaken at 30°C for 30 min to release the $^{14}CO_2$ from the reaction mixture. The filter paper carrying the $^{14}CO_2$ is transferred to a vial containing scintillation cocktail and the radioactivity determined. Under standard reaction conditions, approximately 1:1 stoichiometry is observed between the hydroxylation of hyoscyamine and the decarboxylation of 2-oxoglutarate (Hashimoto and Yamada, 1986). Although 2-oxoglutarate-dependent dioxygenases may catalyse decarboxylation of 2-oxoglutarate without subsequent hydroxylation, the rate of such uncoupled decarboxylation is usually less than a few percent of the rate in the complete reaction (e.g. Myllylä *et al.*, 1984).

C. Purification

H6H has been purified about 380-fold to homogeneity by a five-step process from cultured roots of *Hyoscyamus niger* (Yamada *et al.*, 1990). The purification sequence includes ammonium sulphate fractionation, open-column chromatography on Butyl-Toyopearl and DEAE-Toyopearl, and HPLC on Phenyl-Superose and hydroxyapatite. To stabilise the enzyme, it is important to include 30% (v/v) glycerol in the chromatography buffers. From 500 g (fresh weight) of cultured roots, about 20 μg of homogeneous H6H has been obtained.

Immunoaffinity column chromatography has also been used to purify H6H. Monoclonal antibody mAb5 interacts with H6H near the active site and recognises specifically the H6H proteins, either in native or denatured forms, from various scopolamine-producing plants (Hashimoto *et al.*, 1991). Purified mAb5 antibody is covalently attached to Sepharose CL-4B resin by cyanogen bromide activation and relatively crude enzyme preparations are applied to the antibody affinity column. After washing the column with a phosphate buffer, enzymatically active, nearly homogeneous H6H is specifically eluted with 10 mM borate buffer (pH 10.5). Recovery of H6H protein (220 μg from 750 g roots) is considerably higher than the conventional purification procedures, but the specific activity of the immunoaffinity-purified H6H is lower than that achieved by the conventional methods because of significant inactivation of the enzyme during the elution with the alkaline buffer.

D. Properties

The molecular weight of *H. niger* H6H deduced from the cDNA sequence is 39 kDa (Matsuda *et al.*, 1991). Because the native enzyme exhibits a molecular weight of 41 kDa by gel filtration (Hashimoto and Yamada, 1987), H6H must act catalytically as a monomer.

H6H belongs to the enzyme family of 2-oxoglutarate-dependent dioxygenases that require 2-oxoglutarate, Fe^{2+}, ascorbate and molecular oxygen for catalysis. The amino acid sequence of *H. niger* H6H has several regions homologous to the sequences of several other proteins including another member of this enzyme family, deacetoxy-cephalosporin C synthase (Matsuda *et al.*, 1991).

The most intriguing property of H6H is the multiple reactions it catalyses. Besides the 6β-hydroxylation of various hyoscyamine derivatives, the enzyme incorporates molecular oxygen into the double bond of 6,7-dehydrohyoscyamine, thereby ep-oxidising the synthetic alkaloid to scopolamine, as well as catalysing, although at a low rate, the formation of scopolamine by the removal of 7β-hydrogen from 6β-hydroxyhyoscyamine (Hashimoto and Yamada, 1987a,b,c,d; Yamada and Hashimoto, 1989; Hashimoto *et al.*, 1989; our unpubl. res.). These three types of reactions (Fig. 14.2) may be explained by postulating a highly reactive, ferryl enzyme inter-mediate, which may be formed when H6H oxidatively decarboxylates 2-oxoglutarate (Step 1). As reported for other dioxygenases of this class, H6H is expected to catalyse a low level of decarboxylation of 2-oxoglutarate without subsequent oxidation reaction in Step 2, and ascorbate may be required as a specific alternative oxygen acceptor in such uncoupled reaction cycles. It has been suggested that the oxidising ferryl species facilitates the homolytic abstraction of the proton (Hanauske-Abel and Günzler, 1982). With hyoscyamine as the substrate, the enzyme preferentially cleaves the C_6-H_β bond, followed by the insertion of the ferryl oxygen, resulting in hydroxyla-tion at the C-6 (Pathway A). The carbon radical produced from the homolytic cleavage of the C_7-H_β bond of 6β-hydroxyhyoscyamine would be attacked by the adjacent 6β-hydroxy oxygen to give the epoxide, scopolamine (Pathway C). Cleavage of the C_7-H_β bond would take place slowly because the ferryl oxygen of H6H must be positioned much closer to the C-6 than to the C-7 of the substrate. Another route to epoxide formation would be direct insertion of the ferryl oxygen into the 6,7-double bond of 6,7-dehydrohyoscyamine, which also would produce scopolamine (Pathway B).

Usually low accumulation of 6β-hydroxyhyoscyamine in scopolamine-producing plants raises an important question as to whether *in vitro* low epoxidation activity catalysed by H6H (Pathway C) is solely, or primarily, responsible for scopolamine formation *in vivo*. It has been recently demonstrated, however, that the *H. niger* H6H gene expressed in tobacco under the control of the cauliflower mosaic virus 35S promoter enables efficient conversion of exogenously supplied hyoscyamine and 6β-hydroxyhyoscyamine to scopolamine (our unpubl. res.). It therefore seems plausible that H6H acts *in vivo* as a bifunctional enzyme that catalyses two sequential oxidation reactions from hyoscyamine to scopolamine.

E. Regulation

H6H mRNA is abundant in cultured roots and present in plant roots, but absent in leaves, stems and cultured cells of *H. niger* (Matsuda *et al.*, 1991). Immunohisto-chemical studies using antibody mAb5 further localised H6H protein to the pericycle of the root (Hashimoto *et al.*, 1991). This pericycle-specific localisation of H6H protein provides an anatomical explanation for the high biosynthetic potential of cultured roots, and also suggests that alkaloids produced in the pericycle cells of the branch root are moved to the neighbouring xylem and then translocated to the aerial part where alkaloids accumulate.

REFERENCES

Couladis, M. M., Friesen, J. B., Landgrebe, M. E. and Leete, E. (1991). *Phytochemistry* **30**, 801–805.

Dräger, B., Hashimoto, T. and Yamada, Y. (1988). *Agric. Biol. Chem.* **52**, 2663–2667.

Evans, W. C. (1980). *In* "Proceedings of the 4th Asian Symposium on Medicinal Plants and Spices", Bangkok, Sept. 1980, pp. 125–135.

Feth, F., Arfmann, H.-A., Wray, V. and Wagner, K. G. (1985). *Phytochemistry* **24**, 921–923.

Fodor, G., Romeike, A., Janzso, G. and Koczor, I. (1959). *Tetrahedron Lett.* **7**, 19–23.

Friesen J. B. and Leete, E. (1990). *Tetrahedron Lett.*, **31**, 6295–6298.

Goldmann, A., Milat, M.-L., Ducrot, P.-H., Lallemand, J.-Y., Maille, M., Lepingle, A., Charpin I. and Tepfer, D. (1990). *Phytochemistry* **29**, 2125–2127.

Hanauske-Abel, H. M. and Günzler, V. (1982). *J. Theor. Biol.* **94**, 421–455.

Hashimoto, T. and Yamada, Y. (1986). *Plant Physiol.* **81**, 619–625.

Hashimoto, T. and Yamada, Y. (1987). *Eur. J. Biochem.* **164**, 277–285.

Hashimoto, T. and Yamada, Y. (1989). *Agric. Biol. Chem.* **53**, 863–864.

Hashimoto, T. and Yamada, Y. (1991). *In* "Methods in Plant Biochemistry" (K. Hostettmann, ed.). Academic Press, London (in press).

Hashimoto, T., Kohno, J. and Yamada, Y. (1987). *Plant Physiol.* **84**, 144–147.

Hashimoto, T., Kohno, J. and Yamada, Y. (1989a). *Phytochemistry* **28**, 1077–1082.

Hashimoto, T., Yamada, Y. and Leete, E. (1989b). *J. Am. Chem. Soc.* **111**, 1141–1142.

Hashimoto, T., Yukimune, Y. and Yamada, Y. (1989c). *Planta* **178**, 123–130.

Hashimoto, T., Yukimune, Y. and Yamada Y. (1989d). *Planta* **178**, 131–137.

Hashimoto, T., Mitani, A. and Yamada, Y. (1990). *Plant Physiol.* **93**, 216–221.

Hashimoto, T., Hayashi, A., Amano, Y., Kohno, J., Iwanari, S., Usuda, S. and Yamada, Y. (1991). *J. Biol. Chem.* **266**, 4648–4653.

Hashimoto, T., Nakajima, K., Ongena, G. and Yamada, Y. (1992). *Plant Physiol.*, in press.

Hibi, N., Fujita, T., Hatano, M. Hashimoto, T. and Yamada, Y. (1992). *Plant Physiol.*, in press.

Jindra, A. and Staba, E. J. (1968). *Phytochemistry* **7**, 79–82.

Kaczkowski, J. (1964). *Bull. Acad. Polonaise Sci.* **12**, 375–378.

Koelen, K. J. and Gross, G. G. (1982). *Planta Med.* **44**, 227–230.

Leete, E. (1964). *Tetrahedron Lett.* **24**, 16129–1622.

Leete, E. (1979). *Planta Med.* **36**, 97–112.

Leete, E. (1991). *Planta Med.* **56**, 339–352.

Leete, E. and Lucast, D. H. (1976). *Tetrahedron Lett.* **38**, 3401–3404.

Matsuda, J., Okabe, S., Hashimoto, T. and Yamada, Y. (1991). *J. Biol. Chem.* **266**, 9460–9464.

Mizusaki, S., Tanabe, Y., Noguchi, M. and Tamaki, E. (1971). *Plant Cell Physiol* **12**, 633–640.

Mizusaki, S., Kisaki, T., Noguchi, M. and Tamaki, E. (1972). *Phytochemistry* **11**, 2757–2762.

Myllylä, R., Majamaa, K., Günzler, V., Hanauske-Abel, H. M and Kivirikko, K. I. (1984). *J. Biol. Chem.* **259**, 5403–5405.

Robins, R. J., Parr, A. J. and Walton, N. J. (1991). *Planta* **183**, 196–201.

Romeike, A. (1978). *Bot. Notiser* **131**, 85–96.

Tiburcio, A. F. and Galston, A. W. (1986). *Phytochemistry* **25**, 107–110.

Tiburcio, A. F., Kaur-Sawheny, R., Ingersoll, R. B. and Galston, A. W. (1985). *Plant Physiol.* **78**, 323–326.

Waller, G. R. and Nowacki, E. K. (1979). "Alkaloid Biology and Metabolism in Plants". Plenum Press, New York.

Yamada, Y. and Hashimoto, T. (1989). *Proc. Japan Acad. Sci., Ser. B* **65**, 156–159.

Yamada, Y., Okabe, S. and Hashimoto, T. (1990). *Proc. Japan Acad. Sci., Ser.B* **66**, 73–76.

15 Abscisic Acid Metabolism

ANDREW D. PARRY

Department of Biological Sciences, University College of Wales, Aberystwyth, Dyfed SY23 3DA, Wales UK*

1. INTRODUCTION

Abscisic acid (ABA; Fig. 15.1) is a plant growth regulator perhaps best known for its function in the response of higher plants to environmental stresses such as drought and waterlogging (Addicott, 1983; Zeevaart and Creelman, 1988). Wright and Hiron (1969) found that the levels of ABA in wheat (*Triticum vulgare*) leaves increased rapidly and substantially following wilting. After the discovery that externally applied ABA could initiate stomatal closure in both intact leaves and epidermal strips (Mittel-heuser and Van Steveninck, 1969; Mansfield and Jones, 1971), and that stomatal functioning was impaired in 'wilty' ABA-deficient mutants (Tal and Imber, 1970), ABA was ascribed a role as an endogenous 'anti-transpirant'. Following water stress there is a rapid redistribution of ABA and an increase in its rate of biosynthesis that

*Current address: Department of Biological Sciences, University of Durham, DH1 3LE, UK.

METHODS IN PLANT BIOCHEMISTRY Vol. 9
ISBN 0–12–461019–6

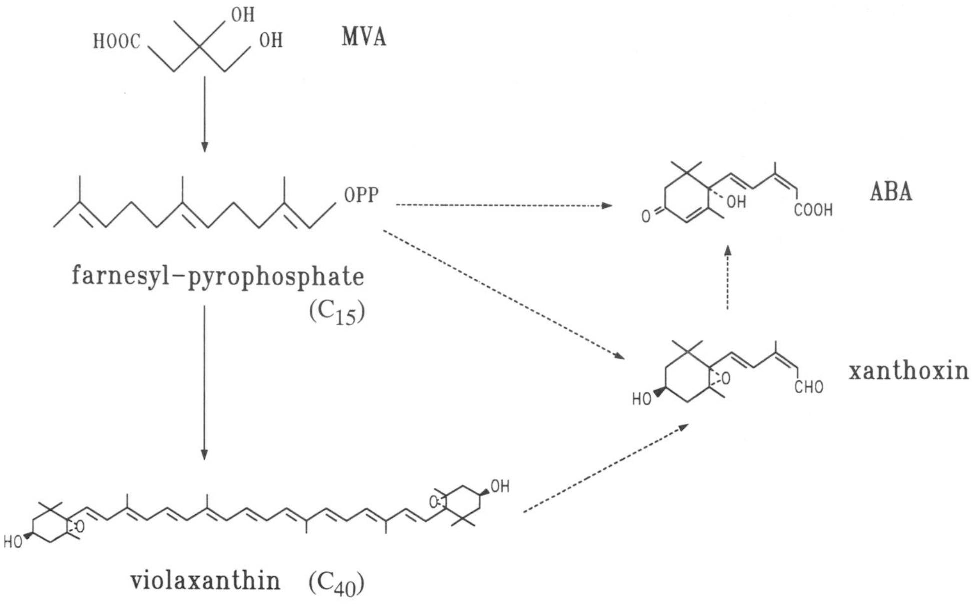

FIG. 15.1. Alternative pathways for the biosynthesis of ABA.

leads to stomatal closure (Pierce and Raschke, 1980, 1981; Cornish and Zeevaart, 1985). Roots may serve as an 'early warning system', sensing stress well before any loss of leaf turgor and producing ABA which acts as a chemical messenger, being transported to the shoots where it reduces transpiration (see Davies *et al.*, 1990). Maintaining turgor may also be facilitated by ABA increasing the hydraulic conductivity of the roots (e.g. Glinka, 1980).

In addition to these so-called 'rapid' effects, relying on alterations of ion flux, ABA has a key role in developmental events (such as the maturation of seeds) and physiological processes (such as the ability to tolerate desiccation, high and low temperatures, freezing, high salinities, etc.) through its ability to induce and suppress the expression of specific genes (Zeevaart and Creelman, 1988; Skriver and Mundy, 1990).

To date at least 27 clones of ABA-responsive genes have been isolated (Cohen and Bray, 1990; Kurkela and Franck, 1990; Skriver and Mundy, 1990). While the function of many of the products of these genes is unknown, evidence from amino acid sequences and homologies with known proteins suggests possible roles as seed storage proteins, osmoprotectants and regulatory proteins (e.g. inhibitors of amylase and protease).

Only five years ago a discussion of the enzymology of the biosynthesis of ABA would have been impossible. Such a discussion may still be premature but an almost complete pathway for ABA biosynthesis has now been elucidated, and likely control points identified (Parry and Horgan, 1991a). ABA levels can also be controlled via catabolism, and this has been extensively studied through the use of ^{14}C-labelled ABA.

II. ABSCISIC ACID BIOSYNTHESIS

ABA was first purified and chemically identified in 1965. It had the structure of a sesquiterpenoid and as such was expected to be derived from the fusion of three isopentenyl pyrophosphate (IPP) residues (see Addicott, 1983). Noddle and Robinson (1969) confirmed that ABA shared the normal terpenoid pathway by feeding [^{14}C]mevalonic acid (MVA), the precursor of IPP, to ripening fruit of avocado (*Persea gratissima*) and tomato (*Lycopersicon esculentum*), and finding incorporation of ^{14}C into ABA.

A. Direct Versus Indirect Pathway

There are two main routes whereby MVA could be converted to ABA (Fig. 15.1). The first is the so-called direct or C_{15} pathway involving the cyclisation of a C_{15} precursor, such as farnesyl pyrophosphate, and its conversion to ABA. The second indirect or C_{40} pathway (Fig. 15.1) involves the cleavage of a carotenoid, such as violaxanthin, to yield a C_{15} ABA precursor such as xanthoxin (Xan).

Despite considerable effort little progress was made over the next 20 years, MVA remained the only identified precursor of ABA and the stereochemistry of its biosynthesis was compatible with the operation of both direct and indirect pathways. The mid- to late 1980s to the present have been described as the 'renaissance' for research into ABA biosynthesis (Zeevaart and Creelman, 1988). Overwhelming evidence in support of an indirect pathway has now accumulated. Evidence against the operation of a carotenoid pathway in avocado fruit (see Milborrow, 1983) has been criticised and alternative explanations presented for the results obtained (Parry and Horgan, 1991a). The major evidence in support of an indirect pathway will be briefly discussed.

1. Heavy oxygen labelling experiments

ABA extracted from stressed leaves of *Phaseolus vulgaris* and *Xanthium strumarium* incubated in the presence of $^{18}O_2$ was labelled predominantly with only one ^{18}O, located in the carboxyl group (Creelman and Zeevaart, 1984). Incorporation into the ring oxygens also occurred but to a much lesser degree (Creelman *et al.*, 1987). The use of H_2 ^{18}O proved the other carboxyl oxygen to be derived from water (Creelman *et al.*, 1987). These findings were consistent with ABA arising from an apo-carotenoid. The oxygenase-mediated cleavage of a xanthophyll such as violaxanthin would give rise to Xan labelled with ^{18}O in the aldehyde group. Dehydrogenase-mediated oxidation of Xan to ABA, utilising water as an oxygen source, would produce ABA labelled with one ^{18}O atom in the carboxyl group. The ring oxygens of the xanthophylls are derived from molecular oxygen (see Britton, 1982, 1988) but as large pools of xanthophylls such as violaxanthin and neoxanthin exist in light-grown leaves, with low turnover rates, incorporation into the ABA-precursor ring oxygens would be slow. ABA extracted from other tissues incubated in the presence of $^{18}O_2$, such as non-stressed leaves, roots, fruit and embryos, showed a similar labelling pattern, suggesting the existence of a universal ABA biosynthetic pathway (Creelman *et al.*, 1987; Gage *et al.*, 1989; Zeevaart *et al.*, 1989).

Li and Walton (1987) manipulated the xanthophyll cycle in *Phaseolus* leaves to replace the epoxide oxygens of 40–45% of the violaxanthin with ^{18}O *in vivo*. Analysis of the extracted stress-induced ABA revealed that 10–15% was labelled in the 1'-hydroxyl group, and had therefore been synthesised from the labelled violaxanthin.

2. Carotenoid-deficient plants.

The best characterised carotenoid-deficient mutants are the viviparous (*vp*) ones of maize (*Zea mays*). These germinate precociously on the ear and some (*vp2, vp5, vp7* and *vp9*) have blocks in the carotenoid biosynthetic pathway between phytoene and the xanthophylls (see Koornneef, 1986). Neill *et al.* (1986) showed that, in addition to their xanthophyll deficiency, the mutant embryos had a much reduced ABA content (6–16%) compared to the wild-type. Chemical inhibition of carotenogenesis in normal plants, by fluridone and norflurazon, can produce phenotypic copies of mutants such as *vp5*, causing ABA deficiency and inducing vivipary (Fong *et al.*, 1983).

The ABA-deficient *aba* mutant of *Arabidopsis thaliana* contains abnormally low levels of antheraxanthin, violaxanthin and neoxanthin, presumably due to an inability to convert zeaxanthin, which accumulates in the mutant, to antheraxanthin (J. A. D. Zeevaart, MSU-DOE Plant Res. Lab., MI, USA, pers. commun.; I. B. Taylor, University of Nottingham School of Agriculture, Leics., UK, pers. commun.).

3. Correlative changes in ABA and xanthophylls

In light-grown leaves it has proved impossible to correlate changes in the levels of ABA with those of the xanthophylls, due to the large ratio of xanthophylls:ABA (at least 50:1; Norman *et al.*, 1990; Parry *et al.*, 1990a). Gamble and Mullet (1986) observed a significant decrease in the levels of violaxanthin, comparable with increases in ABA,

TABLE 15.1. The effects of water stress on xanthophyll and ABA metabolism in leaves of etiolated *Phaseolus vulgaris* and roots of hydroponically-grown *Lycopersicon esculentum*.

Compound	*Phaseolus* leaves		*Lycopersion* roots	
	Non-stressed	Stressed	Non-stressed	Stressed
	(nmol g^{-1} fresh weight)		(nmol g^{-1} fresh weight)	
t-Neoxanthin[a]	1.6	1.6	0.188	0.120
c-Neoxanthin[b]	4.1	1.3	0.043	0.022
t-Violaxanthin	26.0	17.6	0.157	0.093
c-Violaxanthin	0.6	0.5	0.013	0.007
t-Antheraxanthin	7.1	6.3	0.010	0.007
Lutein	24.9	24.3	0.040	0.046
Xanthophylls cleaved:	12.6		0.156	
ABA	0.06	4.20	0.004	0.096
PA	0.22	4.69	0.001	0.001
DPA	0.57	0.75	0.008	0.033
ABA synthesised:	8.8		0.118	

[a] *t* = all-*trans*.
[b] *c* = 9/9'-*cis*.

following dehydration of fluridone-treated etiolated *Hordeum vulgare* seedlings. Li and Walton (1990a), and subsequently Parry *et al.* (1990a), found that etiolated leaves of *Phaseolus vulgaris* had much reduced xanthophyll levels (15–20% of light-grown) but retained the ability to synthesise large amounts ABA in response to water stress. Under these circumstances it was possible to show a stoichiometric relationship between the amount of ABA made, measured as increases in the levels of ABA plus its two main metabolites phaseic acid and dihydrophaseic acid, and the losses of certain xanthophylls (all-*trans*- and 9-*cis*-violaxanthin and 9'-*cis*-neoxanthin). Similar correlations have now been obtained in both soil-grown and hydroponic roots of *Lycopersicon esculentum* (Parry *et al.*, 1991a; Table 15.1). These experiments are discussed in more detail below.

4. *Heavy water labelling experiments*

The extent of deuteration of violaxanthin, neoxanthin, lutein and ABA extracted from etiolated leaves of *Phaseolus* grown on 50% D_2O was very similar (12–18%; Parry *et al.*, 1990a). Whilst not excluding the possibility of a direct pathway, these results were consistent with the operation of an indirect one. Additional experiments aimed at investigating the interrelationship of the xanthophylls and the origin of ABA (Parry and Horgan, 1991b) are discussed below.

It appears, therefore, that in higher plants there is a universal pathway for the biosynthesis of ABA, involving the cleavage of specific xanthophylls. The biosynthesis of the carotenes from MVA is well understood (see Goodwin, 1979; Britton, 1988), however little is known about the biosynthesis of the major leaf xanthophylls from the carotenes, except that the introduction of oxygen groups is believed to occur as the final stage of their biosynthesis (Britton, 1982, 1988). The interconversion of all-*trans*-violaxanthin, antheraxanthin and zeaxanthin via the xanthophyll cycle has been intensively studied (Yamamoto, 1979), but it is not clear whether or not the *de novo* synthesis of violaxanthin and antheraxanthin takes place through the functioning of this cycle (Britton, 1982). Almost nothing is known about the biosynthesis of allenic xanthophylls such as neoxanthin, but a possible route from zeaxanthin via epoxides such as violaxanthin has been proposed (Goodwin, 1980). In a cell-free system from the alga *Amphidinium carterae* [¹⁴C]zeaxanthin was converted to all-*trans*-neoxanthin but no intermediates were isolated (Swift *et al.*, 1982). For the purposes of this chapter the ABA biosynthetic pathway shall be assumed to begin with those xanthophylls shown to be converted into ABA.

B. Xanthophyll ABA precursors

The similarity between ABA and the end-groups of certain xanthophylls was first noted by Taylor and Smith (1967). Photo-oxidation of those xanthophylls possessing 5,6 epoxy- and 3 hydroxy-groups (such as violaxanthin and neoxanthin) gave rise to a neutral compound later identified as Xan which inhibited cress seed germination (Taylor, 1968; Taylor and Burden, 1970a). Xan was present in a variety of plant extracts together with its geometric isomer 2-*trans*-xanthoxin (*t*-Xan; Taylor and Burden, 1970b; Firn *et al.*, 1972). Xan is active in a range of bioassays, probably as a result of its conversion into ABA (see Section II.E), whereas *t*-Xan possesses little or no biological activity (Taylor and Burden, 1972; Addicott, 1983).

Taylor and Burden (1972) extracted xanthophylls from various tissues and cleaved them using mild chemical oxidation. Violaxanthin, neoxanthin, antheraxanthin and lutein epoxide all gave rise to Xan and *t*-Xan, and under these experimental conditions it appeared that the ratio of *t*-Xan:Xan obtained was determined by the ratio of all-*trans* to 9-*cis* xanthophyll isomers extracted. Feeding experiments utilising [13]C- and ^{14}C-labelled Xan and *t*-Xan have shown that *t*-Xan cannot act as a precursor to Xan or ABA (Taylor and Burden, 1973; Parry *et al.*, 1988). Similar results have been obtained using cell-free systems capable of synthesising ABA from Xan (Sindhu and Walton, 1987). It seems likely, therefore, that Xan and *t*-Xan have independent origins, and that *in vivo* isomerisation of *t*-Xan to Xan does not occur.

Thus, any endogenous Xan precursor would have to possess a *cis* configuration in the position corresponding to the 2,3 double bond of Xan (e.g. the 9,10 double bond of carotenoids). In photosynthetic tissues carotenoids are present mainly in the thermodynamically most stable all-*trans* configuration—leaf violaxanthin for example is 95–99% all-*trans*, but 9-*cis*-violaxanthin has been identified as a minor component (Li and Walton, 1990a; Parry *et al.*, 1990a). The 9,9'-di-*cis*-violaxanthin isomer has the potential to generate two molecules of Xan but its presence has not been detected in any leaf or root extracts (Parry *et al.*, 1990a). In contrast neoxanthin occurs predominantly as the 9'-*cis* isomer, with all-*trans*-neoxanthin making up only *c.* 2–5% of the total. The relative composition of leaf xanthophylls is similar in most higher plants (Goodwin, 1980). While the same xanthophylls are present in etiolated leaves and roots, their distribution is much more varied (Parry and Horgan, 1991b; Parry *et al.*, 1991a; Table 15.2).

These two potential Xan, and therefore ABA precursors, i.e. 9-*cis*-violaxanthin and 9'-*cis*-neoxanthin, and no others, have been identified in extracts of light-grown and

TABLE 15.2. Xanthophyll composition in leaves and roots of a variety of plants.

	t-Neo	*c*-Neo	*t*-Viola	*c*-Viola	Anthera	Lutein	Total
Light-grown leaves							
Lycopersicon	1.5	85.3	91.1	0.6	4.2	254.6	438
esculentum	(<1)	(19)	(21)	(<1)	(1)	(58)	
Phaseolus	0.7	84.1	94.5	0.3	10.2	246.7	436
vulgaris	(<1)	(19)	(22)	(<1)	(2)	(56)	
Pisum	4.8	138.5	168.9	1.3	7.2	355.2	675
sativum	(1)	(20)	(26)	(<1)	(1)	(52)	
Etiolated leaves							
Phaseolus	1.6	4.1	25.4	0.6	6.8	23.3	62
vulgaris	(3)	(7)	(41)	(<1)	(11)	(38)	
Lycopersicon	4.8	7.0	37.7	1.8	0.8	61.0	113
esculentum	(4)	(6)	(33)	(2)	(1)	(54)	
Roots							
Lycopersicon	0.80	0.04	0.29	0.02	0.01	0.03	1.2
esculentum	(67)	(3)	(24)	(2)	(<1)	(3)	
Zea mays	0.02	0.09	0.75	0.02	0.04	0.04	1.0
	(2)	(9)	(78)	(2)	(4)	(4)	

Given as nmol g^{-1} fresh weight with % of total xanthophyll, given in parentheses ( ).

etiolated leaves and roots of a variety of species. In addition 9'-*cis*-neoxanthin was identified together with ABA in undifferentiated callus of *Vinca rosea* (A. D. Parry and R. Horgan, unpubl. data).

The levels of 9-*cis*-violaxanthin in both light-grown and etiolated leaves are comparable with those of ABA, but while levels of ABA increase by 3 to 40-fold during stress, those of 9-*cis*-violaxanthin remain constant or show only minor decreases (Li and Walton, 1990a; Parry *et al.*, 1990a). If 9-*cis*-violaxanthin were an ABA precursor, either being cleaved directly to generate Xan or via its conversion to 9'-*cis*-neoxanthin which is cleaved, then it must have a rapid turnover rate (20–25 times an hour) to explain the observed increases in ABA levels. The possibility that 9-*cis*-violaxanthin occurs only as a result of artefactual isomerisation from the more abundant all-*trans* isomer cannot be discounted.

C. The Conversion of Violaxanthin to Neoxanthin

The stoichiometric decreases in xanthophyll levels which accompany ABA synthesis in water-stressed etiolated *Phaseolus* leaves are due to decreases in 9'-*cis*-neoxanthin (20–30% of total xanthophyll 'lost') and 9-*cis*-violaxanthin (<5%), as might be expected, but more importantly to decreases in all-*trans*-violaxanthin (65–80%). The relationship between these xanthophylls was investigated by Parry and Horgan (1991b).

Transfer of etiolated seedlings to the light results in a general stimulation of carotenoid synthesis, although the rates of accumulation for individual carotenoids

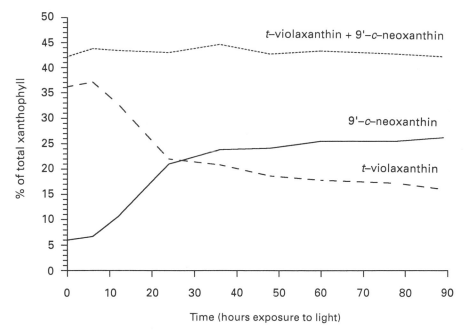

FIG. 15.2. Conversion of all-*trans*-violaxanthin to 9'-*cis*-neoxanthin in fluridone-treated etiolated seedlings of *Phaseolus vulgaris* following transfer to the light.

vary widely. This accumulation of carotenoids is controlled both by phytochrome and by the availability of chlorophyll and intact thylakoids (see Britton, 1988). Fluridone was used to prevent the synthesis of new carotenoids following the transfer of etiolated seedlings to light. After a time lag of 6–25 h a decrease occurred in the percentage of all-*trans*-violaxanthin concomitant with an increase in the percentage of 9′-*cis*-neoxanthin. The combined percentage of these two xanthophylls remained virtually constant (42.5 ± 1.3, *Lycopersicon*; 42.6 ± 1.4, *Phaseolus*) while the ratio of all-*trans*-violaxanthin:9′-*cis*-neoxanthin changed from *c.* 5:1 to *c.* 1:1 (Fig. 15.2).

These results suggest that during the transition of etioplasts into chloroplasts all-*trans*-violaxanthin acts as a precursor of 9′-*cis*-neoxanthin. This is consistent with previous hypotheses regarding neoxanthin formation (Goodwin, 1980). There were no significant changes in the percentage of either all-*trans*-neoxanthin or 9-*cis*-viola-xanthin during the conversion of all-*trans*-violaxanthin to 9′-*cis*-neoxanthin, making it impossible to predict which, if either, was an intermediate in this conversion.

In *Lycopersicon* roots carotenoids are present at levels only 0.1–0.2% of those in light-grown leaves and the major xanthophyll is all-*trans*-neoxanthin (65–80% of the total; Parry *et al.*, 1991a). In soil-grown plants dehydration lead to a mean increase in root ABA levels of $0.57 \, \text{nmol g}^{-1}$ fresh weight, and a mean decrease in xantho-phyll levels of $0.54 \, \text{nmol g}^{-1}$ fresh weight, most (90%) of which was due to all-*trans*-neoxanthin. Roots of hydroponically grown seedlings also synthesised ABA following stress and showed a comparable reduction in levels of all-*trans*-violaxanthin, 9′-*cis*- and all-*trans*-neoxanthin (Table 15.1). This is good evidence that, at least in roots, all-*trans*-neoxanthin is the intermediate between all-*trans*-violaxanthin and 9′-*cis*-neoxanthin.

To further investigate the relationship between violaxanthin, neoxanthin and ABA, extracts were made of etiolated *Phaseolus* seedlings grown on 50% D_2O, and then either kept in the dark on 50% D_2O or transferred to the light and fed with water. Following transfer a reduction in the amount of labelling occurred (Table 15.3). The extent of deuteration of lutein and violaxanthin fell by 44–50% compared to a fall of 32% for neoxanthin, even though the levels of neoxanthin had increased by over six-fold compared to increases of only 1.5 to 1.7-fold for the other xanthophylls (Parry and Horgan, 1991b). This indicated that violaxanthin and lutein were being synthesised from newly formed precursors while neoxanthin was derived from an existing, labelled

TABLE 15.3. Extent of deuteration of xanthophylls, ABA and PA extracted from *Phaseolus vulgaris* leaves, grown on 50% D_2O in the dark for 5 days and then either kept in the dark on 50% D_2O or transferred to the light and fed with water.

Treatment, after 5 days in the dark grown on 50% D_2O	Violaxanthin		Neoxanthin		Lutein		ABA	PA
	%D	Conc.[a]	%D	Conc.	%D	Conc.	%D	%D
Dark, 3 days, 50% D_2O	18	147	19	30	16	199	15	16
Light, 3 days, H_2O	9	256	13	188	9	299	11	n.d.

[a] nmol g^{-1} fresh weight.
n.d., not determined.
Adapted from Parry and Horgan (1991b).

compound, proposed to be violaxanthin. The extent of deuteration of ABA fell by 27% following transfer, consistent with it being formed from a cleavage product of neoxanthin.

The early part of the biosynthetic pathway of ABA can therefore be considered to be the conversion of all-*trans*-violaxanthin to 9'-*cis*-neoxanthin (Fig. 15.3). The nature of the enzymes responsible for this conversion (allene formation and isomerisation) are unknown, but may exist in a dual-enzyme complex as all-*trans*-neoxanthin does not accumulate during the conversion of all-*trans*-violaxanthin to 9'-*cis*-neoxanthin in developing chloroplasts, or decrease following water stress. The time-scale of the former conversion is a matter of days whilst water stress can lead to over a third of the violaxanthin in etiolated *Phaseolus* leaves being converted to ABA, via neoxanthin, in under 7 h.

Following exposure of etiolated leaves to the light the conversion of violaxanthin to neoxanthin may be stimulated directly by light, via phytochrome, or result from the formation of chloroplasts and the accumulation of chlorophyll. During water stress the stimulus may be direct or be due to a reduction in neoxanthin levels. Inhibitors of the cleavage reaction are required to investigate this further. This conversion of violaxanthin to neoxanthin must take place within the plastid, the exclusive subcellular

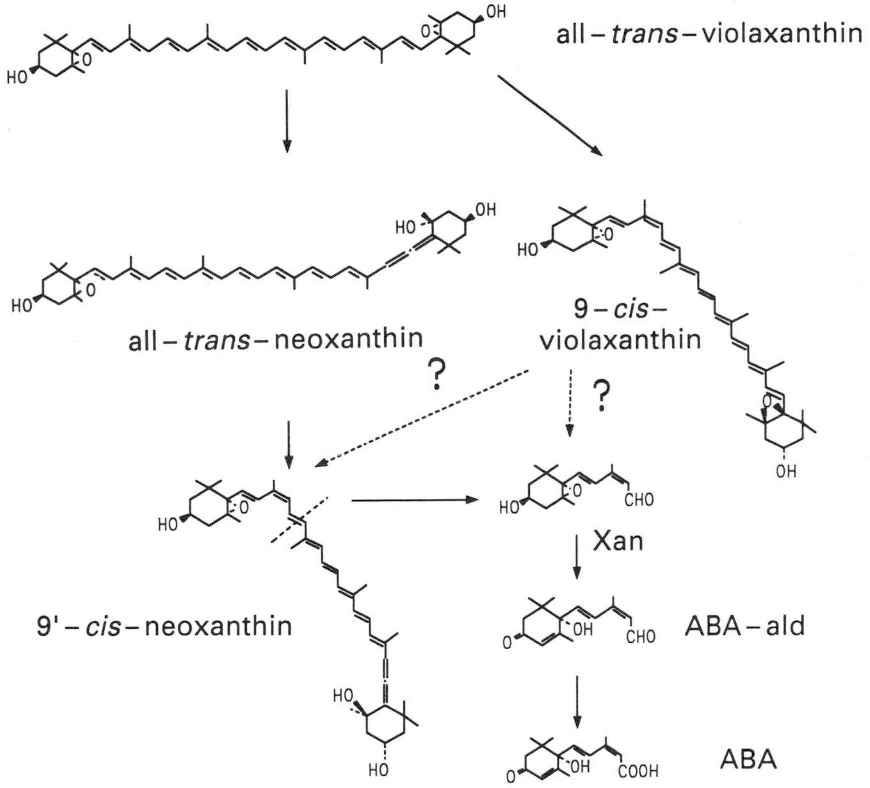

FIG. 15.3. Pathway of ABA biosynthesis, showing the main (only?) route from all-*trans*-violaxanthin, via all-*trans*-neoxanthin, to 9'-*cis*-neoxanthin.

site of carotenoid location (Goodwin, 1980). Preliminary experiments suggest that chloroplast envelopes are enriched in violaxanthin but that the actual levels are much lower than those necessary to generate the stress-induced ABA, suggesting that some thylakoidal xanthophylls are also utilised (A. D. Parry and R. Horgan, unpubl. data).

D. The Cleavage of 9′-cis-Neoxanthin

Taylor and Burden (1972) fed Xan to turgid shoots of *Lycopersicon* and observed 70-fold increases in ABA levels. Sindhu and Walton (1987) extracted 'Xan oxidising' activity from a variety of tissues (see later) and measured rates of conversion of Xan to ABA of up to 7000 ng $h^{-1} g^{-1}$ fresh weight, much greater than the observed *in vivo* rates. The amount of activity extracted was unaffected by water stress. These authors also commented that their attempts to load *Phaseolus* leaves with Xan failed due to the rapid conversion of Xan to ABA. Levels of ABA in non-stressed *Lycopersicon* leaves fed with ^{13}C-Xan were higher than those in non-fed stressed leaves (Parry *et al.*, 1988; Table 15.4). In the fed non-stressed leaves over 75% of the extracted ABA was labelled with ^{13}C, while in fed stressed leaves the ^{13}C content fell to 20–30%. Parry (1989) concluded that in stressed leaves the labelled Xan was being diluted by the increased synthesis of natural Xan.

Collectively the above results indicate that *in vivo* the rate of ABA biosynthesis is limited not by the conversion of Xan to ABA but by the production of Xan. A constitutive ability to convert Xan to ABA would explain the very low levels of Xan found in plant extracts (Parry *et al.*, 1990b). It seems most likely that the cleavage of 9′-cis-neoxanthin is the rate-limiting step.

In response to water stress ABA biosynthesis is accelerated and the stimulus for this appears to be a loss of cell turgor (Pierce and Raschke, 1980, 1981; Cornish and Zeevaart, 1985). Nothing is known about the transduction mechanism but this accumulation of ABA can be eliminated by inhibitors of nuclear transcription such as cordycepin and actinomycin D and by inhibitors of protein translation such as cycloheximide (e.g. Guerrero and Mullet, 1986; Stewart *et al.*, 1986). Treatment of turgid *Phaseolus* leaves with cycloheximide lead to a 50% reduction in ABA levels within 36 h, which was not the result of increased catabolism (Li and Walton, 1990b). The amount of 'Xan oxidising' activity isolated from plant tissues by Sindhu and Walton (1987) was unaffected by pre-treatment with cycloheximide. It appears that the

TABLE 15.4. The conversion of [^{13}C]Xan and [^{13}C]*t*-Xan to ABA in non-stressed and stressed leaves of *Lycopersicon esculentum*.

Treatment	[ABA] (nmol g^{-1} fresh weight)	% ^{13}C	% Incorporation
Non-stressed	1.2	–	–
Stressed	3.4	–	–
Non-stressed + ^{13}C-Xan	5.5	75.8	28.3
Stressed + ^{13}C-Xan	3.5	29.0	14.7
Non-stressed + ^{13}C-*t*-Xan	1.4	14.6	1.5
Stressed + ^{13}C-*t*-Xan	4.1	4.2	1.6

Adapted from Parry *et al.* (1988).

biosynthesis of ABA is regulated by the activity of an inducible enzyme, probably a dioxygenase which acts specifically to cleave 9'-*cis*-neoxanthin, and has a high turnover rate. Re-watering stressed plants leads to a rapid decrease in the rates of ABA synthesis (see Zeevaart and Creelman, 1988).

The alteration of plant gene expression in response to environmental stress has received much attention but until recently this was mainly restricted to the effects of anaerobic and high temperature stress (Sachs and Ho, 1986). The research effort being directed into drought-influenced gene expression is now substantial and many such changes have been observed (Quarrie, 1990; Skriver and Mundy, 1990). Changes in gene expression as a result of water stress may be due directly to water stress, such as those leading to increased levels of ABA, or may be secondary effects such as those observed in response to increased ABA levels (Ho, 1983; Skriver and Mundy, 1990).

Quarrie (1990) has made use of a rice genotype having a high capacity for water-stress-induced ABA synthesis in an attempt to identify proteins responsible for this increased synthesis, which would include the cleavage enzyme. ABA accumulation in this genotype begins approximately 30 min after the imposition of stress and therefore only proteins that are synthesised prior to this are of interest. Preliminary results indicate that a 38.8 kDa protein accumulates following stress, and is not induced by ABA. Cloning of mRNAs produced in the first 30 min after stress has begun and future work could involve the use of labelled oligonucleotide probes to investigate ABA biosynthesis or antisense technology to determine the function of any gene isolated.

The location of the cleavage enzyme has not been established, neoxanthin exists exclusively within the plastid while the enzymes responsible for the conversion of Xan to ABA are cytosolic (Sindhu and Walton, 1987). It seems likely that the enzyme will exist within the chloroplast, perhaps associated with the envelope, but it could also be attached to the outer chloroplast membrane, cleaving neoxanthin molecules that by some (stress-related?) mechanism have become exposed. Parry and Horgan (1991b) have concluded that the neoxanthin pool available to the cleavage enzymes is in equilibrium with the bulk of the neoxanthin produced by light-induced *de novo* synthesis, although it is likely to be spatially separate.

The stoichiometry observed by Li and Walton (1990a) and Parry *et al.* (1990a, 1991a) implies that one mole of xanthophyll is converted to one mole of ABA, and indicates that the cleavage reaction is a specific one, where 9'-*cis*-neoxanthin is cleaved across the 11', 12' double bond to generate Xan. If this is so then in addition to Xan a C_{25} allenic apo-aldehyde should also be produced (Fig. 15.4). Parry and Horgan (1991b) synthesised a range of allenic and epoxy apo-aldehydes ranging from C_{15} to C_{27} in an attempt to confirm the specificity of the cleavage. However none of these apo-aldehydes, or their related alcohols, acids or conjugates, were found in any extracts of light-grown and etiolated leaves or roots. In one experiment the levels of neoxanthin and violaxanthin in etiolated *Phaseolus* leaves fell by over 20 nmol in the 60 min after stress but still no apo-carotenoids were identified, with a limit of detection of *c*. 0.02 nmol.

Numerous possible apo-carotenoids have been identified in extracts of plant and animal tissues but knowledge of the biochemistry of carotenoid catabolism is extremely limited (Enzell, 1985). The majority of these apo-carotenoids are volatile (C_9–C_{13}) compounds but others range in size up to C_{30}, such as apo-8'-violaxanthal which has been extracted from orange peel (Molnar and Szabolcs, 1980). The bleaching of

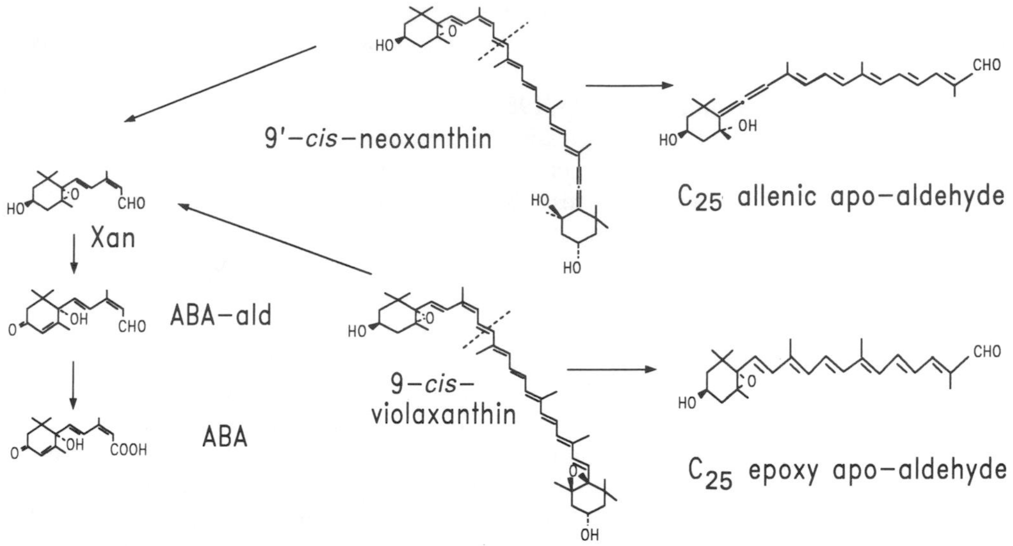

FIG. 15.4. Specific cleavage of 9'-*cis*-neoxanthin would, in addition to Xan, produce a C_{25} allenic apo-aldehyde. Similarly a C_{25} epoxy apo-aldehyde would be produced from 9-*cis*-violaxanthin.

carotenoid solutions by lipoxygenase (LOX) has been described frequently but the products of such degradations have not usually been identified/reported, although Xan and *t*-Xan were isolated from incubations of LOX and violaxanthin (Firn and Friend, 1972). Pigment bleaching by cytochrome *c* and peroxidases also takes place *in vitro* (Ben-Aziz *et al.*, 1971; Matile and Martinoia, 1982).

The bleaching by these enzymes is due to the coupled oxidation (co-oxidation) of the carotenoids with primary hydroperoxidation (using molecular oxygen) of poly-unsaturated fatty acids possessing a *cis,cis*-1,4-pentadiene moiety, like linoleic acid (Galliard and Chan, 1980). *In vitro* LOX rapidly cleaved neoxanthin and violaxanthin into small fragments ($<C_{13}$). If this reaction was interrupted by the antioxidant *tert*-butylated hydroxyquinoline (TBHQ) then low yields of C_{15} (including Xan and *t*-Xan), C_{25} and C_{27} apo-aldehydes were found (Parry and Horgan, 1991b). The relevance of such observations to the specific or non-specific degradation of caro-tenoids *in vivo* is uncertain.

In an attempt to measure *in vivo* bleaching activity, plant extracts were assayed for linoleic acid-dependent oxygen use, using a Clark-type oxygen electrode, and zeaxan-thin cleavage. These techniques are not specific for LOX but should give estimates of cleavage capacity. The results of these experiments are shown in Table 15.5. Etiolated leaves contain far more activity than light-grown ones, but light-grown ones and roots still possess far in excess of the activity necessary to account for the observed *in vivo* xanthophyll cleavage or apo-aldehyde disappearance. The precise enzymic nature of this activity is unknown, as is its subcellular localisation. LOX activity has been found in extracts from seeds, fruit, leaves and roots of many species (Galliard and Chan, 1980). Similarly peroxidase isozymes are widely distributed (Butt, 1980). Peroxidases are usually located within the cell wall, the cytosol and the vacuole, but have also been detected within the chloroplast (Butt, 1980). Little is known about the subcellular

TABLE 15.5. Oxygen utilisation and zeaxanthin cleavage by extracts of *Phaseolus vulgaris* leaves and *Lycopersicon esculentum* roots in the presence of linoleic acid.

Tissue			O_2 use (μmol min^{-1} g^{-1} fresh weight)	Zeaxanthin cleavage (nmol min^{-1} g^{-1} fresh weight)	Rate of xanthophyll cleavage *in vivo* (nmol min^{-1} g^{-1} fresh weight)
Etiolated *Phaseolus* leaves	(a)	Non-stressed	47.9 ± 5.3	506 ± 3	0.02–0.07
	(b)	Stressed	43.3 ± 4.9	395 ± 2	
Light-grown *Phaseolus* leaves	(a)	Non-stressed	0.50 ± 0.09	5.2	≤ 0.02
	(b)	Stressed	0.81 ± 0.19	7.4	
Lycopersicon roots			0.33	3.0	≤ 0.001

distribution of LOX, and there is conflicting evidence as to whether LOX activity is associated with the chloroplast (Galliard and Chan, 1980; Vernooy-Gerritsen *et al.*, 1984).

The possible role of LOX in ABA biosynthesis or apo-carotenoid degradation is being further investigated through the use of inhibitors. TBHQ has been used to inhibit LOX *in vitro* (Klein *et al.*, 1984; Parry and Horgan, 1991b) but as a free radical scavenger it is too non-specific to be of use *in vivo*. ETYA (5,8,11,14-eicosatetraynoic acid; Kuhn *et al.*, 1984) is an acetylenic fatty acid and acts as a suicidal substrate for LOX, irreversibly inactivating the enzyme. Parry and Horgan (unpubl. data) found that 20 μM ETYA inhibited 70–80% of the linoleic acid-dependent oxygen use and 35–55% of the zeaxanthin cleavage by extracts of *Phaseolus* leaves. Preliminary results from feeding ETYA to detached *Phaseolus* leaves have revealed an inhibition of only *c.* 20% of the potential cleavage capacity, which did not affect ABA or apo-carotenoid metabolism. Such inefficient inhibition may have resulted from poor uptake or rapid metabolism to non-inhibitory compounds.

A model system for the cleavage of 9'-*cis*-neoxanthin should be the formation of retinal from β-carotene. Surprisingly although the product/precursor relationship of β-carotene and retinal was discovered over 60 years ago (Ganguly, 1989), there is little known about this reaction other than that it makes use of molecular oxygen. Again there are two alternative cleavage mechanisms, specific (central) cleavage generating two moles of retinal per mole of carotene, and non-specific (excentric) cleavage where the carotene is cleaved at random, and retinal is just one of the products formed. A large amount of effort has gone into studying this reaction *in vitro* but the stoichiometry is still the subject of debate. In a recent book Ganguly (1989) has summed up the available evidence and on balance suggests that an excentric cleavage reaction is more likely, given the range of apo-carotenoids that have been isolated from animal tissues. The carotene cleavage enzyme has been partially purified and its specificity examined. It cleaves a range of apo-carotenals and apo-carotenols but shows little activity against carotenoids or apo-carotenoids possessing 5,6 epoxy groups (see Olson, 1983).

E. The Conversion of Xan to ABA

To summarise what has already been discussed about Xan, it can be formed by the cleavage of specific xanthophylls *in vitro*, has been identified in extracts of a variety of tissues, and can be converted to ABA by plants. If 9'-*cis*-neoxanthin is the immediate pre-cleavage precursor of ABA, as now seems likely (Parry and Horgan, 1991a), then Xan must be the *in vivo* C_{15} post-cleavage intermediate. Sindhu and Walton (1987) made cell-free systems from leaves of *Phaseolus vulgaris*, *Vigna radiata*, *Zea mays*, *Cucurbita maxima* and *Pisum sativum* and roots of *Phaseolus vulgaris* which had the ability to convert Xan to ABA. This 'Xan oxidising' activity was associated exclusively with cytosolic fractions and in potassium phosphate buffer (50 mM) had a pH optimum of between 7.0 and 7.5. Following acetone precipitation (4 vols) or dialysis there was a loss of activity, but if NAD or NADP were added then a three-to four-fold increase in activity compared to the crude extract was seen. This suggested that an inhibitor was present in the crude extract, and fractionation indicated that it had a chloroplastic origin.

Further experiments with leaves and cell-free systems of the wild-type and three ABA-deficient mutants of *Lycopersicon* showed that two of them, *flacca* (*flc*) and *sitiens* (*sit*), were unable to convert Xan to ABA (Parry *et al.*, 1988; Sindhu and Walton, 1988). To obtain 'Xan oxidising' activity from *Lycopersicon* leaves it was necessary to include 7.5 mM dithiothreitol (DTT) in the extraction buffer, and in contrast to other extracts dialysis or acetone precipitation led to a loss of activity that could not be regained by addition of NAD or NADP. Both *flc* and *sit* are single gene mutants, but with their mutations located on different chromosomes (see Rick, 1980). Combining cell-free extracts of *flc* and *sit* did not overcome their inability to convert Xan to ABA, suggesting that in some way the two mutations were affecting the same enzyme or multienzyme complex (Sindhu and Walton, 1988). The actual step of the pathway affected, and the immediate precursor to ABA, were identified by feeding a range of related compounds to these cell-free systems. Extracts of *flc* and *sit* were incapable of converting ABA-aldehyde to ABA, showing the last step of the pathway to involve the simple oxidation of the side-chain aldehyde to the carboxylic acid (Fig. 15.5).

Sindhu *et al.* (1990) have now fractionated the 'Xan oxidising' activity from *Phaseolus* leaves, by differential precipitation with acetone. The two activities have been termed xanthoxin oxidase (Xan oxidase), which catalyses the two-step reaction of Xan to ABA-ald, and abscisic aldehyde oxidase (ABA-ald oxidase), which converts ABA-ald to ABA. Xan oxidase has an absolute requirement for NAD/NADP while ABA-ald oxidase is the activity affected by the endogenous inhibitor. Preliminary kinetic characteristics have been obtained (apparent K_m and V_{max} for Xan oxidase, 3.6 μM and 0.19 nmol ABA-ald formed min^{-1} mg^{-1} protein; ABA-ald oxidase,

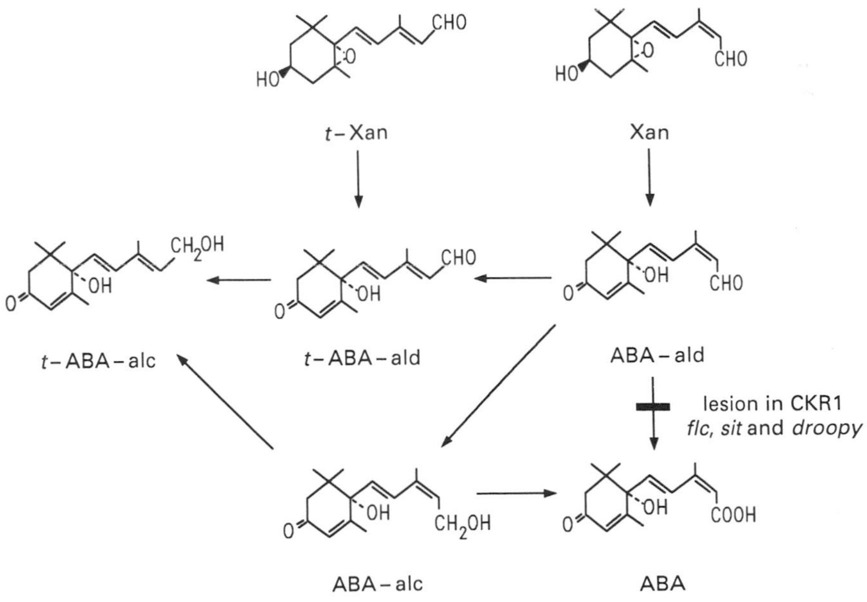

FIG. 15.5. The latter, C_{15}, part of the ABA biosynthetic pathway. In normal plants Xan is converted via ABA-ald to ABA, whereas mutants such as *flc* accumulate *t*-ABA-alc.

2.5–3.0 μM and 1.0–1.3 nmol ABA formed min^{-1} mg^{-1} protein). ABA-ald oxidase was estimated to have been purified 145-fold. ABA-ald oxidase from *Lycopersicon* has also been partially purified (I. B. Taylor, pers. commun.), by ion-exchange, gel filtration and affinity chromatography to give an active fraction with a molecular weight in the 100–200 kDa range.

Extracts of *flc* and *sit* leaves were, as expected, shown to be deficient in ABA-ald oxidase and not Xan oxidase (Sindhu *et al.*, 1990). When ABA-ald oxidase extracted from *Phaseolus* leaves was added to cell-free systems from *flc* and *sit* conversion of Xan to ABA was observed. Walker-Simmons *et al.* (1989) identified an ABA-deficient mutant of *Hordeum vulgare* (Az34) whose mutation is located in a gene controlling a molybdenum cofactor. This mutation results in the pleitropic deficiency in at least three molybdoenzymes, nitrate reductase, xanthine dehydrogenase and aldehyde oxidase. The mutant was found to lack aldehyde oxidase activity with a number of substrates including ABA-ald. Sindhu *et al.* (1990) failed to detect ABA-ald oxidase from either turgid or stressed leaves of Az34, indicating that ABA-ald oxidase may in fact be a molybdoenzyme.

In leaves of *flc* and *sit* a compound identified as 2-*trans*-ABA-alcohol (*t*-ABA-alc) was found to accumulate, especially after water stress, and was proposed to be an ABA precursor (Linforth *et al.*, 1987). Later feeding studies with [^{2}H]ABA-ald revealed that while wild-type and leaves converted ABA-ald to ABA those of *flc*, *sit* and the *droopy* potato mutant reduced and isomerised ABA-ald to *t*-ABA-alc (Taylor *et al.*, 1988; Duckham *et al.*, 1989; Fig. 15.5). This was further evidence for ABA-ald being the immediate ABA precursor. Parry *et al.* (1991b) characterised a new ABA-deficient mutant of *Nicotiana plumbaginifolia* (CKR1) as being similarly unable to convert ABA-ald to ABA. Wild-type *Lycopersicon* and *Nicotiana* leaves converted Xan to ABA and *t*-Xan to *t*-ABA-alc while leaves of *flc* and CKR1 converted both Xan and *t*-Xan to *t*-ABA-alc. Both ABA-alc and *t*-ABA-ald are converted to *t*-ABA-alc and so the order of the reduction and isomerisation is unclear (Linforth *et al.*, 1990; Fig. 15.5).

Mutants such as *flc*, *sit* and CKR1 are all ABA-deficient but not ABA-lacking; such a condition may well be lethal. An explanation for this leakiness may have been found by Rock and Zeevaart (1990a). Using ^{18}O$_2$ they have obtained evidence that some ABA-ald is reduced to ABA-alc which is then re-oxidised by a monooxygenase (this reaction can be inhibited by carbon monoxide) to ABA. This may operate as a minor pathway in normal plants as well as mutants blocked in the conversion of ABA-ald to ABA.

III. ABSCISIC ACID CATABOLISM

In contrast to the biosynthetic precursors of ABA, which have only recently been identified, the major catabolites of ABA were discovered in the first decade after its structural characterisation (Fig. 15.6). This resulted from the availability and extensive use of ^{14}C-labelled ABA. However, knowledge about the enzymology and regulation of ABA catabolism is scant, and little progress has been made since the publication of several major discussions of this subject (Milborrow, 1983; Loveys and Milborrow, 1984; Zeevaart and Creelman, 1988). The catabolism of ABA, and its resulting inactivation, takes place in two main ways, through oxidation and conjugation.

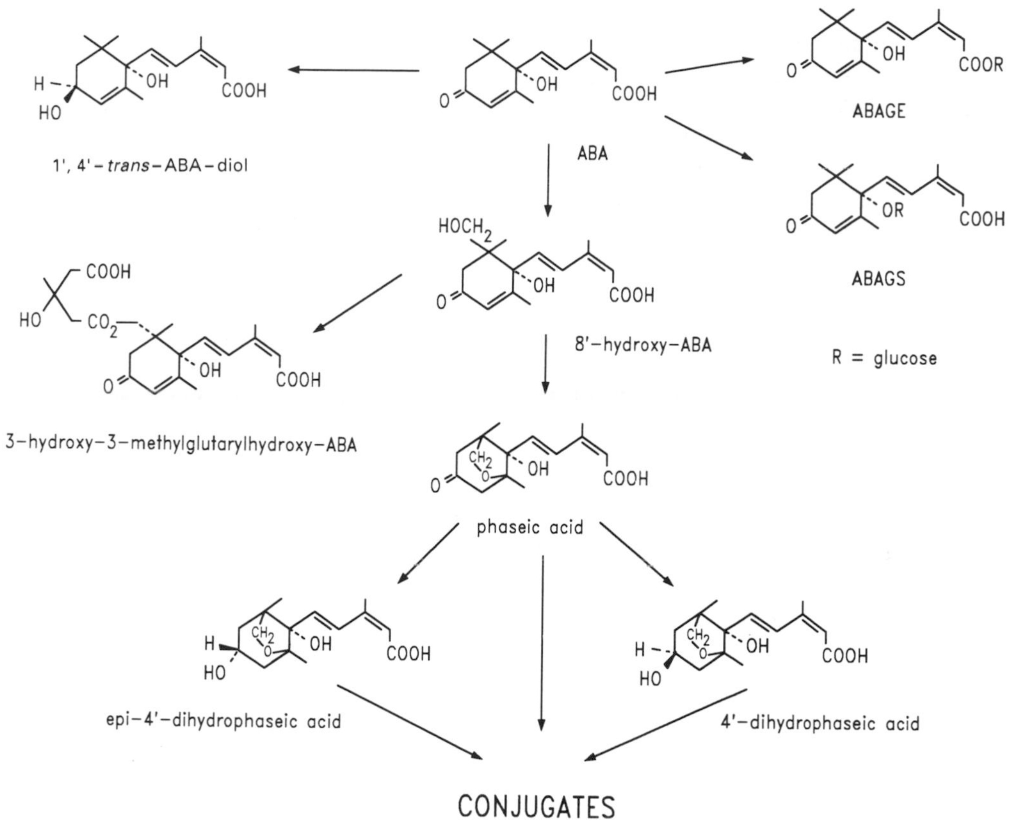

FIG. 15.6. Pathways of ABA catabolism.

A. Oxidation

Milborrow (see Loveys and Milborrow, 1984) was the first to feed [¹⁴C]ABA to plants and in 1969 identified 8-hydroxy-ABA as a metabolite of ABA. This compound has only rarely been identified in plant extracts because it spontaneously rearranges *in vitro* to phaseic acid (PA). This rearrangement to PA *in vivo* is probably carried out enzymically as the insertion of a proton at C-3′ of 8-hydroxy-ABA during cyclisation is stereospecific (Milborrow *et al.*, 1988).

Two acidic metabolites were isolated from excised axes of *Phaseolus* fed with [¹⁴C]ABA, one was identified as PA but the main one was dihydrophaseic acid (DPA), which accumulated to 100-fold higher levels than ABA (Walton and Sondheimer, 1972; Tinelli *et al.*, 1973). PA was suggested to be an intermediate between ABA and DPA, the latter being the major inactivation product (Tinelli *et al.*, 1973). Gillard and Walton (1976) obtained cell-free preparations from the liquid endosperm of *Echinocystis lobata* which were capable of metabolising ABA. The reaction products were again identified as PA and DPA, but acetylation of short-term incubations 'trapped' 8′-hydroxy ABA and provided further evidence for its role as an intermediate between ABA and PA. This crude preparation was fractionated by centrifugation to yield a particulate ABA-hydroxylating activity (converted ABA to PA) and a soluble PA-reducing activity (PA to DPA). The former activity showed requirements for molecular oxygen and NADPH, and was inhibited by carbon monoxide. These characteristics are similar to those of cytochrome P-450 monooxygenases found in animals. In addition the hydroxylating enzyme showed a high substrate specificity for ABA. Creelman and Zeevaart (1984) confirmed that this ABA-hydroxylating enzyme was an oxygenase by incubating leaves of *Xanthium*, which had undergone a cycle of dehydration and rehydration to stimulate ABA catabolism, in the presence of $^{18}O_2$. The extracted PA was labelled with one ^{18}O atom, located in the 8'-hydroxy group.

As well as DPA its epimer, 4'-epi-DPA, can be formed by reduction of PA and occurs in plant extracts at 2–50% of the levels of DPA (Loveys and Milborrow, 1984). Milborrow (1983) has commented on the unusual occurrence of two epimeric forms of the same compound in one tissue at the same time. Two possibilities were proposed to account for their biosynthesis: either one enzyme was capable of accepting PA in two positions at the active site, or two separate enzymes were present. Preliminary data were said to favour the latter theory.

The reduction of ABA to the 1',4'-ABA-diols has also been reported (e.g. Vaughan and Milborrow, 1987; Rock and Zeevaart, 1990b) and both 1',4'-*cis*- and 1',4'-*trans*-ABA-diols are present in plant tissues, usually at much lower levels than ABA (Vaughan and Milborrow, 1987; Parry *et al.*, 1988; Rock and Zeevaart, 1990b). The relationship between ABA and the ABA-diols is complicated as the *trans*-ABA-diol can also act as an ABA precursor (Vaughan and Milborrow, 1987; Parry *et al.*, 1988). It seems unlikely that the reduction of ABA to the diols serves as a major inactivation route.

B. Conjugation

The conjugation of ABA involves its covalent attachment to glucose. The first such conjugate was isolated from yellow lupin pods and identified as ABA-β-D-glucosyl ester (ABA-GE; Koshimizu *et al.*, 1968). A second conjugate, 1'-*O*-ABA-β-D-glucoside (ABA-GS), has been extracted from tomato shoots (Loveys and Milborrow, 1981). Both ABA-GE and ABA-GS are hydrolysed by alkali and estimates of total conjugated ABA are often made from the amount of ABA released by alkaline hydrolysis of an extract. The conjugation of ABA is an irreversible event (see Zeevaart and Creelman, 1988) with conjugates being sequestered in the vacuole (Bray and Zeevaart, 1985; Lehmann and Glund, 1986). Lehmann and Schutte (1980) partially purified (18-fold) an enzyme capable of glucosylating ABA from suspension cultures of *Macleaya microcarpa*. In phosphate buffer (33 mM) maximal activity was at pH 5.0, consistent with its operating within the vacuole. Conjugates of 8-hydroxy-ABA, PA, DPA and the 1',4,-ABA-diols have also been identified (Zeevaart and Creelman, 1988; Milborrow, 1990).

C. Regulation of Catabolism

Water stress results not only in an increase in ABA levels, but also in those of PA and DPA (e.g. Li and Walton, 1990a; Parry *et al.*, 1990a). The increase in ABA levels is due to a balance between increased biosynthesis and catabolism. Following rehydration the conversion of ABA to PA is further stimulated, resulting in a rapid return to pre-stress ABA levels (Pierce and Raschke, 1981; Zeevaart and Creelman, 1988). The rate of conjugation is not greatly affected by stress or recovery from stress (Zeevaart and Creelman, 1988). Oxidative catabolism appears to be a specific mechanism plants have evolved for the rapid inactivation of ABA, while conjugation is a general mechanism for getting rid of unwanted acids.

Cowan and Railton (1987) claimed that cycloheximide treatment of *Hordeum* leaves led to the inhibition of catabolism (both oxidative and conjugative), suggesting that continued protein synthesis was required. However the ABA levels in cycloheximide-treated non-stressed leaves of *Phaseolus* decreased due to the inhibition of biosynthesis and continued catabolism (Li and Walton, 1990b). The conversion of ABA to PA in aleurone layers of barley and wheat is stimulated by addition of ABA (Uknes and Ho, 1984). This stimulation can be blocked by inhibitors of transcription and translation, and it has been proposed that ABA induces the monooxygenase responsible for its hydroxylation to PA. Such an effect was not observed in other tissues (Uknes and Ho, 1984).

Oxidative catabolism of ABA is localised within the cytosol (Hartung *et al.*, 1982). The subcellular distribution of ABA is determined by pH gradients. During the light most leaf ABA is found within the chloroplast, and is therefore protected from the enzymes of catabolism (Zeevaart and Creelman, 1988). Stress and darkness both cause

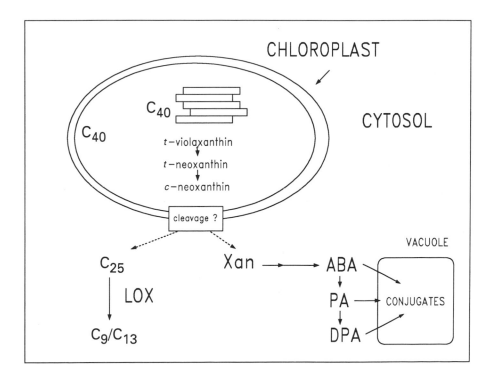

FIG. 15.7. Simplified scheme showing the subcellular localisation of ABA metabolism.

a decrease in stromal pH and consequently a re-distribution of ABA and an increase in the rates of catabolism. Therefore the increased rates of catabolism could simply be due to the increased availability of ABA.

IV. CONCLUSIONS

The metabolism of ABA, from all-*trans*-violaxanthin to conjugates PA and DPA, has been described (see Fig. 15.7). The key regulatory step on this pathway appears to be the cleavage of 9'-*cis*-neoxanthin to produce Xan. This is catalysed by an inducible oxygenase with high turnover and specificity (9'-*cis*-neoxanthin-11',12'-dioxygenase?). Xan is converted to ABA by constitutive enzymes in the cytosol. The absence of any apo-carotenoid by-products of 9'-*cis*-neoxanthin cleavage could result from the very high endogenous 'bleaching capacity' found in plant extracts. The conversion of all-*trans*-violaxanthin to 9'-*cis*-neoxanthin may be stimulated by alterations in neoxanthin levels or be directly affected by stress. Similarly the oxidative catabolism of ABA might be directly stimulated by stress, or be limited by substrate availability. Much of the work described in this chapter, especially that on ABA biosynthesis, has been done in the last five years. It is hoped that with the combination of biochemical and molecular techniques now available further rapid progress can be made into the understanding of how ABA metabolism is controlled.

REFERENCES

Addicott, F. T. (1983). "Abscisic Acid". Praeger, New York.

Ben-Aziz, A., Grossman, S., Ascorelli, I. and Budowski, P. (1971). *Phytochemistry* **10**, 1445–1452.

Bray, E. A. and Zeevaart, J. A. D. (1985). *Plant Physiol.* **79**, 719–722.

Britton, G. (1982). *Physiol. Veg.* **20**, 735–755.

Britton, G. (1988). *In* "Plant Pigments" (T. W. Goodwin, ed.), pp. 133–182. Academic Press, London.

Butt, V. A. (1980). *In* "The Biochemistry of Plants", Vol. 2 (D. D. Davies, ed.), pp. 81–123. Academic Press, London.

Cohen, A. and Bray, E. A. (1990). *Planta* **182**, 27–33.

Cornish, K. and Zeevaart, J. A. D. (1985). *Plant Physiol.* **78**, 1029–1035.

Cowan, A. K. and Railton, I. D. (1987). *Plant Physiol.* **84**, 157–163.

Creelman, R. A. and Zeevaart, J. A. D. (1984). *Plant Physiol.* **75**, 166–169.

Creelman, R. A., Gage, D. A., Stults, J. T. and Zeevaart, J. A. D. (1987). *Plant Physiol.* **85**, 726–732.

Davies, W. J., Mansfield, T. A. and Heterington, A. M. (1990). *Plant Cell Environ.* **13**, 709–719.

Duckham, S. C., Taylor, I. B., Linforth, R. S. T., Al-Naieb, R. J., Marples, B. A. and Bowman, W. R. (1989). *J. Exp. Bot.* **217**, 901–905.

Enzell, C. (1985). *Pue Appl. Chem.* **57**, 693–700.

Firn, R. D. and Friend, J. (1972). *Planta* **103**, 263–266.

Firn, R. D., Burden, R. S. and Taylor, H. F. (1972). *Planta* **102**, 115–126.

Fong, F., Koehler, D. E. and Smith, J. D. (1983). *In* "The Third International Symposium on Pre-harvest Sprouting in Cereals" (J. E. Kruger and D. E. La Berge, eds), pp. 188–196. Westview Press, Boulder, CO, USA.

Gage, D. A., Fong, F. and Zeevaart, J. A. D. (1989). *Plant Physiol.* **89**, 1039–1041.

Galliard, T. and Chan, H. W-S. (1980). *In* "The Biochemistry of Plants", Vol. 4 (P. K. Stumpf and E. E. Conn, eds), pp. 131–161. Academic Press, New York.

Gamble, P. E. and Mullet, J. E. (1986). *Eur. J. Biochem.* **160**, 117–121.

Ganguly, J. (1989). "Biochemistry of Vitamin A". CRC Press, Cleveland, OH.

Gillard, D. F. and Walton, D. C. (1976). *Plant Physiol.* **58**, 790–795.

Glinka, Z. (1980). *Plant Physiol.* **65**, 537–540.

Goodwin, T. W. (1979). *Ann. Rev. Plant Physiol.* **30**, 369–404.

Goodwin, T. W. (1980). "The Biochemistry of the Carotenoids", Vol. 1. Chapman and Hall, London and New York.

Guerrero, F. and Mullet, J. E. (1986). *Plant Physiol.* **80**, 588–591.

Hartung, W., Gimmler, H., Heilmann, B. and Kaiser, G. (1982). *In* "Plant Growth Substances 1982" (P. F. Wareing, ed.), pp. 325–333. Academic Press, London.

Ho, T-H. D. (1983). *In* "Abscisic Acid" (F. T. Addicott, ed.), pp. 147–169. Praeger, New York.

Klein, B. P., Grossman, S., King, D., Cohen, B. S. and Pinsky, A. (1984). *Biochim. Biophys. Acta.* **793**, 72–79.

Koornneef, M. (1986). *In* "A Genetic Approach to Plant Biochemistry" (A. D. Blonstein and P. J. King, eds), pp. 35–54. Springer-Verlag, Wien and New York.

Koshimizu, K., Inui, M., Fukui, H. and Mitsui, T. (1968). *Agric. Biol. Chem.* **32**, 789–791.

Kuhn, H., Holzhutter, H.-G., Schewe, T., Hiebsch, C. and Rapoport, S. M. (1984). *Eur. J. Biochem.* **139**, 577–583.

Kurkela, S. and Franck, M. (1990). *Plant Mol. Biol.* **15**, 137–144.

Lehmann, H. and Glund, K. (1986). *Planta* **168**, 559–562.

Lehmann, H. and Schutte, H. R. (1980). *Z. Pflanzenphysiol.* **96**, 277–280.

Li, Y. and Walton, D. C. (1987). *Plant Physiol.* **85**, 910–915.

Li, Y. and Walton, D. C. (1990a). *Plant Physiol.* **92**, 551–559.

Li, Y. and Walton, D. C. (1990b). *Plant Physiol.* **93**, 128–130.

Linforth, R. S. T., Bowman, W. R., Griffin, D. A., Marples, B. A. and Taylor, I. B. (1987). *Plant Cell. Environ.* **10**, 599–606.

Linforth, R. S. T., Taylor, I. B., Duckham, S. C., Al-Naieb, R. J., Bowman, W. R. and Marples, B. A. (1990). *New Phytol.* **115**, 517–521.

Loveys, B. R. and Milborrow, B. V. (1981). *Austral. J. Plant Physiol.* **8**, 571–589.

Loveys, B. R. and Milborrow, B. V. (1984). *In* "The Biosynthesis and Metabolism of Plant Hormones" (A. Crozier and J. R. Hillman, eds), pp. 71–104. Cambridge University Press, Cambridge.

Mansfield, T. A. and Jones, R. J. (1971). *Planta* **101**, 147–158.

Matile, P. and Martinoia, E. (1982). *Plant Cell Rep.* **1**, 244–246.

Milborrow, B. V. (1983). *In* "Abscisic Acid" (F. T. Addicott, ed.), pp. 79–111. Praeger, New York.

Milborrow, B. V. (1990). *In* "Plant Growth Substances 1988" (R. P. Pharis and S. B. Rood, eds), pp. 241–253. Springer-Verlag, Berlin.

Milborrow, B. V., Carrington, N. J. and Vaughan, G. T. (1988). *Phytochemistry* **27**, 757–759.

Mittelheuser, C. J. and Van Steveninck, R. F. M. (1969). *Nature* **221**, 281–282.

Molnar, P. and Szabolcs, J. (1980). *Phytochemistry* **19**, 633–637.

Neill, S. J., Horgan, R. and Parry, A. D. (1986). *Planta* **169**, 87–96.

Noddle, R. C. and Robinson, D. R. (1969). *Biochem. J.* **112**, 547–548.

Norman, S. M., Maier, V. P. and Pon, D. L. (1990). *J. Agric. Food Chem.* **38**, 1326–1334.

Olson, J. A. (1983). *In* "Biosynthesis of Isoprenoid Compounds", Vol. 2 (J. W. Porter and S. L. Spurgeon, eds), pp. 371–412. John Wiley and Sons, New York.

Parry, A. D. (1989). "Abscisic Acid Biosynthesis in Higher Plants". PhD Thesis, University College of Wales.

Parry, A. D. and Horgan, R. (1991a). *Physiol. Plant.* **82**, 320–326.

Parry, A. D. and Horgan, R. (1991b). *Phytochemistry* **30**, 815–821.

Parry, A. D., Neill, S. J. and Horgan, R. (1988). *Planta* **173**, 397–404.

Parry, A. D., Babiano, M. J. and Horgan, R. (1990a). *Planta* **182**, 118–128.

Parry, A. D., Neill, S. J. and Horgan, R. (1990b). *Phytochemistry* **29**, 1033–1039.

Parry, A. D., Griffiths, A., Tomos, A. D., Jones, H. G. and Horgan, R. (1991a). *J. Exp. Bot.* **42**, Suppl. 14.

Parry, A. D., Blonstein, A. D., Babiano, M. J., King, P. J. and Horgan, R. (1991b). *Planta* **183**, 237–243.

Pierce, M. and Raschke, K. (1980). *Planta* **148**, 174–182.

Pierce, M. and Raschke, K. (1981). *Planta* **153**, 156–165.

Quarrie, S. A. (1990). *In* "Importance of Root to Shoot Communication in the Responses to Environmental Stress" (W. J. Davies and B. Jeffcoat, eds), pp. 13–27. B.S.P.G.R., Bristol.

Rick, C. (1980). *Rep. Tomato Genet. Coop.* **30**, 2–17.

Rock, C. D. and Zeevaart, J. A. D. (1990a). *Plant Physiol.* **93**, suppl. 5.

Rock, C. D. and Zeevaart, J. A. D. (1990b). *Plant Physiol.* **93**, 915–923.

Sachs, M. M. and Ho, T.-H. D. (1986). *Ann. Rev. Plant Physiol.* **37**, 363–376.

Sindhu, R. K. and Walton, D. C. (1987). *Plant Physiol.* **85**, 916–921.

Sindhu, R. K. and Walton, D. C. (1988). *Plant Physiol.* **88**, 178–182.

Sindhu, R. K., Griffin, D. H. and Walton, D. C. (1990). *Plant Physiol.* **93**, 689–694.

Skriver, K. and Mundy, J. (1990). *The Plant Cell* **2**, 505–512.

Stewart, C. R., Voetberg, G. and Rayapati, P. J. (1986). *Plant Physiol.* **82**, 703–707.

Swift, I. E., Milborrow, B. V. and Jeffrey, S. W. (1982). *Phytochemistry* **21**, 2859–2864.

Tal, M. and Imber, D. (1970). *Plant Physiol.* **46**, 373–376.

Taylor, H. F. (1968). *S.C.I. Monograph* **31**, 22–33.

Taylor, H. F. and Burden, R. S. (1970a). *Phytochemistry* **9**, 2217–2223.

Taylor, H. F. and Burden, R. S. (1970b). *Nature* **227**, 302–304.

Taylor, H. F. and Burden, R. S. (1972). *Proc. R. Soc. Lond. B.* **180**, 317–346.

Taylor, H. F. and Burden, R. S. (1973). *J. Exp. Bot.* **24**, 873–880.

Taylor, H. F. and Smith, T. A. (1967). *Nature* **215**, 1513–1514.

Taylor, I. B., Linforth, R. S. T., Al-Naieb, R. J., Bowman, W. R. and Marples, B. A. (1988). *Plant Cell Environ.* **11**, 739–745.

Tinelli, E. T., Sondheimer, E., Walton, D. C., Gaskin, P. and MacMillan, J. (1973). *Tetrahedron Lett.* **2**, 139–140.

Uknes, S. J. and Ho, T-H. D. (1984). *Plant Physiol.* **75**, 1126–1132.

Vaughan, G. T. and Milborrow, B. V. (1987). *Austral. J. Plant Physiol.* **14**, 593–604.

Vernooy-Gerritsen, M., Leunissen, J. L. M., Veldink, G. A. and Vliegenthart, J. F. G. (1984). *Plant Physiol.* **76**, 1070-1079.

Walker-Simmons, M., Kudrna, D. A. and Warner, R. L. (1989). *Plant Physiol.* **90**, 728–733.

Walton, D. C. and Sondheimer, E. (1972). *Plant Physiol.* **49**, 285–289.

Wright, S. T. C. and Hiron, R. W. P. (1969). *Nature* **224**, 719–720.

Yamamoto, H. Y. (1979). *Pure Appl. Chem.* **51**, 639–648.

Zeevaart, J. A. D. and Creelman, R. A. (1988). *Ann. Rev. Plant Physiol., Plant Mol. Biol.* **39**, 439–473.

Zeevaart, J. A. D., Heath, T. G. and Gage, D .A. (1989). *Plant Physiol.* **91**, 1594–1601.

16 Enzymes of Gibberellin Synthesis

THEODOR LANGE and JAN E. GRAEBE

Pflanzenphysiologisches Institut und Botanischer Garten, Universität Göttingen, D-3400 Göttingen, Germany

I. INTRODUCTION

A. The Gibberellin Pathway

The gibberellins (GAs) are tetracyclic diterpenoid compounds, some of which are recognised as hormones in higher plants. Their biosynthesis (Graebe, 1987) and their structures (N. Takahashi *et al.*, 1986; Hedden, 1987; Beale and Willis, 1991) have

METHODS IN PLANT BIOCHEMISTRY Vol. 9
ISBN 0–12–461019–6

recently been reviewed in detail, so that the following overview is intended only to orient the reader and define our scopes. Gibberellin biosynthesis first follows the normal terpenoid pathway from mevalonic acid (MVA) to geranylgeranyl pyrophosphate. This part of the pathway is catalysed by soluble enzymes, which are presumably located in plastids. Soluble enzymes of unknown cellular localisation then catalyse the two-step cyclisation of geranylgeranyl pyrophosphate to *ent*-kaurene. This hydrocarbon is subsequently oxidised in several steps to GA_{12}-aldehyde and, at least in *Cucurbita maxima*, to GA_{12} by microsomal cytochrome P-450 type enzymes, which are associated with the endoplasmic reticulum. Finally, GA_{12}-aldehyde and GA_{12} are oxidised to C_{20}- and C_{19}-GAs, by soluble enzymes, which show the properties of 2-oxoglutarate-requiring dioxygenases. The localisation of these enzymes in the cell is unknown. Interestingly, the reactions on the borderline between lipophilic and more soluble substrates, e.g. the C-7 oxidation of GA_{12}-aldehyde to GA_{12}, may be catalysed by both microsomal and soluble enzymes. In such cases the cofactor requirements of the microsomal enzymes are of the cytochrome P-450 type, whereas those of the soluble enzymes are of the 2-oxoglutarate-dependent dioxygenase type.

The enzymes catalysing the reactions from MVA to *ent*-kaurene are treated in Chapter 9 of this volume and the cytochrome P-450 type enzymes catalysing *ent*-kaurene oxidation are treated in Chapter 10. We will deal with these parts of the pathway in connection with the preparation of labelled precursors only. The main part of this chapter deals with the soluble 2-oxoglutarate-dependent enzymes, catalysing the steps beyond GA_{12}-aldehyde. Since these enzymes have not yet been rigorously classified as dioxygenases (EC 1.14.11.-), they are variously referred to as oxidases and hydroxylases in the literature.

B. Short History

After Birch *et al.* (1959) had established the diterpenoid character of GA biosynthesis in the fungus *Gibberella fujikuroi*, cell-free systems from higher plants capable of GA biosynthesis were sought as a prerequisite for enzyme purification and characterisation. The first successful cell-free system came from the almost liquid endosperm of immature *Marah macrocarpus* seeds (then named *Echinocystis macrocarpa*; Graebe *et al.*, 1965). This system became instrumental in elucidating the pathway from MVA to *ent*-7α-hydroxykaurenoic acid (Dennis and West, 1967; Lew and West, 1971) and in the purification of the enzymes involved in *ent*-kaurene synthesis and oxidation (Chapters 9 and 10). Further success was obtained with a cell-free system from the endosperm of immature *Cucurbita maxima* seeds (first incorrectly identified as *C. pepo*; Graebe, 1969, 1972), which catalysed the whole pathway from MVA to C_{20}-GAs and one C_{19}-GA (Graebe *et al.*, 1972, 1974a, b, c). Other useful cell-free systems were obtained from immature seeds of *Pisum sativum* (Kamiya and Graebe, 1983), *Phaseolus coccineus* (Turnbull *et al.*, 1985) and *Phaseolus vulgaris* (M. Takahashi *et al.*, 1986) as well as from leaves of *Spinacea oleracea* (Gilmour *et al.*, 1986). The first attempt to purify a GA 2β-hydroxylase from germinating bean seeds was unsuccessful because the cofactor requirements were not known (Patterson *et al.*, 1975). After it became clear that the enzymes had the cofactor requirements of 2-oxoglutarate-dependent dioxygenases (Hedden and Graebe, 1982), 2β-hydroxylases (Smith and MacMillan, 1984, 1986; Griggs *et al.*, 1991), 3β-hydroxylases (Kwak *et al.*,

1988; Smith *et al.*, 1990) and C-20 oxidases (Gilmour *et al.*, 1987; Lange and Graebe, 1989; Graebe *et al.*, 1991) have been partially purified and characterised. The most active enzyme preparations are obtained from immature seeds, but the ultimate goal is to characterise the GA biosynthesis enzymes in tissues which are involved in the regulation of stem growth. This has been partially realised in the case of photoperiodically regulated GA C-20 oxidases from spinach leaves (Gilmour *et al.*, 1986, 1987; Section III.B).

II. PREPARATION OF LABELLED SUBSTRATES

Labelled GAs are needed as substrates for monitoring enzyme activity during purification, since the GA molecules do not have a characteristic UV absorption or other useful properties for their detection. Some research workers use tritiated substrates and measure the release of tritiated water as a rapid and convenient assay for enzyme activity (Section III.A.5). This method is suited for monitoring the activity during enzyme purification, but the use of substrates, tritiated at the site of reaction, is not suitable for enzyme characterisation because (1) the large tritium isotope effect causes underestimation of enzyme activity and distorts product ratios and kinetic parameters, (2) the high specific activity usually necessitates working at very low substrate concentrations, and (3) the isotope label cannot be detected by gas chromatography–mass spectrometry (GC-MS) for conclusive product identification. These disadvantages are avoided by working with substrates that are double-labelled with tritium and a stable isotope at sites that are remote from the reaction, such as in $[17\text{-}^{13}C\ ^3H_2]GA_{20}$. Here the ^3H-label is used for detecting the products during purification and the ^{13}C-label to identify the products by GC-MS (Smith *et al.*, 1990). Beale and Willis (1991) have reviewed the synthesis of isotopically labelled GAs. Some ^3H-labelled GAs are available from Amersham International plc. We prefer to use biosynthetically prepared ^{14}C-labelled GA substrates with high enough specific activity to allow the detection of the isotopic peaks by GC-MS. Once the identities of the products in a particular reaction have been conclusively established by this method, the enzyme activity can be monitored by radio-high performance liquid chromatography (HPLC) or radio-thin layer chromatography (TLC) of the extracted products.

The ^{14}C-labelled precursors are produced with the *Cucurbita maxima* endosperm cell-free system, which is particularly suitable for this purpose and has been extensively studied (Graebe, 1969, 1972; Graebe *et al.*, 1972, 1974a,b,c, 1980, 1991; Hedden and Graebe, 1982; Turnbull *et al.*, 1985; Graebe and Lange, 1990). Section II.A describes the use of this system for the preparation of ^{14}C-labelled GA_{12}-aldehyde, GA_{12} and GA_{15}. Section II.B describes the 13-hydroxylation of GA_{12} to GA_{53} with a microsomal system from immature pea cotyledons, and Section II.C describes the conversion of $[^{14}C]GA_{12}$ or $[^{14}C]GA_{53}$ to more oxidised GAs with high-speed supernatants from *C. maxima* endosperm and cotyledons.

A. $[^{14}C]GA_{12}$-Aldehyde, $[^{14}C]GA_{12}$ and $[^{14}C]GA_{15}$

If a dialysed low-speed supernatant of the pumpkin endosperm is incubated with $[2\text{-}^{14}C]MVA$ and cofactors needed to activate the isoprenoid pathway, the kaurene synthetase and the microsomal cytochrome P-450 part of the pathway, the main products are ^{14}C-labelled GA_{12}-aldehyde, ent-6α,7α-dihydroxykaurenoic acid, and GA_{12}. With very active preparations, GA_{15} is also obtained, but the presence of Mn^{2+} and the absence of Fe^{2+} in the incubation mixture prevent the formation of more highly oxidised GAs. Gibberellin A_{12}-aldehyde is a general GA precursor. It can be used to assay the activity of C-7 oxidases (Section III.E), or as a precursor of GAs. Gibberellin A_{12}, the least oxidised GA, is used to assay the activity of C-20 oxidases (Section III.B) or to produce more highly oxidised GAs (Section II.C). Gibberellin A_{15} can be used to assay the activity of 3β-hydroxylases (Section III.C). ent-6α,7α-Dihydroxykaurenoic acid is not on the pathway to GAs.

1. Cucurbita maxima *endosperm preparation*

Seeds of field grown *Cucurbita maxima* L. (we use 'Riesenmelone, gelb genetzt', in the UK 'Big max' has been used) are harvested when the fruits are fully grown or almost so, but the seeds are still immature. We recognise this stage by the size and colour of the fruits and the condition of the exocarp, but others prefer to cut sterile holes in the fruits and sample the seeds (Birnberg *et al.*, 1986). All the seeds in a fruit are usually at the same stage, and seeds in which the cotyledons have grown to maximally 50% the length of the seed lumen give the best results. The original publication (Graebe, 1972), which also has an illustration of the seeds, states that maximal activity was obtained with seeds 13–46% the length of the lumen, but good activities are obtained with younger stages as well. More endosperm and higher activity are obtained in warm weather than in cold, and growing the fruits in transparent tents increases the activity in cold climates.

The seeds are removed from the fruits, cut at the distal end relative to the point of attachment and the endosperm is squeezed out and collected with tweezers in a beaker on ice. The combined endosperm of seeds from one pumpkin (up to 100 ml) is gently homogenised with 4 to 5 strokes in a Potter-type glass homogeniser, which has been ground to some clearance. The resulting homogenate is filtered through three layers of cheesecloth, centrifuged for 5 min at 2000 × g and dialysed three times for 1 h at 4°C against 1 l 50 mM potassium phosphate buffer (pH 8.0), containing 2.5 mM $MgCl_2$. The endosperm preparation, which contains about 4 mg protein ml^{-1}, should be stored in liquid N_2 or in a freezer at −80°C. The preparations from different fruits should be kept apart and it is useful to freeze them in droplets or aliquots so that they can be sampled without defrosting the main part. Note that the dialysis is not complete and low molecular weight compounds are still present in the preparations, but either omitting the dialysis or attempting gel filtration results in loss of activity. A slightly modified method for the preparation of the pumpkin cell-free system (Birnberg *et al.*, 1986) is more expensive and yields less active extracts in our hands.

2. *Identifying active endosperm preparations*

Since individual fruits yield extracts of very different activities for the cytochrome P-450 part of the pathway (Graebe, 1972), each preparation must be tested for activity on a small scale. The endosperm preparation (83.4 µl) is mixed with 8.3 µl of a cofactor solution to give (final concentrations) 5 mM $MgCl_2$, 1 mM $MnCl_2$, 5 mM ATP, 5 mM phosphoenol pyruvate (optional) and 0.5 mM NADPH. The reaction is started by adding 8.3 μl of a solution, containing 3×10^{-8} mol R-[2-^{14}C] mevalonate, previously prepared by hydrolysing R-[2-^{14}C]MVA lactone (1.96×10^{12} Bq mol^{-1}, Amersham-Buchler, Braunschweig) for 15 min at 30°C with two equivalents of KOH and diluting it to the appropriate concentration. After incubation in open tubes with shaking for 2.5 h at 30°C the reaction is stopped by adding 10 μl 0.5 N HCl (to pH 3.0) and 100 μl acetone to ensure contact with the organic phase in the following extraction.

3. Extraction and separation

The incubation mixture is diluted with 0.5 ml H_2O, adjusted to pH 3.0 with HCl and the products are extracted with 3×1 ml ethyl acetate. The combined organic phases are washed once with 1 ml dilute acetic acid (pH 3.0) and dried under a gentle stream of N_2. The residues are dissolved in 50 μl acetone-methanol (1:1; v/v) and 3 μl are applied to a silica gel thin layer (TLC) plate, which is developed with chloroform–ethyl acetate–acetic acid (80:20:1 to 65:35:1; v/v, depending on the make of plates and atmospheric conditions). The radioactivity is best located with a radioscanner. By this method, several preparations can be tested more rapidly than by HPLC. The remaining (47 μl) portions of the extracts from several experiments are pooled and the labelled products recovered by HPLC as described in the next section.

The interpretation of the scans requires experience, since the more active the cytochrome P-450 type enzymes are, the further along the pathway the biosynthesis proceeds (Fig. 16.1). The peaks for *ent*-6α,7α-dihydroxykaurenoic acid and GA$_{12}$-aldehyde, which are usually the predominant ones, are often about equally high. Gibberellin A$_{15}$, when present, chromatographs like *ent*-7α-hydroxykaurenoic acid. The oxidation products near the origin belong to branches of the pathway (see Section 3.3.6 in Hedden, 1983). The reader may want to consult the complete schemes of this part of the pathway in one of the reviews (Graebe, 1982, 1987; Hedden, 1983).

4. Preparative-scale production

Endosperm preparations with the desired activity are used for the larger scale production of GAs as described for the small-scale incubations, but in 250 ml Erlenmeyer flasks and with 60 times greater amounts of all components. After incubation for 2.5 h at 30°C in a shaking water bath the reaction is stopped with 5 N HCl (*c.* 60 μl, final pH 3.0) and 6 ml acetone.

The incubation products are extracted with 3×6 ml ethyl acetate. The combined organic phases are washed once with 6 ml dilute acetic acid (pH 3.0) and dried under reduced pressure in a rotatory evaporator. The products are purified on an analytical C_{18} reverse-phase HPLC column (e.g. 15×0.39 cm, 5 μm, Radial-Pak Liquid Chromatography Cartridge, Waters) using isocratic or gradient elution systems and on-line radiation detection or aliquot scintillation counting. Isocratic elution at 1 ml min^{-1} in 65% methanol yields GA$_{15}$ (about 21 min), *ent*-6α,7α-dihydroxykaurenoic acid (36 min), *ent*-7α-hydroxykaurenoic acid (44 min), GA$_{12}$ (61 min), and GA$_{12}$-aldehyde (86 min). The retention times differ considerably for different columns and must be

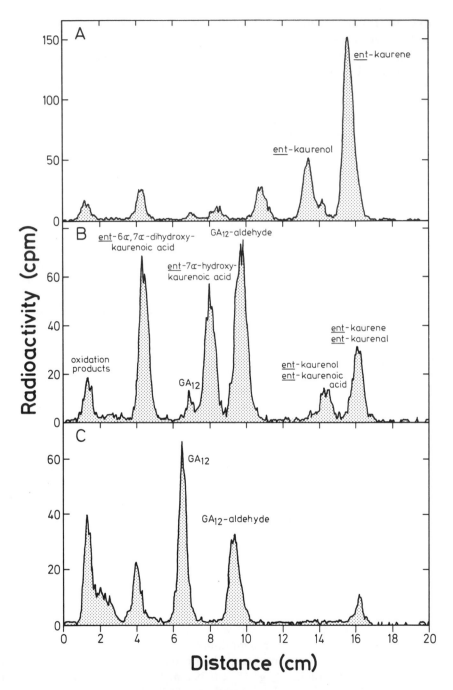

FIG. 16.1. TLC product patterns obtained by incubating endosperm preparations from different fruits of *Cucurbita maxima* with [2-^{14}C]MVA and cofactors. (A) Weak cytochrome P-450 activity: main products are *ent*-kaurene and *ent*-kaurenol. (B) Higher cytochrome P-450 activity: main products are GA$_{12}$-aldehyde, *ent*-6α,7α-dihydroxykaurenoic acid and *ent*-7α-hydroxykaurenoic acid, the neutral intermediates are almost used up. (C) High cytochrome P-450 activity: main product is GA$_{12}$, the precursors are almost gone and the oxidation products at the origin have increased.

determined for each column. More hydrophobic kaurenoids (from *ent*-kaurene to *ent*-kaurenoic acid) are eluted with 100% methanol. Preparative incubations give about the same product patterns as the small-scale incubations, but usually have lower enzyme activity. Active preparations convert 100% of the added [2-^{14}C]MVA, whereby ^{14}C-labelled GA$_{12}$-aldehyde, GA$_{12}$ and GA$_{15}$ account for 30–50% of the products. Their specific radioactivity is about 7×10^{12} Bq mol^{-1} by the method of Bowen *et al.* (1972). All identities must be confirmed by GC-MS.

The *C. maxima* endosperm may also be used for the preparation of *ent*-kaurene (Graebe, 1969) by incubating as described above, but using the 200 000 × *g* instead of the low-speed supernatant and omitting NADPH from the incubation mixture. After 4 vols of methanol have been added to the reaction mixture, *ent*-kaurene may be extracted with petroleum ether and further purified by HPLC (Grosselindemann *et al.*, 1991).

B. 13-Hydroxylation, [^{14}C]GA$_{53}$

[^{14}C]Gibberellin A$_{53}$ is needed as a precursor for the preparation of 13-hydroxylated GAs (Section II.C) and as a substrate to measure GA C-20 oxidase activity (Section III.B). It is prepared from [^{14}C]GA$_{12}$ with microsomal preparations from immature pea cotyledons, which are high in 13-hydroxylating activity (Kamiya and Graebe, 1983; Lange and Graebe, 1989). Gibberellin A$_{12}$-aldehyde is also converted to GA$_{53}$ by the pea system, but the yield is only one-fifth as large as with GA$_{12}$.

1. *Extraction of the cotyledons*

We have used different field grown pea cultivars, including 'Torsdag' (cv. WB 2157, Weibullsholms, Landskrona, Sweden), but we have also successfully used peas diverted from the production line of a local pea grower before being blanched. Frozen peas from the market have been boiled and do not function. The immature seeds are harvested 21 days after anthesis or when they have reached about 40% of their maximal fresh weight. The testa are removed and the cotyledons (about 300 mg per seed) are frozen in liquid N$_2$. Portions of 100 g are pulverised in a Waring blender, the powder is thawed in 100 ml 0.2 M Tris-HCl, pH 7.9, containing 8 mM dithiothreitol (DTT), and the homogenate is centrifuged at 2000 × *g* for 15 min. The resulting supernatant is centrifuged again for 1 h at 200 000 × *g*. Almost all (95%) of the 200 000 × *g* supernatant is carefully removed and discarded, but the supernatant next to the pellet is high in 13-hydroxylating activity and should be left. The pellet is resuspended in 0.1 M Tris-HCl (pH 7.9 at 4°C), containing 4 mM DTT, and centrifuged again at 200 000 × *g* for 1 h, whereafter 95% of the supernatant is discarded as before. This washing step will remove most of the endogenous low molecular weight compounds. The high-speed pellet is carefully resuspended in the remaining supernatant, frozen by dripping it into liquid N$_2$, and stored.

2. *Incubation and separation*

To find the most economical concentrations, 48, 24, 12 and 6 µl microsomal preparation in 5-ml test tubes are mixed with 1 µl aqueous NADPH (final concentration

1 mM) and 3500 Bq [^{14}C]GA$_{12}$ dissolved in 1 μl methanol (final concentration 10 μM). The total volume is adjusted to 50 μl with 100 mM Tris-HCl, pH 7.0 (at 30°C), and the mixture is incubated in a shaking water bath for 4 h at 30°C. Another 1 μl NADPH-solution is added after 2 h. The incubation is terminated by acidifying to pH 3.2 with 5 μl acetic acid. The incubation products, primarily [^{14}C]GA$_{12}$ and [^{14}C]GA$_{53}$, are extracted using C$_{18}$ reverse-phase Sep Pak cartridges (Section II.C.4) and identified by TLC on silica gel developed with chloroform–ethyl acetate–acetic acid (40:60:1; v/v; R_f of GA$_{53}$ c. 0.4), or by C$_{18}$ reverse-phase HPLC (Section II.C.4).

The preparative production of [^{14}C]GA$_{53}$ is done in 250 ml Erlenmeyer flasks, using 60 times the amounts of all components and the lowest enzyme concentration giving complete conversion of GA$_{12}$ to GA$_{53}$, plus 50% to account for the scale-up loss in activity. The reaction is stopped by adding 3 ml H$_2$O and lowering the pH to 3.0 with about 40 μl 5 N HCl. The products (mainly GA$_{53}$) are extracted by liquid–liquid extraction (Section II.A.4). Pure [^{14}C]GA$_{53}$ with about the same specific activity as that of the substrate is obtained by C$_{18}$ reverse-phase HPLC (Section II.C.4). The identity should be confirmed by GC-MS.

C. C-20 Oxidation and Insertion of 3β- and 12α-Hydroxyl Groups

The oxidation of [^{14}C]GA$_{12}$ and [^{14}C]GA$_{53}$ to further GAs is accomplished by using enzymes of the 2-oxoglutarate-requiring dioxygenase type, which are contained in the high-speed supernatants of extracts from endosperm and cotyledons of immature pumpkin seeds. By varying the precursors, the extracts and the protein concentrations, the system can be fine tuned to produce specific GAs in high yields (Tables 16.1 and 16.2). The relations are rather complex (Lange, 1989; Graebe et al., 1991). The endosperm system has high C-20 oxidasing and 3β-hydroxylating activities, while the cotyledon system is high in 12α-hydroxylase activity. The presence of a 13-hydroxy

TABLE 16.1. Approximate protein concentrations for obtaining maximal yields of selected ^{14}C-labelled GAs in incubations of endosperm preparations with [^{14}C]GA$_{12}$ or [^{14}C]GA$_{53}$ for 5 min (T. Lange, P. Hedden and J. E. Graebe, unpubl. res.).

	GA$_{12}$ as a substrate						GA$_{53}$ as a substrate			
						12α-OH				
	GA$_{15}$	GA$_{24}$	GA$_{25}$	GA$_{13/36}$	GA$_{43}$	GA$_{43}$	GA$_{44}$	GA$_{23}$	GA$_1$	GA$_{28}$
Protein (μg ml^{-1})	20	20	40	80	400	>2000	100	1500	8000	>8000
Yield (%)	45	15	35	80	70	>80	70	60	35	>30

TABLE 16.2. Approximate protein concentrations for obtaining maximal yields of selected ^{14}C-labelled GAs by incubation of cotyledon preparations with [^{14}C]GA$_{12}$ or [^{14}C]GA$_{53}$ for 2 h (T. Lange, P. Hedden and J. E. Graebe, unpubl. res.).

	GA$_{12}$ as a substrate						GA$_{53}$ as a substrate			
				12α-OH						
	GA$_{15}$	GA$_{24}$	GA$_{25}$	GA$_{25}$	GA$_{13/36}$	GA$_{39}$	GA$_{44}$	GA$_{19}$	GA$_{17}$	GA$_{28}$
Protein (μg ml^{-1})	20	20	80	320	320	>1300	40	80	320	>8000
Yield (%)	50	40	60	60	15	>20	70	60	35	>30

group in GA_{53} results in the ready production of a C_{19}-GA (GA_1) by the endosperm system. Finally, the use of higher protein concentrations yields GAs further down the respective pathways. Gel filtration chromatography (Sephadex G-100, Section III) separates the 3β- and C-20 hydroxylases, which simplifies the selective preparation of the corresponding GAs. In contrast, 12α- and C-20 hydroxylases co-elute on Sephadex G-100.

1. Enzyme extraction

Immature pumpkin seeds (with cotyledons less than 50% the length of the seed lumen) are harvested and split open length-wise with a pair of scissors. Endosperm and cotyledons are separated in the cold, and the endosperm is frozen by dripping it into liquid N_2. The cotyledons are washed with distilled water (4°C) and carefully dried with paper tissue to remove adhering endosperm before they are frozen in liquid N_2. Frozen endosperm (10 g) or cotyledons (10 g) are pulverised with mortar and pestle, and 10 ml of 0.2 M Tris-HCl, pH 7.9 (at 4°C), containing 8 mM DTT is added to the powder. After thawing, the homogenate is centrifuged at $20\,000 \times g$ for 20 min and the resulting supernatant is centrifuged at $200\,000 \times g$ for 1 h. The high-speed supernatant is gel filtered over Sephadex PD-10 columns equilibrated with 100 mM Tris-HCl, pH 7.9, containing 4 mM DTT. The gel filtration step removes most of the endogenous GAs, thus preventing subsequent dilution of the radioactivity of the incubation products. The endosperm preparation contains about 2 mg, the cotyledon preparation about 10 mg protein ml^{-1}. The gel-filtered high-speed supernatant can be concentrated by ultrafiltration to improve activity. The enzyme preparation should be stored in liquid N_2 or at -80°C in a freezer.

2. Surveying the activity

Tables 16.1 and 16.2 show the range of GAs which can be produced and the approximate protein concentration giving maximal yields with cell-free systems from endosperm and cotyledons, respectively. Some five to ten small-scale incubations with different protein concentrations (e.g. 1/6, 1/4, 1/2, 1/1, 2, 4 and 6 times the concentrations suggested in Table 16.1 or 16.2) are done for every enzyme preparation. The incubations are done in 5- to 10-ml centrifuge tubes, to which 12.5 μl cofactor solution, containing (final concentrations) 2-oxoglutarate (4 mM), ascorbate (4 mM), $FeSO_4$ (0.5 mM), bovine serum albumin (BSA; 2 mg ml^{-1}) and catalase (0.1 mg ml^{-1}), are added. The total volume is adjusted to 250 μl with 100 mM Tris HCl, pH 7.0 (at 30°C). This mixture is pre-incubated at 30°C for 5 min and the reaction is started by adding about 3500 Bq [^{14}C]GA_{12} or [^{14}C]GA_{53} (final 2 μM) in 2 μl methanol, depending on which GA products are required. The mixtures are incubated from 5 min to 2 h at 30°C in a shaking water bath, depending on the activity of the individual enzyme preparations (Tables 16.1 and 16.2). The reaction is terminated by lowering the pH to about 3.0 with 25 μl acetic acid.

3. Incubations on a preparative scale

After the optimal protein concentration has been determined, preparative incubations are done in 250 ml Erlenmeyer flasks on a 60 times larger scale, using the same ingredients and the same concentrations, except for the protein, which is increased by 50%. At the end of the incubation time, pH is lowered to about 3.0 with 1.5 ml acetic acid.

4. *Extraction and purification of incubation products*

The following extraction and purification scheme is used for both small-scale and large-scale incubations. Sep-Pak C_{18} reverse-phase cartridges (Waters Ass., Milford, MA, USA) are first conditioned with 5 ml methanol and equilibrated with 5 ml 10 mM acetic acid, pH 3.2. The incubation mixtures are loaded onto the cartridges, which are

TABLE 16.3. C_{18} reverse-phase HPLC retention times for [^{14}C]GAs. Solvent system and flow rate in all cases: methanol–10 mM acetic acid (pH 3.2), at 1 ml min^{-1}. Gradient programme number 1: first 20 min linear gradient from 25% to 62.5% methanol, followed by a 5 min step to 62.7% methanol, a 1 min step to 63.2% and seven 2 min steps to 64.3%, 67.2%, 70.5%, 74.9%, 81%, 89% and 100% methanol, respectively. Programme number 2: first 5.4 min isocratic flow at 75% methanol, followed by 0.9 min steps to 76%, 78.6%, 85.2% and 100% methanol, respectively. Programme number 3 was an 8 min linear gradient from 57.5% to 60%, followed by a 1 min step to 100% methanol.

Gibberellin	R_t (min) for gradient programme no.		
	1	2	3
12α-Hydroxy-GA$_{43}$	7′08″		
Putative 2β-hydroxy-GA$_{28}$	7′42″		
GA$_8$	9′32″		
GA$_{39}$	11′30″		
GA$_{23}$	13′24″		
GA$_{28}$	13′58″		
GA$_{38}$	14′00″		
GA$_1$	15′40″		
GA$_{43}$	21′06″	3′42″	
GA$_{20}$	23′34″		
GA$_{46}$	23′36″		
GA$_{36}$	24′00″		
GA$_{13}$	24′02″	4′09″	
Putative 12α-hydroxy-GA$_{25}$	24′48″		
GA$_{44}$	25′00″	4′17″	9′31″
GA$_{19}$	26′09″	4′19″	10′46″
GA$_{34}$	26′21″		
GA$_{17}$	26′24″		
GA$_{37}$	27′09″		13′00″
Putative 12α-hydroxy-GA$_{12}$	28′21″		
GA$_4$	30′48″		
GA$_{53}$	33′27″	6′06″	13′47″
GA$_{24}$	35′39″	7′36″	
GA$_{25}$	36′27″	7′37″	
GA$_9$	37′34″		
GA$_{15}$	37′36″	8′23″	14′10″
ent-7α-Hydroxykaurenoic acid	40′57″		
GA$_{12}$	42′21″	12′57″	
GA$_{12}$-aldehyde	43′18″	13′36″	

washed with at least 10 ml 10 mM acetic acid. This washing step will avoid the formation of GA artefacts due to DTT in the incubation mixture (Graebe and Lange, 1990). The eluates are discarded and the bound GAs are eluted with 5 ml methanol. The methanol is removed under a gentle stream of N_2 or under reduced pressure in a Speed-Vac centrifugal concentrator. This extraction method offers high recoveries of all GAs, although it can only be used with incubation mixtures of relatively low protein concentration lest the cartridge gets plugged up. The extracted incubation products are purified by C_{18} reverse-phase HPLC (Radial-Pak Cartridge as described in Section II.A.4), using the gradient programme number 1, Table 16.3 (gradient programmes no. 2 and 3 will be referred to later). The identity of the products must be confirmed by GC-MS.

III. ENZYMES OF GIBBERELLIN METABOLISM

The following sections describe the purification and characterisation of the soluble, 2-oxoglutarate-dependent oxidases of GA-biosynthesis, which catalyse the steps after GA_{12}-aldehyde in the pathway. General aspects of these topics are discussed first.

A. General Aspects

1. Choice of plant material

The amounts of GA-biosynthetic enzymes in plant tissues are often very low. The careful selection of plant material at specific stages of development is therefore important for obtaining enzymes of reasonable purity in a limited number of steps. Immature pumpkin seeds contain very high concentrations of GA C-7 oxidase(s), C-20 oxidase(s), 2β-hydroxylases, 3β-hydroxylase(s) and 12α-hydroxylase(s), making them a prime source of these enzymes (Lange, 1989; Graebe and Lange, 1990). Seeds of *Marah macrocarpus* provide another good source of GA enzymes, particularly those catalysing the conversion of GA_9 to GA_7 and GA_3 (Albone *et al.*, 1990), which do not occur in *Cucurbita maxima*. However, *Marah macrocarpus*, supposedly, cannot be cultivated but must be collected wild, which limits its usefulness. Other workable sources are pea and bean seeds as described in the following sections. Most GA enzymes are located in the embryos and endosperm, but not in the testa, of immature seeds.

2. Stability

Gibberellin enzymes are very labile and the first purification steps must be done in the cold. As they become purer, the enzymes are often stable even at 30°C. Values of pH below pH 7.0 usually destroy the activity (Smith and MacMillan, 1986; Smith *et al.*, 1990; Lange, 1989). DTT, DTE or (more seldom) 2-mercaptoethanol must be added to the crude extract and during all purification procedures to preserve activity, although it is usually not needed during incubations. The presence of trace amounts of DTT during the extraction of incubation products can cause artefacts (Graebe and Lange, 1990). Further protective agents may be used, such as ascorbate, polyvinyl-pyrrolidone (PVP) and EDTA. Protease activity can be inhibited by phenylmethyl-sulphonylfluoride (about 0.1 mM). Enzyme stability decreases during purification, presumably because of dilution and removal of stabilising proteins. The addition of sucrose, glycerol or BSA helps maintaining enzyme stability during dilution, but must be left out during the purification.

3. Extraction

Methods that were originally developed for preparing cell-free systems to investigate GA metabolism are also very useful for the initial steps in enzyme purification. The extraction buffer most often used is Tris-HCl, but phosphate and HEPES buffers can also be used. The kind of buffer used may influence the Fe^{2+} requirement of the purified enzyme. All buffers should contain enzyme-stabilising agents. Enzymes from soft tissues, such as the endosperm from *Cucurbita maxima* or *Marah macrocarpus*, can be extracted in buffer with a glass homogeniser. Firmer material, such as cotyle-dons from pumpkin or pea, or young leaves from spinach, may be ground in extraction buffer with a mortar and pestle, but it is better to freeze the plant material in liquid N_2 and grind it to a fine powder in a mortar or Waring blender. Extraction buffer is then added to the powder.

4. Strategies for the purification

The homogenate is filtered through several layers of cheesecloth and the filtrate centrifuged at $20\,000 \times g$ for 20 min to remove cell debris and intact organelles. The microsomes are removed by centrifugation at $200\,000 \times g$ for 1 h. Centrifugation results in up to a 1.3-fold increase in enzyme-specific activity. The high-speed supernatant is then usually fractionated by precipitation with $(NH_4)_2SO_4$ or chilled methanol. These precipitation techniques result in another up to 6-fold increase in specific activity. Finally various gels (e.g. ion-exchange, hydrophobic interaction, hydroxylapatite) can be used batch-wise to bind and remove unwanted compounds.

Following these initial procedures the extract is purified by column chromatography. Ion-exchange chromatography is a useful technique at this stage, because of its high protein-binding capacity and resolution. A disadvantage is that organic solvents and salts, remaining after the precipitation step, must first be removed by dialysis. Alternatively, gel filtration can be used as the first column chromatography step, simultaneously removing salts and organic solvents. A disadvantage of *this* method is the long separation time caused by low flow rates ($1-2\,\mathrm{ml\,min^{-1}}$) and the very voluminous columns needed (2–3 l).

Protein purification methods that imply changes to low pH values, such as preparative isoelectric focusing or chromatofocusing, tend to inactivate GA enzymes and are therefore less useful (Volger and Graebe, 1985; Gilmour *et al.*, 1987). Hydro-phobic interaction column chromatography is seldom used as an early purification step because of its low protein-binding capacities and problems arising from the strong binding of certain GA oxidases (Smith and MacMillan, 1986; Smith *et al.*, 1990). The use of affinity column chromatography has also been of limited use in GA oxidase purification, although attempts have been made to link GA substrates to affinity gels (Beale *et al.*, 1986). The development of monoclonal antibodies using partially purified enzyme material for immunoaffinity purification has been tried (for methods, see Harlow and Lane, 1988), but the selection of positive cell-lines from mice hybridomas by immunoprecipitation of enzyme activity were not successful (D. L. Griggs and J. A. D. Zeevaart, pers. commun., 1991). The development of specific monoclonal antibodies against a 3β-hydroxylase from beans has been mentioned (Smith *et al.*, 1990), but a conclusive characterisation of these antibodies has not been published.

5. Assays for GA-oxidases

C-7 Oxidases can be assayed by monitoring the formation of [^{14}C]GA$_{12}$ from [^{14}C]GA$_{12}$-aldehyde (Hedden and Graebe, 1982). C-20 oxidases and 13-hydroxylases are assayed by using [^{14}C]GA$_{12}$ as a substrate (Gilmour *et al.*, 1987; Lange, 1989). Certain C-20 oxidases can be assayed with [^{14}C]GA$_{53}$, especially if 13-hydroxylation otherwise complicates the product pattern (Gilmour *et al.*, 1987; Lange and Graebe, 1989). For other C-20 oxidases, [^{14}C]GA$_{44}$ is a useful substrate (Gilmour *et al.*, 1986, 1987). Enzymes that catalyse the C$_{20}$ to C$_{19}$ conversion can be assayed directly with C-20 aldehydes, e.g. [^{14}C]GA$_{19}$ (Gilmour *et al.*, 1986, 1987). 3β-Hydroxylase activity in pumpkin endosperm has been measured by using [^{14}C]GA$_{15}$ (Volger and Graebe, 1985). [2,3-^{3}H$_2$]- or [1,2,3-^{3}H$_3$]GA$_{20}$ or [2,3-^{3}H$_2$]GA$_9$ can be used as substrates for studying 3β- and 2β-hydroxylases (Kwak *et al.*, 1988; Smith *et al.*, 1990, Griggs *et al.*, 1991), whereas [1,2-^{3}H$_2$]GA$_1$ or [1,2-^{3}H$_2$]GA$_4$ are substrates for 2β-hydroxylases only (Smith and MacMillan, 1984, 1986; Griggs *et al.*, 1991). Several ^{14}C-labelled GAs (e.g. GA$_1$, GA$_{13}$, GA$_{39}$) can be used as substrates for 2β-hydroxylases. Finally, [^{14}C]GA$_{43}$ is a suitable substrate for 12α-hydroxylases (Lange, 1989).

The reaction mixture (e.g. 100–400 μl) contains enzyme preparation, diluted in incubation buffer, usually 100 mM Tris-HCl buffer or 50 mM phosphate buffer, adjusted to pH 7.0 at 30°C, 2-oxoglutarate, ascorbate and FeSO$_4$. Enzyme activity can often be stimulated by the addition of BSA and catalase, especially if highly purified enzyme preparations are used. Indeed, highly purified enzyme preparations should never be diluted without the addition of BSA (2 mg ml^{-1}) to the dilution buffer (Graebe and Lange, 1990).

If true initial rates are to be obtained, the reaction mixtures must be equilibrated at 30°C for 5 min. The reaction is initiated by addition of the appropriate substrate (c. 1 μM ^{14}C-labelled or 0.01 μM ^{3}H-labelled) in methanol, whereby the final concentration of methanol should not exceed 2% (v/v). The incubation time is kept short (5–10 min), but if the enzyme activity is low, up to 2 h may be necessary to get measurable amounts of the products. It is often not practicable to use saturating substrate concentrations, which must be taken into account when kinetic parameters are determined (e.g. Glick *et al.*, 1979).

Reactions, in which ^{14}C-labelled GAs are used as substrates, are stopped by lowering the pH to 2.5–3.2 with acetic acid to 10% (v/v) or 0.1 N HCl. The incubation products are extracted by solvent partitioning (Section II.A.3,4) or solid-phase extraction using C$_{18}$ reverse-phase Sep-Pak cartridges (Section II.C.4). Qualitative and quantitative analysis by C$_{18}$ reverse-phase radio-HPLC follows. Jensen *et al.* (1986) gives an overview of elution conditions for separating individual GAs by C$_{18}$

reverse-phase HPLC, and programmes can be devised to separate GAs in less than 30 min. Table 16.3 shows two short-gradient programmes which are suitable for the separation of products obtained with GA_{12}-aldehyde C-7 oxidases and GA_{12} C-20 hydroxylases (programme 2), or those obtained with GA_{53} C-20 hydroxylases and GA_{15} 3β-hydroxylases (programme 3), respectively. HPLC equipment with column-selection and valve-switching facilities makes it possible to extract and analyse the incubation products in one procedure, thus eliminating the need for solvent partition-ing or solid phase extraction of the incubation mixture (T. Lange, D. A. Ward and P. Hedden, unpubl. res.).

Enzyme reactions with ^{3}H-labelled substrates are stopped by the addition of charcoal slurry (50 mg ml^{-1} in 25 mM EDTA, pH 7.0, at least 2:1; v/v) to the incubation mixture. The GAs bind to the charcoal and, after centrifugation (3000 × g, 5 min), aliquots of the aqueous supernatant are counted for ^{3}H$_2$O in a liquid scintillation counter (Smith and MacMillan, 1984; Kwak et al., 1988; Griggs et al., 1991). The disadvantages of this very convenient method were mentioned in Section II.

B. C-20 Oxidases

GA C-20 oxidases catalyse the successive oxidation of C-20 from methyl to aldehyde via the alcohol (Scheme 16.1, Steps 1 and 2). The aldehyde can either be further oxidised to a carboxylic acid, resulting in a tricarboxylic GA (Step 3a), or the C-20 can be removed as CO_2 (Kamiya et al., 1986), resulting in a C_{19}-GA (Step 3b). Only C_{19}-GAs are active as hormones, whereas tricarboxylic GAs have no known function and are not converted to C_{19}-GAs (Graebe et al., 1980; Kamiya and Graebe, 1983). Our results indicate that one protein catalyses all four steps in immature seeds (Lange, 1989; Graebe and Lange, 1990), but several C-20 oxidases with different functions were separated in an extract from spinach leaves (Gilmour et al., 1987). The pathway branch point might have some regulatory function by controlling the relative amounts of physiologically active C_{19}-GAs and physiologically inactive GA tricarboxylic acids formed; this is not proven. In spite of their key function in GA-production, very little is known about the enzymes that catalyse the C_{20} to C_{19} conversions.

1. Purification

A GA_{12} C-20 oxidase from pumpkin endosperm (Lange, 1989; Graebe et al., 1991), a GA_{53} C-20 oxidase from immature pea cotyledons (Lange and Graebe, 1989) and a GA_{44} C-20 oxidase from spinach leaves (Gilmour et al., 1986, 1987) have been partially purified.

(a) C-20 oxidases from pumpkin endosperm and pea cotyledons. A GA_{12} C-20 oxidase from pumpkin endosperm and a GA_{53} C-20 oxidase from pea cotyledons were partially purified by a combination of $(NH_4)_2SO_4$ precipitation, gel filtration and anion-exchange chromatography (Lange, 1989; Lange and Graebe, 1989; Graebe and Lange, 1990; Graebe et al., 1991). This resulted in 270-fold purification of the GA_{53} C-20 oxidase from pea with a yield of 15% (3.7 mg protein). The GA_{12} C-20 oxidase from pumpkin endosperm was further purified by hydrophobic interaction chromatography, the overall purification was 70-fold, the yield 0.5% (190 µg protein).

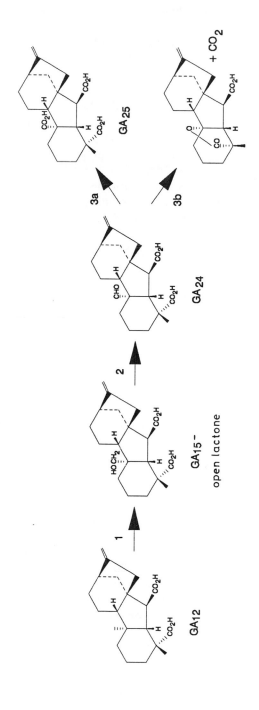

SCHEME 16.1.

In spite of the lower apparent purification of the pumpkin enzyme, the final preparation was much purer than that of the pea enzyme because of the cleaner starting material. In the following description, values for the pea enzyme purification are given in brackets.

Enzyme extraction. 250 g of frozen pumpkin endosperm (or 120 g of frozen immature pea cotyledons) were pulverised in a Waring blender with 1 vol (w/v) of 200 mM Tris-HCl, pH 7.9 (8.6), containing 8 mM DTT. After centrifugation for 20 min at $20\,000 \times g$ (plus another 1 h at $200\,000 \times g$ for the pea preparation), the supernatant was fractionated between 50 and 70% (35 and 55%) saturated $(NH_4)_2SO_4$. After centrifugation, the precipitate was resuspended in 50 ml (25 ml) 100 mM Tris-HCl, pH 7.9 (8.3), containing 4 mM DTT.

Gel filtration chromatography. The crude enzyme preparation, containing 120 mg (700 mg) protein, was applied to an 80×7 cm (49×6 cm) Sephadex G-100 column, which had been equilibrated with and was eluted with the resuspension buffer at a flow rate of 100 ml h^{-1} (120 ml h^{-1}). The GA_{12} C-20 oxidase from pumpkin endosperm eluted between 1.2 and 1.4 l, and the GA_{53} C-20 oxidase from pea cotyledons between 0.52 and 0.63 l.

Anion-exchange chromatography. The $(NH_4)_2SO_4$ precipitation and gel filtration steps were done four times with pumpkin endosperm and twice with pea cotyledons, and the active fractions from the gel filtration were pooled and concentrated by ultrafiltration. The pumpkin enzyme preparation was applied in five runs (each 35 mg protein) to a DEAE-silica 5PW HPLC column (Waters, 7.5×0.8 cm) equilibrated at 24°C with 50 mM Tris-HCl, pH 7.2, containing 4 mM DTT. After elution of the excluded material with equilibration buffer, a linear gradient from 0 to 63 mM NaCl at 1 ml min^{-1} for 60 min eluted the GA_{12} C-20 oxidase between 20 and 50 mM NaCl at 24°C. The pea enzyme preparation (100 mg) was applied to a DEAE-Sephacel anion-exchange column (Pharmacia, 25×2.6 cm) equilibrated at 4°C with 100 mM Tris-HCl, pH 8.3, containing 4 mM DTT. After elution of the excluded material a linear gradient from 100 to 200 mM Tris-HCl at 50 ml h^{-1} for 12 h eluted the GA_{53} C-20 oxidase between 160 and 190 mM Tris-HCl (all at 4°C).

Hydrophobic interaction chromatography was done with the pumpkin enzyme preparation only. Active fractions from five anion-exchange runs were pooled and concentrated by ultrafiltration. The sample (7 mg protein) was adjusted to 2 M NaCl in 100 mM Tris-HCl, pH 7.2 at 24°C, containing 4 mM DTT and loaded on a Phenyl-silica 5PW HPLC column (Waters, 7.5×0.8 cm) equilibrated in the same buffer. After elution of the excluded material, a linear gradient from 2 M to 1 M NaCl was applied for 10 min, and followed by a linear gradient from 1 M to 0 M NaCl for 30 min (both at 1 ml min^{-1} and at 24°C). The GA_{12} C-20 oxidase eluted in two peaks, a narrow one between 1 M and 0.9 M NaCl and a second, broader one, between 0.9 and 0 M NaCl. The first enzyme peak was the purest and most active. No further purification could be obtained by gel filtration HPLC (200SW column, Waters, 30×0.8 cm), equilibrated with 50 mM Tris-HCl, pH 7.2 at 24°C, containing 4 mM DTT and 250 mM NaCl, and only one protein band was found on SDS-PAGE by Ag-staining.

(b) GA$_{44}$ C-20 oxidase from spinach leaves.   Gilmour *et al.* (1987) partially purified GA_{44} C-20 oxidase from 1 kg of spinach leaves using a combination of $(NH_4)_2SO_4$

precipitation, anion-exchange chromatography, hydrophobic interaction chromatography and gel filtration. This resulted in about 100-fold purification and 0.3% yield (140 μg protein).

Extraction. 1 kg of frozen pulverised spinach leaves were extracted in a mortar and pestle with 3 l of extraction buffer (50 mM HEPES buffer, pH 7.0, containing 5 mM DTT and 5 mM ascorbate) and 300 g PVP. After filtration through cheesecloth and centrifugation for 20 min at 14 000 × g, the supernatant was fractionated between 45 and 55% saturated $(NH_4)_2SO_4$. After centrifugation, the precipitate was resuspended in about 20 ml of extraction buffer and dialysed three times against 1 l of extraction buffer for 1 h.

Anion-exchange chromatography. The crude enzyme preparation (0.8 g protein) was applied to a preparative anion-exchange HPLC column (25 × 1.2 cm) packed with Synchroprep AX-300 (SynChrom, Linden, IN) and equilibrated with 50 mM HEPES, pH 7.0, containing 5 mM DTT. After elution of excluded material with the same buffer, a linear gradient from 0 to 0.7 M NaCl in the same buffer was applied at 3 ml min^{-1} for 120 min. GA_{44} oxidase was eluted between 0.3 and 0.4 M NaCl.

Hydrophobic interaction chromatography. Active fractions from the anion-exchange chromatography were pooled and concentrated by ultrafiltration. The sample was adjusted to 10% (w/v) with $(NH_4)_2SO_4$ in 50 mM HEPES, pH 7.0 at 4°C, containing 5 mM DTT and loaded onto a Phenyl-Sepharose CL-4B column, equilibrated with the same buffer. After elution of the excluded material, the column was eluted stepwise at a flow rate of 48 ml h^{-1} with 50 mM HEPES (pH 7.0), containing (1) 5 mM DTT and 0.5 M NaCl, (2) 5 mM DTT and 0.1 M NaCl and (3) 5 mM DTT only. GA_{44} oxidase activity was eluted in Step (2).

Gel filtration chromatography. Active fractions from the hydrophobic interaction chromatography were pooled and concentrated by ultrafiltration. Separation of up to 4 mg of protein in 200 μl was performed on three columns (30 × 0.8 cm) in series. The first column was packed with NuGel 6.5 GP 300 and the other two with NuGel 10 GP 200 (Separation Industries, Metuchen, NJ). A flow rate of 0.4 ml min^{-1} of 50 mM HEPES, pH 7.0, containing 5 or 10 mM DTT and 0.1 M Na_2SO_4 was used.

2. Properties

(a) Substrate specificities and catalytic properties. The purified C-20 oxidase from pumpkin endosperm catalyses the two series of oxidations, from GA_{12} to GA_{25} via GA_{15}-open lactone (20-hydroxy-GA_{12}, GA_{15}-hydroxy acid) and GA_{24}, and from GA_{53} to GA_{17} via GA_{44}-open lactone and GA_{19} (Lange, 1989). The partially purified GA_{53} C-20 oxidase from pea cotyledons catalyses at least the first two oxidation steps from GA_{53} to GA_{19} via GA_{44}-open lactone (Lange and Graebe, 1989). The formation of GA_{17} is low in peas *in vivo* and was not observed in cell-free systems of this species (Kamiya and Graebe, 1983). Neither enzyme accepts GA_{15} or GA_{44} in their lactonic form as substrates (Graebe *et al.*, 1974a; Kamiya and Graebe, 1983). Given the right substrate, the C-20 oxidase from pumpkin endosperm also catalyses the conversion of C_{20}-GAs to C_{19}-GAs. Thus, GA_{53} is converted to GA_{20} in better yields than GA_{12} is converted to GA_4 in the corresponding 13-deoxy reaction (Lange, 1989). The relative activity of the conversion of C_{20}- to C_{19}-GAs in pea cotyledons is much higher than in pumpkin endosperm (Kamiya and Graebe, 1983). Gilmour *et al.* (1987) separated

two C-20 oxidases from spinach leaves: a GA_{53} C-20 oxidase, which catalyses the oxidation of GA_{53} to GA_{44}, and a GA_{44} C-20 oxidase, which oxidises GA_{44} (in its lactonic form) to GA_{19}. The conversion of the lactonic form is found exclusively in vegetative tissues, as has been shown for spinach, barley and pea (Gilmour et al., 1986; Grosselindemann, 1990; Zander, 1990), but not in seed tissues. Also in spinach, the C_{20}- to C_{19}-GA conversion activity was co-purified with the C-20 oxidase and both activities are under photoperiodic control (Gilmour et al., 1986). K_m values given for GA-substrates vary between 0.1 and 0.7 μM depending on preparation, substrate and incubation conditions (Gilmour et al., 1986; Lange, 1989; Lange and Graebe, 1989). The specific activities of the purified enzymes differ considerably. The GA_{44}-oxidase partially purified from spinach leaves has a specific activity of about 10 nmol h^{-1} mg^{-1} (Gilmour et al., 1987). For the partially purified GA_{53} C-20 oxidase from pea cotyledons it is 10 times higher (Lange and Graebe, 1989) and for the pure enzyme from pumpkin endosperm it is 5100 times higher (Lange, 1989).

(b)   Cofactor requirements and the influences of other additives.   All the C-20 oxidases require 2-oxoglutarate (4–5 mM is optimal), ascorbate (4–5 mM), Fe^{2+} (8–500 μM) and O_2 for optimal activity. An exception is the GA_{44} C-20 oxidase from spinach leaves, which is not stimulated by ascorbate (Gilmour et al., 1986). Other additives (BSA at 2–6 mg ml^{-1}, catalase at 0.1–1 mg) often stimulate purified C-20 oxidases. The K_m with respect to 2-oxoglutarate is 100–180 μM for the pea and the pumpkin enzymes (Lange, 1989). DTT (4 mM) strongly inhibits the C-20 oxidases from pea and pumpkin in the presence of added Fe^{2+} (0.5 mM) (Lange and Graebe, 1989; Lange, 1989) or stainless steel, as in normal HPLC columns. This inhibition can be overcome by high concentrations (1 mg ml^{-1}) of catalase (T. Lange, D. A. Ward, P. Hedden and J. E. Graebe, unpubl. res.). The purified GA_{12} C-20 oxidase from pumpkin endosperm is denatured by low concentrations of $(NH_4)_2SO_4$ (half-life 14 h) or its cofactors in the absence of its GA-substrate (half-life 6 h; Lange, 1989).

(c)   pH Optima, pI and molecular weights.   The pH optimum for the GA_{53} C-20 oxidase from pea cotyledons and spinach leaves and for the GA_{12} C-20 oxidase from pumpkin endosperm is 7.0 (Gilmour et al., 1986; Lange and Graebe, 1989; Lange, 1989). The GA_{44} C-20 oxidase from spinach leaves has a pH optimum at 8.0 (Gilmour et al., 1986). The pI values for the C-20 oxidases from peas and pumpkin endosperm are about 5.5 (Lange and Graebe, 1989; Lange, 1989). The molecular weights of the C-20 oxidases from pea, pumpkin and spinach lie between 43 and 45 kDa, as determined by gel filtration. Only the GA_{44} C-20 oxidase from spinach has a slightly lower molecular weight (38 kDa) (Gilmour et al., 1987; Lange and Graebe, 1989; Lange, 1989). The molecular weight of the pumpkin enzyme was verified by SDS-PAGE.

(d) C_{20} to C_{19} conversion.   In at least two cases, purified C-20 oxidases catalyse the C_{20}- to C_{19}-GA conversion (Scheme 16.1, Reaction 3b). In spinach leaves this activity ($GA_{19} \rightarrow GA_{20}$) has a slightly lower pH optimum (pH 6.5) than the initial C-20 oxidasing activity ($GA_{53} \rightarrow GA_{44}$; pH 7.0). Both activities have the same cofactor requirements and are associated with the same approximate molecular weight (40–43 kDa). At least in the pumpkin endosperm, both activities probably reside in the

same enzyme, since all their properties are the same and they are not separated when the enzyme becomes very pure.

(e) Conclusions. The C-20 oxidases studied to date are all very similar. Substrate specificity, cofactor requirements, molecular weight and pH optima are comparable, except for GA$_{44}$ C-20 oxidase from spinach leaves. However, differences occur in the ratio of the C$_{19}$- and C$_{20}$-GAs produced as end-products with enzymes from different sources. Whereas the pumpkin enzyme produces more tricarboxylic acid (Scheme 16.1, Reaction 3a), the enzymes from both spinach and pea produce more C$_{19}$-GA (Reaction 3b).

C. 3β-Hydroxylases

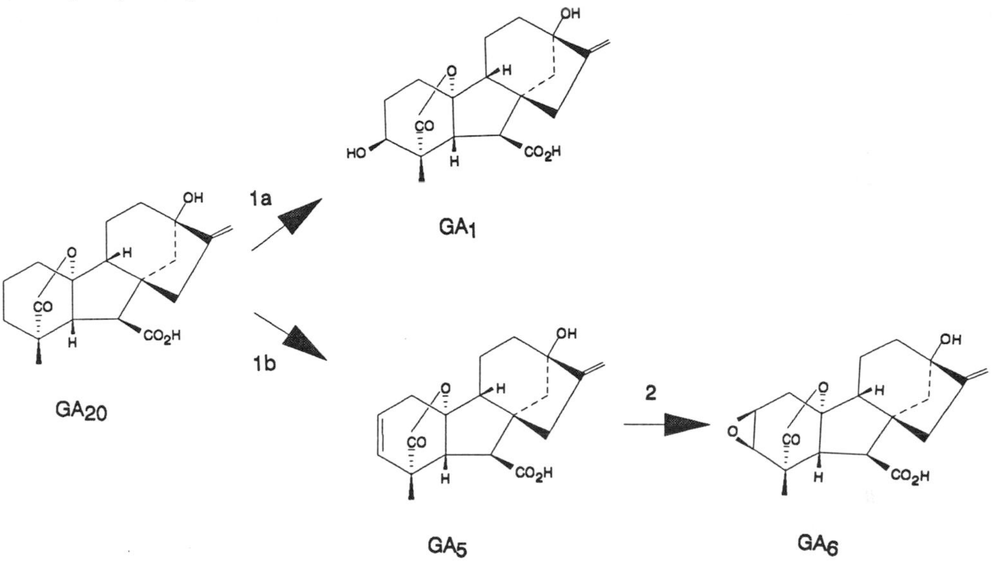

SCHEME 16.2.

Gibberellin 3β-hydroxylases primarily catalyse 3β-hydroxylation (Step 1a), but 2,3-desaturation (Step 1b), followed by epoxidation (Step 2), and even 2β-hydroxylation, have been found associated with the preparations. None of the preparations is pure enough to exclude the possibility that these activities are due to contaminating GA enzymes. 3β-Hydroxylation of C$_{19}$-GAs strongly enhances their physiological activity and may even be essential for internode elongation, which makes this reaction eminently important for plant development.

1. Purification

Kwak *et al.* (1988) and Smith *et al.* (1990) independently purified GA$_{20}$ 3β-hydroxylases from immature bean embryos. The method by Kwak *et al.* (1988) is described below, because it resulted in the higher purification (313-fold) and better recovery (1.7%; 86 µg).

(a) Methanol precipitation. Immature bean embryos (220 g) were extracted with mortar and pestle in (1:1; w/v) 20 mM phosphate buffer, pH 8.0. After filtration through cheesecloth and centrifugation for 1 h at 200 000 × g, methanol (−20°C) was added to the supernatant to a final concentration of 15% (v/v). After centrifugation, the supernatant was dialysed twice for 1 h against 10 vols of 50 mM phosphate buffer, pH 8.0, containing 30% (v/v) glycerol, 0.2 M sucrose and 2 mM DTT (buffer A).

(b) Hydrophobic interaction chromatography. The dialysed enzyme solution (1200 mg protein in about 200 ml) was loaded onto a preparative hydrophobic interaction column (20 × 2.2 cm) packed with Butyl-Toyopearl 650S (Toso, Tokyo) and equilibrated with buffer A, containing 20% saturated $(NH_4)_2SO_4$. After elution of excluded material, a linear gradient from 20% to 0% saturated $(NH_4)_2SO_4$ in buffer A at 3 ml min^{-1} for 67 min, followed by an isocratic flow in only buffer A for 100 min, eluted the enzyme at about 5% saturated $(NH_4)_2SO_4$.

(c) Anion-exchange chromatography. Active fractions from the hydrophobic interaction chromatography were pooled and $(NH_4)_2SO_4$ was removed by ultrafiltration and subsequent dilution with buffer A three times. The sample (49 mg protein in 21 ml) was applied to a DEAE-Toyopearlpak 650S column (Toso, Tokyo, 20 × 2.2 cm) equilibrated with buffer A. After elution of excluded material, a linear gradient from 0 to 0.25 M NaCl in buffer A at 2 ml min^{-1} for 50 min eluted the GA_{20} 3β-hydroxylase between 0.12 and 0.20 M NaCl.

(d) Gel filtration. Active fractions from the anion-exchange chromatography were pooled and concentrated by ultrafiltration. Separation of up to 1.4 mg protein in 400 μl was achieved with a TSK gel G3000SW HPLC column (60 × 0.75 cm, Toso, Tokyo) equilibrated in buffer A. The retention volume of the enzyme was about 17 ml at a flow rate of 0.3 ml min^{-1}. After SDS-PAGE, only one major protein band occurred, which had the same molecular weight as determined by gel filtration HPLC for the 3β-hydroxylase. However, studies with monoclonal antibodies show that this major protein band was not identical with the GA_{20} 3β-hydroxylase (Abe *et al.*, 1991)

2. Properties

(a) GA-substrate specificities and catalytic properties. The partially purified GA_{20} 3β-hydroxylase from immature bean embryos catalyses the conversions of GA_{20} to GA_1 and GA_9 to GA_4 (Kwak *et al.*, 1988; Smith *et al.*, 1990). In addition, Kobayashi *et al.* (1991) found that an epoxidation of GA_5 to GA_6 was associated with the preparation, whereas Smith *et al.* (1990) found conversion of GA_{20} to GA_5 (2,3-desaturation) and low yields of GA_{29} (2β-hydroxylation), but no epoxidation of GA_5. Full certainty as to which of these activities are actually associated with the 3β-hydroxylase must await the pure proteins. Inhibition studies with unlabelled GAs indicate that GA_{15} and GA_{44} are possible substrates, but not GA_1, GA_3, GA_4, GA_6, GA_8, GA_{12}, GA_{19}, GA_{22}, GA_{24}, GA_{29} or GA_{53} (Kwak *et al.*, 1988; Smith *et al.*, 1990). Kwak *et al.* (1988) found that unlabelled GA_5 inhibited their GA_{20} 3β-hydroxylase, and that unlabelled GA_{20} inhibited their GA_5-epoxidase. Smith *et al.* (1990) found that unlabelled GA_5 inhibited all the activities of their preparation (3β-hydroxylation,

2,3-desaturation and 2β-hydroxylation). These results were seen as an indication that the activities are catalysed by a single enzyme. The K_m value of the 3β-hydroxylase from beans for $[2\beta,3\beta\text{-}^3H_2]GA_{20}$ is 0.29 μM and for $[2\beta,3\beta\text{-}^3H_2]GA_9$ 0.33 μM (Kwak et al., 1988). The K_m value determined by Smith et al. (1990) for the overall conversion of $[1\beta,2\beta,3\beta\text{-}^3H_3]GA_{20}$ to the three products was 1.5 μM. These values are affected by the tritium isotope effect. Partially purified preparations from pumpkin endosperm convert GA_{12}-aldehyde, GA_{12}, GA_{15}, GA_{19}, GA_{20}, GA_{24}, GA_{25}, GA_{44}, GA_{46} and GA_{53} to their 3β-hydroxylated derivatives with highest affinity for GA_{15}, GA_{24} and GA_{25} (Volger and Graebe, 1985; Lange, 1989).

(b) Cofactor requirements and the influences of other additives. The cofactor requirements for all the reactions described above are typical for 2-oxoglutarate-dependent dioxygenases: 2-oxoglutarate (optimal concentration between 4 and 5 mM), $FeSO_4$ (0.5 mM), O_2 and ascorbate (4–5 mM). The K_m for 2-oxoglutarate of the GA_{15} 3β-hydroxylase from pumpkin endosperm is 80 μM, which is similar to the values obtained with the C-7 and C-20 oxidases from the same material (Sections III.E and III.B, respectively). Gibberellin A_{20} 3β-hydroxylase from immature bean embryos has a K_m of 250 μM for 2-oxoglutarate (Kwak et al., 1988). The preparation of Smith et al. (1990) is very dependent on added Fe^{2+} and ascorbate. BSA (2 mg ml^{-1}) and catalase (0.5–1 mg ml^{-1}) stimulate the activity of purified 3β-hydroxylases.

Smith et al. (1990) doubt the hypothesis that the GA enzymes, and in any case GA_{20} 3β-hydroxylase from beans, are 2-oxoglutarate-dependent dioxygenases. They base this view on their failure to detect ^{14}C-labelled and ^{18}O-labelled succinate after incubation of 2-oxo-[U-^{14}C]glutarate and [$^{18}O_2$]oxygen, respectively, with their partly purified enzyme preparations. However, the cofactor requirements of the GA enzymes (see Graebe and Lange, 1990) as well as the facts that at least the oxidation of GA_{12}-aldehyde and GA_{12} requires O_2 (Hedden and Graebe, 1982; Kamiya and Graebe, 1983), that [^{14}C]succinate is formed from 2-oxo-[U-^{14}C]glutarate when GA_{12}-aldehyde is converted to GA_{12} (Hedden and Graebe, 1982), that the formation of $^{14}CO_2$ from 2-oxo-[U-^{14}C]glutarate is stimulated by GA_{12}-aldehyde in the same reaction and that hexadione, an analogue of 2-oxoglutarate, strongly inhibits GA_9 3β-hydroxylases and GA_5 epoxidases from beans (Nakayama et al., 1990), all indicate that 2-oxoglutarate is directly involved in the catalytic reaction as demanded for 2-oxoglutarate-dependent dioxygenases.

(c) pH Optimum and molecular weight. The properties of 3β-hydroxylases from pumpkin endosperm and immature bean embryos are quite distinct. The pumpkin GA_{25} 3β-hydroxylase has its maximal activity between pH 6.5 and 7.0, whereas the GA_{20} 3β-hydroxylase from beans has its pH optimum between 7.0 and 8.0. The apparent molecular weight for the GA_{15} 3β-hydroxylase from pumpkin (about 60 kDa by gel filtration) is higher than for most other GA enzymes, whereas the molecular weight for bean GA_{20} 3β-hydroxylase (42–45 kDa by gel filtration and SDS-PAGE) lies within the normal range. A further difference between the properties of the pumpkin and bean 3β-hydroxylases is the strong competetive inhibition of the latter, but not of the former, by deoxy-GA C (DGC) and 16-deoxy-DGC (Saito et al., 1991).

D. 2β-hydroxylases

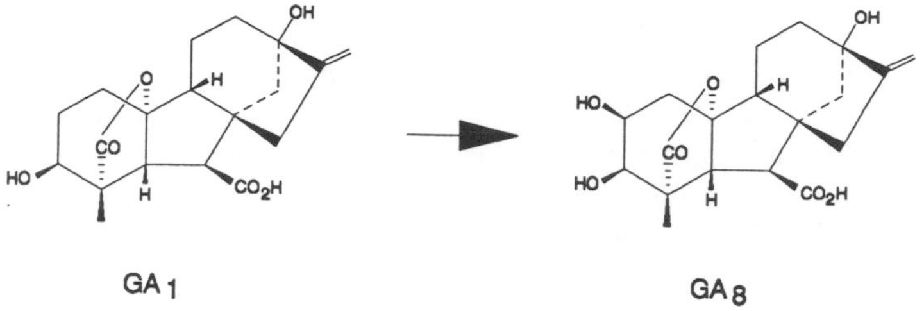

SCHEME 16.3.

2β-Hydroxylation occurs at the end of the GA pathways and results in physiologically inactive GAs. 2β-Hydroxylases therefore have an important function in the regulation of physiological activity. These enzymes also have higher substrate specificities than C-20 oxidases and 3β-hydroxylases.

1. Purification

Gibberellin 2β-hydroxylases have been partially purified from cotyledons of mature *Phaseolus vulgaris* (Smith and MacMillan, 1984; Griggs *et al.*, 1991) and *Pisum sativum* seeds (Smith and MacMillan, 1986).

(a) GA$_1$ and GA$_9$ 2β-hydroxylases from mature bean cotyledons. Griggs *et al.* (1991) partially purified two different 2β-hydroxylases, a GA$_1$ and a GA$_9$ 2β-hydroxylase, from 150 g of mature bean cotyledons by a combination of batch anion-exchange, $(NH_4)_2SO_4$ precipitation, cation-exchange chromatography, gel filtration, and again cation-exchange chromatography. The purification was about 220-fold and the yield 0.5% (0.75 mg) for the GA$_1$ 2β-hydroxylase, the corresponding figures for the GA$_9$ 2β-hydroxylase being 270-fold and 4.5% (5.9 mg). Smith and MacMillan in an earlier publication (1984) reported a similar purification method but did not succeed in separating the two 2β-hydroxylases.

Batch ion-exchange and $(NH_4)_2SO_4$ precipitation. Bean cotyledons (150 g) from imbibed mature seeds were frozen in liquid N_2, pulverised in a coffee grinder and extracted in a Polytron homogeniser with two vols (w/v) of 50 mM Tris-HCl, pH 7.2, containing 20 mM $MgSO_4$ (1:2; w/v). The extract was centrifuged at $16\,000 \times g$ and $45\,000 \times g$ for 45 min each, and 1 vol of DEAE-anion-exchange gel (DE 52, Whatman International Ltd.), equilibrated with and suspended in extraction buffer, was added to the supernatant. After centrifugation at $16\,000 \times g$ for 45 min, the supernatant was fractionated between 25 and 40% saturated $(NH_4)_2SO_4$. After centrifugation, the pellet was resuspended, dialysed three times for 1 h against 2 l of 100 mM NaOAc, pH 6.0, and diluted to 300 ml with the same buffer.

Cation-exchange chromatography. The dialysed enzyme preparation (2300 mg protein) was applied to a preparative cation-exchange column (2.5 × 9.5 cm) packed with S-Sepharose fast flow (Pharmacia) and equilibrated with 100 mM NaOAc, pH 6.0. After elution of non-retained protein a linear gradient of 0 to 350 mM NaCl in the same buffer at 5 ml min^{-1} was applied over 85 min, followed by a linear gradient to 1 M NaCl at 5 ml min^{-1} over 40 min. The GA$_1$ 2β-hydroxylase was eluted between 100 and 200 mM NaCl and the GA$_9$ 2β-hydroxylase between 170 and 250 mM NaCl. The enzymes were subsequently purified separately.

Gel filtration. The active fractions of each enzyme were pooled and dialysed three times for 1 h against 2 l of 20 mM NaOAc buffer, pH 6.0, concentrated by lyophilisation and dissolved in 2 ml H$_2$O. The GA$_1$ and GA$_9$ 2β-hydroxylase preparations (240 mg and 95 mg protein, respectively) were each applied to Superose-12 columns (1.6 × 50 cm, Pharmacia), which were equilibrated and eluted with 50 mM MES buffer, pH 6.0, at a flow rate of 0.4 ml min^{-1}. GA$_1$ 2β-hydroxylase eluted between 86 and 90 ml and GA$_9$ 2β-hydroxylase had a retention volume of 84–88 ml.

Cation-exchange chromatography. The active fractions from the gel filtration of GA$_1$ and GA$_9$ 2β-hydroxylase (16.3 and 24.4 mg protein, respectively) were pooled and applied to Mono S FPLC cation-exchange columns (Pharmacia, HR 5/5), equilibrated with 50 mM MES buffer, pH 6.0. After elution of the excluded material, a linear gradient of 0 to 430 mM NaCl in the same buffer at 1 ml min^{-1} over 70 min eluted the GA$_1$ and GA$_9$ 2β-hydroxylases at similar NaCl concentrations as those described above for the S-Sepharose fast flow cation-exchange chromatography.

(b) 2β-Hydroxylases from mature pea cotyledons. Smith and MacMillan (1986) partially purified 2β-hydroxylases from pea cotyledons. The purification was eight-fold and the yield 37% (170 mg protein). The existence of two different enzymes (GA$_1$/GA$_4$ and GA$_9$/GA$_{20}$ 2β-hydroxylase) was deduced but they were not separated by the methods used.

Methanol precipitation. Cotyledons from about 180 imbibed mature pea seeds were pulverised in liquid N$_2$ and extracted with 150 ml 100 mM Tris-HCl buffer, pH 7.6, containing 2 mM dithioerythritol (DTE). After filtration through cheesecloth and centrifugation for 45 min at 18 000 × g, methanol (final concentration, 20%, v/v) was added to the supernatant. After centrifugation, the supernatant was dialysed against 50 mM Tris-HCl, pH 7.6, for 2 h and concentrated by further dialysis against 50 mM Tris-HCl, pH 7.6, containing 2 mM DTE and 40% (w/v) polyethylene glycol.

Anion-exchange chromatography. The 2β-hydroxylases from pea (2050 mg protein in 40 ml) did not bind to an anion-exchange column (30 × 2.2 cm) packed with DEAE-cellulose (DE-52, Whatman), equilibrated with 50 mM Tris-HCl, pH 7.6, containing 2 mM DTE, and eluted with the same buffer at a flow rate of 20 ml h^{-1}.

Gel filtration. The excluded material from anion-exchange chromatography was concentrated by dialysis against 100 mM Tris-HCl, pH 7.6, containing 2 mM DTE and 40% (w/v) polyethylene glycol. The enzyme preparation (610 mg protein in 18 ml) was applied to a Sephadex G-100 column (60 × 3.2 cm), equilibrated with 100 mM Tris-HCl, pH 7.6, containing 2 mM DTE. The 2β-hydroxylases were eluted between 190 and 210 ml at a flow rate of 17 ml h^{-1}.

2. Properties

(a) Substrate specificities and catalytic properties. Two distinct 2β-hydroxylases occur in cotyledons of mature seeds from both peas and beans. The GA$_1$ 2β-hydroxylases prefer 3β-hydroxylated substrates (GA$_1$ and GA$_4$), whereas the GA$_9$

2β-hydroxylases prefer a 3β-deoxy substrate (GA_9). Gibberellin A_{20} is a poor substrate for both enzymes. In bean cotyledons the GA_9 2β-hydroxylase has low activity, whereas in pea cotyledons both enzymes have similar specific activities (Smith and MacMillan, 1984, 1986; Griggs et al., 1991). The purified 2β-hydroxylases from mature bean seeds do not 3β-hydroxylate or 2,3-desaturate GA_{20} (Griggs, 1991; cf. the 2β-hydroxylation activity suggested for GA_{20} 3β-hydroxylase from immature seeds from the same species, Section III.C.2). Different 2β-hydroxylases occur in different organs of pumpkin seeds: those from the endosperm readily hydroxylate C_{20}-GAs, such as GA_{39}, GA_{28} and GA_{43}, while those from the cotyledons preferentially hydroxylate C_{19}-GAs, such as GA_1 and GA_4. 2β-Hydroxylation of GA_{20} has not been found in pumpkin seeds (Lange, 1989).

K_m values for $[1,2-^3H_2]GA_1$, $[1,2-^3H_2]GA_4$, $[2,3-^3H_2]GA_9$ and $[1,2-^3H_2]GA_{20}$ with bean GA_1-2β-hydroxylase are 0.10, 0.06, 0.54 and 0.30 μM, respectively, and with GA_9 2β-hydroxylase 0.05, 0.14, 0.03 and 12 μM, respectively (uncorrected for the isotope effect; Griggs et al., 1991).

(b) Cofactor requirements. All 2β-hydroxylases require 2-oxoglutarate (0.5–2.5 mM), Fe^{2+} (0.5–1 mM), O_2 and ascorbate (5 mM) for activity. Enzyme activity is stimulated by catalase (0.05 mg ml^{-1}). K_m for 2-oxoglutarate with the GA_1 and the GA_9 2β-hydroxylases from beans are 12 and 4.5 μM, respectively (Griggs et al., 1991). Thus, 2β-hydroxylases have the highest affinity for 2-oxoglutarate of all GA-enzymes studied to date.

(c) pH-Optimum and molecular weight. Optimal activity of 2β-hydroxylases from peas is between pH 7.4 and 7.8 (Smith and MacMillan, 1986). The pH optimum for a GA_{13} 2β-hydroxylase from pumpkin is 7.0–8.0 (Lange, 1989). In contrast to other GA enzymes, 2β-hydroxylases from beans are fairly stable even under acidic conditions (pH 6.0) and do not require a thiol reagent during column chromatography. The GA_1 2β-hydroxylase from beans is less stable during purification than the GA_9 2β-hydroxylase (Griggs et al., 1991), whereas the GA_1 2β-hydroxylase from peas is more stable than the GA_9 2β-hydroxylase from the same source, according to kinetic measurements of thermal denaturation (Smith and MacMillan, 1986). The apparent molecular weight is 44 kDa for both 2β-hydroxylases from pea (Smith and MacMillan, 1986), whereas the two 2β-hydroxylases from beans are different with molecular weights 26 kDa for the GA_1 2β-hydroxylase and 42 kDa for the GA_9 2β-hydroxylase, respectively (Griggs et al., 1991). As in the case of GA_{15} 3β-hydroxylase, the GA_{13} 2β-hydroxylase from pumpkin has the highest molecular weight of all (about 60 kDa; T. Lange, D. A. Ward, P. Hedden and J. E. Graebe, unpubl. res.)

E. Other Gibberellin Oxidases

The activities of the enzymes described in this section are found in both microsomal and soluble enzyme preparations of some species. Thus, C-7 oxidases and 12α-hydroxylases occur both as microsomal and soluble enzymes in pumpkin endosperm and cotyledons (Graebe et al., 1974a; Hedden et al., 1984; Lange, 1989), and a soluble 13-hydroxylase occurs in spinach leaves (Gilmour et al., 1986, 1987), although 13-hydroxylation is usually microsomal (Section II.B). Only the soluble 2-oxoglutarate-dependent varieties are considered in this chapter.

1. GA C-7 oxidases

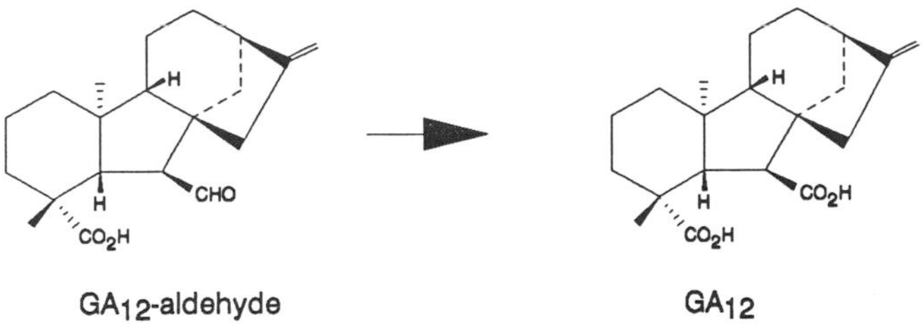

GA$_{12}$-aldehyde $\longrightarrow$ GA$_{12}$

SCHEME 16.4.

The oxidation of the aldehyde group at C-7 to a carboxylic acid is the first step in the biosynthesis of GAs from GA$_{12}$-aldehyde. The soluble, 2-oxoglutarate dependent variety has so far only been found in pumpkin endosperm and cotyledons. Several substrates are oxidised: GA$_{12}$-aldehyde is metabolised to GA$_{12}$, GA$_{14}$-aldehyde to GA$_{14}$ and 12α-hydroxy-GA$_{14}$-aldehyde to 12α-hydroxy-GA$_{14}$ (Graebe and Hedden, 1974; Hedden et al., 1984). However, clear differences in rates of conversion indicate a certain substrate specificity, although this field has been little investigated. The GA$_{12}$-aldehyde C-7 oxidase from pumpkin endosperm is very unstable, and attempts to purify it have been unsuccessful (Volger and Graebe, 1985). The enzyme has absolute requirements for 2-oxoglutarate (optimal concentration 5 mM), Fe^{2+} (0.5 mM) and O$_2$ and the activity is stimulated by ascorbate (5 mM). The K_m for 2-oxoglutarate is 70–80 μM. The release of ^{14}CO$_2$ from 2-oxo[U-^{14}C]glutarate was enhanced six-fold in the presence of the substrate GA$_{12}$-aldehyde (Hedden and Graebe, 1982). These results provide good evidence that the GA$_{12}$-aldehyde C-7 oxidase belongs to the group of 2-oxoglutarate-dependent dioxygenases.

The enzyme has an unusually low pH optimum between pH 5 and 7 and enzyme activity decreases rapidly at higher pH values (Lange, 1989). The molecular weight of the GA$_{12}$-aldehyde C-7 oxidase is about 30 kDa as determined by gel filtration (T. Lange, D. A. Ward, P. Hedden and J. E. Graebe, unpubl. res.). This is significantly lower than most of the other known GA-oxidases and makes the separation of the C-7 oxidase by gel filtration very effective.

2. GA 13-hydroxylases

13-Hydroxylation is an early step in GA-biosynthesis in many plants (Graebe, 1987). It is usually catalysed by microsomal NADPH-requiring enzymes, but one soluble GA$_{12}$ 13-hydroxylase with the properties of a 2-oxoglutarate-dependent dioxygenase has been partially purified from spinach leaves (Gilmour et al., 1986, 1987). Its molecular weight, as determined by gel filtration chromatography, is 28 kDa, which is very low and similar to that of pumpkin C-7 oxidase. Its binding characteristics to

GA$_{12}$ GA$_{53}$

SCHEME 16.5.

an anion-exchange column (Synchroprep AX-300 support, SynChrom, Linden, IN), equilibrated and pre-eluted with 50 mM HEPES, pH 7.0, containing 4 mM DTT, are similar to those of the GA$_{44}$ C-20 oxidase isolated from the same plant tissue (Section III.B). The enzyme elutes in a broad peak with 0.3–0.5 mM NaCl of a 0–0.7 M NaCl gradient in eluting buffer.

3.   GA 12α-hydroxylases

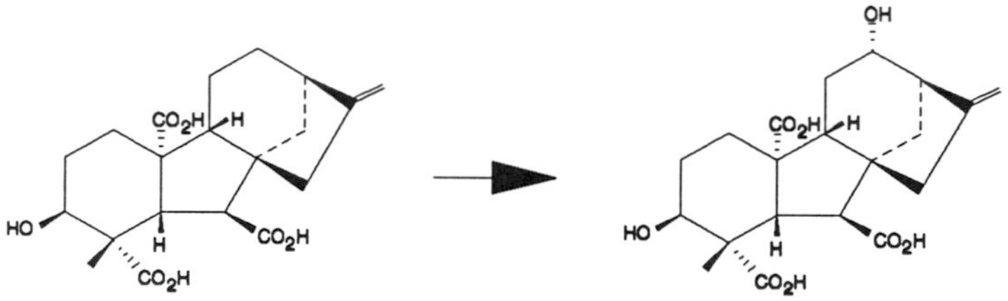

GA$_{13}$ GA$_{39}$

SCHEME 16.6.

The significance of 12α-hydroxylation is unknown. In standard bioassays (stem growth; α-amylase induction), 12α-hydroxylated GAs have much less activity than the corresponding 12α-deoxy GAs. Soluble, 2-oxoglutarate-dependent 12α-hydroxylases introduce 12α-hydroxy groups in several different GAs, e.g. GA$_{12}$, GA$_{25}$, GA$_{13}$ and GA$_{43}$. The enzymes from *C. maxima* endosperm and cotyledons have the typical cofactor requirements of 2-oxoglutarate-dependent dioxygenases, including a pronounced dependency on added FeSO$_4$ for full activity. Their pH optimum is 7.0 and their molecular weight around 45 kDa, as determined by gel filtration chromatography (Lange, 1989).

ACKNOWLEDGEMENTS

We wish to thank Mr Dennis A. Ward, Dr Peter Hedden and Professor Jake MacMillan for fruitful discussions and constructive criticism. Work in our laboratory was supported by the Deutsche Forschungsgemeinschaft.

REFERENCES

Abe, H., Kamiya, Y. and Sakurai, A. (1991). *In* "Abstracts of the 14th International Conference on Plant Growth Substances", Abstract TU-C8-PI6, p. 63. Agricultural University, Wageningen.

Albone, K. S., Gaskin, P., MacMillan, J., Phinney, B. O. and Willis, C. L. (1990). *Plant Physiol.* **94**, 132–142.

Beale, M. H. and Willis, C. L. (1991). *In* "Methods in Plant Biochemistry", Vol. 7 (D. V. Banthorpe and B. C. Charlwood, eds), pp. 289–330. Academic Press, London.

Beale, M. H., Hooley, R. and MacMillan, J. (1986). *In* "Plant Growth Substances 1985" (M. Bopp, ed.), pp. 65–73. Springer-Verlag, Berlin, Heidelberg and New York.

Birch, A. J., Richards, R. W., Smith, H., Harris, A. and Whalley, W. B. (1959). *Tetrahedron* **7**, 241–251.

Birnberg, P. R., Maki, S. L., Brenner, M. L., Davies, G. C. and Carnes, M. G. (1986). *Anal. Biochem.* **153**, 1–8.

Bowen, D. H., MacMillan, J. and Graebe, J. E. (1972). *Phytochemistry* **11**, 2253–2257.

Dennis, D. T. and West, C. A. (1967). *J. Biol. Chem.* **242**, 3293–3300.

Gilmour, S. J., Zeevaart, J. A. D., Schwenen, L. and Graebe, J. E. (1986). *Plant Physiol.* **82**, 190–195.

Gilmour, S. J., Bleecker, A. B. and Zeevaart, J. A. D. (1987). *Plant Physiol.* **85**, 87–90.

Glick, N., Landman, A. D. and Roufogalis, B. D. (1979). *Trends Biochem. Sci.* **4**, N82–N83.

Graebe, J. E. (1969). *Planta* **85**, 171–174.

Graebe, J. E. (1972). *In* "Plant Growth Substances 1970" (D. J. Carr, ed.), pp. 151–157. Springer-Verlag, Berlin, Heidelberg and New York.

Graebe, J. E. (1982). *In* "Plant Growth Substances 1982" (P. F. Wareing, ed.), pp. 71–80. Academic Press, London.

Graebe, J. E. (1987). *Ann. Rev. Plant Physiol.* **38**, 419–465.

Graebe, J. E. and Hedden, P. (1974). *In* "Biochemistry and Chemistry of Plant Growth Regulators" (K. Schreiber, H. R. Schütte and G. Sembder, eds), pp. 1–16. Academy of Science of the German Democratic Republic, Institute of Plant Biochemistry, Halle (Saale).

Graebe, J. E. and Lange, T. (1990). *In* "Plant Growth Substances 1988" (R. P. Pharis and S. B. Rood, eds), pp. 314–321. Springer-Verlag, Berlin, Heidelberg and New York.

Graebe, J. E., Dennis, D. T., Upper, C. D. and West, C. A. (1965). *J. Biol. Chem.* **240**, 1847–1854.

Graebe, J. E., Bowen, D. H. and MacMillan, J. (1972). *Planta* **102**, 261–271.

Graebe, J. E., Hedden, P., Gaskin, P. and MacMillan, J. (1974a). *Phytochemistry* **13**, 1433–1440.

Graebe, J. E., Hedden, P. and MacMillan, J. (1974b). *In* "Plant Growth Substances 1973", pp. 260–266. Hirokawa, Tokyo.

Graebe, J. E., Hedden, P., Gaskin, P. and MacMillan, J. (1974c). *Planta* **120**, 307–309.

Graebe, J. E., Hedden, P. and Rademacher, W. (1980). *In* "Gibberellins – Chemistry, Physiology and Use" (J. R. Lenton, ed.), Monograph 5, pp. 31–47. British Plant Growth Regulation Group, Wantage.

Graebe, J. E., Lange, T., Pertsch, S. and Stöckl, D. (1991). *In* "Gibberellins" (N. Takahashi, B. O. Phinney and J. MacMillan, eds), pp. 51–61. Springer-Verlag, New York.

Griggs, D. L. (1991). Ph.D. Thesis, University of Bristol, UK.

Griggs, D. L., Hedden, P. and Lazarus, C. M. (1991). *Phytochemistry* **30**, 2507–2512.

Grosselindemann, E. (1990). Ph.D. Thesis, Göttingen University, Germany.

Grosselindemann, E., Graebe, J. E., Stöckl, D. and Hedden, P. (1991). *Plant Physiol.*, **96**, 1099–1104.

Harlow, E. and Lane, D. (1988). "Antibodies, A Laboratory Manual". Cold Spring Harbor Laboratory, New York.

Hedden, P. (1983). *In* "The Biochemistry and Physiology of Gibberellins", Vol. 1 (A. Crozier, ed.), pp. 99–149. Praeger, New York.

Hedden, P. (1987). *In* "The Principles and Practice of Plant Hormone Analysis", Vol. 1 (L. Rivier and A. Crozier, eds), pp. 9–110. Academic Press, London.

Hedden, P. and Graebe, J. E. (1982). *J. Plant Growth Reg.* **1**, 105–116.
Hedden, P., Graebe, J. E., Beale, M. H., Gaskin, P. and MacMillan, J. (1984). *Phytochemistry* **23**, 569–574.
Jensen, E., Crozier, A. and Monteiro, A. M. (1986). *J. Chromatogr.* **367**, 377–384.
Kamiya, Y. and Graebe, J. E. (1983). *Phytochemistry* **22**, 681–689.
Kamiya, Y., Takahashi, N. and Graebe, J. E. (1986). *Planta* **169**, 524–528.
Kobayashi, M., Kwak, S. S., Kamiya, Y., Yamane, H., Takahashi, N. and Sakurai, A. (1991). *Agric. Biol. Chem.* **55**, 249–251.
Kwak, S. S., Kamiya, Y., Sakurai, A., Takahashi, N. and Graebe, J. E. (1988). *Plant Cell Physiol.* **29**, 935–943.
Lange, T. (1989). Ph.D. Thesis, Göttingen University, Germany.
Lange, T. and Graebe, J. E. (1989). *Planta* **179**, 211–221.
Lew, F. T. and West, C. A. (1971). *Phytochemistry* **10**, 2065–2076.
Nakayama, I., Kamiya, Y., Kobayashi, M., Abe, H. and Sakurai, A. (1990). *Plant Cell Physiol.* **31**, 1183–1190.
Patterson, R. J., Rappaport, L. and Breidenbach, R. W. (1975). *Phytochemistry* **14**, 363–368.
Saito, T., Kwak, S. S., Kamiya, Y., Yamane, H., Sakurai, A., Murofushi, N. and Takahashi, N. (1991). *Plant Cell Physiol.* **32**, 239–245.
Smith, V. A. and MacMillan, J. (1984). *J. Plant Growth Reg.* **2**, 251–264.
Smith, V. A. and MacMillan, J. (1986). *Planta* **167**, 9–18.
Smith, V. A., Gaskin, P. and MacMillan, J. (1990). *Plant Physiol.* **94**, 1390–1401.
Takahashi, M., Kamiya, Y., Takahashi, N. and Graebe, J. E. (1986). *Planta* **168**, 190–199.
Takahashi, N., Yamaguchi, I. and Yamane, H. (1986). *In* "Chemistry of Plant Hormones" (N. Takahashi, ed.), pp. 57–151. CRC Press, Boca Raton, FL.
Turnbull, C. G. N., Crozier, A., Schwenen, L. and Graebe, J. E. (1985). *Planta* **165**, 108–113.
Volger, H. and Graebe, J. E. (1985). *In* "Book of Abstracts, 12th International Conference on Plant Growth Substances, Heidelberg", Abstract 12, p. 10.
Zander, M. (1990). Ph.D. Thesis, Göttingen University, Germany.

17   Ethylene Metabolism

WING-KIN YIP and SHANG FA YANG

Department of Vegetable Crops, University of California, Davis, CA 95616, USA

I. INTRODUCTION

Ethylene is a plant hormone which regulates many aspects of plant growth and development, ranging from seed germination to organ senesence (Abeles, 1973). Ethylene exerts its effects at concentrations between 0.01 and 10 ppm (Abeles, 1973; Burg, 1973). Because of the spectacular effects of ethylene, such as the triple response in pea seedlings and the initiation of ripening in fruits, interest in ethylene research

METHODS IN PLANT BIOCHEMISTRY Vol. 9
ISBN 0-12-461019-6

was initiated early in this century (Neljubov, 1901; Crocker *et al.*, 1932). Research into the action of ethylene was, however, slow until the late 1950s when the technique of gas chromatography (GC) was developed. With the advent of the flame ionisation detector (FID), the limit of detection of ethylene became as low as 0.01 ppm in a 1-ml sample. At this sensitivity the endogenous ethylene production of intact plant tissues can be monitored (Burg and Stolwijk, 1959). Since then, the knowledge about ethylene in higher plant tissues has rapidly advanced, and ethylene has been clearly identified as an endogenous growth regulator.

The discovery of 1-aminocyclopropane-l-carboxylic acid (ACC) as the immediate precursor of ethylene by Adams and Yang (1979) was a major step toward the understanding of ethylene biosynthesis. This finding led to the identification of two enzymes, ACC synthase and ACC oxidase, which act in series to convert *S*-adenoylmethionine to ACC and ACC to ethylene. The characterisation of ACC synthase and ACC oxidase, and the elucidation of their primary structures by DNA sequence analysis, constitute the major research efforts in recent years. Those research efforts have led to a better understanding of the regulation of ethylene biosynthesis and, more importantly, opened the possibility of manipulating ethylene biosynthesis via biotechnology.

In this chapter we shall describe: general methods for ethylene and ACC analysis; the enzymology and primary structure of ACC synthase; and the characterisation of ACC oxidase.

II. ANALYSIS OF ETHYLENE AND ACC

A. Measurement of Ethylene

Gas chromatography using an alumina-packed column equipped with a FID is the device most commonly used to analyse ethylene (Abeles, 1973). A short column can be easily prepared by packing 40–60 mesh alumina (or Porapak) in a 0.5 m × 3 mm i.d. copper tube. The column must be tightly packed to ensure good performance. For best results, it is often necessary to bake the column for 1 or 2 days at around 150–180°C with a slow flow of carrier gas (nitrogen, 20 ml min^{-1}). Once the column has been conditioned, it is advisable to maintain the oven temperature at about 80°C and fix the carrier gas flow rate at around 40 ml min^{-1} for operation. The flow rates of hydrogen and oxygen are usually set at 20 ml min^{-1} and 10 ml min^{-1}, respectively, to maintain the flame for the FID. Under these conditions, the retention time for ethylene should be about 0.6–0.9 min, and the detection limit is about 0.01 ppm. Normally 1–2 ml of gas sample can be injected.

B. Trapping and Releasing of Ethylene

Plant tissues are usually kept in closed containers for various periods of time to allow accumulation of ethylene in the head space before samples are taken for ethylene analysis. In some cases, if the build-up of ethylene in the container becomes a concern, ethylene can be first absorbed into mercuric perchlorate (0.25 M mercuric oxide in 2 M perchloric acid; Young *et al.*, 1952) during the incubation period. This mercuric perchlorate solution is then transferred into a small closed container or test tube and an equal volume of 4 M lithium chloride is injected to release ethylene for analysis by (GC). This trapping method is especially useful if radioactive compounds such as methionine or ACC are employed as ethylene precursors. Radioactive ethylene can be adsorbed directly into a small filter paper disc (1 cm in diameter) which is moistened with 50–100 μl of mercuric perchlorate reagent and hung in the reaction flask. At the end of the incubation period the filter paper is transferred into a scintillation vial and the radioactivity is assayed by scintillation counting. The counting efficiency of this method is about 30% (Liu *et al.*, 1985). A detailed description of the measurement and handling techniques for ethylene has been documented by Saltveit and Yang (1987).

C. Measurement of ACC

Ethylene precursor ACC, a non-protein amino acid, was first isolated from perry pears and cider apples by Burroughs (1957) after yeast fermentation. Burroughs also found that during the storage process the level of ACC in pears increased and he suggested that changes in this compound were in some way related to fruit ripening. Despite its seemingly fragile cyclopropane structure, ACC is quite stable. ACC standard solutions can be stored in 10 mM HCl at 0°C for several months without degradation. Like glycine or α-aminoisobutyric acid, ACC does not possess a chiral carbon. However, it has four different stereoisomers when one of the four hydrogens attached to C-2, C-3 of ACC is substituted. There are many methods to assay ACC, including high performance liquid chromatography (HPLC) for free ACC (Grady and Bassham, 1982), or its 2,4-dinitrophenyl-, phthalimido- or phenythiocyanate derivatives (Savidge *et al.*, 1983; McGaw *et al.*, 1985; Lanneluc-Sanson *et al.*, 1986); GC-MS for the methyl ester of phthalimido-ACC (McGaw *et al.*, 1985) or the propyl ester of benzoyl ACC (Hall *et al.*, 1989); and colorimetric analysis after enzymatic degradation of ACC to 2-oxobutanoic acid by ACC deaminase (Honma, 1983). The method we routinely employ is GC after chemical oxidation of ACC to ethylene (Boller *et al.*, 1979; Lizada and Yang, 1979) because of its simplicity and sensitivity. The indirect method described by Lizada and Yang (1979) is the most widely used. In this method ACC is converted to ethylene with NaOCl (commercial bleach) in the presence of Hg^{2+} at 0°C, as shown below in Scheme 17.1.

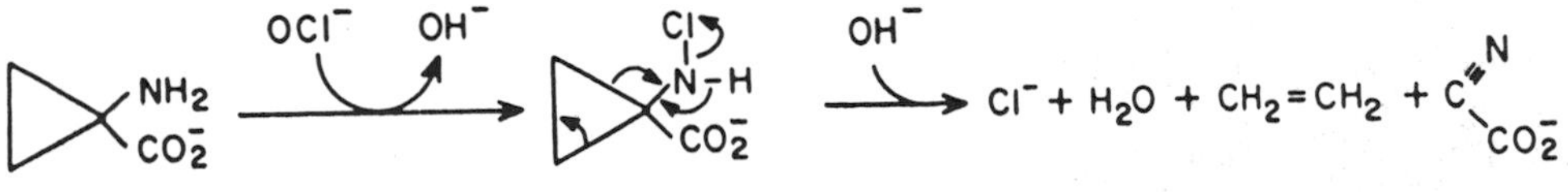

SCHEME 17.1.

Due to the high efficiency of the conversion of ACC to ethylene and the high sensitivity of GC for ethylene, the detection limit of this method is around 5 pmol of ACC and can be completed within 10 min. The following procedure for ACC assay is adopted from that described by Lizada and Yang (1979), with some modifications.

1. Extraction of ACC from plant tissues

ACC has been extracted from plant tissues using 80% ethanol, 80% methanol, 5% sulphosalicylic acid, 5% perchloric acid or water. We routinely isolate ACC from plant tissue (1–10 g) with two extractions of 5 vols of hot 80% ethanol (50–70°C) for 30 min. After the ethanol is evaporated *in vacuo*, the residue is dissolved in a known amount of water (this is referred to as the sample extract). ACC in the extract can be assayed directly or further purified by passage through a cation-exchange column. In the extraction of ACC from leafy tissues, the removal of pigments from the sample extract by extraction with an equal volume of chloroform improves the conversion efficiency of ACC to ethylene with the NaOCl reagent. If the ACC concentration in the tissue is low (> 0.1 nmol g^{-1}), a larger amount of sample extract is required. In order to improve the conversion of ACC to ethylene, it is necessary to further purify the sample. The sample extract is passed through a small column (disposable micropipette tip) containing 1 ml cation-exchange resin Dowex-50 (H$^+$ form) at a flow rate of 0.5 ml min^{-1}. The column is then washed with 10 ml of water and the ACC retained in the column is eluted with 5 ml of 2 N NH$_4$OH. After the removal of ammonia *in vacuo*, the residue is dissolved in a known volume of water and is ready for ACC assay. The extraction efficiency can be easily estimated by adding a known amount of [^{14}C]ACC to the medium prior to tissue extraction, and determining the radioactivity after the purification procedures.

2. Assay conditions

Sample solutions containing plant extracts or standard ACC solutions are assayed in 15-ml test tubes at 0°C. To each tube containing a sample solution is added 5 μmol HgCl$_2$ and the volume is brought up to 0.9 ml with H$_2$O. These reaction tubes are kept on ice. After the addition of 0.1 ml of a cold mixture of 5% NaOCl and saturated NaOH (2:1; v/v), which contains about 45 μmol NaOCl, the test tube is immediately sealed with a serum stopper. The test tube is then vortexed for 10 s and returned to ice for 2–5 min. Prior to gas sampling, the reaction mixture is vortexed for another 5–10 s, and a 1-ml gas sample is then withdrawn for ethylene analysis by GC.

The conversion efficiency of ACC to ethylene in each plant extract is determined separately with a replicate sample containing an appropriate amount of ACC as an internal standard. The amount of ACC added as an internal standard should be at least three times the endogenous ACC in the sample. Normally the amount of plant extract employed is adjusted so that it contains no more than 0.7 nmol ACC. Thus, we routinely employ 2.0 nmol of ACC as an internal standard. The efficiency of ACC to ethylene conversion in a plant extract is calculated as follows: [(ethylene released from a sample with 2 nmol of ACC as an internal standard) − (ethylene released from a replicate sample without internal standard)] ÷ 2 nmol. The amount of ethylene (nmol) released in the reaction tube is calculated as ppm (nl ml^{-1}) of C$_2$H$_4$ in one ml gas sample × 14 ml ÷ 24 nl nmol^{-1}. The amount of ACC (nmol) in the sample is calculated by the following equation: the amount of ACC in the sample = the amount of ethylene from the sample ÷ conversion efficiency of the sample. It should be noted that the conversion efficiencies of standard ACC solution should be at least 50%. A lower efficiency indicates that there must be something wrong in the assay. To get reasonable accuracy, it is important that the conversion efficiency is greater than 50%. For those plant extracts which yield low conversion efficiencies, purification by cation-exchange resin, as described above, prior to the ACC assay is required.

3. Specificity and interferences

The present ACC assay method is vulnerable if any compounds other than ACC are present which evolve ethylene under the assay conditions. Lizada and Yang (1979) found no such material in avocado fruit extracts after Dowex-50 purification. Knee (1984) has observed an uncharacterised compound in apple fruit that is not ACC but which produces small amounts of ethylene under ACC assay conditions. The interference is small (1–2%) in ripe apples; however, in unripe apple fruit tissue where endogenous ACC is extremely low (< 0.1 nmol g^{-1}), ACC could be overestimated by up to 20%. To ascertain the specificity of this assay method for ACC, Lizada and Yang (1979) and Knee (1984) subjected their plant extracts to paper chromatograhy or paper electrophoresis, and each zone was then analysed for ethylene production in the ACC assay system. The extent of ethylene production from the non-ACC zones was found to be insignificant. Knee (1984) has also tested a number of amino acids for their activity to produce ethylene in the ACC assay system and found that certain natural amino acids such as homoserine, 2,4-diaminobutyrate, and homocysteine yield traces of ethylene. However, he concluded that physiological levels of these compounds were too low to be a significant source of interference.

Although ethanol alone produces no ethylene with the NaOCl reagent, an ethanol extract of etiolated pea seedlings yields considerably more ethylene than the same extract in which ethanol has been removed by evaporation (Nieder et al., 1986). This observation indicates that this extra ethylene is derived from ethanol. Various amines including homoserine, amino acids, ammonia, and mono-, di- and tri-alkylamines are active in facilitating ethylene production from ethanol in the presence of NaOCl. The above-mentioned observation cautions that when ethanol is used for ACC extraction, it is important that ethanol be evaporated before assay. If it is impractical to remove all ethanol, we recommend using acetone or methanol as the extraction medium, which does not yield ethylene upon reaction with NaOCl in the presence of ammonia/amines.

Although Lizada and Yang's method is both simple and sensitive, it does suffer several deficiencies. In addition to the interfering compounds mentioned above, low and variable conversion efficiencies could render the method less accurate. Although the low conversion efficiencies can be corrected by the internal standards as described above, the accuracy of this correction becomes poorer as the conversion efficiency decreases. Thus, we recommend that the conversion efficiencies should be at least 50%. Sitrit et al. (1987) observed that phenolic compounds existing in avocado fruit pedicel extracts greatly reduced the conversion efficiency. Removal of phenolic compounds from the extract by polyvinylpolypyrrolidone (PVP) or anion-exchange resin, or precipitation of phenolic compounds with lead acetate, or purification of ACC in the extract by Dowex-50 cation-exchange resin as described above, resulted in significant increases in the conversion efficiency. Nieder et al. (1986) observed that ammonia, primary amines (including amino acids), and secondary amines caused low yields of ethylene from ACC in this assay. The presence of high concentrations of ammonia and amino acids in plant extracts or amine bufffers in enzyme mixtures could lower the conversion efficiency. In these cases, the amount of plant extracts to be assayed should be reduced, and the use of primary and secondary amine buffers should be avoided (see Section III.B).

Although other direct analyses of ACC by HPLC or GC-MS have not been widely adopted for routine analysis of ACC, as these procedures are time consuming and require expensive equipment, these methods are useful when improved precision is required.

III. ENZYMOLOGY AND PRIMARY STRUCTURE OF ACC SYNTHASE

Ethylene biosynthesis in plant tissue is normally very low but is greatly promoted at certain developmental stages, such as fruit ripening, by auxin treatment or under certain environmental stresses (Yang and Hoffman, 1984). In all cases the increased ethylene biosynthesis results from increased synthesis of ACC from AdoMet catalysed by ACC synthase (S-adenosyl-L-methionine methylthioadenosine-lyase, EC 4.4.1.14). Based on the observations that the enzyme activity is inhibited by pyridoxal 5'-phosphate (PLP) inhibitors such as aminoethoxyvinylglycine and aminoxyacetic acid, and that PLP is required for maximal activity, ACC synthase is thought to be a PLP-utilising enzyme. ACC synthase has strict substrate specificity; the naturally occurring isomer $(-)$-Ado-L-Met acts as its substrate with a K_m of 20 μM; $(+)$-Ado-L-Met is a potent inhibitor with a K_i of 15 μM; while $(-)$- and $(+)$-Ado-D-Met are inactive as substrates (Khani-Oskouee et al., 1987).

A. Assay Method

For assaying ACC synthase activity from various plant tissues, the following isolation procedures were found to be satisfactory. Typically, 2 g of plant tissue is homogenised with a mortar and pestle in 2 ml of ice-cold buffer containing 400 mM potassium phosphate (pH 8.5), 10 μM PLP, 0.5% 2-mercaptoethanol, and 20% glycerol. In some tissues such as apple and pear, 1% Triton X-100 is required during homogenisation. The homogenate is then passed through four layers of laboratory tissue paper (Kimwipe) and centrifuged at 14 000 $\times$ g in 4°C for 5 min. The resulting supernatant (2.5 ml) is then filtered through a G-25 Sephadex column (10 ml bed volume) which has been equilibrated with 20 mM potassium phosphate (pH 8.5), 10 μM PLP, 2 mM 2-mercaptoethanol, and 20% glycerol. The protein fraction in the void volume (3.5 ml) is collected for ACC synthase activity assay. Aliquots of enzyme preparation are incubated at 30°C for 15 min with 50 mM K-HEPES (pH 8.0), 10 μM PLP and 200 μM AdoMet in a total volume of 0.6 ml. The reaction is initiated by adding AdoMet and stopped by adding 300 μl of 10 mM HgCl$_2$. ACC synthase activity is assayed by following the formation of ACC. The amount of ACC formed is determined according to Lizada and Yang (1979) as described above (Section II.C).

B. Interference

In the course of purification and activity measurement of ACC synthase, we noticed that certain buffers or compounds interfered with Lizada and Yang's method for ACC assay by causing poor conversion of ACC to ethylene. We have examined the influence of various salts and buffers at 0.1 M concentration on the conversion efficiency of ACC to ethylene (Nieder *et al.*, 1986). Ammonia and aliphatic primary and secondary amines greatly decrease the conversion efficiency. Pyridine, tertiary and quaternary amines, however, show little effect. The concentrations of amines to give 50% inhibition are estimated to be 2.5, 32 and 37 mM for ammonia, glycine and Tris, respectively. In the course of ACC synthase purification, 70–80% saturated ammonium sulphate is generally used to precipitate the enzyme from the crude extract. Therefore, it is important that ammonium sulphate be removed in order to have accurate assay of ACC synthase activity by Lizada and Yang's method. Since Tris and glycine also interfere with ACC assay at relatively low concentration, it is recommended that phosphate or HEPES (a tertiary amine) buffer is used for ACC synthase extraction and purification.

C. Enzyme Sources for Purification

Fruit tissues, such as apple (Yip *et al.*, 1991), tomato (Bleecker *et al.*, 1986), zucchini (Sato and Theologis, 1989) and winter squash (Nakajima *et al.*, 1988) are enzyme sources for purification of ACC synthase. Because of the low abundance of ACC synthase protein (i.e. 0.001% in wound tomato pericarp, 0.003% in ripening apple), and its apparent lability, the purification of the enzyme has been slow. Using conventional chromatographic techniques, a partially purified ACC synthase preparation can be obtained. Monoclonal antibodies against ACC synthase can then be raised and screened. Using these specific antibodies, an immunoaffinity column was prepared, which has shown to be a very powerful and useful tool to purify and to characterise ACC synthase. As an example, we shall briefly descibe the procedures we employed to purify ACC synthase from apple fruit (Yip *et al.*, 1991; Dong *et al.*, 1991).

D. Purification of ACC Synthase from Apple Fruit

1. Plant material

Apples (*Malus sylvestris* Mill. var Golden Delicious) that had been stored at 0–4°C were placed at 24°C for about 5 days until their ethylene production rate reached around 10 nmol $g^{-1} h^{-1}$, at which time their skin colour turned slightly yellow. These fruits were then transferred to 0°C and stored for at least 3 days before extraction. For an unknown reason, apple fruits immediately after warming yield relatively low enzyme activity.

2. Pellet preparation and solubilisation

Unlike ACC synthase isolated from other sources, apple ACC synthase is associated with the pellet fraction and can be solubilised in active form with Triton X-100. Apples that had been stored at 0°C were peeled, cut into small slices (2 cm × 2 cm), and homogenised at maximum speed for 2 min in a Waring blender with an equal volume of homogenisation buffer (w/v) containing 400 mM potassium phosphate (pH 8.5), 1 mM EDTA, 0.5% 2-mercaptoethanol and 10 μM PLP. The homogenate was squeezed through four layers of cheesecloth and centrifuged at $28\,000 \times g$ for 30 min. The pellet fraction was collected and stored at $-20°C$ until use. Pellet preparation stored at $-20°C$ was thawed and resuspended in buffer A (20 mM potassium phosphate at pH 8.5, 1 mM EDTA, 2 mM 2-mercaptoethanol, 10 μM PLP and 20% glycerol) to a concentration of 2 mg protein ml^{-1}. To the pellet suspension was added with stirring an equal volume of the solubilisation buffer, which consisted of buffer A plus 0.2% (v/v) Triton X-100. After incubation for 30 min, the mixture was centrifuged at $28\,000 \times g$ for 20 min. The resulting supernatant, which contained ACC synthase activity, was collected and further purified. All procedures were carried out at $1°C$.

3. Partial-purification of the solubilised ACC synthase

The solubilised ACC synthase was further purified using conventional chromatographic media including DEAE-Sepharose, hydroxyapatite, phenyl-Sepharose and aminohexyl-Sepharose as shown in Table 17.1. This scheme allowed us to purify the enzyme over 5000-fold with 30% yield. The specific activity of the enzyme at this partially purified stage is 100 000 units mg^{-1} protein (1 unit is defined as the catalytic formation of 1 nmol ACC per h) and its purity is estimated to be about 20%.

Table 17.1. Purification of ACC synthase from the homogenate of 35-kg apple fruits.

Purification step	Total activity (units)	Protein (mg)	Specific activity (units mg^{-1})	Recovery (%)	Purification (-fold)
Pellet	385 500	19 670	19.6	100	—
DEAE-Sepharose	258 000	2 407	107.2	66.9	5.5
Hydroxylapatite	195 000	110.9	1 758	50.6	90
Phenyl-Sepharose	172 000	9.08	18 940	44.6	966
AH-Sepharose	146 200	1.378	106 100	37.9	5410

Reproduced from Yip *et al.* (1991).

4. Production and screening of monoclonal antibodies

A partially purified ACC synthase preparation (60 000 units mg^{-1} protein) was injected into two mice, and after two subsequent boosts, sera from both mice were tested for anti-ACC synthase activity. Spleen cells were isolated from one of the two mice that demonstrated anti-ACC synthase activity and were fused with myeloma cells to generate hybridoma. Hybridoma were then screened by their ability to secrete antibody that could precipitate ACC synthase activity. Monoclonal antibodies were obtained by limiting dilution of the positive clones and by repeated subcloning. Finally, eight different monoclonal hybridoma cell lines were selected; two of the eight monoclones recognised ACC synthase in Western blot analysis. The monoclonal line 6A10, which tested positive in Western blots, was chosen to generate anti-ACC synthase ascitic fluid for further experiments.

5. *Immunoaffinity purification of ACC synthase*

An anti-ACC synthase ascitic fluid (monoclonal line 6A10) was incubated with protein A-agarose and the adsorbed IgG antibodies were covalently conjugated to the protein A-agarose with coupling reagent dimethyl pimelimidate (Reeves *et al.*, 1981). This immunoaffinity agarose gel had a binding capacity of 200 units μl^{-1} and was used to purify ACC synthase from a relatively crude enzyme preparation (specific activity of about 100 units mg^{-1} protein). Before loading onto a small column containing 50–100 μl of immunoaffinity agarose gel, the enzyme preparation was dialysed to remove 2-mercaptoethanol, which would cause the leakage of IgG subunits from the gel matrix. The loading rate was kept under 0.1 ml min^{-1}, at which rate over 95% of ACC synthase was adsorbed. After washing, the ACC synthase was eluted with 2% SDS in H$_2$O. As a result, a single 48 kDa protein band was eluted from the immuno-affinity column. The use of 2% SDS completely eluted all adsorbed protein but resulted in total inactivation of the enzyme activity. However, before elution, the native apple ACC synthase bound on the immunoaffinity column exhibited approximately one-third of its initial activity, a result similar to that reported for the tomato enzyme (Bleecker *et al.*, 1988). The use of immunoaffinity purification for apple ACC synthase enabled us to obtain enough material for structure analysis, as will be described later in this chapter.

E. Suicidal Inactivation of ACC Synthase

An important characteristic of ACC synthase is that its substrate, AdoMet, serves as a suicide inactivator. Satoh and Yang (1988) demonstrated that when a partially purified ACC synthase preparation isolated from tomato fruit was incubated with Ado[3,4-^{14}C]Met and the resulting protein was analysed by SDS/PAGE, only one radioactive protein was observed. This protein was judged to be ACC synthase based on the observations that its molecular weights was 50 kDa and it was specifically bound to a monoclonal antibody against ACC synthase prepared by Bleecker *et al.* (1986). Since ACC synthase can be radiolabelled with Ado[carboxyl-^{14}C]Met but not with Ado[methyl-^{14}C]Met, Satoh and Yang (1989) suggest that AdoMet-induced inactivation of ACC synthase involves a covalent linkage of a portion of AdoMet, probably the 2-aminobutyric acid moiety, into the active site of ACC synthase. Such an AdoMet-dependent radiolabelling has been employed as a tool for the identification of ACC synthase (Van Der Streaaten *et al.*, 1989). To radiolabel ACC synthase with AdoMet, a partially purified ACC synthase preparation in 100 mM HEPES-KOH (pH 8.5) and 10 μM PLP is incubated for 6 h at 30°C with 10 μCi of Ado[carboxyl-^{14}C]Met (55 mCi mmol^{-1}, in 0.1 N H$_2$SO$_4$), which has been pre-heated at 100°C for 7 min (Satoh and Yang, 1989). To maintain the AdoMet concentration above 200 μM, the reaction mixture volume is kept under 200 μl. After incubation, the reaction mixture is passed through a Sephadex G-25 column (10 ml bed volume), that has been equilibrated with 10 mM HEPES-KOH buffer (pH 8.0), and the resulting protein fraction is concentrated to 50 μl using a Centricon-30 (Amicon) before SDS-PAGE analysis and fluorography.

F. Characterisation of the Active Site Structure Using Affinity Labelling

In addition to the radiolabelling of ACC synthase with AdoMet, ACC synthase can also be radiolabelled with NaB^3H_4 by reduction of the PLP enzyme aldimine to the hydrolytically stable secondary amine. While radiolabelling with AdoMet is specific for ACC synthase in crude enzyme preparations (Satoh and Yang, 1988), radio-labelling with NaB^3H_4 is not expected to be specific for ACC synthase, because there are many other pyridoxal phosphate enzymes in the crude enzyme preparation. These radiolabelling techniques have been shown to be useful tools for the probing of the active site of ACC synthase.

To probe the active site of apple ACC synthase with NaB^3H_4 immunoaffinity agarose gels prepared from anti-ACC synthase monoclonal antibodies (Yip et al., 1990) were used to specifically adsorb ACC synthase (50 000 units) from a partially purified apple enzyme preparation (purity c. 100 units mg^{-1} protein). After the unbound proteins were washed off, ACC synthase was radiolabelled with 25 mCi NaB^3H_4 (14 Ci $mmol^{-1}$) while it was still bound to the immunoaffinity gel (0.5 ml). This labelling technique is possible presumably because the monoclonal antibodies used in these experiments bind to an epitope of the enzyme that is different from the substrate and PLP-binding sites, and the bound enzyme is still catalytically active (Bleecker et al., 1988; Yip et al., 1990). SDS-PAGE analysis of the apple enzyme preparation after immunoaffinity gel purification and NaB^3H_4 reduction revealed only one radioactive band which coincided with the major protein band eluted from the immunoaffinity gel (Yip et al., 1990). This protein was judged to be ACC synthase based on the observation that its molecular weight was 48 kDa and it was specifically radiolabelled with AdoMet, and specifically bound to a monoclonal antibody against apple ACC synthase. The radiolabelled ACC synthase eluted from the immunoaffinity column with 2% SDS was then lyophilised, and suspended in acetone–acetic acid–triethyla-mine–H_2O (86:5:5:4; v/v). In this buffer system, SDS forms an ion pair with tri-ethylamine that dissolves in acetone, but causes the protein to precipitate (Henderson et al., 1979). The protein was extacted twice with acetone and dried under vacuum. The dried ACC synthase (c. 60–100 μg) was suspended in 2 M urea containing 50 mM NH_4HCO_3 (pH 8.0), and TPCK-treated trypsin was added to yield a final 5% trypsin-to-protein ratio. The total volume of the digestion mixture was kept under 100 μl, and the mixture was incubated at 37°C for 12 h. The tryptic peptides were separated on a C_4 reverse-phase HPLC column with a linear acetonitrile gradient with a flow rate of 1 ml min^{-1}: the first dimension was 10–40% acetonitrile in 20 mM ammonium acetate (pH 6.0) for 80 min; the second dimension was 10–40% acetonitrile in 0.1% CF_3COOH (pH < 2.0) for 50 min; and the third dimension was 10–40% acetonitrile in 0.1% CF_3COOH (pH < 2.0) for 80 min. The profile of HPLC elution was monitored by a UV detector equipped with a 214 nm filter, and the radioactivity in each fraction, collected manually, was determined by liquid scintillation counting. Only one radiolabelled peptide was isolated. Edman degradation of this labeled peptide revealed a 12-amino acid sequence, Ser-Leu-Ser-Xaa-Asp-Leu-Gly-Leu-Pro-Gly-Phe-Arg, where Xaa in cycle 4 was radioactive and did not match any known standard. It is logical to assume that Xaa is N^ε-phosphopyridoxyllysine. To ascertain this, the tritiated ACC synthase was hydrolysed in 6 M HCl at 110°C for 12 h, and the resulting radioactive product was analysed by paper co-electrophoresis and co-chromatography.

Table 17.2. Sequence homology of the active-site peptide of apple and tomato ACC synthase.

Source	Sequence
Apple	H$_2$N–Ser–Leu–Ser–Lys^a–Asp–Leu–Gly–Leu–Pro–Gly–Phe–Arg–COOH
Tomato	H$_2$N–Ser–Leu–Ser–Lys^a–Asp–Met–Gly–Leu–Pro–Gly–Phe–Arg–COOH
	H$_2$N–Ser–Leu–Ser–Lys^a–Asp–Leu–Gly–Leu–Pro–Gly–Phe–Arg–COOH

^{a} Location of bound radioactivity when the enzyme was radiolabelled with NaB^{3}H$_4$ or Ado[*carboxyl-*^{14}C] Met.
Reproduced from Yip *et al.* (1990).

Indeed, it yielded only a single radioactive compound, which co-migrated with the authentic N^ε-pyridoxyllysine in both paper chromatography and paper electrophoresis. These data indicate that residue 4 of the tryptic peptide is a radioactively modified lysine (Table 17.2).

In order to elucidate to which amino acid residue the 2-aminobutyrate moiety of AdoMet links during suicidal inactivation, we have similarly radiolabelled apple ACC synthase with AdoMet. Following trypsin digestion and HPLC, only one radiolabelled peptide was detected. Edman degradation of this labelled peptide revealed the same dodecapeptide, Ser–Leu–Ser–Xbb–Asp–Leu–Gly–Leu–Pro–Gly–Phe–Arg, where the unidentified radioactive PTH-amino acid derivative (Xbb) was also released in cycle 4.

The above results indicate that the same lysyl residue (position 4 of the tryptic peptide) in the active site participates in binding the PLP in native enzyme and in linking covalently to the 2-aminobutyrate portion of AdoMet during the inactivation process.

G. Elucidation of the Primary Structure of ACC Synthase by DNA Sequence Analysis

One of the major objectives of purifying ACC synthase is to reveal the DNA gene sequence that encodes the protein. To screen positive clones from cDNA libraries prepared from mRNA isolated from plant tissues which express ACC synthase activity, researchers have used antibodies raised against purified protein or oligonucleotides derived from the partial amino acid sequence as probes. Two ACC synthase cDNA clones, zucchini (Sato *et al.*, 1991) and winter squash (Nakajima *et al.*, 1990), have been isolated using antibodies as probes. Two tomato clones (Van Der Streaten *et al.*, 1990) have been isolated using oligonucleotide screening. By employing oligonucleotides probes derived from the active site and two other tryptic peptides from apple enzyme, we have recently obtained an ACC synthase cDNA clone from a cDNA library prepared from ripening apple fruits.

From the cDNA nucleic acid squences, primary amino acid sequences are deduced. From the full-length cDNAs isolated from tomato, winter squash and zucchini, ACC synthases of 485, 493 and 493 amino acids, respectively, have been predicted. The native apple ACC synthase is thought to consist of 463 amino acid (Dong *et al.*, 1991). While the ACC synthase cDNAs isolated from tomato, winter squash and zucchini predict molecular weights of 53–58 kDa, which are equivalent to the *in vitro* translation products of mRNAs, purified ACC synthases from these tissues show molecular weights of 45–50 kDa on SDS-PAGE. It is suggested that part of the *C*-terminal amino acid sequence of ACC synthase from these tissues is cleaved during the purification procedures (Nakajima *et al.*, 1990; Van Der Streaten *et al.*, 1990). In apple fruit, however, the size of the translation product of ACC synthase mRNA is similar to that of the purified mature protein on SDS-PAGE, indicating that post-translational processing of ACC synthase does not occur.

The deduced amino acid sequences of ACC synthase from tomato, winter squash, zucchini and apple show a high degree of conservation (> 50% among any two sequences) and overall similarity (> 80%). There are seven highly conserved regions including the active site, containing at least eight amino acid residues and showing greater than 95% identity. The similarity of these deduced amino acid sequences from various tissues may provide some explanation for the results previously obtained from the immunological studies. For instance, polyclonal antibodies raised against winter squash ACC synthase showed some degree of cross-reactivity with tomato enzyme, but failed to recognise apple enzyme. It is to be noted that sequence identity between apple and tomato is 52%, between tomato and winter squash it is 62%, and between winter squash and zucchini it is as great as 96%.

IV. CHARACTERISATION OF ACC OXIDASE

The conversion of ACC to ethylene in higher plants is carried out by an oxidative enzyme system which is generally referred to as ethylene forming enzyme (EFE), or ACC oxidase. In most plant tissues ACC oxidase is constitutive, because application of exogenous ACC to these plant tissues results in marked increase in ethylene production (Cameron *et al.*, 1979). Exceptions are immature fruit and flower tissues, where both ACC oxidase and ACC synthase activity are limiting (Yang and Hoffman, 1984). Many previous efforts to isolate active ACC oxidase independent of intact cellular material had been unsuccessful. Intact protoplasts and vacuoles retain EFE activity but their activity disappears completely when membrane integrity is ruptured. It is therefore believed that ACC oxidase requires membrane integrity. Nevertheless, much information about EFE has been gained from experiments carried out *in vivo* by supplying tissues with ACC.

A. *In Vivo* Study of ACC Oxidase

1. *Effect of substrate*

The effect of substrate (ACC and oxygen) concentrations on ethylene production rate by plant tissues was investigated (Yip *et al.*, 1988). The K_m value for O_2 in ethylene production varied greatly depending on the internal ACC content. When ACC levels in the tissues are low (below its K_m value), the concentration of O_2 giving half-maximal ethylene production rate ($[S]_{0.5}$) ranged from 5 to 7%, and was similar among different tissues. As the concentration of ACC is increased (> its K_m value), $[S]_{0.5}$ for O_2 decreased markedly. In contrast, the K_m value for ACC was not as dependent on O_2 concentration, but varied greatly among different plant tissues, ranging from 8 μM in apple fruit tissue to 120 μM in etiolated wheat leaves. These studies nevertheless indicate that while the apparent K_m for ACC *in vivo* is difficult to determine accurately because of compartmentation of both ACC and EFE, the true value is probably no more than 0.1 mM. These kinetic data are consistent with the view that ACC oxidase follows an ordered binding mechanism in which ACC oxidase binds first to oxygen and then to ACC.

2. The fate of the carbons of ACC

Trace studies reveal that ethylene comes from C-2 and C-3 of ACC. It was not until 1984 that HCN and CO_2 were established as the other two co-products of ACC during its biological oxidation to ethylene (Peiser *et al.*, 1984). When mungbean hypocotyls were fed with [carboxyl-^{14}C]ACC, $^{14}CO_2$ was released. When [1-^{14}C]ACC was administered, the label was found to be located at C-4 of asparagine in mungbean hypocotyls, but at β-cyanoalanine moiety of γ-glutamyl-β-cyanoalanine in common vetch (*vicia sativa*), in the same amount as that of ethylene production from ACC. It has been well established that HCN is metabolised by β-cyanoalanine synthase, which is widely distributed in all plant tissues (Miller and Conn, 1980). The enzyme catalyses the conversion of HCN and cysteine to H_2S and β-cyanoalanine; the latter is then hydrated to asparagine in mungbean or further metabolised to γ-glutamyl-β-cyanoalanine in common vetch. By employing [^{13}C-1]ACC and nuclear magnetic resonance (NMR) techniques, Pirrung (1985) also found that the ^{13}C was incorporated into C-4 of asparagine in mungbean hypocotyls. Thus the oxidation of ACC *in vivo* can be represented by the following equation:

$$ACC + \tfrac{1}{2}[O] \rightarrow C_2H_4 + HCN + CO_2 + H_2O$$

Since no free HCN was detected in plant tissues which were actively producing ethylene, Peiser *et al.* (1984) suggested that those plant tissues possess ample capacity to detoxify HCN (presumably by β-cyanoalanine synthase) generated from the oxidation of ACC. By employing an isotope dilution method, Yip and Yang (1988) estimated that the steady state concentration of HCN was below 0.2 μM in ripening fruits and in auxin-treated mungbean hypocotyls, both of which produced ethylene at high rates. Thus, plant tissues are capable of maintaining HCN at such a low level that it does not cause any significant inhibition of cytochrome oxidase ($K_i = 10\ \mu M$), which is vital for the respiration of plant tissues.

3. Selectivity toward stereoisomers of 1-amino-2-ethylcyclopropane-1-carboxylic acid (AEC)

Many cell-free EFE systems have been reported including pea homogenate (Konze and Kende, 1979), the lipoxygenase system from oat leaf (Bousquet and Thimann, 1984), microsomal preparations from pea (Adam and Mayak, 1984) and from carnation petal (Mayak and Adam, 1984), and IAA oxidase from olive leaf (Vioque *et al.*, 1981). However, those systems are judged to be non-physiological, because they all exhibit low affinity for ACC and are unable to discriminate between AEC isomers. In contrast, Hoffman *et al.* (1982) have shown that plant tissues can only convert the (1*R*, 2*S*)-AEC { (+)-allocoronamic acid} to 1-butene, but not other AEC isomers. The four stereomers of AEC comprise two pairs of enantiomers: $(+)$- and $(-)$-allocoronamic acid, and $(+)$- and $(-)$-coronamic acid. Since individual stereoisomers are more difficult to prepare, racemic mixtures of allocoronamic acid and coronamic acid (1 mM) are incubated with the putative 'in vitro' ACC oxidase to test for its authenticity.

B. Recent Advances in the Research on ACC Oxidase

In studying the physiology of ripening and wounding in tomato, Grierson and his colleague (Smith et al., 1986) isolated a cDNA clone (pTOM13) from tomato that corresponds to an mRNA whose expression is correlated with ethylene production in ripening fruits and in wounded leafy tissues. Subsequent experiments using transgenic plants in which the accumulation of pTOM13 mRNA was inhibited with an antisense gene confirmed its role in ethylene synthesis and suggest it may encode a protein involved in the conversion of ACC to ethylene (Hamilton et al., 1990). Interestingly, the predicted amino acid sequence of the pTOM13 polypeptide shows close homology with that of flavanone-3-hydroxylase (EC 1.14.11.9) from Antirrhinum majus. Using extraction and assay procedures similar to those that had been shown previously to preserve in vitro the activity of flavanone-3-hydoxylase from Petunia hybrida petals (Britsch and Grisebach, 1986), Ververidis and John (1991) have recently reported a crude enzyme extracted from ripening melon that exhibited authentic ACC oxidase activity. For enzyme preparation 2 g of melon pericarp tissue was frozen in liquid N_2 and homogenised with 4 ml of buffer containing 0.1 M Tris-HCl (pH 7.2) and 10% glyerol which was degassed before used. The extract was then filtered through 2 layers of Miracloth under N_2. ACC oxidase assay of the extract was performed in a 3 ml reaction mixture containing 2.7 ml extraction buffer, 1 mM ACC, 30 mM sodium ascorbate, and 0.1 mM $FeSO_4$. After 1-2 h incubation at 25°C, a gas sample was withdrawn for ethylene analysis by GC. The enzyme activity appeared in the supernatant, exhibited saturable kinetics for ACC ($K_m = 85 \, \mu M$), and discriminated AEC isomers for 1-butene production. In agreement with in vivo studies, melon ACC oxidase can be inhibited by cobalt ions. Although both Fe^{2+} and ascorbate are required, unlike other dioxygenase enzymes, EFE in melon extract does not require 2-oxoglutarate for maximal activity. These studies indicate that the present EFE system resembles flavanone-3-hydroxylase. When this soluble enzyme is proven to be the authentic EFE, many studies with respect to the purification and characterisation of the enzyme will follow.

REFERENCES

Abeles, F. B. (1973). "Ethylene in Plant Biology". Academic Press, New York.

Adam, Z. and Mayak, S. (1984). FEBS Lett. 172, 47-50.

Adams, D. O. and Yang, S. F. (1979). Proc. Natl. Acad. Sci. USA 70, 170-174.

Bleecker, A. B., Kenyon, W. H., Somerville, S. C. and Kende, H. (1986). Proc. Natl. Acad. Sci. USA 83, 7755-7759.

Bleecker, A. B., Robinson, G. and Kende, H. (1988). Planta 173, 385-390.

Boller T., Herner, R. C. and Kende, H. (1979). Planta 145, 293-303.

Bousquet, J. F. and Thimann, K. V. (1984). Proc. Natl. Acad. Sci. USA 81, 1724-1727.

Britsch, L. and Grisebach, H. (1986). *Eur. J. Biochem.* **155**, 322–327.

Burg, S. P. (1973). *Proc. Natl. Acad. Sci. USA* **70**, 591–597.

Burg, S. P. and Stolwijk, J. A. J. (1959). *J. Biochem. Microbiol. Technol. Eng.* **1**, 245–259.

Burroughs, L. F. (1957). *Nature* **179**, 360–361.

Cameron, A. C., Fenton, C. A. L., Yu, Y. B., Adams, D. O. and Yang, S. F. (1979). *HortSci.* **14**, 178–80.

Crocker, W. Zimmerman, P. W. and Hitchcock, A. E. (1932). *Contrib. Boyce Thompson Inst.* **4**, 177–218.

Dong, J.-G., Yip, W.-K. and Yang, S. F. (1991). *Plant Cell Physiol.* **32**, 25–31.

Dong, J.-G., Kim, W.-T., Yip, W.-K., Thompson, G. A., Li, L., Bennett, A. B. and Yang, S. F. (1991). *Planta* **185**, 38–45.

Grady, K. L. and Bassham, J. A. (1982). *Plant Physiol.* **70**, 919–921.

Hall, K. C., Pearce, D. M. E. and Jackson, M. B. (1989). *Plant Growth Reg.* **8**, 297–307.

Hamilton, A. J., Lycett, G. W. and Grierson, D. (1990). *Nature* **346**, 284–287.

Henderson, S., Oroszlan, S. and Konigberg, W. (1979). *Anal. Biochem.* **93**, 153–157.

Hoffman, N. E., Yang, S. F., Ichihara, A. and Sakamura, S. (1982). *Plant Physiol.* **70**, 195–199.

Honma, M. (1983). *Agric. Biol. Chem.* **47**, 617–618.

Khani-Oskouee, S., Rammlingam, K., Kalvin, D. and Woodard, R. W. (1987). *Bioorg. Chem.* **15**, 92–99.

Knee, M. (1984). *J. Exp. Bot.* **35**, 1794–1799.

Konze, J. R. and Kende, H. (1979). *Planta* **146**, 293–301.

Lanneluc-Sanson, D., Phan, C. T. and Granger, R. L. (1986). *Anal. Biochem.* **155**, 322–327.

Liu, Y., Hoffman, N. E. and Yang, S. F. (1985). *Planta* **164**, 565–568.

Lizada, M. C. C. and Yang, S. F. (1979). *Anal. Biochem.* **100**, 140–145.

Mayak, S. and Adam, Z. (1984). *Plant Sci. Lett.* **33**, 345–352.

McGaw, B. A., Horgan, R. and Heald, J. K. (1985). *Anal. Biochem.* **149**, 130–135.

Miller, J. M. and Conn, E. E. (1980). *Plant Physiol.* **65**, 1199–1202.

Nakajima, N., Nakagawa, N. and Imaseki, H. (1988). *Plant Cell Physiol.* **29**, 989–998.

Nakajima, N., Mori, H., Yamazaki, K. and Imaseli, H. (1990). *Plant Cell Physiol.* **31**, 1021–1029.

Neljubov, D. (1901). *Pflanzen. Bot. Centralbl. Beih.* **10**, 128–139.

Nieder, M., Yip, W.-K. and Yang, S. F. (1986). *Plant Physiol.* **81**, 156–160.

Peiser, G. D., Wang, T.-T., Hoffman, N. E., Yang, S.F. and Walsh, C.T. (1984). *Proc. Natl. Acad. Sci. USA* **81**, 3059–3063.

Pirrung, M. C. (1985). *Bioorg. Chem.* **13**, 219–226.

Reeves, H. C., Heeren, R. and Malloy, P. (1981). *Anal. Biochem.* **115**, 194–196.

Saltveit, M. E. Jr. and Yang, S. F. (1987). In "Principles and Practice of Plant Hormone Analysis", Vol. 2 (L. Rivier and A. Crozier, eds), pp. 367–401. Academic Press, London.

Sato, T. and Theologis, A. (1989). *Proc. Natl. Acad. Sci. USA* **86**, 6621–6625.

Sato, T., Oeller, P. W. and Theologis, A. (1991). *J. Biol. Chem.* **266**, 3752–3759.

Satoh, S. and Yang, S. F. (1988). *Plant Physiol.* **88**, 109–114.

Satoh, S. and Yang, S. F. (1989). *Arch. Biochem. Biophys.* **271**, 107–112.

Savidge, R. A., Mutumba, G. M. C., Heald, J. K. and Wareing, P. F. (1983). *Plant Physiol.* **71**, 434–436.

Sitrit, Y., Rivo, J. and Blumenfeld, J. (1987). *Plant Physiol.* **86**, 13–15.

Smith, C. J. S., Slater, A. and Grierson, D. (1986). *Planta* **168**, 94–100.

Van Der Streaten, D., Van Wiemeersch, L., Goodman, H. M. and Van Montagu, M. (1989). *Eur. J. Biochem.* **182**, 639–647.

Van Der Streaten, D., Van Wiemeersch, L., Goodman, H. M. and Van Montagu, M. (1990). *Proc. Natl. Acad. Sci. USA* **87**, 4859–4863.

Ververidis, P. and John, P. (1991). *Phytochemistry* **30**, 725–727.

Vioque, A., Albi, M. A. and Vioque, B. (1981). *Phytochemistry* **20**, 1473–1475.

Yang, S. F. and Hoffman, N. E. (1984). *Ann. Rev. Plant. Physiol.* **35**, 155–189.

Yip, W.-K. and Yang, S. F. (1988). *Plant Physiol.* **88**, 473–476.

Yip, W.-K., Jiao, X.-Z. and Yang, S. F. (1988). *Plant Physiol.* **88**, 553–558.

Yip, W.-K., Dong, J.-G., Kenny, J. W., Thompson, G. A. and Yang, S.F. (1990). *Proc. Natl Acad. Sci. USA* **87**, 7930–7934.

Yip, W.-K., Dong, J.-G. and Yang, S. F. (1991). *Plant Physiol.* **95**, 251–257.

Young, R. E., Pratt, H. K. and Biale, J. B. (1952). *Anal. Chem.* **24**, 551–555.

Note Added in Proof

Recent work confirmed that the corrected pTOM13 and related sequences confer EFE activity when expressed in yeast [Hamilton, A. J., Bouzayen, M. and Grierson, D. (1991). *Proc. Natl Acad. Sci. USA* **88**, 7434–7437] or *Xenopus* oocytes [Spanu, P., Reinhardt, D. and Boller, T. (1991). *EMBO J.* **10**, 2007–2013]. Authentic EFE activity has since been isolated from avocado [McGarvey, D. J. and Christofferson, R. E. (1992). *J. Biol. Chem.* **267**, 5964–5967] and apple [Kuai, J. and Dilley, D. R. (1992). *Postharvest Biol. Technol.* **1**, 203–211; Fernandez, J. C. and Yang, S. F. (1992). *Plant Physiol.* **99**, 571–574] fruits, besides melon fruits [Smith, J. J., Ververidis, P. and John, P. (1992). *Phytochemistry* **31**, 1485–1494].

18 Polyamine Biosynthesis

HELENA BIRECKA and MIECZYSLAW BIRECKI

Department of Biological Sciences, Union College, Schenectady, NY 12308, USA

METHODS IN PLANT BIOCHEMISTRY Vol. 9
ISBN 0–12–461019–6

I. INTRODUCTION

The term polyamines (PAs) is used here as a collective term for diamines, 1,3-diamino-propane (Dap), putrescine (Put) and cadaverine (Cad); the triamine spermidine (Spd); and the tetramine spermine (Spm). Put, Spd and Spm are very common in higher plants, whereas Dap, the product of Spd and/or Spm oxidation by polyamine oxidases, and Cad, the product of lysine (Lys) decarboxylation by Lys decarboxylase (LDC), have not been frequently detected. PAs, such as homospermidine (homoSpd), norspermidine and norspermine, naturally occurring in microorganisms, have been found only sporadically in higher plants (Birecka et al., 1984; Smith, 1985; Rodriguez-Garay et al., 1989). A putative pathway of norspermidine and norspermine bio-synthesis from Dap and decarboxylated S-adenosylmethionine (d-SAM) has been presented by Kuehn et al., (1990).

Biosynthesis of Spd, Spm and homoSpd is initiated by the formation of Put from either arginine (Arg) or ornithine (Orn). Figure 18.1 presents the most common bio-synthetic pathway of these PAs. Orn gives rise to Put directly after decarboxylation by Orn decarboxylase (ODC), whereas Arg yields Put via three steps: decarboxylation by Arg decarboxylase (ADC) produces agmatine (Agm), which in turn is deiminated by Agm iminohydrolase producing N-carbamoylPut; the latter is then converted to Put by N-carbamoylPut amidohydrolase. An alternative pathway leading from Agm to Put, catalysed by a multifunctional enzyme called Put synthase, has been found in Lathyrus sativus and cucumber and is also reported to be present in pea and corn (Srivenugopal and Adiga, 1981; Prasad and Adiga, 1986a, b).

Spd and Spm are formed by the transfer of an aminopropyl group from d-SAM to Put and Spd, respectively, the reaction being catalysed by Put aminopropyl-transferase and Spd aminopropyltransferase, respectively; d-SAM is the product of S-adenosylmethionine decarboxylation by S-adenosylmethionine decarboxylase (SAMDC). HomoSpd, which may be of interest as an intermediate in pyrrolizidine alkaloid biosynthesis (Birecka et al., 1984), is formed from two molecules of Put via 4-aminobutyraldehyde in Lathyrus sativus and sandal tree (Srivenugopal and Adiga, 1980a); the reaction is catalysed by a NAD^+-dependent enzyme.

ODC and/or ADC activities have been considered as the rate-limiting step in biosynthesis of common PAs. In most cases both enzymes can be detected in a plant extract. However, many reported activities, especially those for ODC, may represent highly overestimated — as discussed below — values. It has been claimed that in tobacco ADC produces high levels of free Agm and Put, whereas ODC contributes only to biosynthesis of Put conjugates, thus implying a possibility of subcellular compartment action of the two enzymes and their metabolites (Burtin et al., 1989); it is worth noting, however, that in corn cell lines conjugates were formed only by Arg-derived Put (Hiatt, 1989).

II. AMINO ACID DECARBOXYLASES

The most frequently employed assays, especially for ODC and ADC, monitor CO_2 by trapping and counting $^{14}CO_2$ released from $1\text{-}^{14}C$- or $U\text{-}^{14}C$-labelled substrates in crude or dialysed-only enzyme extracts; rarely ammonium sulphate fractionation has

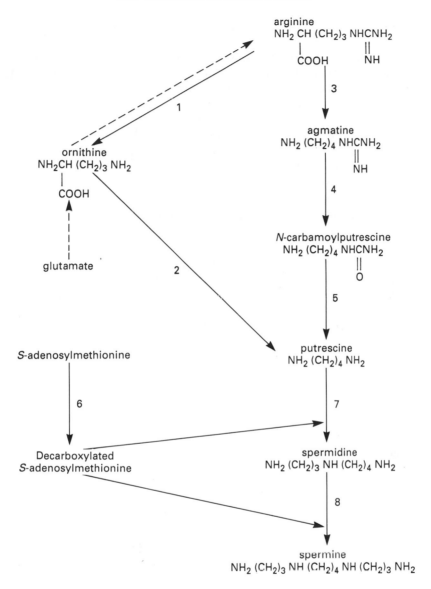

FIG. 18.1 Biosynthesis of commom polyamines. Key: 1, arginase; 2, ornithine decarboxylase; 3, arginine decarboxylase; 4, agmatine iminohydrolase; 5, N-carbamoylputrescine amidohydrolase; 6, S-adenosyl-methionine decarboxylase; 7, putrescine aminopropyltransferase; 8, spermidine aminopropyltransferase.

been applied. In such assays the CO_2 released can be many times greater than Put and/or Agm produced in the extract (Birecka *et al.*, 1985a,b). The artefactual CO_2 production can be due to several factors:

(1) The presence of high levels of phenolics, which create a danger of not only affecting the enzymes directly, but also of oxidising the substrates which are expected to undergo non-oxidative decarboxylation. Removal of phenolics can

be accomplished by adding polyvinylpyrrolidone to crude extracts (Loomis, 1974).

(2) The presence of diamine oxidases, which by virtue of their broad specificity may oxidise Arg, Lys, Orn, Spd, or Spm in addition to Put, Cad or Agm. Moreover, Orn, Arg, Lys and/or SAM can undergo decarboxylation in the presence of H_2O_2 liberated during amine oxidase-mediated catabolism of the diamines (Suresh and Adiga, 1977; Adiga and Prasad, 1985). Thus, diamine oxidases in plant extracts may lead to an overestimation of the activity when based on CO_2 evolution and to its underestimation when based on PAs formed. Amino-guanidine, a potent diamine oxidase inhibitor, at 0.1–0.2 mM proved to be very effective in eliminating the amine oxidase interference (Birecka et al., 1985a).

(3) Possible oxidative decarboxylation of Orn, Arg and Lys, unrelated to diamine oxidase activities, under conditions employed in ODC, ADC and LDC assays (Smith and Marshall, 1988a, b, c).

(4) The presence of high arginase and urease activities in the extract, which may lead to a significant conversion of Arg into Orn as well as release of guanido-C CO_2. The latter artefact is of special importance when U-^{14}C-labelled Arg is used as a substrate. Orn, which is a relatively effective inhibitor of arginase (Wright et al., 1981; Carvajal et al., 1982), at 20 mM significantly but not completely suppressed arginase activity in extracts from *Heliotropium* plants (Birecka et al., 1985b).

Thus, CO_2 release will furnish good estimates of decarboxylase activities only in conjunction with precise stoichiometry of amine formation. However, the time-consuming and more costly separation of PAs from the reaction mixture can be avoided if no significant release of CO_2 is observed in enzyme extracts pre-incubated with appropriate irreversible inhibitors, such as DL-α-difluoromethylornithine (DFMO), DL-α-difluoromethylarginine (DFMA) or DL-α-difluoromethyllysine (DFML), specific for ODC, ADC and LDC, respectively, at 0.5–1.0 mM for about 20–30 min. One should add that arginase at high activities can hydrolyse DFMA to DFMO, thus affecting not only ADC but also ODC activity (Slocum et al., 1988; Birecka et al., 1991). New ADC inhibitors, α-monofluoromethylagmatine and [E]-α-monofluoromethyldehydroarginine proved to be even more potent than DFMA and at the same time do not serve as substrates for arginase or Agm-metabolising enzymes (Bitonti et al., 1987).

III.   ORNITHINE DECARBOXYLASE

Ornithine decarboxylase (EC 4.1.1.17) is located in the cytosol, but it has been also found as chromatin-bound in the nuclei of barley, corn, bean and pea cells (Foudouli and Kyriakidis, 1989); in barley, both the cytosolic and chromatin-bound ODC can form an inactive complex with antizyme, a protein non-competitive inhibitor. ODC activity has been also detected in chloroplasts and mitochondria (Torrigiani et al., 1986).

ODC requires pyridoxal phosphate (PLP) and a thiol reagent, dithiothreitol (DTT), 2-mercaptoethanol (2-ME) or ascorbate for maximum activity. The reported K_m Orn values for tobacco and tomato (Heimer and Mizrahi, 1982), wheat (Christ *et al.*, 1989) and barley (Koromilas and Kyriakidis, 1988a,b) are 0.14, 0.25 and 0.36 mM, respectively; the optimum pH ranges between 7.5 and 8.5. Only in two cases has DFMO been mentioned as an ineffective inhibitor, namely for tobacco cell lines with a genetically controlled ODC resistance to DFMO (Hiatt and Malmberg, 1988) and for purified jute embryo ODC when applied at 1–10 mM (Pandit and Ghosh, 1988).

Tris-HCl or phosphate buffers at 40–250 mM (pH 7.5–8.3) containing 100 μM EDTA, 20–50 μM PLP and 5–10 mM DTT have been usually used for tissue homogenisation. A proteinase inhibitor, phenylmethylsulphonylfluoride (PMSF), is sometimes added at 0.05–1.0 mM (Altman *et al.*, 1983; D'Orazi and Bagni, 1987; Foudouli and Kyriakidis, 1989). Crude extract obtained after centrifugation at 10 000 $\times$ g for 15–30 min at 4°C is assayed directly or after overnight dialysis. Partial purification is achieved by precipitating the ODC fraction with 50–70% ammonium sulphate followed by centrifugation at 20 000 $\times$ g; the pellet is suspended in a small volume of the assay buffer and desalted either by dialysis or by filtration over a Sephadex G-25 column pre-conditioned with the assay buffer. The enzyme extracts are stored in microvials at −20°C or −80°C; unused thawed extracts should be discarded.

A. Assay

The most widely used assay for ODC—as mentioned above—is the radiometric one in which the $^{14}CO_2$ released from labelled L-Orn is trapped and its radioactivity measured and/or Put, formed in the reaction mixture from [U-^{14}C]Orn, is isolated and counted. The reaction mixture of 0.5–1.0 ml (the higher the ODC activity the lower the total volume) contains 100 mM Tris-HCl (pH 8.0), 2 mM DTT, 50 μM PLP, 50 μM EDTA, 0.2–1.0 mM Orn and the enzyme extract. A preliminary assessment of the enzyme activity at least at two different substrate concentrations is advisable. The radioactivity may range between 0.2 and 1.0 μCi of L-Orn in the reaction mixture; for low levels of activity, the higher values of specific radioactivity should be employed. The blank may contain a boiled enzyme extract or buffer substituting for the extract volume. However, a test with an extract pre-incubated with DFMO is a prerequisite. The reaction is carried out in small vials with rubber stoppers pierced by a pin on which a piece of filter paper (4 $\times$ 25 mm, folded) impregnated with about 20 μl methylbenzethonium hydroxide to absorb CO_2 is suspended. The vials are incubated at 37°C for 30–45 min, then placed on ice and the reaction is immediately stopped by injection of 10–50 μl 25% trichloroacetic acid (TCA) with a microsyringe (exact volume is required when Put is recovered from the mixture). After an additional 40–60 min at room temperature to allow the $^{14}CO_2$ evolving from the acidified mixture to be absorbed by the filter paper, the vials are opened. The filter paper is removed and placed in a scintillation fluid and the radioactivity measured in a counter.

A slight modification of this basic procedure is presented by Gaines *et al.* (1989) for animal ODC. A microcentrifuge tube with 100 μl of the reaction mixture (pH 7.1) is placed in a 20 ml glass scintillation vial with a polypropylene centre well containing 200 μl of a trapping agent. The vial is closed with a silicone septum-lined plastic screw cap. After incubation 100 μl of 0.67 N HCl is injected through the septum into the tube, then after an additional 60 min the cap is removed, the tube discarded and 15 ml of scintillation fluid added.

When the DFMO test indicates a relatively significant artefactual release of CO_2 and therefore cannot be used as a blank, isolation of Put from the reaction mixture containing [U-^{14}C]- or [5-^{14}C]Orn is necessary and the need for aminoguanidine should be checked.

A very good separation of Put as well as of other PAs is obtained by thin layer electrophoresis (TLE) carried out according to Seiler (1983); however, the pyridine acetate buffer applied is acidified to pH 4.2 instead of 4.8 (Birecka et al., 1985a). The reaction mixtures are transferred into microtubes and centrifuged at 15 000 × g. Aliquots of 20–150 μl, depending on the total mixture volume, are applied on pre-coated 20 × 20 cm silica gel 60 (250 μm) plates as repeated streaks at a distance of c. 3 cm from one edge and of c. 1.5 cm between samples. Markers are applied on both sides of the plate; three to four samples can be run on one plate. The plate is evenly sprayed with the buffer and portions of about 20 ml of the same buffer are poured into the troughs of a TLE cooled plate apparatus before each run. The plate and troughs are connected by buffer-soaked filter paper strips. The separation is carried out at 600 V for about 60 min. After the plate is dried the marker spots are developed with 0.4% ninhydrin in acetone, the silica gel bands corresponding to Put are scraped off, transferred into scintillation vials, and their radioactivity measured. The total recovery of the radioactivity applied ranged between 90 and 108%. The R_f values related to Put ($R_f = 1$) were as follows: 0.08, 0.27, 0.40, 0.42, 0.51, 0.57, 0.60, and 0.75 for citrulline, Spm, γ-aminobutyric acid, Arg, Orn, Spd, N-carbamoylPut, and Agm, respectively.

Labelled Put can be separated from Orn using ion-exchange chromatography (Weber, 1987). Amberlite CG-50 resin (H$^+$) washed with 0.2 M ammonium acetate (pH 6.5) containing 1 mg ml^{-1} EDTA is filled into glass wool-plugged Pasteur pipettes to form 3 cm high columns. Each column is washed with 2 ml 0.2 M ammonium acetate (without EDTA). The complete assay mixture (2 ml) containing [2,3-^{3}H]Orn, specific radioactivity 20 μCi μmol^{-1}, is applied to the column, the unreacted Orn being removed by washing the column with 12 ml 0.2 M ammonium acetate followed by 2 ml 0.2 M acetic acid. [1,2-^{3}H] Put is eluted with 3 ml 8 M formic acid directly into scintillation vials and measured after addition of 15 ml scintillation cocktail. The reported detection limit is 50 pmol at the ^{3}H counting efficiency of about 30%. A freshly purified [^{3}H]Orn blank gave counting rates below 0.1% of the applied ^{3}H-label.

[1,2-^{3}H]Put has been separated from [2,3-^{3}H]Orn by an isocratic reversed-phase HPLC method which does not require a separate derivatisation step (Beeman and Rossomondo, 1989). Put is separated from the substrate with a mobile phase containing 0.05 M phosphate (pH 3.9), 0.01 M SDS and 36% acetonitrile; the sample injection volume is 50 μl. Orn (R_t 4–5 min) and Put (R_t 7–8 min) are collected in 0.5 ml fractions—at the eluant flow rate of 1 ml min^{-1}—directly in scintillation vials. The detection limit for [^{3}H]Put was reported to be 2.5 pmol. However, one wonders whether under the assay conditions, when 1 ml of the reaction mixture contains 0.4 μCi L-[2,3-^{3}H]Orn and 100 mM L-Orn, such a low detection limit for labelled Put in a 50 μl sample can be achieved.

Reversed-phase HPLC has been used to assay Put formed from non-labelled Orn; pre-column derivatisation with o-phthalaldehyde is required (Krannes and Flatmark, 1987). The Put derivative, whose half-life was increased to 3 h in darkness by adding the HPLC mobile phase, had R_t of 11.6 min at a flow rate of 1.2 ml min^{-1}. It was quantified fluorometrically with excitation at 360 nm and emission 420–540 nm. The reported detection limit is 0.5 pmol Put at a signal-to-noise ratio of 3.

And finally, the ODC activity can be estimated spectrophotometrically using 2,4,6-trinitrobenzene sulphonic acid which reacts with Put but not with Arg giving a coloured product soluble in 1-pentanol. Its molar extinction coefficient at 420 nm is about $1.9 \times 10^3 \, cm^{-1}$ and the detection limit about 10 nmol Put (Ngo et al., 1987).

B. Purification

Ornithine decarboxylase has been purified from germinated barley seeds (Koromilas and Kyriakidis, 1988a,b) and embryos of jute seeds (Pandit and Ghosh, 1988). The barley cytosolic and chromatin-bound ODC were isolated from a homogenate in an isotonic grinding medium containing 0.25 M sucrose and buffer A, consisting of 50 mM Tris-HCl (pH 8.5), 0.3 mM EDTA, 50 μM PLP, 2.5 mM DTT, 50 μM PMSF and 5 mM NaF, a phosphatase inhibitor. After filtration through several layers of gauze, the suspension was centrifuged at $4000 \times g$ for 30 min and the supernatant was further centrifuged at $100\,000 \times g$ for 60 min to yield the cytosolic ODC fraction. The $4000 \times g$ pellet was used to isolate chromatin by the method of Bonner (1976).

The crude cytosolic fraction was applied to a DEAE Bio-Gel A column equilibrated with buffer A and eluted with 0.15 M NaCl in buffer A. The combined active fractions were applied to a (triethylaminoethyl)cellulose column equilibrated with buffer A, washed with the same buffer and the enzyme was eluted with a linear gradient of 0–0.3 M NaCl in buffer A. The pooled active fractions concentrated by ultrafiltration were passed twice through a Bio-Gel P-150 column; before the second chromatography the preparation was treated with 10% (w/v) ammonium sulphate. The overall activity recovered was 48% with a purification factor of 77.

The bound ODC was released from chromatin by three times repeated freezing and thawing, each time extracting it with buffer A followed by centrifugation at $100\,000 \times g$. The purification involved chromatography using DEAE-Bio-Gel A, (triethylaminoethyl)cellulose, phenyl-Sepharose CL-4B, hydroxylapatite and Sephadex G-200 columns. The latter was needed to separate the ODC antizyme released from chromatin. The purification factor and activity yield were 2670 and 20%, respectively. The molecular weight of the enzyme is 55 kDa as determined by SDS-PAGE. The bound antizyme extracted from the chromatin by 2 M NaCl has been also purified (Panagiotidis and Kyriakidis, 1985; Koromilas and Kyriakidis, 1988b).

A jute embryo ODC has been purified to homogeneity (773-fold purification with 33% recovery); ammonium sulphate fractionation and DEAE-cellulose chromatography were applied (Pandit and Ghosh, 1988). The apparent molecular weight of the enzyme is 39 kDa, as indicated by Sephadex G-100 chromatography.

IV. ARGININE DECARBOXYLASE

Arginine decarboxylase (EC 4.1.1.19) has been found only in the cytoplasm; it requires PLP and a thiol reagent for maximum activity. The enzyme has been purified from various sources and shows significant differences in terms of molecular structure, pH and temperature optima as well as K_m for Arg. The optimum pH ranges between 7.0

for oat (Smith, 1979) and 8.5 for broad bean (Matsuda, 1984), the temperature optimum varies from 32°C for oat to 60°C for avocado (Winer *et al.*, 1984), and the K_m varies from 0.03 mM for oat to about 1.8 mM for avocado and *Lathyrus sativus* (Ramakrishna and Adiga, 1975). ADC purified from cucumber seedlings was reported to display ODC activity as well (Prasad and Adiga, 1986a,b). Usually ADC is extracted from plant tissues using the same procedure as described for ODC.

A. Assay

Like ODC, ADC has been assayed in most cases by the $^{14}CO_2$-trapping method using similar reaction mixtures and conditions. Usually L-[U-^{14}C]Arg has been applied as the source of label. Pre-incubation of the enzyme extract with DFMA at 0.5–1.0 mM as a control is a prerequisite. If the artefactual release of CO_2 is due to arginase activity and cannot be suppressed by 20 mM Orn, additional control with [guanido-^{14}C]Arg may allow to account for this interference (Rabe and Lovatt, 1984).

Under the assay conditions Agm can be partially converted into Put (Birecka *et al.*, 1985b). Therefore, both PAs should be taken into account when ADC activity is assessed on the basis of its products in the mixture. They can be successfully separated and measured using the TLE method described for ODC.

There are several assays for ADC, in which non-labelled Arg is used as the substrate. Manometric estimation of CO_2 in a Warburg apparatus has been applied to broad bean ADC. A spectrophotometric method described by Smith and Barker (1988) is based on the oxidation of Agm by a purified pea diamine oxidase to produce H_2O_2, which in turn is used by peroxidase to form a stable red pigment from chromogens antipyrine/3,5-dichloro-2-hydroxybenzene sulphonic acid (AAP/DCHBS) with a molar extinction coefficient at 575 nm of about $24 \times 10^3 cm^{-1}$; the substitution of the new chromogens for the previously used guaiacol (Smith, 1983) resulted in a six-fold increase in sensitivity. After incubation the reaction mixture (2.5 ml) containing 0.1 M phosphate buffer and 5 mM Arg—without PLP and DTT—is boiled for 5 min and centrifuged; 0.1 ml of the supernatant is added to a cuvette containing 0.1 ml peroxidase (1 mg ml^{-1}), 10 μl diamine oxidase (10 nkat), AAP and DCHBS at final concentrations of 0.1 and 1 mM, respectively, and 0.1 M phosphate (pH 7.0) to a total volume of 2.5 ml. The change in absorbance is measured after completion of oxidation at 30° (60 min). The possibility of an interference by thiols and PLP—if present in the reaction mixture—is mentioned.

The spectrophotometric estimations of Agm were comparable to those obtained by ion-pairing reversed-phase HPLC using post-column derivatisation and fluorometric detection. Such a method, which can be used not only in ADC but also in ODC and LDC assays, is described by Kochhar *et al.* (1989) for bacterial decarboxylases. The resolution of substrates and products is complete within 15 and 30 min of isocratic elution and the detection limit for the product is about 40 pmol. The method allows one to follow the time-course of the reaction in a single assay. Samples (10 μl) are removed at 10 min intervals from the reaction mixture, mixed with 10 μl of pre-chilled 0.2 M HClO$_4$ and centrifuged at 10 000 × g for 5 min; 10 μl of the supernatant are mixed with 190 μl HPLC buffer (30 mM heptafluorobutyrate), degassed by sonication and injected (40 μl) onto the Nucleosil C$_{18}$ reversed-phase column. The column is eluted isocratically at a flow rate of 1 ml min^{-1}, detection is by post-column derivatisation with o-phthaldialdehyde reagent (flow rate 1 ml min^{-1}). The R_t values (min) for Orn and Put; Arg and Agm; and Lys and Cad were 8.25 and 12.9; 21.8 and 33.0; and 11.25 and 18.8, respectively.

B. Purification

1. Oat

Oat leaf ADC has been purified by Smith (1979). It was extracted with 50 mM Na$_2$HPO$_4$; the protein fraction precipitated with 20–50% ammonium sulphate was suspended in 0.1 M Tris-HCl (pH 6.7) and dialysed for 18 h. Acetone at −15°C (50 ml per 100 ml enzyme) was added at 0°C, the precipitate after centrifugation discarded, and a new portion of acetone (150 ml per 100 ml enzyme) added. After centrifugation at 5000 × g the precipitate was dissolved in 0.1 M Tris-HCl and again centrifuged at 15 000 × g. The supernatant was applied to a column of Bio-Gel A-1.5 m equilibrated with 50 mM Tris-HCl (pH 7.0) containing 0.1 M KCl and eluted with this buffer. Two active fractions, A and B, were obtained; they were concentrated by ultrafiltration. As indicated by gel chromatography, fractions A and B have apparent molecular weights of 196 kDa and 118 kDa, respectively. Further purification of fraction A on a DEAE-cellulose column, from which the enzyme was eluted with a linear gradient of KCl (0 to 1 M), resulted in a 3530-fold final purification with a 37% yield.

2. Rice

Rice embryo ADC has been purified by Choudhuri and Ghosh (1982). It was extracted with 50 mM Na$_2$HPO$_4$, containing 5 mM 2-ME and 20μM PLP; the protein fraction was precipitated with 25–55% ammonium sulphate, suspended in 100 mM Tris-HCl (pH 7.0) and then dialysed. The treatment with acetone was similar to that used for oat ADC. The precipitate was dissolved in the buffer, dialysed and applied to a Sepharose 4B column. When eluted with the Tris-HCl buffer containing 0.01 M KCl, it yielded two active fractions. One with molecular weight of 88 kDa had a low activity. The second fraction on further chromatography using a DEAE-cellulose colum yielded an ADC purified to homogeneity with molecular weight 174 kDa (700-fold purification with 12% yield); the latter is considered to be a dimer of the first ADC fraction.

3. Cucumber

The cucumber ADC has been purified by Prasad and Adiga (1985). The crude extract in 20 mM Tris-HCl (pH 7.6), containing 5 mM 2-ME and 20 μM PLP, was treated with MnCl$_2$ (final concentration 7.5 mM) and stirred for 60 min to precipitate nucleo-proteins. After centrifugation the supernatant was applied to a DEAE-cellulose column equilibrated with 10 mM Tris-HCl (pH 7.4) containing 2 mM 2-ME and 10 μM PLP. After washing off the unbound proteins with the buffer, the column was eluted with a linear 0 to 0.4 M KCl gradient. The pooled active fraction was precipitated with 90% ammonium sulphate to concentrate the protein. The clarified fraction was loaded onto a Sephadex G-150 column equilibrated with the buffer containing 0.2 M KCl. ADC activity was exclusively associated with the protein that eluted with the void volume; the enzyme was 19-fold purified with 18% yield. Polyacrylamide gel electro-phoresis (PAGE) at pH 8.3 showed the presence of a single band. When subjected to SDS-PAGE under reducing conditions, the protein resolved in three discrete bands of apparent molecular weights of 48 kDa, 44 kDa and 15 kDa. In the absence of 2-ME the enzyme exhibited only a single band of 150 kDa. Evidence has been obtained that all three peptides were held together by disulphide bonds in the final preparation eluted from Sephadex G-150 and that the preparation possesses an intrinsic or associated proteolytic activity.

When purified to homogeneity from *Lathyrus sativus*, ADC proved to exist as a homohexamer of subunits with a molecular weight of 36 kDa (Ramakrishna and Adiga, 1975).

V. LYSINE DECARBOXYLASE

When assayed using [U-^{14}C]Cad, and recovered by thin layer chromatography (TLC) from the reaction mixture as the activity indicator, LDC (EC 4.1.1.18) has been detected in 47 plant species (Wink and Hartmann, 1981; Schoofs *et al.*, 1983) although at extremely low levels, in most cases below 7 nmol h^{-1} g^{-1} acetone powder. Much higher activities assessed by the ^{14}CO$_2$-trapping method have been reported for leaves, stems and roots of *Nicotiana glauca* (Bagni *et al.*, 1986), soybean embryonic axis (Lin, 1984) and especially for ethylene-treated pea seedlings (Icekson *et al.*, 1986). The enzyme, isolated from both *Lupinus polyphyllus* (Hartmann *et al.*, 1980) and *N. glauca*, required 1.0–2.5 mM Fe^{2+} for maximum activity. However, a significant inhibition of LDC activity by Fe^{2+} was reported to occur in crude extracts from pea seedlings (Bakhanashvili *et al.*, 1985) and *Heimia salicifolia* cultures (Pelosi *et al.*, 1986). The lupin LDC required PLP and a thiol reagent; its pH optimum and K_m for Lys were 8.0 and 0.76 mM, respectively. LDC isolated from *Lathyrus sativus* seedlings catalysed not only Lys but also homoarginine decarboxylation, the K_m values being 0.88 and 3.33 mM, respectively. The enzyme with optimum pH of 8.4 shows an absolute requirement for Mn^{2+} or Fe^{2+} (1 mM) (Ramakrishna and Adiga, 1976).

A. Assay

The enzyme is extracted with a buffer solution similar to that used for ODC and ADC; 1 mM dithioerythritol or 2 mM 2-ME may substitute for DTT in the extraction or assay buffer.

The plant LDC has been assayed only by radiometric methods using [U-^{14}C]Lys. The reaction mixture (50 μl to 2.0 ml) — with or without FeSO$_4$ — contains 100 mM Tris-HCl or phosphate buffer (pH 8.0–8.4), PLP and EDTA at concentrations as for the other decarboxylases; 0.1–0.2 mM aminoguanidine or 10 mM diethylthiocarbamate is added to suppress diamine oxidase activity. The concentration and radioactivity of U-^{14}C-labelled L-Lys range between 0.1 and 3.0 mM and 0.2 and 5 μCi per reaction mixture, respectively, depending on the enzyme activity. The incubation time (at 30–37°C) may be prolonged to 4–6 h. A test with DFML for the possible artefactual release of ^{14}CO$_2$ is necessary. It may be added that DFMO, which in pea seedlings *in vivo* caused a decrease in LDC activity as well as in Cad accumulation, *in vitro* had no effect on the extracted enzyme (Birecka *et al.*, 1991). The LDC activity can be assessed on the basis of $^{14}CO_2$ or ^{14}C-Cad isolated by TLE as previously described, or by TLC (Hartmann *et al.*, 1980). An ion-exchange cartridge, Baker SPE sulphonic acid column, pre-conditioned with water, has been recently used in bacterial LDC assays for separating the labelled product from substrate (Heerze *et al.*, 1990). The reaction mixture of 300 μl containing 0.5–20 mM Lys, 200 000 dpm L-[U-^{14}C]Lys in a buffer (pH 6.0), is quenched after incubation with 1 ml H_2O; samples are loaded on the cartridge and washed with 20 ml 0.03 M sodium citrate (pH 6.9) to remove unreacted Lys. [^{14}C]Cad is eluted with 12 ml phosphate (pH 8.0). It is doubtful whether this method can be applied to plant LDC with very low activities in extracts.

There are several methods for bacterial LDC using non-labelled Lys, besides the reversed-phase HPLC described above for ADC. However, either their sensitivity for plant LDC is too low or the pH conditions are inappropriate (Tonelli *et al.*, 1981; Phan *et al.*, 1982; Scriven *et al.*, 1988).

B. Purification

Only the LDC from *Lathyrus sativus* has been purified. The enzyme extract in 50 mM Na_2HPO_4 containing 2 mM 2-ME and 50 μM PLP was treated in $MnCl_2$ (7.5 mM final concentration) to precipitate nucleoproteins and then centrifuged. The supernatant adjusted to pH 7.0 was treated with solid ammonium sulphate. The proteins precipitated with 25–55% saturation were collected, dissolved in 5 mM phosphate (pH 7.5) containing PLP and thiol reagent, and then fractionated with acetone. The 30–50% fraction was dialysed and absorbed on alumina C_7 and then eluted with 5 mM buffer. The active fractions were concentrated over aquacide and loaded onto a DEAE-Sephadex column washed with 50 mM phosphate buffer. The enzyme was eluted with 0.1–0.4 M KCl in the same buffer; a 114-fold purification was achieved with a 14% yield.

VI. S-ADENOSYLMETHIONINE DECARBOXYLASE

S-Adenosylmethionine decarboxylase (EC 4.1.1.50) is located in the cytosol, although some of its activity has been detected in isolated chloroplasts and mitochondria (Torrigiani *et al.*, 1986). It has been partially purified from *Lathyrus sativus* (Suresh and Adiga, 1977), corn (Suzuki and Hirasawa, 1980), and tobacco (Hiatt *et al.*, 1986). The reported pH optimum ranges between 7.5 for *L. sativus* to 8.6 for the corn SAMDC. K_m values of 5, 17 and 200 μM SAM have been reported for the enzyme from corn, wheat (Christ *et al.*, 1989) and pea (Ickeson *et al.*, 1985), respectively. The enzyme has been extracted with phosphate, Tris-HCl, or HEPES buffer containing 2.5–10 mM DTT (or up to 2 mM 2-ME) and 0.05–10 mM EDTA; PLP at 20–40 μM is sometimes added.

A. Assay

The enzyme has been assayed mostly in crude extracts by the $^{14}CO_2$-trapping method using *S*-adenosyl-L-[carboxyl-^{14}C]methionine. The reaction mixture (100–500 μl) and incubation conditions are similar to those applied in ODC assays; occasionally 5 mM Mg^{2+}, which stimulated *L. sativus* SAMDC, is added to the mixture. The concentration and radioactivity of SAM may range between 40 and 500 μM and 0.1 and 0.25 μCi per reaction mixture, respectively. The reported SAMDC activities range in most cases from 0.2 to 600 pmol h^{-1} mg^{-1} protein. A test for the artefactual $^{14}CO_2$ release is necessary. Methylglyoxal *bis*(guanylhydrazone) (MGBG), a carbonyl group reagent, at 0.1–0.5 mM is a very effective inhibitor of SAMDC. However, SAMDC from a mutant of *Nicotiana tabacum* cells displayed *in vitro* a high resistance of MGBG (Malmberg and Rose, 1987). Recently, when tested on mammalian and bacterial SAMDC, 5'-{[(*Z*)-4-amino-2-butenyl]methylamino}-5'-adenosine proved to be an extremely potent new specific enzyme-activated inhibitor (Danzin *et al.*, 1990).

B. Purification

1. Lathyrus sativus

The enzyme was extracted from 3-day-old seedlings with 50 mM Na_2HPO_4, containing PLP and 2-ME. After centrifugation the supernatant was applied onto an immuno-affinity *p*-hydroxymercuribenzoate-Sepharose column; the column was prepared by coupling antibodies, raised in rabbits against diamine oxidase of *L. sativus*, to CNBr-activated Sepharose 4B. The column was equilibrated with 50 mM Tris-HCl (pH 7.6) containing 0.9% NaCl, 20 μM PLP and 2 mM 2-ME. The crude extract was passed through the column twice. The resultant effluent showed to detectable diamine oxidase activity. The SAMDC active fractions were further fractionated on a DEAE-Sephadex A-50 column. After removal of the unadsorbed proteins with 50 mM Tris-HCl buffer (pH 7.6) containing PLP and 2-ME, the ion-exchanger was washed successively with above buffer supplemented with 0.1, 0.3 and 0.5 M KCl. SAMDC activity was found in the 0.3 M KCl eluate. At this stage the enzymes was purified 42-fold with a yield of 18%. The purified enzyme showed no increase in activity in the presence of Put.

2. Corn

Etiolated seedlings were extracted with Tris-HCl (pH 7.5) containing 1 mM 2-ME and 0.1 mM EDTA. After centrifugation at 10 000 × g the supernatant was fractionated stepwise with ammonium sulphate, and the fraction collected between 20 and 50% saturation was centrifuged; the pellet was dissolved in the Tris-HCl buffer and dialysed overnight. After centrifugation the dialysate was applied onto a column of trimethylamino-2- hydroxypropyl-cellulose equilibrated with the buffer. The pooled active fractions were applied onto an affinity MGBG-Sepharose column which was prepared according to Pegg (1974). The column was equilibrated with the Tris-HCl buffer, washed with the same buffer supplemented with 0.3 M NaCl, and finally the enzyme was eluted with the buffer containing 0.3 M NaCl and 1 mM MGBG and dialysed to remove MGBG. The enzyme fraction was concentrated to 5 ml in a collodion bag; a 500-fold purification with a yield of 30% has been achieved. The molecular weight of the enzyme is about 25 kDa as indicated by Sephadex G-200 and Bio-Gel P-150 chromatography.

3. Tobacco

Cells of a wild-type *Nicotiana tabacum* cv. Xanthis were homogenised with 100 mM Tris-HCl (pH 8.0) containing 100 mM EDTA, 10 mM DTT, 40 μM PLP and 0.5 mM Put. After centrifugation at 100 000 $\times$ g, the supernatant was treated with ammonium sulphate. The protein precipitated by 35–55% saturation was resuspended in the buffer and dialysed against 50 mM HEPES (pH 8.0) containing 100 μM EDTA, 10 mM DTT, 40 μM PLP and 0.5 mM Put. After centrifugation at 5000 $\times$ g, the supernatant was applied onto a DEAE-Sepharose column, washed with the dialysis buffer and the enzyme was eluted with a gradient of 0–1.5 M NaCl. The active fractions were transferred onto a MGBG-Sepharose column prepared as described by Pegg and Poso (1983); the column was washed with the dialysis buffer and the enzyme was eluted with the buffer containing 1 mM MGBG. Each fraction was precipitated with 60% ammonium sulphate, resuspended in the dialysis buffer and dialysed overnight against the buffer supplement with 0.3 NaCl to remove MGBG. A 4540-fold purification with a 16% recovery was achieved. The SDS-PAGE profile of the purified SAMDC indicated a molecular weight of 35 kDa.

VII. AGMATINE IMINOHYDROLASE

The cytosolic enzyme (EC 3.5.3.12), which converts Agm to *N*-carbamoylPut and ammonia, was detected in extracts from oat, sunflower, cabbage, Brussels sprouts and *Passiflora* (Smith, 1969; Desai and Mehta, 1985) and has been purified from groundnut (Sindhu and Desai, 1979), corn (Yanagisawa and Suzuki, 1981) and rice (Choudhuri and Ghosh, 1985). The reported optimum pH values were 5.5–8.5, 6.5–7.5 and 6.0, respectively and the K_m values for Agm 0.76, 0.19 and 15 mM, respectively. The enzyme was extracted with 5 mM phosphate (pH 7.0) or with 0.1 M Na_2HPO_4 followed by dialysis against 0.1 M phosphate (pH 7.0).

A. Assay

The enzyme activity has been assayed by measuring either the ammonia released from Agm in Conway microdiffusion units or the *N*-carbamoylPut formed by the procedure described by Archibald (1944). The reaction mixture (1.0–2.0 ml) containing phosphate buffer, 4–15 mM Agm sulphate and 0.1 ml toluene is placed in the outer compartment of a Conway unit (1 mg streptomycin and 1 mg chloramphenicol—without toluene—may be added); 2 ml 0.01 M HCl (or 0.5 ml 0.1 M HCl) is placed in the central compartment. The unit is sealed and incubated for 2–18 h at 28–37°C; then 1 ml of saturated K_2CO_3 is added to the outer compartment. After an additional 3 h at room temperature, the ammonia absorbed in the inner compartment is determined with Nessler's reagent. The mercuric hydroxyiodimide complex formed is measured at 430 nm.

N-CarbamoylPut can be separated from Agm either by TLC, high voltage paper electrophoresis (3000 V, 15 min in 1 M acetic acid) or TLE as in the case of the decarboxylases and measured by the diaacetylmonoxime-*p*-aminodiphenylamine method (Dunninghake and Grisolia, 1966). When the enzyme activity is assessed only on the basis of N-carbamoylPut produced, the reaction mixture is placed in a small vial or a test tube and after 1–2 h incubation the reaction is terminated by addition of 0.5 ml 10% TCA.

B. Purification

1. Groundnut

Acetone powder, obtained from cotyledons of germinated seeds, was suspended in 10 ml phosphate (pH 7.0) and centrifuged at 10 000 × g. The supernatant was acidified to pH 5.0 with acetic acid, allowed to stand for 40 min and centrifuged. The supernatant with pH adjusted to 7.0 was mixed with $Ca_3(PO_4)_2$ gel (20 mg dry wt ml^{-1}) at a gel : enzyme volume ratio of 2:5, stirred for 30 min and centrifuged at 6000 × g. The supernatant was mixed with alumina C_7 gel (15 mg dry wt ml^{-1}) at a gel : enzyme vol. ratio of 3:2. After stirring for 30 min the mixture was centrifuged and the absorbed enzyme fraction was applied to a DEAE-cellulose column equilibrated with the buffer. The column was washed with 0.1 M phosphate (pH 7.0) and eluted with 0.2 M phosphate (pH 7.0). The enzyme, purified 375-fold with a recovery of 60%, was completely free of amine oxidase activity.

2. Corn

The enzyme was extracted from the shoots with 50 mM phosphate (pH 6.5), containing 5 mM 2-ME and 0.1 mM EDTA, in the presence of polyvinylpyrrolidone (0.1 g per g fresh weight). The homogenate was filtered through cheesecloth and clarified by centrifugation at 10 000 × g. The supernatant was stirred with DEAE-cellulose equilibrated with the buffer (DEAE-cellulose/supernatant = 50 g per 1000 ml) for 1 h and the slurry transferred onto a Buchner funnel. After being washed with the buffer, the DEAE-cellulose was stirred with the buffer supplemented with ammonium sulphate at 30% saturation and again transferred onto the funnel. This operation was repeated with ammonium sulphate concentration increased to 80% saturation and after 30 min the precipitate was collected by centrifugation, dissolved in a small volume of the buffer and passed through a Sephadex G-25 column equilibrated with the buffer. After desalting, the enzyme fraction was applied to a DEAE-cellulose column and eluted with a pH gradient established with the extraction buffer and 0.5 M KH_2PO_4, containing 5 mM 2-ME and 0.1 mM EDTA, at a 1:1 (v/v) ratio. The active fractions (pH 6.2–5.7) were concentrated and filtered through a Sephadex G-100 column. After dialysis the active fraction was subjected to Agm-SA-AM-Sepharose 6B affinity chromatography. The column was washed with 0.05 mM phosphate (pH 8.5) containing 5 mM 2-ME and 0.1 mM EDTA. The enzyme was eluted with the same buffer supplemented with 10 mM Agm sulphate. The pooled active fractions were dialysed against the extracting buffer; the purification factor and yield were 7300 and 23%, respectively. The enzyme molecular weight, estimated by chromatography on a Bio-Gel P-200 column, was 85 kDa; SDS-PAGE indicated that the enzyme was a dimer with identical subunits of 43 kDa.

3. Rice

The enzyme extract from seedlings in 50 mM phosphate (pH 6.0) containing 5 mM 2-ME and 0.1 mM EDTA was centifuged at 8000 × g. The supernatant was treated with $MnCl_2$ (7.5 mM final concentration) for 30 min and the precipitate was removed by centrifugation. The protein fraction precipitated with ammonium sulphate at 30–70% saturation was suspended in the buffer, dialysed for 24 h and chromatographed on a DEAE-cellulose column as described for the corn enzyme. The pooled active fractions were precipitated with 70% ammonium sulphate, dissolved in the buffer, dialysed, applied to a second DEAE-cellulose column and eluted as above. The enzyme was 717-fold purified with a yield of about 9%.

VIII. *N*-CARBAMOYLPUTRESCINE AMIDOHYDROLASE

The cytosolic enzyme (EC 3.5.1.-) converts *N*-carbamoylPut to Put, CO_2, and NH_3. It has been detected in leaves of barley, wheat, rye, oat, corn, pea, radish and sunflower (Smith, 1965) and purified from corn (Yanagisawa and Suzuki, 1982). Its optimum pH ranges from 6.5 to 8.0 and K_m for the substrate is about 90 μM (barley). The enzyme has been extracted with 0.1 M Na_2HPO_4 or 50 mM Tris-HCl containing 5 mM 2-ME, 0.1 M EDTA and 20% glycerine in the presence of polyvinylpyrrolidone (corn).

A. Assay

The amidohydrolase activity — like that of Agm iminohydrolase — is determined by measuring the NH_3 released from the substrate in Conway microdiffusion units. The reaction mixture (1.25–2.0 ml) consists of the enzyme extract and 5.0–7.5 mM *N*-carbamoylPut in either 100 mM phosphate (pH 7.2) or 50 mM Tris-HCl (pH 7.0) containing 5.0 mM 2-ME and 0.1 mM EDTA. Stoichiometry of the reaction can be verified by determining either CO_2 manometrically or Put, separated from the substrate, by the colorimetric method of Dunnighake and Grisolia (1966).

B. Purification

The homogenate of corn shoots in the glycerine-containing Tris-HCl buffer was centrifuged at 10 000 × g and the supernatant fractionated with ammonium sulphate; the 40–65% ammonium sulphate precipitate was dissolved in the buffer, its pH adjusted to 4.5 and centrifuged. The supernatant adjusted to pH 7.0 was brought to 70% saturation with ammonium sulphate. After centrifugation the precipitate was dissolved in a small volume of Tris-HCl without glycerine (buffer B) and desalted on Sephadex G-25. The void fraction of the eluate was ultracentrifuged at 100 000 × g for 1 h and the supernatant concentrated in a cellophane tube in contact with solid polyethylene glycol. After centrifugation at 10 000 × g the enzyme solution was applied onto a Sephadex G-150 column equilibrated with buffer B and eluted with the same buffer. The enzyme was purified 70-fold with a yield of 31%. Its molecular weight was 125 kDa as determined by gel filtration on Sephadex G-200.

IX. PUTRESCINE SYNTHASE

This multifunctional enzyme converts stoichiometrically Agm and Orn to Put and citrulline, respectively; it exhibits Agm iminohydrolase, Put transcarbamylase and Orn transcarbamylase activities, and, in addition it shows also carbamate kinase activity. The overall reaction is as follows:

$$Agm + Orn + H_2O + Pi \rightarrow ATP + CO_2 + NH_3$$

and the additional reaction:

$$Carbamoylphosphate + ADP + H_2O \rightarrow ATP + CO_2 + NH_3$$

The enzyme has been purified from *Lathyrus sativus* (Srivenugopal and Adiga, 1981) and cucumber (Prasad and Adiga, 1986a, b); its optimum pH is 8.8. The enzyme is extracted with 20 mM Tris-HCl (pH 8.5) containing 5.0 mM 2-ME and 0.5 mM $MnCl_2$.

A. Assay

The overall reaction catalysed by the enzyme is assayed by measuring citrulline production with either Agm or *N*-carbamoylPut as the substrate. The reaction mixture (0.5 ml), containing 50 mM Tris-HCl (pH 8.8), 3 mM DTT, 10 mM $MgSO_4$, 5 mM Orn, 2.5 mM Na_2HPO_4 and 5 mM Agm or *N*-carbamoylPut, is incubated with the enzyme at 37°C for 1 h; the reaction is stopped with 0.1 ml 20% $HClO_4$. During each assay a blank containing the boiled enzyme is regularly included. The denatured proteins are removed by brief centrifugation and the reaction mixture is applied to a Dowex50W [H^+] column (1.0 × 4 cm). After washing the column with 5 ml H_2O, citrulline is selectively eluted with 3 ml 2 M NH_4OH. The amine substrates, which would otherwise interfere with the determination, are preferentially retained on the column. An aliquot of the eluate is subjected to the colour reaction by the method of Prescott and Jones (1969). The amount of citrulline formed is assessed by the difference between the active assay mixture and the blank.

B. Purification

The seedlings of *Lathyrus sativus* were homogenised with 50 mM imidazole-Cl buffer (pH 8.0) containing 5 mM 2-ME. After centrifugation the supernatant was treated with 7.5 mM $MnCl_2$ (final concentration) and stirred for 30 min. The supernatant obtained after centrifugation was adjusted to pH 7.0 and fractionated with ammonium sulphate. The protein fraction precipitated at 40–85% saturation was collected by centrifugation, dissolved in 5 mM imidazole-HCl (pH 7.5) containing 2 mM 2-ME, and dialysed against the same buffer. The enzyme fraction was subjected to affinity chromatography on a Put-CH Sepharose column, which was prepared according to March *et al.* (1974) and Cuatrecasas (1970), and equilibrated with the imidazole buffer without 2-ME. After washing off the unadsorbed proteins (monitored by A_{280}) the enzyme was eluted with the same buffer supplemented with 2 mM 2-ME and 2 mM Put. The fractions were collected directly into test tubes containing 10 μg BSA to stabilise the enzyme. The pooled concentrated active fractions were applied onto a DEAE-Sephadex column and Put synthase was eluted with 0.5 M KCl; the purification factor was 230 with a 25% recovery of the activity. The enzyme is a single protein, with a molecular weight of 55 kDa as indicated by SDS-gel electrophoresis.

The cucumber Put synthase was purified in the same way as that from *L. sativus* and resolved by PAGE at pH 8.3 into two bands, one representing an aggregated enzyme and the other the enzyme monomer; two-dimensional gel electrophoresis and SDS-PAGE indicated that both bands comprised identical polypeptide chains. The enzyme moved as a single 150 kDa protein under non-reducing conditions during SDS-PAGE. Unlike Put synthase in *L. sativus*, the enzyme isolated from cucumber also harbours ADC activity (Prasad and Adiga, 1987).

X. PUTRESCINE AMINOPROPYLTRANSFERASE AND SPERMIDINE AMINOPROPYLTRANSFERASE

By transferring the propylamine moiety from d-SAM to Put and Spd, putrescine aminopropyltransferase (Spd synthase; EC 2.5.1.16) and spermidine aminopropyltransferase (Spm synthase) produce Spd and Spm, respectively, and in both cases 5'-methylthioadenosine and a proton. In contrast to other enzymes involved in PA biosynthesis in plants, the two synthases have not attracted much attention. Spd synthase has been found in *Lathyrus sativus* extracts and has been separated from SAMDC by affinity chromatography (Suresh and Adiga, 1977); it has been also detected in corn (Smith, 1985) and wheat (Christ *et al.*, 1989). Spd synthase and Spm synthase were assayed in extracts from cabbage; relatively high activities of the former were also found in spinach leaves (Sindhu and Cohen, 1984). Spd synthase is present in the cytosol as well as in chloroplasts of cabbage leaves and it has been partially purified (Cohen *et al.*, 1981).

The *L. sativus* Spd synthase was extracted together with SAMDC as described previously; the wheat enzyme was extracted together with ODC, ADC and SAMDC using 200 mM phosphate buffer containing PLP, DTT and EDTA, and the cabbage synthases were extracted with 10 mM glycine-NaOH (pH 8.8).

A. Assay

Animal and bacterial synthases have been assayed using ^{14}C-labelled Put (Spd); also, methyl-labelled d-SAM has been used with the subsequent recovery of 5'-methylthioadenosine (Raina *et al.*, 1983; Pegg, 1983; Tabor and Tabor, 1983; Anton, 1986). Plant synthases have been assayed employing either ^{14}C-labelled Put (Spd) or non-labelled substrates and measuring the product Spd by a fluorometric method (Suzuki *et al.*, 1981); d-SAM may be generated from SAM by the SAMDC present in the extract (as in the cases of *L. sativus* and wheat) or it may be added as such.

The *L. sativus* enzyme extract devoid of diamine oxidase activity was assayed in a reaction mixture (1 ml) containing 100 mM Tris-HCl (pH 7.6), 0.1 mM SAM, 0.5 mM Mg^{2+} and 5 mM DTT. After 2 h incubation at 37°C the reaction was terminated with

0.2 ml 1 M KOH; 100 nmol each of carrier Spd and Spm were added and the mixture heated at 100°C for 30 min to degrade the sulphonium compounds. The PAs were separated by paper chromatography, located with ninhydrin and counted in a scintillation counter.

The wheat Spd synthase, fractionated with 70% ammonium sulphate, was assayed in a mixture (100 μl) containing 100 mM phosphate (pH 8.0), 0.5 mM SAM, 0.1 mM [1,4-^{14}C]Put (0.5 μCi), 50μM EDTA, 50 μM PLP and 2 mM DTT. After a 40 min incubation the reaction was stopped by adding 25 μl of a solution containing 10 mM Spd and 30% HClO$_4$. The PAs were dansylated, extracted with 0.25 ml benzene and separated by TLC using silica plates and chloroform and triethylamine (5:1; v/v) as solvents; they were visualised under UV (254 nm), scraped off the plates and transferred into scintillation vials.

The synthases of Chinese cabbage and spinach leaves were assayed in a reaction mixture (0.325 ml) containing the crude extract, 150 mM glycine-NaOH (pH 8.8), 37 μM ^{14}C-labelled Put (1 μCi) for Spd synthase or 37μM ^{14}C-labelled Spd (1 μCi) and 25 μM d-SAM (blanks lacked either d-SAM or the enzyme). The mixture was incubated for 1 h at 37°C and the reaction was stopped with 1 ml 5% HClO$_4$. After standing for 30 min at 4°C the mixture was centrifuged, the precipitate washed twice with 1 ml 3% HClO$_4$ and 30 nmol Spd or Spm was added as carrier to the combined supernatant fluids. The solution was loaded on a Dowex-50 W [H$^+$] column prepared as described by Inoué and Mizutani (1973). The labelled substrate was eluted with 2.3 M HCl for Spd synthase or 3.3 M HCl for Spm synthase; the labelled products were eluted with 30 ml 6 M HCl with a recovery of about 90%; the eluates were dried at 45°C *in vacuo*, dissolved in 0.2 ml 30 mM HCl, dansylated and analysed as described by Cohen *et al.* (1981).

The fluorometric method was applied to partially purified Spd synthase from Chinese cabbage leaves. The method is based on the oxidation of Spd by a polyamine oxidase yielding H$_2$O$_2$ that is subsequently used by peroxidase to oxidise homo-vanillic acid. The assay mixture (0.6 ml) contains the enzyme, 33 μM phosphate (pH 7.4), 50 μl of oat seedling polyamine oxidase solution (1 mg ml^{-1}), 40 μl of horseradish peroxidase solution (1 mg ml^{-1}), 50 μl of homovanillic acid solution (1 mg ml^{-1}), 100μM Put and 25μM d-SAM; the reaction is started by addition of d-SAM. After incubation at 37°C for 1 h the reaction is stopped with 50 μl 1 M NaOH and the fluorescence measured with excitation at 323 nm and emission at 426 nm. The applied method is reliable within the range from 1 to 10 nmol of Spd. However, its detectability threshold may be decreased to 0.1 nmol by substituting the fluorogenic 3-(p-hydroxyphenyl) propionic acid for the homovanillic acid (Zaitsu and Ohkura, 1980).

B. Purification

The partially purified Chinese cabbage Spd synthase was obtained by ammonium sulphate fractionation (35–65% saturation) followed by acetone treatment. The acetone powder was dissolved in 25 mM phosphate (pH 7.2) containing 0.1 M KCl and filtrated through a Sephadex G-100 column equilibrated with the same buffer. The 160-fold purified enzyme showed a molecular weight of 81 kDa and K_m values for d-SAM and Put 6.7 and 32 μM, respectively; its optimum pH was 8.8.

In *L. sativus* an alternative pathway of Spd biosynthesis involving carboxy-Spd synthase has been found (Srivenugopal and Adiga, 1980b). The NADPH-dependent enzyme catalyses the reaction between Put and aspartate semialdehyde yielding carboxy-Spd which in turn — after decarboxylation — gives rise to Spd. The enzyme was tested in an assay mixture (1.0 ml) containing 50 mM phosphate (pH 7.5), 5 mM L-threonine, 3 mM DTT, 0.25 mM [1,4-^{14}C]Put (75 nCi), 10 mM aspartic β-semialdehyde and 0.5 mM NADPH. After incubation at 37°C for 1 h and enzyme inactivation with 0.1 ml 20% HClO$_4$, the supernatant was adjusted to pH 6.0, the carboxy-Spd was fractionated on phosphocellulose, separated by paper electrophoresis and quantified by ninhydrin reaction or by measuring its radioactivity.

Decarboxylation of carboxy-Spd was assayed in a reaction mixture (1.0 ml) containing 30 mM Tris-HCl (pH 8.4), 5 mM DTT, 20 μM PLP, 5 mM MgCl$_2$ and 0.5 mM substrate (5000 cpm). After 2 h incubation at 37°C the pH of the reaction mixture was brought to 12 and the labelled Spd was extracted wit butanol. The acidified butanol was evaporated *in vacuo* and the radioactivity measured.

XI. HOMOSPERMIDINE SYNTHASE

The NAD$^+$-dependent thiol enzyme, which catalyses the formation of *sym*-homo-Spd from two molecules of Put (Srivenugopal and Adiga, 1980b), is assayed in a reaction mixture (1.0 ml) containing 50 mM Tris-HCl (pH 8.4), 2 mM DTT, 10 mM KCl, 0.5 mM [^{14}C]Put (0.1 μCi) and 1 mM NAD$^+$. After the reaction is stopped with 0.1 ml 20% HClO$_4$, the amines are purified on a Dowex W50 [H$^+$] column as in the case of cabbage Spd synthase and separated by paper chromatography or electrophoresis.

The *L. sativus* homoSpd synthase was 100-fold purified by affinity chromatography on a Blue Sepharose column equilibrated with 50 mM phosphate (pH 7.5) containing 2 mM 2-ME. The enzyme was eluted with 5 mM NAD$^+$ in the equilibration buffer. The homoSpd synthase has a molecular weight of 75 kDa as indicated by filtration on a Sepharose S-200 column; its K_m for Put is 3 mM.

REFERENCES

Adiga, P. R. and Prasad, G. L. (1985). *Plant Growth Reg.* 3, 205–226.
Altman, A., Friedman, R. and Levin, N. (1983). *In* "Advances in Polyamine Research", Vol. 4 (U. Bachrach, A. Kaye, and R. Chayen, eds), pp. 395–408. Raven Press, New York.
Anton, D. L. (1986). *Anal. Biochem.* 156, 43–47.
Archibald, R. M. (1944) *J. Biol. Chem.* 156, 121–126.
Bagni, N., Creus, J. and Pistocchi, R. (1986), *J. Plant Physiol.* 125, 9–15.
Bakhanashvili, M., Icekson, I. and Apelbaum, A. (1985). *Plant Cell Rep.* 4, 297–299.
Beeman, C. S. and Rossomondo, E. F., (1989). *J. Chromatogr.* 496, 101–110.
Birecka, H., Dinolfo, T. E., Martin, W. B. and Frohlich, M. W. (1984). *Phytochemistry* 23, 991–997.
Birecka, H., Bitonti, A. J. and McCann, P. P. (1985a). *Plant Physiol.* 79, 509–514.
Birecka, H., Bitonti, A. J. and McCann P. P. (1985b). *Plant Physiol.* 79, 515–519.
Birecka, H., Birecki, M., Bitonti, A. J. and McCann, P. P. (1991). *Phytochemistry* 30, 99–103.
Bitonti, A. J., Casara, P. J., McCann, P. P. and Bey, P. (1987). *Biochem. J.* 242, 69–74.

Bonner, J. (1976). *In* "Plant Biochemistry", (J. Bonner and J. E. Warner, eds), pp. 37–40. Academic Press, New York and London.

Burtin, D., Martin-Tanguy, J., Paynot, M. and Rossin, N. (1989). *Plant Physiol.* **89**, 104–110.

Carvajal, N., Acoria, M., Rodriguez, J. P., Fernandez, M. and Martinez, J. (1982). *Biochem. Biophys. Acta* **701**, 146–148.

Choudhuri, M. M. and Ghosh, B. (1982). *Agric. Biol. Chem.* **46**, 739–743.

Choudhuri, M. M. and Ghosh, B. (1985). *Phytochemistry* **24**, 2433–2435.

Christ, M., Felix, H. and Harr, J. (1989). *Z. Naturforsch* **44c**, 49–54.

Cohen, J. S., Balint, R. and Lindhu, R. K. (1981). *Plant Physiol.* **68**, 1150–1155.

Cuatrecasas, P. (1970). *J. Biol. Chem.* **245**, 3059–3065.

Danzin, C., Marchal, P. and Casara, P. (1990). *Biochem. Pharmacol.* **40**, 1499–1503.

Desai, H. V. and Mehta, A. R. (1985). *J. Plant Physiol.* **119**, 45–53.

D'Orazi, D. D. and Bagni, N. (1987). *Physiol. Plant.* **71**, 177–183.

Dunninghake, D. and Grisolia, S. (1966). *Anal. Biochem.* **16**, 200–205.

Foudouli, A. C. and Kyriakidis, D. A. (1989). *Plant Growth Regul.* **8**, 233–242.

Gaines, D. W., Friedman, L. and Braunberg, R. C. (1989). *Anal. Biochem.* **178**, 82–56.

Hartmann, T., Schoofs, G. and Wink, M. (1980). *FEBS Lett.* **115**, 35–38.

Heimer, Y. M. and Mizrahi, Y. (1982). *Biochem. J.* **201**, 373–376.

Heerze, L. D., Kang, Y. J. and Palcic, M. M. (1990). *Anal. Biochem.* **185**, 201–205.

Hiatt, A. (1989). *Plant Physiol.*, 1378–1381.

Hiatt, A. and Malmberg, R. L. (1988). *Plant Physiol.* **86**, 441–446.

Hiatt, A. C., McIndoo, J. and Malmberg, R. L. (1986). *J. Biol. Chem.* **261**, 1293–1298.

Icekson, I., Goldlust, A., and Apelbaum, A. (1985). *J. Plant Physiol.* **119**, 335–345.

Icekson, I., Bakhanashvili, M. and Apelbaum, A. (1986). *Plant Physiol.* **82**, 607–609.

Inoue, H. and Mizutani, A. (1973). *Anal. Biochem.* **56**, 408–416.

Kochhar, S., Mehta, P. K. and Christen, P. (1989). *Anal. Biochem.* **179**, 182–185.

Koromilas, A. E. and Kyriakidis, D. A. (1988a). *Physiol. Plant.* **72**, 718–724.

Koromilas, A. E. and Kyriakidis, D. A. (1988b). *Phytochemistry* **27**, 989–992.

Krannes, J. and Flatmark, T. (1987). *J. Chromatogr.* **419**, 291–295.

Kuehn, G. D., Rodriguez-Garay, B., Bagga, S. and Phillips, G. C. (1990). *Plant Physiol.* **94**, 855–857.

Kyriakidis, D. A., Panagiotidis, C. A. and Georgetsos, J. G. (1983). *In* "Methods in Enzymology", Vol. 94 (H. Tabor, and C. W. Tabor, eds), pp. 162–166. Academic Press, New York and London.

Lin, P. P. (1984). *Plant Physiol.* **76**, 372–380.

Loomis, W. D. (1974). *In* "Methods in Enzymology", Vol. 31 (S. Fleischer and L. Packer, eds), pp. 528–544. Academic Press, New York and London.

Malmberg, R. L. and Rose, D. J. (1987). *Mol. Gen. Genet.* **207**, 9–14.

March, S. C., Parikh, I. and Cuatrecasas, P. (1974). *Anal. Biochem.* **60**, 149–152.

Matsuda, H., (1984). *Plant Cell Physiol.* **25**, 523–530.

Ngo, T. T., Brillhart, K. L., Davis, R. H., Wong, R. C., Bovaird, J. H., Digangi, J. J., Ristow, J. L., Marsh, J. L., Phan, A. P. H. and Lenhoff, H. H. (1987). *Anal. Biochem.* **160**, 290–293.

Panagiotidis, C. A., and Kyriakidis, D. A. (1985). *Plant Growth Regl.* **3**, 247–255.

Panagiotidis, C. A., Georgatsos, J. G. and Kyriakaidis, D. A. (1982). *FEBS Lett.* **146**, 193–196.

Pandit, M. and Ghosh, B. (1988). *Phytochemistry* **27**, 1609–1610.

Pegg, A. E. (1974). *Biochem. J.* **141**, 581–583.

Pegg, A. E. (1983). *In* "Methods in Enzymology", Vol. 94 (H. Tabor and C. W. Tabor, eds), pp. 260–265. Academic Press, New York and London.

Pegg, A. E. and Poso, H. (1983). *In* "Methods in Enzymology", Vol. 94 (H. Tabor and C. W. Tabor, eds), pp. 234–239. Academic Press, New York and London.

Pelosi, L. R., Rother, A. and Edwards, J. M. (1986). *Phytochemistry* **25**, 2315–2319.

Phan, A. P. H., Ngo, T. T. and Lenhoff, H. M. (1982). *Anal. Biochem.* **120**, 193–199.

Prasad, G. L. and Adiga, P. R. (1985). *J. Biosci.* **7**, 331–343.

Prasad, G. L. and Adiga, P. R. (1986a). *J. Biosci.* **10**, 203–213.

Prasad, G. L. and Adiga, P. R. (1986b). *J. Biosci.* **10**, 373–391.

Prasad, G. L. and Adiga, P. R. (1987). *J. Biosci.* **11**, 571–579.
Prescott, L. M. and Jones, M. E. (1969). *Anal. Biochem.* **32**, 409–419.
Rabe, E. and Lovatt, C. J. (1984). *Plant Physiol.* **76**, 747–752.
Raina, A., Eloranta, T. and Pajula, R. C. (1983). *In* "Methods in Enzymology", Vol. 94 (H. Tabor and C. W. Tabor, eds), pp. 257–259. Academic Press, New York and London.
Ramakrishna, S. and Adiga, P. R. (1975). *Eur. J. Biochem.* **59**, 377–386.
Ramakrishna, S. and Adiga, P. R. (1976). *Phytochemistry* **15**, 83–86.
Rodriguez-Garay, B., Phillips, G. C. and Kuehn, G. D. (1989). *Plant Physiol.* **89**, 525–529.
Schoofs, G., Teichman, S., Hartmann, T. and Wink, M. (1983). *Phytochemistry* **22**, 65–69.
Scriven, F., Wlasichuk, K. B. and Palcic, M. M. (1988). *Anal. Biochem.* **170**, 367–371.
Seiler, N. (1983). *In* "Methods in Enzymology", Vol. 94 (H. Tabor and C. W. Tabor, eds), pp. 3–9. Academic Press, New York and London.
Sindhu, R. K. and Cohen, S. S. (1984), *Plant Physiol.* **74**, 645–649.
Sindhu, R. K. and Desai, H. V. (1979). *Phytochemistry* **18**, 1937–1938.
Slocum, R. D., Bitonti, A. J., McCann, P. P. and Feifer, R. P. (1988). *Biochem. J.* **255**, 197–202.
Smith, T. A. (1965). *Phytochemistry* **4**, 599–607.
Smith, T. A. (1969). *Phytochemistry* **8**, 2111–2117.
Smith, T. A. (1979). *Phytochemistry* **18**, 1447–1452.
Smith, T. A. (1983). *In* "Methods in Enzymology", Vol. 94 (H. Tabor and C. W. Tabor, eds), pp. 176–180. Academic Press, New York and London.
Smith, T. A. (1985). *Ann. Rev. Plant Physiol.* **36**, 117–143.
Smith, T. A. and Barker, J. H. A. (1988). *Adv. Exp. Med. Biol.* **250**, 573–587.
Smith, T. A. and Marshall, J. H. A. (1988a). *Biochem. Soc. Trans.* **16**, 972–973.
Smith, T. A. and Marshall, J. H. A. (1988b). *Phytochemistry* **27**, 703–710.
Smith, T. A. and Marshall, J. H. A. (1988c). *Phytochemistry,* **27**, 1611–1613.
Srivenugopal, K. S. and Adiga, P. R. (1980a). *Biochem. J.* **190**, 461–464.
Srivenugopal, K. S. and Adiga, P. R. (1980b). *FEBS Lett* **112**, 260–264.
Srivenugopal, K. S. and Adiga, P. R. (1981). *J. Biol. Chem.* **256**, 9532–9541.
Suresh, M. R. and Adiga, P. R. (1977). *Eur. J. Biochem.* **79**, 511–518.
Suzuki, O., Matsumoto, M., Oya, Y., Katsumata, Y. and Samejima, K. (1981). *Anal. Biochem.* **115**, 72–77.
Tabor, C. W. and Tabor, H. (1983). *In* "Methods in Enzymology", Vol. 94 (H. Tabor and C. W. Tabor, eds), pp. 265–269. Academic Press, New York and London.
Tonelli, D., Budini, K., Gattavecchia, E. and Girotti, S (1981). *Anal. Biochem.* **111**, 189–194.
Torrigiani, P., Serafini-Fracassini, D., Biondi, S. and Bagni, N. (1986). *J. Plant Physiol.* **124**, 23–29.
Weber, L. W. D. (1987). *Experientia* **43**, 176–178.
Winer, L., Winkler, C. and Apelbaum, A (1984). *Plant Physiol.* **76**, 233–237.
Wink, M. and Hartmann, T. (1981). *Z. Pflanzenphysiol.* **102**, 337–344.
Wright, L. C., Brady, C. J. and Hinde, R. W. (1981). *Phytochemistry* **20**, 2641–2645.
Yanagisawa, H. and Suzuki, Y. (1981). *Plant Physiol.* **67**, 697–706.
Yanagisawa, H. and Suzuki, Y. (1982). *Phytochemistry* **21**, 2201–2203.
Zaitsu, K. and Ohkura, Y. (1980). *Anal. Biochem.* **109**, 109–113.

Index